Chambers Concise Encyclopedia of Film and Television

Chambers Concise Encyclopedia of Film and Television

Editor

ALLAN HUNTER

EDINBURGH NEW YORK TORONTO

Published 1991 by W & R Chambers Ltd
43 – 45 Annandale Street, Edinburgh EH7 4AZ, Scotland
95 Madison Avenue, New York, NY 10016, USA

Published in the UK under the title
Chambers Film and Television Handbook

Library of Congress Cataloging-in-Publication Data applied for

ISBN 0-550-17253-X Pbk

Editorial Manager: Min Lee

Printed in England by Butler and Tanner Ltd, Frome, Somerset

CONTENTS

Preface	vii
Explanatory Notes	xi
Contributors	xii
Acknowledgements	xiii
Dictionary A–Z	1
Oscars	379
Berlin Film Festival Awards	391
Cannes Film Festival Awards	395
Venice Film Festival Awards	399

PREFACE

Once, the film industry regarded television as a heinous rival and undoubted threat to its well-being and future survival. If audiences could select between the plethora of entertainment and drama that was available in the home at the touch of a button, what possible incentive could there be for venturing out to those archaic cathedrals of the silver screen known as cinemas?

The industry attempted to fight back with spectacle and scale – wide-screen epics and 3-D gimmicks. Successive generations keep discovering, however, that their strongest card is the cinema-going experience itself. Nothing can replace the hushed expectancy as the lights go down in a cocoon-like auditorium or the communal experience of shared laughter and tears in the dark with a group of like-minded strangers.

Cinema attendances in this country have risen consistently over the past five years and audiences have shown a healthy appetite for not just the latest Hollywood blockbuster but also foreign-language productions like *Cinema Paradiso* (1988) and *Cyrano De Bergerac* (1990), which will have grossed around £2 million by the time its British theatrical distribution has been completed.

Far from killing the film industry, such technological innovations as television, video, cable and satellite have all assisted in the expansion of outlets for the dissemination of film culture and stimulated an ever-increasing appetite for film in which cinema-going is but one experience.

An average of one million video cassettes are rented every day in Britain whilst terrestrial television channels will broadcast between thirty and forty feature films a week, many more at a peak holiday period. A recent week's television viewing figures show the films *The Goonies* (1985) and *Live and Let Die* (1973) among the top ten highest-rated events of the week with audiences of just under 13 million.

The interdependence of all forms of entertainment is well established. A film may be seen by a much smaller audience in the cinema than when it is screened on television, but that's where its reputation will be made. A film can fail to generate a profit on its cinema release alone but will turn a healthy shade of black when revenue from a video deal or television sale is also accrued. Television stars are no longer doomed to remain forever confined to the small screen, and performers like Bruce Willis or Michael J. Fox can readily make the leap from TV hits like *Moonlighting* and *Family Ties* to the global big-screen popularity of *Die Hard* (1988) or *Back to the Future* (1985) which will then have an even greater attraction for the television scheduler who already knows that a loyal home audience will tune in when it is later broadcast.

As attitudes to the 'threat' of television to the film industry have changed, so too has the level of appreciation for television's own heritage. Gone are the days when executives could make callous decisions to wipe the tapes of some of the medium's classic shows forever on the practical grounds that the space was needed. Proper respect for television's achievements and study of its history are only gradually coming to gain the level of application that prevails towards film but *I Love Lucy* or *Coronation Street* are every bit as seminal to the history of a vastly popular medium as *Birth Of A Nation* (1915) or *Citizen Kane* (1941) are to the history of the movies.

This *Chambers Film and Television Handbook* is an attempt to pay due homage to the past landmarks and current achievements of two heavily interrelated popular media.

The book consists of around seven hundred entries covering the careers of some of the most influential and significant performers and directors in both media. There are entries on individual films of note from *The Great Train Robbery* (1903) to *Dances With Wolves* (1990) and key television series from *Coronation Street* to *Twin Peaks* that have helped mould perceptions of that medium. Other entries explore subjects like film noir and neo-realism, provide insight into the history of the Oscar or the censorious Production Code and explain such technical details as the functions of a 'best boy' or the condition called persistence of vision that allows the human eye to see moving pictures.

Throughout, the bias has been towards the contemporary. Reference books abound that will provide the eager fan with details on the careers of Gregory Peck, Charlton Heston and other figures from earlier generations and there seemed more value in covering the likes of Kevin Costner and Pedro Almodovar, hence the contents include a biographical entry on Michael Douglas and not his father Kirk.

Anyone with a passing interest in film or television should find something of interest within these pages whether it is further details and an appreciation of a particular director's work, an explanation of a perplexing technical term or the answer to some niggling trivial pursuit question. The book is designed to appeal to the general reader but is packed with a level of detail, trivia and solid information that should also make it a good read for the armchair fan and a source of reference for the dedicated buff.

Inevitably, in selecting the contents of a volume like this the ultimate choices are subjective. To those who find themselves annoyed or disappointed that their favourite is not included, one can only reply that overwhelming demand on the part of particular individuals may be met in future editions.

Suggestions or comments, related to this subject and any other matter arising from the book, are welcome care of the publishers.

Allan Hunter
April 1991

EXPLANATORY NOTES

Individual biographical entries provide a chronological account of the key events and major films or television series in a career, noting as well all Oscar nominations and Oscar wins, in selected categories, since the first award. Individual film and television entries provide information on the product concerned, with a list of director, cast and, in the case of films, any major international award received e.g. Oscar, Cannes Palme D'Or etc.

Throughout the text all films are referred to by their original language title first with an English-language translation appearing in brackets. Titles covered by individual entries are listed under their best-known title with a cross-reference listing under the English-language or foreign-language equivalent, thus *L'Année Dernière A Marienbad* is listed under *Last Year at Marienbad* whilst *The Children of Paradise* will be found under its much more familiar title of *Les Enfants Du Paradis*.

The dating of film titles and biographical data is notoriously tricky. Throughout the text, films are listed by the year in which they were copyrighted. This may vary from their year of release, particularly in Britain where American films can sometimes take between six and nine months to make the journey across the Atlantic. In the case of birth dates, the most widely accepted date has been the one used, with any area of debate covered by a listing of more than one alternative. When actors or directors are mentioned within the text and do not have an entry of their own, birth and death dates are given in brackets after their names.

Abbreviations have largely been avoided and those that do appear are of a self-evident nature with b/w standing for black-and-white and col. short for colour.

Cross references to other relevant material are indicated by the use of the symbol ▷.

Contributors

The editor is indebted to the work of three colleagues whose contributions were crucial to the compilation of this volume. Entries on television drama and series were researched and written by Alastair McKay, whilst the entries on individual films were written and researched by Kenny Mathieson and Trevor Johnston.

Acknowledgements

Gill Crawford; Rosemary Goring; Jim Hickey; Richard Mowe; Peter Scott; the staff of the British Film Institute, Kobal Collection and Cinémathèque Française.

p 2 Courtesy Kobal Collection
p 4 © 1979 United Artists Corporation
p 16 © Collection Archives Photeb
p 18 © Collection Archives Photeb
p 21 © Palace Pictures
p 23 © Collection Archives Photeb
p 28 Courtesy Kobal Collection © Collection Archives Photeb
p 35 © Collection Archives Photeb
p 38 Courtesy Kobal Collection
p 49 © Collection Archives Photeb
p 61 © Collection Archives Photeb
p 63 © 1987 Paramount Pictures Corporation
p 69 Courtesy Kobal Collection
p 71 Courtesy Kobal Collection
p 73 Courtesy Kobal Collection
p 75 © Collection Archives Photeb
p 77 Courtesy Kobal Collection
p 78 © Collection Archives Photeb
p 82 © Cannon Film Distributors (UK) Ltd.
p 84 © Collection Archives Photeb
p 86 © Warner Bros., Inc.
p 96 © Collection Archives Photeb
p 98 © Collection Archives Photeb
p 100 © United Press International, Inc. © Collection Archives Photeb
p 103 Courtesy Kobal Collection
p 104 Courtesy Kobal Collection
p 105 © Collection Archives Photeb
p 110 Courtesy Kobal Collection
p 113 © 1986 Columbia Pictures Industries, Inc.
p 117 © Collection Archives Photeb
p 119 © Collection Archives Photeb
p 120 Courtesy Kobal Collection
p 123 © 1980 Paramount Pictures Corporation
p 124 © 1985 Warner Bros., Inc.
p 128 © Collection Archives Photeb
p 133 Courtesy Kobal Collection
p 135 © Collection Archives Photeb
p 143 © Collection Archives Photeb
p 145 © Collection Archives Photeb
p 148 © 1982 National Broadcasting Co., Inc.
p 149 © Universal Pictures
p 150 © 1982 Columbia Pictures Industries, Inc.
p 160 © Collection Archives Photeb
p 164 © 1986 Black Snake, Inc.
p 165 © Collection Archives Photeb
p 170 Courtesy Kobal Collection
p 172 Courtesy Kobal Collection
p 174 © Collection Archives Photeb
p 176 Courtesy Kobal Collection
p 177 © Toho Company, Ltd.
p 180 © 1973 United Artists
p 182 Courtesy Kobal Collection
p 183 © Collection Archives Photeb
p 184 Courtesy Kobal Collection
p 186 © 1989 Universal City Studios, Inc.
p 194 Courtesy Kobal Collection
p 195 Courtesy Kobal Collection
p 199 © 1986 De Laurentis Entertainment Group
p 202 © 1986 Columbia Pictures Industries, Inc.
p 216 © Collection Archives Photeb
p 221 Courtesy Kobal Collection
p 224 © Collection Archives Photeb
p 227 © 1984 Paramount Pictures Corporation
p 229 © Warner-Pathé
p 231 Courtesy Kobal Collection
p 233 © Collection Archives Photeb
p 235 © Collection Archives Photeb
p 237 Courtesy Kobal Collection
p 238 © Collection Archives Photeb
p 241 © Columbia Pictures Industries, Inc.
p 253 © The Rank Organisation
p 254 © Collection Archives Photeb
p 260 © Collection Archives Photeb
p 262 © 1960 Shamley Productions, Inc. (Courtesy Paramount Pictures Corporation)
p 272 © Collection Archives Photeb
p 288 © 1984 Orion Pictures Corporation

A BOUT DE SOUFFLE *see* BREATHLESS

ACADEMY AWARDS *see* OSCAR

THE ADVENTURES OF ROBIN HOOD

UK 1955–1959
Richard Greene, Alexander Gauge, Bernadette O'Farrell, Patricia Driscoll, Archie Duncan, Paul Eddington, Alan Wheatley.

Whilst unable to rival the panache of ▷ Errol Flynn's portrayal in *The Adventures of Robin Hood* (1938), Richard Greene (1918–1985) and his merry men kept armchair viewers glued through four years and 143 episodes of one of ▷ Lew Grade's first attempts to crack the American market.

The hearty tales of wholesome combat against Alan Wheatley's villainous Sheriff of Nottingham were popular enough to spawn a whole series of imitators and the influence of the series can be seen on *The Buccaneers* (1956) which starred Robert Shaw (1927–78) as a reformed 17th century pirate, *The Adventures of William Tell* (1957) once described as 'Robin Hood in the Alps' and *Sword of Freedom* (1957) which starred Edmund Purdom (1924–) as a 15th century Italianate variation on the character. Shot mostly in studio, a twenty-foot hollow tree trunk on wheels served as a good deal of the famed Sherwood Forest. Greene repeated his role in the cinema tale *Sword of Sherwood Forest* (1961) and the story of a blacklisted Hollywood writer working on episodes of the series formed the backdrop of the 1989 feature film *Fellow Traveller*.

British television returned to the adventures of Robin Hood with *Robin of Sherwood* (1984–6) in which the legendary outlaw was a younger, more virile figure involved with magic, pagan Gods and ancient folklore. Michael Praed (1960–) starred for the first two series with Jason Connery (1963–) assuming the role thereafter.

ALEXANDER NEVSKY

Sergei Eisenstein (USSR 1938)
Nikolai Cherkasov, Nikolai Okhlopkov, Alexander Abrikosov, Dimitri Orlov.
112 mins. b/w.

While ▷ Sergei Eisenstein's pioneering Twenties work, ▷ *Battleship Potemkin* (1925) and *October* (1928) for example, formulated the innovative editing technique and focused on the Soviet masses as collective hero, the films that were to follow in subsequent decades, *Alexander Nevsky* and the Forties *Ivan The Terrible* (1944–6) diptych, seem much more traditional both in formal terms and in their narrative focus upon one heroic individual. *Alexander Nevsky* is the first flowering of this later style and the filmmaker's efforts should be viewed, not as a retreat from the cerebral cinematic exploration of the earlier period, but as an advance in developing his method to meet the political and technological demands of the time. Dramatising Prince Nevsky's drive to expel the brutal Teutonic knights from thirteenth century Holy Russia, the film draws on established literary and artistic means of representation, and given the extent to which the impressive pageant of action is fused with Prokofiev's stirring score Eisenstein himself called the structure 'symphonic'. To contemporary eyes raised largely on classical American cinema, it thus remains one of his most accessible works, with the breathtaking Battle of The Ice sequence more than a match for many a Hollywood epic set-piece.

THE AFRICAN QUEEN

John Huston (USA 1951)
Katharine Hepburn, Humphrey Bogart, Robert Morley, Peter Bull.
103 mins. col.
Academy Award: *Best Actor* Humphrey Bogart.

Humphrey Bogart in *The African Queen*

In 1951 shooting on location was much less common than it has since become, and so ▷ John Huston's insistence in filming this adaptation of C. S. Forester's novel some 1,100 miles up the Congo was at the time looked upon as near fanatical. As scripted by influential film critic James Agee (1909–55) ▷ Katharine Hepburn's prim English missionary Rose Sayer and ▷ Humphrey Bogart's grizzled Canadian boat captain Charlie Allnut are two ill-matched souls brought together in adversity, navigating hazardous river conditions in World War I East Africa to escape the German forces and conquering initial antipathy to reluctantly fall in love. In retrospect, Huston's gamble came off, for the very real difficulties of the surroundings are lucidly preserved on screen, and combining with the chemistry of the stars, help to overshadow the narrative's many moments of improbability; while the film's sentimentality (as opposed to the cynical worldview expressed in the director's earlier *Treasure of The Sierra Madre* (1947), for example) ensures that it is always fondly remembered. Uncredited co-writer Peter Viertel's 1953 novel *White Hunter, Black Heart*, chronicling Huston's obsessional excesses was creditably filmed in 1990 by actor/director ▷ Clint Eastwood.

L'AGE D'OR

Luis Buñuel (France 1930)
Lya Lys, Gaston Modot, Max Ernst, Pierre Prevert.
60 mins. b/w.

Iconoclastic Spaniard ▷ Luis Buñuel's collaboration with his equally eccentric countryman Salvador Dali (1904–1989) marks a key moment in the development of the European film as art. Following their groundbreaking work together on the celebrated short *Un Chien Andalou* (1928), noted for its opening shot of an eye brazenly sliced open by a straight razor, Buñuel more firmly allied himself to the ▷ surrealist cause and was fortunate to find a wealthy sponsor in the Vicomte de Noailles, who handed the director a million francs and a free hand to go with it. The result, *L'Age D'Or*, caused a riot on its first appearance, when right wing extremists clubbed fellow audience members, and the film remained unshown in France until 1979. Following its own dream logic, the piece focused on two lovers forever kept apart by the constraints of the church and of bourgeois morality, and its typical surrealist juxtapositions included the invasion of a rocky shoreline by Catholic bishops in full regalia and the scandalous appearance of Christ at a Sadean orgy. The themes and method of the film were to become constants throughout Bunuel's long and distinguished career: his particular brand of ecclesiastical satire was later even more apparent in the blasphemous *Viridiana* (1961), while his obsession with amour fou (signified here by a memorable bout of toe-sucking) surfaced most bizarrely in *Belle de Jour* (1967).

L'ALBERO DEGLI ZOCCOLLI

see **TREE OF WOODEN CLOGS**

ALL ABOUT EVE

Joseph L. Mankiewicz (US 1950)
Bette Davis, Anne Baxter, George Sanders, Celeste Holm.
138 mins. b/w.
Academy Award: *Best Picture*; *Best Director*; *Best Supporting Actor* George Sanders; *Best Screenplay* Joseph L. Mankiewicz; *Best Sound Recording*; *Best Costume Design* Edith Head *and* Charles Le Maire.
Cannes Film Festival: *Special Jury Prize*; *Best Actress* Bette Davis.

'Fasten your seatbelts, it's going to be a bumpy night' remains the most famous line from this acerbic tale of back-biting showbusiness folk, its very quotability a testament to both the vigour of ▷ Bette Davis's performance as fading Broadway star Margo Channing and the sheer deftness of writer/director ▷ Joseph L. Mankiewicz's feel for scripting. His films, the 1949 Oscar-winner *A Letter To Three Wives* and 1953 Shakespearean adaptation *Julius Caesar* amongst them, have been characterised and sometimes criticised for their reliance on verbal rather than visual communication, and while *All About Eve*'s chronicle of the vitriolic thespian conflict between the mature Davis and the rising Baxter out to destroy her does little to refute such claims, it does offer a reminder of Mankiewicz's unshowy elan for staging multi-character scenes and getting the best out of all his performers. Davis (making something of a comeback after a period of career doldrums) and Baxter are certainly outstanding, but George Sanders as the sardonic drama critic Addison de Witt also offers a textbook demonstration in the piquancies of cynicism, introducing a young ▷ Marilyn Monroe as 'a graduate of the Copacabana school of acting'.

ALL IN THE FAMILY

USA 1970–9
Carroll O'Connor, Jean Stapleton, Sally Struthers, Rob Reiner

▷ Norman Lear's transplantation of the British sitcom ▷ *Till Death Us Do Part* from the East End of London to lower-class New York revolutionised the form in the US and was the highest rated series on American television for five consecutive years, with an audience of around 50 million.

ABC rejected the show, which was then picked up by CBS network president Robert Wood. Though ready to go into production in 1969, it was delayed while research was carried out on the possible impact of its upfront style, which was at odds with the cosiness of TV comedy of the time.

Archie Bunker (O'Connor), the Alf Garnett character, was a bigoted, loud-mouthed dock foreman who, from the comfort of his easy chair, witnessed the televised evolution of a world with which he was profoundly at odds. Lear said the show was modelled on his childhood in Queens, with the central character based on his Jewish father, yet Bunker was strictly a WASP. The only breadwinner in a house of four adults, he struggled with economic misfortune and railed against a society which he thought promoted the rights of deviants and political insurgents above his own. As with Garnett there were questions about audience identification with the character, but there were limits on the show's realism (blacks were 'coloreds' not 'niggers'). The intended heroes were Bunker's self-determined daughter Gloria (Struthers), and her husband Mike 'Meathead' Stivic (Reiner), the post-Woodstock embodiment of liberal reason.

When the show began to fade it spun off into *Archie Bunker's Place* (in a disastrous piece of upward mobility Bunker was given a bar to run, losing the disgruntled essence of his character). Bunker's chair – the US TV icon of the 1970s – was gifted to the Smithsonian – while his black neighbours were kept alive in *The Jeffersons* and *Maude*, which broke down sitcom apartheid and prepared the way for *The Cosby Show*, the sitcom success of the 1980s.

ALL QUIET ON THE WESTERN FRONT

Lewis Milestone (USA 1930)
Lew Ayres, Louis Wolheim, John Wray, Slim Summerville.
140 mins. b/w.
Academy Award: *Best Picture*; *Best Director*.

The final image here of a hand stretching out to touch a butterfly before receiving a fatal enemy bullet is undoubtedly one of the most famous moments in all cinema, and Lewis Milestone's film even today is still regarded as one of the classic celluloid anti-war statements. Following the pattern set by 1925's silent blockbuster *The Big Parade* and echoed as recently as 1989 in ▷ Oliver Stone's *Born On The Fourth Of July*, this faithful adaptation of Erich Maria Remarque's famous novel follows the fortunes of volunteer recruits to the 1914 German war effort as they have their patriotic illusions of glory shattered by the grim horrors of battle. Although Ayres (1908–) (a conscientious objector during World War II himself) takes top billing, the focus on the group makes it something of an ensemble movie, with the real star perhaps being director Milestone's bravura staging of the action sequences. Taking over a huge California ranch, he was the first to use a giant crane with which to sweep the camera across the bullet-riddled muddy landscapes, and the startling results have only rarely been bettered by the likes of ▷ Stanley Kubrick's work on *Paths of Glory*. With Milestone thus occupied, it's also worth remembering ▷ George Cukor's input rehearsing the actors' performances, which pleased the studio enough to set him off on a lauded, though very different, Hollywood canon.

ALLEN, Woody

Real Name Allen Stewart Konigsberg
Born 1 December 1935, Brooklyn, New York, USA.

The humorous insights of Woody Allen began as one-line jokes which he sold to newspapers and entertainers. Expelled from New York University and the City College of New York, he wrote gags for a number of performers and joined the Writer's Programme at NBC Television where his comic skills were honed as a writer for comedian Sid Caesar (1922–).

In the early 1960s, he began performing in nightclubs as a stand-up comedian and developed his distinctive persona of the bespectacled urban neurotic; eternally priapic but always befuddled by his own romantic inadequacies.

He made his film debut as a writer and performer in *What's New Pussycat?* (1965) and re-dubbed and re-edited the Japanese film *Kagi No Kag* (*Key of Keys*) into *What's Up, Tiger Lily?* (1966) an espionage spoof about the search for an egg salad recipe.

He then embarked on a prolific filmmaking career as writer, director and performer with early efforts that revealed his admiration for ▷ Bob Hope, ▷ Groucho Marx and classic Hollywood slapstick in frenetically-paced outbursts of anarchy and genre parodies like *Bananas* (1971), *Sleeper* (1973) and *Love and Death* (1975).

▷ *Annie Hall* (1977) marked a shift in style and substance to more concentrated autobiographical pieces and won him Oscars for Best Director and for co-writing the screenplay. He has averaged a film per annum since then, exploring his concerns with mortality, human foibles, sexual longing, showbusiness nostalgia, psychoanalysis and urban living in such productions as *Manhattan* (1979), *Broadway Danny Rose* (1984), *The Purple Rose of Cairo* (1985) and *Hannah and Her Sisters* (1986) (Best Original Screenplay Oscar).

Attempts at more serious, contemplative work like *Interiors* (1978) and *September* (1987) received a less warm welcome but he provided a rich and rewarding portrait of an academic's mid-life crisis in *Another Woman* (1988) and expertly blended the comic and tragic in his witty, profound commentary on a Godless society—*Crimes and Misdemeanours* (1989).

Remaining as productive as ever, he followed this with an acting appearance opposite Bette Midler (1945–) in *Scenes from the Mall* (1990), his own film *Alice* (1990) (Best Original Screenplay Oscar nomination) in which a married woman re-examines her life, and an untitled project in which Madonna (1958–) stars as a circus trapeze artist.

His rare appearances for other directors comprise *Casino Royale* (1967), *Play It Again, Sam* (1972) which was adapted from his Broadway success of 1969, *The Front* (1976) and *King Lear* (1987). A frequent contributor to New Yorker, his books include *Getting Even* (1971) and *Without Feathers* (1976).

Woody Allen and Diane Keaton in *Manhattan* (1979)

ALMODOVAR, Pedro
Born 25 September 1951, Calzada de Calatrava, Spain.

Moving to Madrid at the age of 17, Almodovar spent ten years as an employee of Spain's National Telephone Company alternating his day job with an array of extra-curricular activities that included the creation of cartoon strips for such comics and underground magazines as *Star*, *Vibora* and *Vibraciones*.

He has chronicled the continuing memoirs of fictional porn star Patty Diphusa in *La Luna* magazine, published a novella *Fuego en las Entranas* (Guts on Fire) and a 'porno-photo story' *Todo Tuya* (All Yours) as well as forming punk pop group Almodovar and McNamara which he fronted bedecked in leather miniskirt, fishnet stockings and platform shoes.

Whilst appearing with the Los Golliardos theatre group he first met many of the actors who would later form his film repertory company, including Antonio Banderas (1960–) and Carmen Maura (1945–). He also began making short films on Super-8 including *Dos Putas, O Historia De Amor Que Termina En Boda* (Two Whores, Or A Love Story Which Ends in Marriage) (1974), *La Caida De Sodoma* (The Fall of Sodom) (1974) and *Sexo Va, Sexo Viene* (Sex Comes and Goes) (1977).

His first feature *Folle, Folle, Folle Me Tim* (Fuck, Fuck, Fuck Me Tim) (1978) was followed by *Pepi, Luci, Bom Y Otras Chicas Del Monton* (Pepi, Lucy, Bom and A Whole Lot of Other Girls) (1979–80) an elaborate tale of revenge that includes lesbianism, graphic urination and a competition to discover the largest erection among the males present. Taking full advantage of the post-Franco cultural freedoms he had developed a recognisable personal style that consisted of provocative subject matters, iconoclastic attitudes, a flamboyant use of rich colours and a playful perspective on the cliches of popular culture.

His narratives would remain deliberately provocative as he grew technically more sophisticated. His sympathy for society's outsiders, understanding of women and disdain for the established order fuelled such work as *Entre Tinieblas* (Dark Habits) (1983) in which a nightclub singer takes refuge in a convent where the unorthodox nuns glory in such unlikely monikers as Sister Shit and Sister Sin.

Like a naughty schoolboy, he loves to shock but his films often lack the cohesion and substance to support his consistently interesting notions. *Matador* (1986) in which a couple are drawn to each other by an intense and erotic attraction to death represents one of his most complex and successful fusions of story and style.

With the gay melodrama *La Ley Del Deseo* (Law of Desire) (1987) he began to acquire an international reputation and enjoyed a worldwide success with the frenetic farce *Mujeres Al Borde de Un Ataque de Nervous* (Women on the Verge of a Nervous Breakdown) (1988) which won 50 prizes and an Oscar nomination for Best Foreign Film. He followed this with *Atame!* (Tie Me Up! Tie Me Down!) (1990) in which a psychiatric patient kidnaps the object of his affections and holds her captive until she acknowledges the sincerity of his love. A controversial 'romance', it once again failed to completely fulfil the promise of its subject matter.

ALTMAN, Robert

Born 20 February 1925, Kansas City, Missouri, USA.

Altman toiled long and hard on a chequered early career before enjoying any kind of cinematic success. Employed on industrial documentaries for a company in Kansas City, his first attempts at features, *The Delinquents* (1957) and *The James Dean Story* (1957), attracted little attention.

From 1957, he worked in television on such series as *Bonanza* and *Combat*, returning to the cinema with *Countdown* (1967). The critical and commercial success of *M*A*S*H* (1970) firmly established his credentials and inaugurated a prodigious body of work.

Consistently working against the grain of audience expectations, his films have sought to explode myths by exploring cherished American institutions and hallowed film genres. *McCabe and Mrs Miller* (1971) avoided the popular cliches of frontier-taming cowboys to show a more squalid nation founded by cheats, liars and frauds. *The Long Goodbye* (1973) was as far from the stereotypical trenchcoated detective as Elliott Gould (1938–) is from ▷ Humphrey Bogart, while his 24-character bi-centennial fresco *Nashville* (1975) was typically cynical rather than celebratory.

His distinctive style of multi-track sound, overlapping, seemingly spontaneous, dialogue and persistent search for the anti-heroic has lent an experimental air to his work. His audacity and iconoclasm, in *Buffalo Bill and the Indians* (1976) for example, have not often been appreciated at the box-office.

Also interested in portraits of the female psyche like *Images* (1972) and *Three Women* (1977), he has recently concentrated on screen adaptations of plays with variable results; *Come Back to the Five and Dime, Jimmy Dean, Jimmy Dean* (1982), *Streamers* (1983) and *Secret Honor* (1984) were among the more accomplished.

Returning to television, he was acclaimed for *Tanner* (1988) which followed a fictional Presidential campaign through an inventive mix of fact and drama, and his Van Gogh mini-series *Vincent and Theo* (1990) avoided the romantic notions of the tragic, misunderstood artist genre to focus on a more human level of failure, desperation and the daily grinding relationship between art and life.

As the producer of such films as *The Late Show* (1977), *Welcome to L.A.* (1977) and *Remember My Name* (1978) he was instrumental in promoting the directorial careers of Robert Benton (1932–) and ▷ Alan Rudolph.

AMERICA

UK 1972

Presenter Alistair Cooke (1908–)
Producer Michael Gill.

Lavish platform and a sharp-eyed camera crew for the unique stateside musing of journalist Alistair Cooke, whose understanding of the country is matched by his wit and incisive intelligence. Cooke, born in Manchester, UK, first visited the US in 1932, filing impressionistic reports for *The Times* and the *Manchester Guardian*, before embarking, in 1947, on the weekly BBC radio slot *Letter from America* – now the longest-running talk show on radio. Cooke became an American citizen in 1941, but his reflections have the freshness of a first-time visitor. In *America*, television audiences were introduced to his measured approach. From Columbus and Drake through slave trade, revolution and gold rush, to superpower status and the nuclear stalemate of the Cold War, Cooke was on site with pertinent observations. The series won an Emmy, Cooke took a British Academy Award, and the book of the series became a bestseller.

AMERICAN FILM INSTITUTE (AFI) Founded in 1967 with government and private funding the AFI exists to foster an appreciation of all aspects of the cinema through its programmes of education, preservation, retrospective screenings and grants to young filmmakers. Based at the Kennedy Centre for the Performing Arts in Washington, it also runs a Centre for Advanced Film Studies in Beverly Hills, California and publishes the magazine *American Film*. George Stevens Jnr. (1938–) was the founding director and chief executive from 1967 to 1980 and currently serves as co-chairman.

Its most visible international act is the annual presentation of a prestigious Life Achievement Award honouring a career of distinction within the industry. The recipients to date are; 1973 – ▷ John Ford, 1974 – ▷ James Cagney, 1975 – ▷ Orson Welles, 1976 – ▷ William Wyler, 1977 – ▷ Bette Davis, 1978 – ▷ Henry Fonda, 1979 – ▷ Alfred Hitchcock, 1980 – ▷ James Stewart, 1981 – ▷ Fred Astaire, 1982 – ▷ Frank Capra, 1983 – ▷ John Huston, 1984 – ▷ Lillian Gish, 1985 – ▷ Gene Kelly, 1986 – ▷ Billy Wilder, 1987 – ▷ Barbara Stanwyck, 1988 – ▷ Jack Lemmon, 1989 – Gregory Peck (1916–), 1990 – ▷ David Lean and 1991 – Kirk Douglas (1916–).

AN AMERICAN IN PARIS
Vincente Minelli (USA 1951)
Gene Kelly, Leslie Caron, Oscar Levant, Georges Guetary.
113 mins. col.
Academy Award: *Best Picture*; *Best Screenplay* Alan Jay Lerner; *Best Cinematography* Al Gilks, John Alton; *Best Art Direction* Preston Ames, Cedric Gibbons; *Best Musical Scoring* Saul Chaplin, Johnny Green; *Best Costume Design* Orry Kelly, Walter Plunkett, Irene Sharaff.

An American in Paris is an important milestone in the development of the musical as one of the first of the genre to win serious respect from the critical establishment and the American Academy. Sprung from producer Arthur Freed's influential musical unit at MGM, responsible for a string of classics including *Meet Me In St Louis* (1944), ▷ *Singin' in the Rain* (1952) and *The Band Wagon* (1953), the film's origin in an orchestral piece by George Gershwin (1898–1937) and its climactic 17-minute modern ballet sequence, which draws heavily for visual inspiration on famous paintings (Toulouse Lautrec, Van Gogh, Manet, Renoir), announced unprecedented aspirations to higher artistic credibility. While these very same qualities have latterly seen the film criticised for its over-ripe pretension, the high-falutin' trappings are actually quite germane to the typical ▷ Vincente Minnelli approach of telling the story through a combination of song, dance, dialogue and decors as frustrated young Yank artist ▷ Gene Kelly falls for the charms of French discovery Leslie Caron (1931–), and the degree of sophistication in all aspects of the production remains a testament to the studio's high standards of creative excellence right across the board. Most significantly perhaps, the focus on actor-choreographer Kelly's exuberant and contemporary routines greatly expanded the filmic dance vocabulary by integrally linking it to the narrative and thus challenged the received notion that a popular audience would not accept ballet forms in a mainstream entertainment. In some respects, it's a challenge that all too few subsequent Hollywood filmmakers have ventured to take up.

ANAMORPHIC LENS A lens with cylindrical elements giving different magnifications in horizontal and vertical directions. In wide-screen cinematography like ▷ CinemaScope the horizontal axis or width is compressed laterally in the camera to half its size and then expanded to compensate in projection. The vertical axis is left undistorted in both instances resulting in a wide-screen image. The process was developed during World War I by Henri Chrétien (1879–1956) and later used for aerial photography in map-making. Its first use in the cinema is acknowledged to be in *Construire Un Feu* (1926) directed by Claude Autant-Laura (1903–).

ANDERSON, G. M. ('Bronco Billy')

Real Name Max Aranson
Born 21 March 1882, Little Rock, Arkansas, USA.
Died 20 January 1971, South Pasadena, California, USA.

A travelling salesman and relatively unsuccessful actor on the New York stage, Anderson made his film debut in *The Messenger Boy's Mistake* (1902) and played several small roles in the famous ▷ *The Great Train Robbery* (1903), an eleven-minute western also directed by ▷ Edwin S. Porter and often credited as the first film to tell a substantial fictional story.

He subsequently joined the Vitagraph studios as an actor and director and, in 1907, helped form the Essanay Film Manufacturing Company. After experimenting with location filming in Colorado and New Mexico, the company chose California as a site for its outdoor stories and was therefore instrumental in establishing that state as a filmmaking centre.

Convinced of the potential popularity in filmic tales of the old West and deciding that audiences were more likely to identify with a strongly drawn recurring character, he appeared in *Bronco Billy's Redemption* (1910) and would star in an estimated 400 short westerns over the next decade including such inventive titles as *Bronco Billy's Christmas Dinner* (1911), *Bronco Billy and The Squatter's Daughter* (1913) and *Bronco Billy and the Revenue Agent* (1916).

He resigned from Essanay in 1916 and retired from acting in 1920, although still directing the likes of *Ashes* (1922). Absent from the screen for over four decades, he returned for a one-off cameo contribution to *The Bounty Killer* (1965).

In 1958, he received an honorary Oscar in recognition of his status as a 'motion picture pioneer, and for his contributions to the development of motion pictures as entertainment'.

ANDERSON, Gerry

Born 14 April 1929.

A youthful entrant to the British film industry as a trainee with the Colonial Film Unit, Anderson later worked as an assistant editor on such films as *The Wicked Lady* (1945) before co-founding Pentagon Films in 1955.

Initially intent on making commercials, he co-produced and directed such television series as *The Adventures of Twizzle* (1956) and *Torchy, the Battery Boy* (1957). He subsequently enjoyed great success with children's adventure series that combined a range of puppet characters with technologically advanced hardware and special effects. Among the best known are *Fireball XL-5* (1961), *Stingray* (1962–3), *Thunderbirds* (1964–6), *Captain Scarlett* (1967) and *Joe 90* (1968).

He later branched out into films and live-action shows with human actors like *The Protectors* (1971–2) and *Space 1999* (1973–6) before returning to the use of increasingly sophisticated puppetry in *Terrahawks* (1983–4) and *Dick Spanner* (1987).

ANDREI RUBLEV

Andrei Tarkovsky (USSR 1966)
Anatoli Solonitzine, Ivan Lapikov, Nikolai Grinko, Nikolai Bourliaiev.
185 mins. b/w. & col.
Cannes Film Festival: *International Critics Award.*

Although completed as far back as 1965, *Andrei Rublev*, ▷ Andrei Tarkovsky's epic panorama of medieval life, waited until the 1969 Cannes Film Festival before it was first seen in the West and faced a further three year delay before distribution in the USSR. Apparently objecting to the film's violent realism, the Soviet authorities who shelved the film for so long were, paradoxically enough, also responsible for providing the first in a series of lavish budgets Tarkovsky would receive for works of uncompromisingly distinctive artistry, including the ruminative science fiction pieces *Solaris* (1972), *Stalker* (1979), and the highly personal recollections of *Mirror* (1974). Perhaps appropriately, *Andrei Rublev* dramatises the time-honoured dilemma of the artist's relationship with society. The eponymous hero, a monk and icon painter who lived during the 15th century, bears witness to a world of constant misery and brutality as the Tartar hordes plunder all before them, and begins to ponder the worth of his existence until he watches a peasant waif's extraordinarily confident supervision of the casting of a prestigious church bell. The young lad's instinctive drive towards the completion of his project is proof to the doubting protagonist that the God-given gift of creativity flourishes even in times of the profoundest social turmoil, and as Tarkovsky bursts into colour to linger over Rublev's exquisite icons, the parallel with the position of the pre-perestroika Soviet artist is taken as read.

ANDREWS, Julie

Real Name Julia Elizabeth Wells
Born 1 October 1935, Walton-on-Thames, UK.

Born into a showbusiness family, Andrews took singing lessons as a child to develop her four-octave vocal range and made her London debut in the revue *Starlight Roof* (1947). Popular on radio and the stage, she appeared in *Cinderella* (1953) at the London Palladium and then enjoyed repeated success on Broadway in the musicals *The Boyfriend* (1954), *My Fair Lady* (1956–9) and *Camelot* (1960).

She lent her vocal talents to the film *Rose of Baghdad* (1952) but her allegiance to the theatre delayed her entry into films, although she did appear before the cameras in such television specials as *High Tor* (1956) with ▷ Bing Crosby and *Cinderella* (1959).

Passed over in favour of ▷ Audrey Hepburn for the 1964 film version of *My Fair Lady*, she won a Best Actress Oscar for her eventual debut as the magical, flying nanny in *Mary Poppins* (1964) and consolidated her stardom with a further nomination as the novice nun in the phenomenally popular ▷ *The Sound of Music* (1965).

For a while, her wholesome girl-next-door persona, refreshingly unaffected personality, gaiety and impeccable singing voice made her one of the world's best loved personalities. She was effective in straight dramas like *The Americanization of Emily* (1964) and *Hawaii* (1966) but musicals proved her forte as she illustrated in *Thoroughly Modern Millie* (1967) and *Star!* (1968).

The virtual demise of the lavish, old-style Hollywood musical and the financial failure of *Darling Lili* (1970) damaged her career opportunities. Over the past twenty years, she has appeared almost exclusively in the films of her second husband, director Blake Edwards (1922–), making strenuous efforts to break the limitations of her saccharine, somewhat prim, image, notably with her topless appearance in the dyspeptic *S.O.B.* (1981) and as the transvestite 1930s entertainer in *Victor/Victoria* (1982) which earned her another Best Actress Oscar nomination.

Recent films include a rare dramatic role in *Duet for One* (1986) and the comedy *Chin-Chin* (1990) opposite ▷ Marcello Mastroianni.

A recording artist and star of innumerable television specials, she has also written children's fiction including *The Last of the Really Great Whangdoodles* (1973).

ANGELOPOULOS, Theodoros

Born 27 April 1935, Athens, Greece.

A student of law in Athens and at the Sorbonne in Paris, Angelopoulos practised law whilst writing short stories and poetry. The film critic for *Dimoktatiki Allaghi* from 1964 to 1967, he made an abortive attempt to start filmmaking in 1965 and subsequently worked as an actor and line producer before making his directorial debut with *Ekpombi* (The Broadcast) (1968).

He followed this with *Anaparastassi* (Reconstruction) (1970), a tale of infidelity and murder that subverted the standard narrative drive of ▷ film noir with an austere tone and deromanticised portrait of Greece. He then embarked on a trilogy of films exploring recent Greek history.

Meres Tou 36 (Days of '36) (1972) tells a true story of a political prisoner who takes a hostage and demands his freedom in return for the man's life. A tense, claustrophobic tale it rails against injustice and the inertia and incompetence of the authorities. *O Thassios* (The Travelling Players) (1975), generally considered his masterpiece, is a slowly unravelling four-hour epic, redolent of the myth of Mycenae, that conveys the agonising political and personal experiences of an itinerant theatrical troupe between 1939 and 1952. *I Kynighi* (The Huntsmen) (1977) uses the discovery of a corpse to examine memories of the Civil War and its aftermath.

Fond of establishing a slow, meditative mood through long, uninterrupted shots and images of bleak and barren terrain, he has long been regarded as Greece's most eminent film director. In the 1980s he continues to use historical incidents to reflect on contemporary Greece in films like *O Megalexandros* (Alexander The Great) (1980) and *Taxid Sta Kithira* (Voyage to Cythera) (1984). Recent work, including *O Melissokomos* (The *Beekeeper*) (1986) has, if anything, grown more ruminative and pessimistic.

Recently, *Topio Stin Omichli* (Landscape in the Mist) (1989) received the Felix Award as European Film of the Year. Covering a journey from Greece to Germany undertaken by two children, it recounts moments both magical and miserable as they encounter society's confusion, lack of compassion and abandonment of basic values.

His latest film, *The Suspended Step of The Stork* starring ▷ Jeanne Moreau and ▷ Marcello Mastroianni was denounced as 'blasphemous,, unpatriotic and pornographic' just as filming began in December of 1990 and promises to be his most controversial work.

ANGER, Kenneth

Born 1930, Santa Monica, California, USA.

An influential character in the American ▷ avant-garde, Anger was a child of Hollywood who took tap dancing lessons with ▷ Shirley Temple and appeared as the Changeling Prince in *A Midsummer Night's Dream* (1935). He began making his own films with the family's 16mm Kodak camera and his earliest efforts include *Who Has Been Rocking My Dream Boat* (1941), a montage of children at play, *Tinsel Tree* (1942), the ritual decoration and destruction of a Christmas Tree, and *Prisoner of Mars* (1942) a science-fiction version of the Minotaur myth in which he also appeared.

He gained considerable notoriety with *Fireworks* (1947) which details the homoerotic, masochistic fantasies of a lonely adolescent. Admired by ▷ Jean Cocteau, he moved to Europe making the gentler fantasy of *Rabbit's Moon* (1950), *Eaux D'Artifice* (1953) set in the Tivoli water gardens, and the mythical *Inauguration of the Pleasure Dome* (1954).

Interested in the iconography of popular culture, his films are deftly edited works of visual splendour, violent emotions and sexual frankness that explore his obsessions with ritual, mythology and, increasingly, the occult. An anarchic sensibility was at play on his return to America with the creation of *Scorpio Rising* (1962–3) a paean to the motorcycle cult that exposes the homoeroticism and violence inherent in this symbol of American manhood and, implicitly, in American society.

His career has been littered with abortive or unrealised projects and when the completed print of *Lucifer Rising* (1967) was stolen by a member of the Manson gang, he placed an advertisement in *Village Voice* announcing his retirement from filmmaking. However, he used some of the remaining footage to create *Invocation of My Demon Brother* (1969) and new versions of *Lucifer Rising*, a mystical piece shot in Egypt, have appeared in 1974 and 1980.

He has also gained a reputation for his intimate revelations of celebrities' scandalous misdemeanours in *Hollywood Babylon* which was first published in Paris in 1959 and *Hollywood Babylon II* (1984).

ANNAUD, Jean-Jacques

Born 1 October 1943, Draveil, near Paris, France.

An avid cinemagoer who collected old movie projectors as a child, Annaud studied extensively at the IDHEC (Institut Des Hautes Etudes Cinématographiques) and the Sorbonne, where he received a B.A. in Literature.

After completing his compulsory military service in Africa, he returned to France and secured employment as a director of television commercials; averaging 80 per annum and gaining a prizewinning reputation as a versatile and inventive practitioner of this very specific craft.

He made his debut as a feature film director with *La Victoire En Chantant* (Black and White in Colour) (1976) which won the Oscar as Best Foreign Film. Set in a remote West African trading post in 1914, it satirised the muddled responses to the onset of the World War through the actions of French colonialists who attack a German fort, and was particularly commended for its evocative photography of the Ivory Coast.

He followed this with *Hot Head* (1979) and then *Quest for Fire* (1981) covering the misadventures of three members of a peaceable primitive tribe as they go in search of fire, the most valuable commodity then known to man. Using languages created by Anthony Burgess (1917–) and body movements devised by Desmond Morris (1928–), it displayed an entertaining, humanistic approach to the dramatic presentation of ancient history.

He revealed a similar flair for immersing himself in the detail of the past with his painstaking 'palimpsest' of the 14th century mystery *Der name Der Rose* (The Name of the Rose) (1986).

Ever eclectic in his choice of subject matter, he then lavished care and attention on *The Bear* (1988) a compelling documentary-like portrait of ursine dignity set in the 19th century and making expressive use of soundtrack and Italian locations. Effectively illuminating his recurring preoccupation with promoting non-violence and the folly of conflict it proved to be his most commercially successful project to date. His latest film is *L'Amant* (The Lover) (1991).

L'ANNEE DERNIERE A MARIENBAD *see* LAST YEAR AT MARIENBAD

ANNIE HALL

Woody Allen (US 1977)
Woody Allen, Diane Keaton, Tony Roberts, Paul Simon, Sigourney Weaver, Jeff Goldblum.
93 mins. col.
Academy Award: *Best Picture*; *Best Director*; *Best Actress* Diane Keaton; *Best Original Screenplay* Woody Allen, Marshall Brickman.

With *Annie Hall*, ▷ Woody Allen left behind the dominant parodic (*Love and Death*, 1975) or revue-style modes (*Bananas*, 1971) in which the former gag-writer and nightclub comedian's earlier work had been pitched to achieve the first unified work in the self-conscious, East Coast Jewish upper-middlebrow manner now his trademark. Although the earlier films had established the Allen screen persona, the aspirant romantic as existential schmuck, here he managed a more emotionally satisfying narrative core by contextualising his star turn in a directly confessional love story – drawing on his own highly public Seventies liaison with co-star Diane Keaton (1946–). Without totally eschewing the comic tricks of yore – one scene uses subtitles to hilariously reveal the characters' true feelings, another gag utilises an appearance by media guru Marshal MacLuhan for its effect – the film parades Allen's personal neuroses (the transcience of love and happiness) with a blend of aphoristic wit and would-be seriousness. As such it's the pivotal entry in the director's filmography, establishing a template for his later, more accomplished variations on a similar theme in *Hannah and Her Sisters* (1986) and *Crimes and Misdemeanours* (1989), and setting Allen on a path towards recognition as one of America's most respected film talents.

ANSWER PRINT The first complete print from the edited negative of a film combining synchronized sound and image that is supplied by the laboratory to the filmmakers for their approval as to colour grading, quality of light and other technical considerations. If the answer to its acceptability is in the affirmative then the many release prints of a film will be struck from this original. More often, it will undergo numerous alterations and fine-tuning. It is also known as an approval print or trial print.

ANTONIONI, Michelangelo

Born 29 September 1912, Ferrara, Italy.

Educated in economics and commerce at the University of Bolgona, Antonioni worked in a bank and wrote for the newspaper *Il Corriere Padano* before moving to Rome and writing film criticism for a number of periodicals including *Cinema*.

He subsequently served as an assistant on films like *Il Due Foscari* (1942) and *Les Visiteurs Du Soir* (1942) before embarking on a career as a scriptwriter on such diverse productions as *Una Pilota Ritorna* (1942) and *Lo Sceicco Bianco* (The White Sheik) (1951). He made his directorial debut with the short documentary on fishermen *Gente del Po*, which was begun in 1943 and completed in 1947. His feature-length debut followed with *Cronaca Di Un Amore* (Chronicle Of A Love) (1950) which focused on the guilt of an adulterous couple when the untimely accidental death of the duped husband pre-empts their own intention to engineer his demise.

Throughout the 1950s, in films like *I Vinti* (The Vanquished) (1952), *Le Amiche* (Girlfriends) (1955) and *Il Grido* (The Cry) (1957) he continued to explore the emotional state of love and its afterglow, developing a precise, personal style of long, slow camera movements and an eschewal of traditional event-filled linear narratives.

He made an international breakthrough with ▷ *L'Avventura* (1960) which explored the relationship that develops between her fiancé and a friend when a girl goes missing on a Sicilian island. Star Monica Vitti (1931–) also appeared in *La Notte* (The Night) (1961), *L'Eclisse* (The Eclipse) (1962) and *Il Deserto Rosso* (The Red Desert) (1964), elliptical attempts to convey an austere, melancholic search for self-awareness and the futility of meaningful communication between men and women that made expressive use of colour and landscape.

Thereafter, he moved into the international arena with *Blow-Up* (1966), *Zabriskie Point* (1969) and the highly-regarded *Professione: Reporter* (The Passenger) (1975) which have served to comment on all sorts of modern ills, from consumerism to the nature of film-making, whilst remaining within his familiar terrain of disillusioned, alienated individuals adrift from society and any secure sense of self-identity.

More recently, he has experimented with video techniques in *Il Mistero Di Oberwald* (The Oberwald Mystery) (1979) and *Identificazione Di Una Donna* (Identification Of A Woman) (1982). Plagued by ill-health, he has long been announced as the director of the forthcoming *The Crew*, to be produced by ▷ Martin Scorsese.

THE APARTMENT

Billy Wilder (US 1960)
Jack Lemmon, Shirley MacLaine, Fred MacMurray, Ray Walston.
125 mins. b/w.
Academy Award: *Best Picture*; *Best Director*; *Best Original Story and Screenplay* Billy Wilder, I.A.L. Diamond; *Best Editing* Daniel Mandell.

▷ Billy Wilder entered his fourth decade in Hollywood with a reputation as writer and director for his equal ability with hard-nosed drama (*The Lost Weekend*, 1945; *Ace In The Hole*, 1950) and frequently risqué comedy (*The Seven Year Itch*, 1955; *Some Like It Hot*, 1959). In *The Apartment* both aspects are combined, using the approachability of central performers (▷ Jack Lemmon, ▷ Shirley MacLaine, Fred MacMurray (1907–)) primarily associated with light comedy to draw the audience into a dramatically bleak vision of contemporary social relations. The setting is a large and typical office, where lowly insurance clerk Lemmon is setting about bettering his position by letting out his apartment to various superiors in return for favourable chances of promotion. Among his clients is company director MacMurray who uses it to entertain elevator girl MacLaine, until one traumatic Christmas Eve, when she learns the full extent of her married boss's philandering ways and attempts suicide. Offering her a $100 bill for her trouble, he returns home to the bosom of family respectability, leaving the kindly Lemmon to cope with a casualty in his bedroom; the backdrop of traditional festive goodwill counterpointing the hollowness of a bourgeois society maintained by sexual and personal exploitation. Marked by a moving sense of compassion throughout, Wilder here escapes the harsh bitterness of which he was sometimes capable (*Kiss Me Stupid*, 1964) to achieve perhaps the most emotionally resonant film of his career.

APOCALYPSE NOW

Francis Ford Coppola (US 1979)
Martin Sheen, Marlon Brando, Robert Duvall, Frederic Forrest, Dennis Hopper.
Initial 70mm release 141 mins; subsequent 35mm release with added credit sequence 153 mins. col.

Academy Award: *Best Cinematography* Vittorio Storaro; *Best Sound* Walter Murch.
Cannes Film Festival: *Palme D'Or* (shared with Volker Schlöndorff's *The Tin Drum*)

'This isn't a film about Vietnam. This film is Vietnam.' Such was the characteristically flamboyant statement of co-writer/producer/director ▷ Francis Coppola upon the release of *Apocalypse Now*. Coming off the massive critical and commercial success of his Mafia chronicles ▷ *The Godfather* (1972) and *The Godfather, Part Two* (1974), Coppola now self-consciously attempted his magnum opus, a major artistic statement on America's military involvement in South-East Asia. While the much-troubled shooting of the film on location in the Philippines – which saw sets destroyed by tropical storms, star Martin Sheen (1940–) suffering a near-fatal heart attack, and the budget escalating from $12 to $31 million – ironically echoed the real events, Coppola's approach to his country's recent history was a highly allusive one. Taking the story outline and thematic thrust from Joseph Conrad's novel *Heart of Darkness* (the film has Sheen's US army captain journeying deep into the Vietnamese jungle to assassinate ▷ Marlon Brando's megalomanic rogue Colonel Kurtz) and overlaying the 'fisher king' myth of death and regeneration adopted by T. S. Eliot's *The Waste Land*, the overall effect works towards an introspective Jungian archetype of the deep-rooted evil in the human soul, rather than a specific focus on the ideological contours of the Vietnamese conflict. Certainly, war correspondent Michael (*Dispatches*) Herr's wise narration, *The Doors*' contemporary rock, and the impressively marshalled military hardware effectively date the proceedings, but the film's real achievement perhaps lies in moments of hallucinatory carnage where the sensual delirium, the horror *and* and absurd beauty of wartime are disturbingly conveyed.

ARC SHOT A camera movement involving a turn of up to 360 degrees around a particular character or subject.

ARCAND, Denys

Born 25 June 1941, Deschambault, near Quebec City, Canada.

A graduate from the University of Montreal with a master's degree in history, Arcand applied for work at the National Film Board of Canada and began his career writing and directing a number of government documentaries planned to coincide with the Canadian centennial in 1967.

Continually at odds with the political conservatism of the National Film Board, he left and established himself as a successful director of commercials before returning to the documentary form with such controversial works as *On Est Au Coton* (1970), a hard-hitting examination of the Quebec textile industry, and *Quebec: Duplessis Et Après* (1972) a satire on the Quebec political situation that expressed his disillusionment with the interchangeability of those who hold power.

He also tackled a number of fictional subjects beginning with *La Maudite Galette* (The Damned Loot) (1971) about a group of petty criminals. Other features from this period include *Rejeanne Padovani* (1972) a bleak view of corruption in the city of Montreal, and *Gina* (1974) which recounts the story of a rape in the form of a 'lurid parable about the social injustices of life in Quebec'.

The abolition of the Canadian Film Development Corporation and changes in the country's tax laws served to restrict opportunities for native filmmakers and he spent the next decade employed on such television series as *Empire Inc* (1982) and *Le Crime D'Ovide Plouffe* (The Crime of Ovide Plouffe) (1984). He also made the documentary *Le Confort et L'Indifference* (Comfort and Indifference) (1982) on the Quebec independence referendum which reflected his bitterness at the trivialities imposed on the democratic process and his disappointment at the negative outcome of the vote.

He returned to feature films in triumph with *Le Déclin De L'Empire Américain* (The Decline of the American Empire) (1986) which uses the social gatherings of eight faculty friends and colleagues to explore relationships between men and women and reflect a current desire for hedonism and personal gratification that may not result in happiness.

A substantial box-office success on the international market that received an Oscar nomination as Best Foreign Film, he followed this with *Jesus of Montreal* (1989) an ingenious contemporary reinterpretation of the Bible stories that explored the spiritual quest of a generation who now seemed to be in search of more than mere pleasure and personal gain.

ARENA
UK 1976–

Iconoclastic BBC2 arts slot which came of age under the tutelage of producer Alan Yentob, who was put in charge in 1977, after a tentative first year. *Arena* has succeeded by taking a broad canvas – devoting memorable programmes to the work of ▷ Orson Welles, *The Private Life of The Ford Cortina*, an examination of the song *My Way*. Musical programmes – such as those on Eddie Cochran and Woody Guthrie (1912–67), have been a particular strength. In recognition of his contribution Yentob was promoted, in 1985, to head of music and arts at the BBC. In 1988 he was made controller of BBC2.

ARGENTO, Dario
Born 7 September 1940, Rome, Italy.

In his native Italy, the multi-talented Dario Argento has reached the same peak of public recognition shared by ▷ Alfred Hitchcock and ▷ Steven Spielberg in Britain and America, but his artistic achievement is to have done for the horror genre what ▷ Sergio Leone did for the western; adapting a primarily American form to the demands of his own particular sensibility to produce a stylistically and thematically arresting hybrid. Beginning his career as a critic and screenwriter – he actually co-scripted Leone's *C'Era Una Volta Il West* (Once Upon A Time In The West, 1968) – Argento first came to prominence as a director in the early Seventies with a series of elaborate crime thrillers or *gialli* (named after the yellow covers on pulp paperbacks). The finest of these, *Profondo Rosso* (Deep Red, 1975), set the pattern for the distinctive Argento manner, its ingeniously graphic violence and bravura camerawork closely aligned with American filmmaker ▷ Brian De Palma's more Hitchcockian work. Two stunningly-executed ventures into the covert domain of diabolism, *Suspiria* (1977) and *Inferno* (1980), later saw Argento's orchestration of colour, music and mise-en-scène attain a pinnacle of accomplishment. In the Eighties, Argento moved into producing and writing slick commercial fare for younger proteges, including Lamberto Bava's *Demoni* (Demons, 1985), as his own career reached a plateau (the vicious élan of 1987's *Opera* remained unseen in the UK until a 1991 video release), but a reunion with kindred soul ▷ George A. Romero, for whom he'd earlier co-produced and scored *Dawn of the Dead* (1979), in a filmic tribute diptych to horror pioneer Edgar Allen Poe, *Due Occhi Diabolici* (Two Evil Eyes, 1990) showed that he still had much to contribute.

ARLETTY
Real Name Léonie Bathiat
Born 15 May 1898, Courbevoie, France.

The daughter of a miner, Arletty worked in a munitions factory and as a secretary before her radiant beauty brought her modelling assignments and stage work in revues like *Si Que Je Serais Roi* (1922).

Gaining experience in a succession of comedies, operettas and musicals, she made her film debut in *La Douceur D'Aimer* (1930) but her subsequent appearances were negligible. However, her stage renown grew with successes like *L'Ecole Des Veuves* (1936) and *Fric-Frac* (1936) and her screen roles grew in significance to include a droll Queen of Ethiopia in the trilingual *Les Perles De La Couronne* (The Pearls of the Crown) (1937).

She began a fruitful association with director ▷ Marcel Carne on *Hotel du Nord* (1938) a study of the derelicts who inhabit a shabby hotel in which her vivacious and moving portrayal of a whore was well-liked. Their subsequent work includes *Le Jour se Lève* (Daybreak) (1939) a classic of pre-War French fatalism, the sombre medieval fantasy *Les Visiteurs Du Soir* (1942) in which she played an emissary of the Devil, and *Les Enfants Du Paradis* (1944) which contains her most famous role as the enigmatic, mysterious courtesan Garance; a graceful femme fatale whose ethereal beauty invokes the deepest admiration of a quartet of suitors.

A wartime love of a German officer led to her brief incarceration on charges of collaboration and there were several abortive attempts to resume her career before an appearance in the thriller *Portrait D'Un Assassin* (1949).

Among a variable group, her notable later films include *Huis Clos* (No Exit) (1955) in which her embittered lesbian compelled attention, *The Longest Day* (1962) her sole appearance in an American production, and the comedy *Le Voyage A Biarritz* (1963) a reunion with Fernandel (1903–71) that marked her last screen appearance. However, she did find more challenging roles and greater acclaim on stage in such Tennessee Williams (1911–83) plays as *Un Tramway Nommé Desir* (A Streetcar Named Desire) (1950) and *La Descente D'Orphée* (Orpheus Descending) (1958).

An accident with eye-drops rendered her almost blind and, despite further offers, her only credit in thirty years has been as the narrator of the short *Dina Chez Les Lois* (1967).

ART DIRECTOR The individual responsible for designing the sets and sometimes also the costumes and graphics of a particular film, also referred to as a production designer. Capable of contributing enormously to the look and atmosphere of a film, notable art directors include Polish-born Anton Grot (1884–1974) whose work ranged across such Warner Brother classics as *Little Caesar* (1931), *Captain Blood* (1935) and *The Sea Hawk* (1940), William Cameron Menzies (1896–1957) who was responsible for the massive sets used in *Things To Come* (1936) which he also co-directed and Ken Adams (1921–) who created some of the most spectacular interior backdrops for James Bond in *You Only Live Twice* (1967), *The Spy Who Loved Me* (1977) and *Moonraker* (1979). More recently, Anton Furst has conjured up a fantasy world for *The Company of Wolves* (1984), transformed England into Vietnam for *Full Metal Jacket* (1987) and visualised a brooding Gotham City for *Batman* (1989) for which he received an Oscar, and Ferdinand Scarfiotti has proved an able collaborator to the vision of directors ▷ Bernardo Bertolucci and ▷ Paul Schrader on films like *American Gigolo* (1980), *Cat People* (1982) and *The Last Emperor* (1987) for which he won an Oscar.

ARZNER, Dorothy

Born 3 January 1900, San Francisco, USA.
Died 1 October 1979, Los Angeles, USA.

The only major woman director at work in the Hollywood studio system of the 1920s and '30s, Arzner was often relegated to a footnote in histories of the period but has latterly undergone critical re-evaluation as appreciation of her work has grown.

A medical student at the University of Southern California and volunteer ambulance driver during World War 1, she began her film career as a script typist at Famous Players-Lasky in 1919. Diligently learning the craft of filmmaking, she progressed from script supervisor to editor on such important silent features as *Blood and Sand* (1922). Encouraged by director James Cruze (1884–1942), she edited several of his westerns including *The Covered Wagon* (1923) and *Old Ironsides* (1926) which she also wrote.

She made her directorial debut with *Fashions for Women* (1927) and was subsequently entrusted with Paramount's first sound feature *Wild Party* (1929). Working with many of the top female stars of the era, her career of social comedies and melodramas was distinguished by her portraits of strong, independent women and eschewal of the conventional stereotypes. Her best known films include *Merrily We Go To Hell* (1932), *Christopher Strong* (1933) and *Dance, Girl Dance* (1940).

After a serious bout of pneumonia, she retired after making *First Comes Courage* (1943). She directed WAC training films during World War II and later lectured and made commercials for Pepsi-Cola.

During her final years she was honoured by the Director's Guild of America and a variety of international women's festivals who saluted her unique status in the history of Hollywood as a pioneer and feminist role model.

ASNER, Edward
Born 15 November 1929, Kansas City, Missouri, USA.

Active in college drama, Asner made his professional debut with Chicago's Playwrights Theatre Club (1953–55) where he appeared in over twenty plays. Moving to New York, he worked off-Broadway in such productions as *The Threepenny Opera* (1956–8) and *The Tempest* (1959–60). He also worked in television and made his Broadway debut in *Face of A Hero* (1960) starring ▷ Jack Lemmon.

Resident in Los Angeles from 1961, he appeared in numerous episodes of television series like *The Naked City*, *Peter Gunn* and *Ironside* and co-starred in the series *Slattery's People* (1964–5).

He made his feature film debut in *The Satan Bug* (1965) but remained primarily a television actor, finding fame as the irascible Lou Grant in the long-running situation comedy ▷ *The Mary Tyler Moore Show* (1970–7) which won him three Emmy Awards.

During the run of the series he continued to work in the cinema and theatre and in such superior television movies as *Hey, I'm Alive* (1975), *The Gathering* (1977) and *The Life and Assassination of The Kingfish* (1977) in which he portrayed Louisiana politician Huey Long (1893–1935). He won a further Emmy as the head of the Jordache family in the mini-series *Rich Man, Poor Man* (1976) and played the slave captain in ▷ *Roots* (1977).

He embellished his earlier character in the series ▷ *Lou Grant* (1977–81) with the format expanded to hour-long episodes, the setting changed from a television newsroom to a large Los Angeles newspaper and the tenor primarily dramatic. He won three Emmys for the gruff integrity he brought to the rounded portrayal of a hard-driven professional and fallible human being and the series won plaudits for its mature approach to a number of social issues.

A controversial President of the Screen Actors Guild (1980–5) who clashed with President Ronald Reagan (1911–   ) over American foreign policy, his liberal beliefs were said to have brought the abrupt cancellation of his series and to have damaged his career.

However, he re-emerged in the film *Daniel* (1983), such series as *Off The Rack* (1985) and *The Bronx Zoo* (1987–88) and television movies like *Anatomy of An Illness* (1984) and *Vital Signs* (1986).

ASPECT RATIO (AR) Describing the ratio of the width to the height of a reproduced picture in cinema or television, the term is most often expressed with the height as unity. Thus from the silent era to the present day the general international aspect ratio for cinema and television has been 1.33:1. Later experiments with wide-screen processes necessitated changes in this standard ratio and ▷ CinemaScope, for instance, was first shown in 2.55:1. Today, a standard wide-screen ratio in Europe is 1.66:1 whilst the American equivalent is 1.85:1. 70 mm films are screened 2.2:1.

The inability of the television screen to accommodate changes in ratio explains the technical difficulties with small-screen transmission of wide-screen cinema films. Solutions include 'panning and scanning' across the length of the partial image that remains visible or showing them in a letterbox format that uses a band at the top and bottom of the screen to retain a reproduction of the entire original image.

ASTAIRE, Fred
Real Name Frederick Austerlitz.
Born 10 May 1899, Omaha, Nebraska, USA.
Died 22 June 1987, Los Angeles, California, USA.

Encouraged to take dance lessons from the age of five, Astaire was teamed with his elder sister Adele (1898–1981) as a touring vaudeville act, beginning his professional career in 1906. They rose to stardom on Broadway and London in the 1920s in specially-written shows like *Lady Be Good* (1925), *Funny Face* (1927) and *The Band Wagon* (1931) and he is credited with film appearances in the shorts *Fanchon the Cricket* (1915) and *Municipal Bandwagon* (1931).

When Adele married Lord Charles Cavendish in 1932, the dancing partnership was dissolved and Fred went to Hollywood, finding himself the subject of a notoriously dismissive screen-test that commented: 'Can't act, can't sing, can dance a little.' Undeterred, he made his major film debut in *Dancing Lady*

Fred Astaire and Ginger Rogers in *Swing Time* (1936)

(1933). Working with new partner Ginger Rogers (1911–) and choreographer Hermes Pan (1905–90), he revolutionized the film musical with a succession of original and innovative dance-tap routines in films like *The Gay Divorcee* (1934), ▷ *Top Hat* (1935) and *Swing Time* (1936).

A hardworking perfectionist who made his dancing appear effortless, he was noted for his debonair charm, lighter-than-air grace, unassuming personality and distinctively precise singing voice. His many popular musicals include *Broadway Melody of 1940* (1940), *You'll Never Get Rich* (1941) and *Holiday Inn* (1942).

He first announced his retirement after *Blue Skies* (1946), but returned to replace an ailing ▷ Gene Kelly on *Easter Parade* (1948) and experimented further with the boundaries of dance on film in *The Barkleys of Broadway* (1949), *Royal Wedding* (1951), in which he appears to dance on the ceiling, *The Bandwagon* (1953) and *Funny Face* (1956).

He later showed some flair as a dramatic actor in *On the Beach* (1959) and won a Best Supporting Actor Oscar nomination for *The Towering Inferno* (1974) and an Emmy for the television film *A Family Upside Down* (1978). His later musical appearances include *Finian's Rainbow* (1968) and *That's Entertainment, Part 2* (1976).

A frequent performer on television, his work in the medium includes such specials as *An Evening With Fred Astaire* (1958), *Another Evening With Fred Astaire* (1959), guest appearances in the series *It Takes A Thief* (1965–9) and films like *The Over-The-Hill-Gang Rides Again* (1970) and *The Man in the Santa Claus Suit* (1978). Also a radio star and recording artist, he made his final film appearance in *Ghost Story* (1981).

He received a special Oscar in March 1950 for his 'unique artistry and contributions to the technique of musical pictures'. Among numerous other honours, he was the recipient of the ▷ American Film Institute Life Achievement Award in 1981.

His autobiography, *Steps in Time* was published in 1959.

L'ATALANTE

Jean Vigo (France 1934)
Jean Daste, Dita Parlo, Michel Simon, Gilles Margaritis.
Released 1934 as Le Chaland Qui Passe at 82 mins; 1945 cuts reinstated to original 89 mins; 1990 restored 89 min version reissued. b/w.

The recent restored reissue of ▷ Jean Vigo's *L'Atalante*, pieced together from prints in British, French and Belgian archives to conform to the director's original intentions, cemented Vigo's critical status as one of our most cherishable film artists. While his earlier short work, the scathing 'document-ary' *A Propos De Nice* (1929) and the schoolboy anarchy of *Zéro De Conduite* (1932), confirmed a trenchant and idiosyncratic talent at work, when hired by major French studio Gaumont to work on a potentially-hackneyed waterway romance Vigo's response was to respect only the outline of the material, make the best of difficult shooting conditions, and turn in the uniquely quirky hymn to romance that is *L'Atalante*. From the simplest of scenarios – young newlyweds Daste and Parlo honeymoon on a barge with a crusty old captain Simon – Vigo creates a rich cavalcade of ever-changing moods, switching from broad comedy to moments of tension, from surreal visions to celebration of desire. Perhaps the most extraordinary sequence has the fervent lovers momentarily estranged, and as Vigo deftly cuts between separate beds from restless body to restless body he evokes a memorably poetic and palpable sense of erotic need. Faced with the uncategorisable end

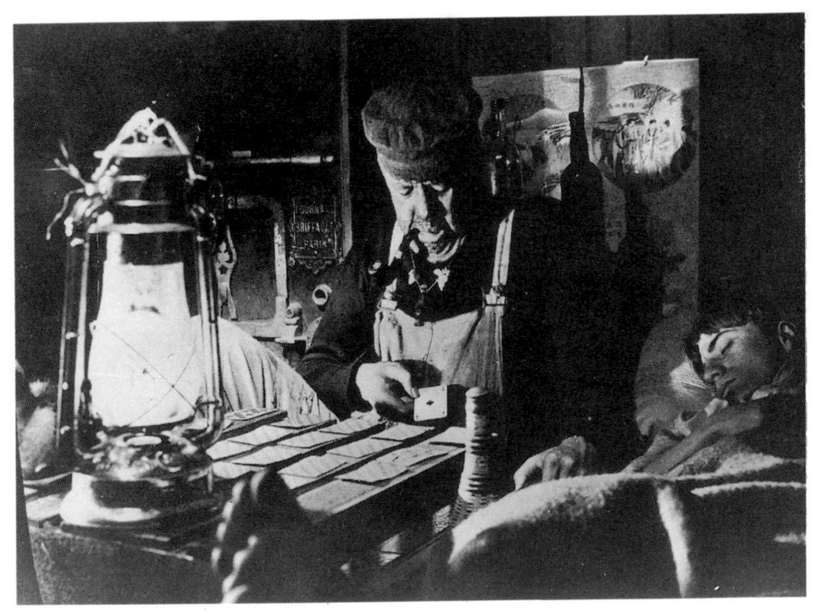

Michel Simon in *L'Atalante* (1934)

result however, Gaumont barred the director from the editing room while they recut the film, replaced some of Maurice Jaubert's dreamy score with a currently popular song, and released the film under another title. Vigo died in Paris of rheumatic septicaemia at the age of 34 a mere two weeks later, unaware that his film would one day be reassembled and recognised for the masterpiece it always was.

ATTENBOROUGH, David Frederick (Sir)

Born 8 May 1926, London, UK.

After gaining an M.A. in zoology and geology from Claire College, Cambridge, Attenborough served as a lieutenant in the Royal Navy and worked as an editorial assistant in an educational publishing house.

Joining the BBC in 1952, he worked as a television producer with responsibilities covering a range of topics from politics to religion and the arts. In 1954, he persuaded the BBC to collaborate with London Zoo in funding a series of zoological and ethnographic expeditions to remote parts of the globe to capture intimate footage of rare wildlife in its natural habitat. The programme, which he later presented, was entitled *Zoo Quest* (1954–64) and set new standards in the presentation of natural history on television, taking him to such diverse locales as Sierra Leone, Indonesia and New Guinea in search of creatures like the bird of paradise and the Komodo dragon lizard.

A very likeable and knowledgeable presenter, his infectious enthusiasm and ability to convey his awe at the marvels of the natural world in inimitable, whispering tones made him one of the medium's most skilled and popular communicators.

In 1965, he was appointed controller of programmes for BBC2 and subsequently served as director of programmes for both channels from 1968 to 1972; proving himself an able administrator and sound advocate of Reithian public service broadcasting principles of making programmes that informed and entertained. Notable successes from this period include *The Forsyte Saga* (1967–68), *Civilisation* (1969) and the popularisation of snooker as a television sport.

He returned to programme-making and has been responsible for some of the most popular, innovative and critically-lauded natural history series including *The Tribal Eye* (1975), which explored the function of art in primitive societies,

Life on Earth (1978), a massively accomplished thirteen-part exploration of the evolution of the species, *The Living Planet* (1984) and *The Trials of Life* (1990).

Knighted in 1985, he has been the recipient of numerous awards and honorary degrees and has served as a trustee of the World Wildlife Fund, the British Museum and as a member of the Nature Conservancy Council since 1975. His many books to accompany the television work have included *Zoo Quest to Guiana* (1956), *The Tribal Eye* (1976) and *The Living Planet* (1984).

ATTENBOROUGH, Richard Samuel (Sir)

Born 29 August 1923, Cambridge, UK.

Drawn to acting from childhood, Attenborough won a scholarship to RADA and made his stage debut in *Ah, Wilderness!* (1941). His film debut followed when Nöel Coward (1899–1973) cast him as the cowardly young seaman in *In Which We Serve* (1942).

He then established a niche, specialising in blustering youths, bristling with empty bravado. His notable early performances include the sadistic gangleader Pinkie in *Brighton Rock* (1947) (a role he had originated on stage in 1943) and the eponymous schoolboy in *The Guinea Pig* (1949).

Gainfully employed among the ranks in many British war films, he also revealed some comic skill in *Private's Progress* (1955). In 1959, he formed Beaver Films with Bryan Forbes (1926–) and enjoyed success with *The Angry Silence* (1959) and *The League of Gentlemen* (1960).

Developing into a versatile character actor, he impressed as the literary agent in *Only Two Can Play* (1961), the browbeaten kidnapper in *Seance On A Wet Afternoon* (1964), the military martinet in *Guns At Batasi* (1964) and as mass murderer John Christie in *10 Rillington Place* (1971).

He made his directorial debut with the lavish, all-star version of the anti-war satire *Oh, What A Lovely War* (1969) and has subsequently concentrated more on a career behind the cameras, gaining a reputation for his technical mastery of large-scale epics and his over-earnest desire to use cinema as a means of delivering great and worthy messages. The variable results include the effective and well-acted supernatural thriller *Magic* (1978), the reverential, simplistic biography *Gandhi* (1982) which won him an Oscar as Best Director and, best of all, *Cry Freedom* (1987) a sincere and persuasive account of a white journalist's growing abhorrence of apartheid in the aftermath of the death of Steve Biko.

His most recent acting appearances are in *Shatranj Ke Khilari* (The Chess Players) (1977) and *The Human Factor* (1979). In 1991, he was scheduled to begin filming his long-planned biography of ▷ Charles Chaplin.

A tireless elder statesman of the British film industry and campaigner for a multiplicity of international humanitarian causes, his ready indulgence in public outbursts of emotion and tendency to long-winded speechifying have rendered him a much-mimicked and much loved celebrity. He was knighted in 1976.

THE AUTEUR THEORY First suggested by French film critic Alexandre Astruc (1923–) in his 1948 article 'Le Camera Stylo', the theory was most widely accepted by fellow French critics writing for *Cahiers du Cinema* and most specifically ▷ Francois Truffaut in his 1954 essay 'A Certain Tendency in French Cinema'. The concept recognises the film director as the sole auteur or author of a film, imposing his or her personal sensibility and artistic vision on every aspect of the production, and building a body of work that can be judged for its thematic consistency and development.

It led to the re-assessment of many directors who had displayed their individuality whilst working within the confines of the Hollywood studio system and figures like ▷ Hitchcock, ▷ Minnelli and ▷ Hawks were championed.

Critic Andrew Sarris (1928–) popularised the theory in English-language circles during the 1960s through his writing in *Film Culture* and the publication of *The American Cinema: Directors and Directions 1929–68* which offered assessments of a vast array of Hollywood directors in the light of the theory.

Whilst a useful tool of critical debate

and still used as a journalistic shorthand in most writings on the film industry, it led to inflated claims for the work of many lesser talents and failed to take into account the collaborative nature of the medium and outside factors of relevance such as the impact of technology or the influence of being under contract to a particular studio.

Given the consistency of their preoccupations and technical mastery directors like ▷ Ingmar Bergman, ▷ Woody Allen or ▷ Luis Buñuel can most readily be analysed through the use of the theory but director Martin Ritt (1919–1990) summed up the feelings of many who recognised its shortcomings when he commented, 'As far as a Martin Ritt Production is concerned, I wouldn't embarrass myself to take that credit. What about the Ravetchs? They wrote it. What about the actors who appear in it? If I ever write one, direct it, and appear in it, then you call it a Martin Ritt Production.'

AVANT-GARDE Used in a general sense to signify any non-commercial film that is experimental or unorthodox in its subject matter or technique, the term is most specifically applied to the group of radical films that appeared in Europe from 1918 until the early 1930s. Significant among these films were the work of the German 'absolute' filmmakers of the 1920s who used animated drawings of lines, shapes or patterns to establish the rhythmic relationship between the separate images, examples include *Rhythmus 21* (1921) and *Diagonale* (1924); the eschewal of linear narrative in favour of the unexpected and spontaneous in the Dadaist efforts of ▷ Rene Clair with *Entr'acte* (1924) and Man Ray (1890–1976) with *Le Retour A La Raison* (1923); and the poetic and shocking imagery deployed by ▷ Surrealists like Germaine Dulac (1882–1942) with *The Seashell and the Clergyman* (1926), ▷ Luis Bunuel with *Un Chien Andalou* (1928) and ▷ Jean Cocteau with *Le Sang D'Un Poète* (The Blood of A Poet) (1930).

Later avant-garde movements of note include the films of ▷ Maya Deren and other underground American filmmakers like ▷ Kenneth Anger, Stan Brakhage (1933–) and ▷ Andy Warhol as well as more recent European practitioners like Jean-Marie Straub.

THE AVENGERS

UK 1961–8.
Patrick MacNee, Honor Blackman, Diana Rigg.

Cultish, witty, fantasy adventure derived from the short-lived series *Police Surgeon*. Initially, square-jawed, bowler-hatted undercover agent John Steed of MI5½ was partnered by men, but the show only really gathered momentum with the introduction of female sidekick Cathy Gale (Blackman). Faced with a role written for a man she gamely donned a catsuit and kinky boots to karate chop her way to immortality and a role as Pussy Galore in the James Bond film *Goldfinger* (1964). Elizabeth Shepherd took over the female lead briefly, but was quickly and memorably replaced by Diana Rigg (1938–) as Emma Peel (the name was a pun on M-Appeal, the 'M' standing for 'man').

The Avengers reflected the confidence and self-consciousness of 1960s Britain, applying inverted logic to the spy format. Key writer Brian Clemens (1931–) has attributed the sparse shooting style to the show's small budgets, whereby a bookshop might be represented by a shot from between the shelves of a single bookcase. Episodes were written round a formula in which a nasty crime or threat to the common good would be thwarted, but not before the heroes had been taunted and dragged into an eerie setting. The quirky scientists and megalomaniacs who did the taunting were played by a succession of guest stars, among them Peter Cushing (1913–), Christopher Lee (1922–), Donald Sutherland (1935–), and Charlotte Rampling (1945–). When Rigg left the show she was replaced by Linda Thorson (1947–), but the moment had passed, and the series was cancelled soon after. In 1976 the rather half-hearted *New Avengers*, was attempted with Clemens still writing. MacNee was present too, with new partners Joanna Lumley (1946–) (who lacked the knowingness of Blackman and Rigg) and Gareth Hunt (1943–). As evidence of *The Avengers*' continuing camp appeal, MacNee and Blackman had a pop hit in 1990, with a revival of the novelty song *Kinky Boots*.

AVERY, Tex

Real Name Frederick Bean Avery
Born 26 February 1907, Taylor, Texas, USA.
Died 26 August 1980, USA.

Graduating from North Dallas High School in 1927, Avery's ambition was to create a newspaper cartoon strip. Failing in this, he joined the Walter Lantz (1900–) animation studio in 1929.

Moving to Warner Brothers, he directed his first cartoon *Golddiggers of '49* (1935) featuring Porky Pig and went on to create the character of Daffy Duck in *Porky's Duck Hunt* (1937), and assist in the development of Bugs Bunny in *A Wild Hare* (1940). It is Avery who is credited with coining Bugs most famous phrase 'What's up, Doc?' which was a common expression in his Texas birthplace.

A highly distinctive talent, his cartoons are tightly choreographed masterpieces of mayhem that entertain with a breathless succession of visual jokes that use exaggerated violence, sexual innuendo and a healthy degree of irreverence to comic effect. An inventive ▷ Surrealist, his style influenced future generations of animators and even live-action directors like Joe Dante (1946–).

Moving to M-G-M, he created Droopy in *Dumb Hounded* (1943), Screwy Squirrel in *Screwball Squirrel* (1944), George and Junior in *Henpecked Hoboes* (1946). Other classic cartoons include *Red Hot Riding Hood* (1943), *King Size Canary* (1947) and *Bad Luck Blackie* (1949).

He left M-G-M in 1955, making his last cartoon in 1958 with *Polar Pests*. He was active thereafter in the field of television commercials winning many awards for his work including a First Prize at the Venice Publicity Festival for *Calo-Tiger* (1958).

He spent the last year of his life working for ▷ Hanna-Barbera.

L'AVVENTURA

Michelangelo Antonioni (Italy 1960)
Monica Vitti, Gabriele Ferzetti, Lea Massari, Dominique Blanchar.
145 minutes. col.
Cannes Film Festival *Special Jury Prize.*

Greeted with a slow handclap and catcalls at its first Cannes screening, ▷ Michelangelo Antonioni's characteristically languorous evocation of contemporary alienation elicited both admiration and dismay from the critics and would go on to bemuse and irritate many an international audience on its surprisingly successful box office release. Although obviously presaged by the director's earlier work (*Le Amiche*, 1955; *Il Grido*, 1957), a combination of distinctive thematic and stylistic elements were to establish *L'Avventura* as a contentious, must-see picture on the then burgeoning art house circuit. Disregarding the usual narrative expectations, the storyline involves the disappearance of a young woman (Massari) on holiday in Sicily, but after a perfunctory search she's nowhere to be found, and the bulk of the running time charts a desultory affair between her lover (Ferzetti) and her friend (Vitti). Any potential audience identification with their plight however is barred by Antonioni's achingly slow pacing, his distanced camera, and by the pair's halting attempts at communication, their joyless, if not pointless, sexuality. Such overwhelming social malaise is of course a common current in much twentieth century European art, but Antonioni's work is significant for the way in which the alienation of individuals from each other is conveyed not through the shaping of a narrative but by a certain formal astringency and the visual motif of the characters' alienation from their surroundings. It was an art he was to memorably develop against a variety of backdrops from the Rome stock exchange (*L'Eclisse*, 1962) to Swinging London (*Blow-Up*, 1966), from hippy California (*Zabriskie Point*, 1969) to arid North Africa (*The Passenger*, 1975).

BABENCO, Hector

Born 7 February 1946, Buenos Aires, Argentina.

As a teenager, Babenco followed a peripatetic lifestyle around Europe finding work as a house-painter, Bible salesman and extra in the plethora of ▷ spaghetti westerns and war movies being produced during the 1960s at the Cinecitta Studios in Rome.

Returning to Latin America, he chose

William Hurt in Hector Babenco's *Kiss of the Spiderwoman* (1985)

Sao Paulo in Brazil as his base and made his feature-length debut with *Rei Da Noite* (King of the Night) (1975).

Drawn to controversial themes and portraits of outsiders threatened by their precarious existences on the periphery of 'normal' society, he achieved a major popular success with *Lucio Flavio-Passagerio Da Agonia* (1978) the fact-based story of a notorious Brazilian 'Robin Hood' figure. This tale of a professional thief who is murdered when he threatens to expose police corruption proved so politically sensitive that Babenco received numerous death threats and discovered his home attacked by machine-gun fire.

Wider international exposure followed with *Pixote* (1980), a sensationalistic portrayal of juvenile delinquents and their squalid, hopeless struggle for survival on the streets of Sao Paulo.

He then progressed to the English-language *Kiss of the Spiderwoman* (1985) which tells of the mutual admiration that develops between two initially hostile cellmates; one a flamboyant, homosexual window-dresser, convicted on a morals charge, the other a committed revolutionary held without trial and brutally tortured. A labour of love, it proved to be one of the sleeper hits of the year and won an Oscar for actor ▷ William Hurt.

Working in America, he made a large-budget version of *Ironweed* (1987) which was a surprising box-office failure despite its well-acted examination of the underside of the American Dream when hobo ▷ Jack Nicholson returns home for the first time in twenty years to make peace with the ghosts of his blighted past.

In 1991 he will release *At Play In The Fields of the Lord.*

BABY SPOT A small incandescent spotlight with a 1,000 watt bulb, frequently used for close-ups. A kick light, designed to highlight a character or object from an angle opposite to that of the main light, uses a 750-watt bulb called a light baby or seven-fifty, similarly a 500-watt bulb is a weak baby or five hundred.

BACK PROJECTION (or REAR PROJECTION) A form of motion picture composite photography in which the actors perform in front of a translucent screen on which the scenic background is projected. Seen in the cinema as early as the western *The Drifter* (1913) this cost-saving illusion of glamorous locations or hazardous endeavours was most widely utilised during the heyday of the Hollywood studio system. Thus Tarzan stalked the fiercest jungles of the African sub-continent without ever leaving the comfy confines of Culver City in California. The essential phoniness of the effect made it increasingly unacceptable to audiences though ▷ Hitchcock favoured the process as late as *The Family Plot* (1976).

BAIRD, John Logie

Born 13 August 1888, Helensburgh, Dumbartonshire, UK.

Died 14 June 1946, Bexhill, Sussex, UK.

A student of electrical engineering at the Royal Technical College in Glasgow, Baird continued his studies at Glasgow

University until the outbreak of the war. Unfit for service, he served as a superintendent engineer at the Clyde Valley Electric Power Company. Persistent poor health forced him to abandon engineering and pursue a career in business but, after a physical and nervous breakdown, he retired to Hastings in Sussex (1922).

He diligently researched the possibilities for television transmission and, in 1924, transmitted the image of a Maltese cross over a distance of several feet. On 26 January, 1926 he gave the first public demonstration of a television image to a group of scientists of 'Noctovision' a form of infra-red television. Further breakthroughs followed; the world's first transatlantic television transmission from London to New York (1928), the first television images in natural colour (1928) and experimentation with stereoscopic television.

His 30-line mechanically-scanned system was adopted by the BBC in 1929 and the Derby from Epsom was televised for the first time in 1931. In 1936 the Corporation adopted his improved 240-line system, but the following year chose a rival 405-line system with electronic scanning made by Marconi-EMIO.

He continued his research into colour, three-dimensional images, and big-screen projection until the time of his death, and was the first British subject to receive the gold medal of the International Faculty of Science in 1937.

BALCON, Michael Elias (Sir)

Born 19 May 1896, Birmingham, UK. *Died* 17 October 1977, Upper Parrock, Sussex, UK.

Rejected for military service because of a flaw in his left eye, Balcon spent World War I in the employment of the Dunlop Rubber Company. Shortly afterwards, he formed a modest distribution company with Victor Saville (1897–1979) and moved into production with advertising films, the documentary *The Story of Oil* (1921) and the popular melodrama *Woman to Woman* (1923).

In 1924, he helped found Gainsborough Pictures and was responsible for such silent successes as *The Rat* (1925), ▷ Alfred Hitchcock's *The Lodger* (1926) and *Blighty* (1927).

In 1931, he took charge of production at Gaumont-British determined to encourage the making of British pictures that could compete in the international marketplace and specifically in America. Notable productions include *Rome Express* (1932), *The Good Companions* (1932), *Man of Aran* (1934) and *The Thirty-Nine Steps* (1935).

A brief spell as head of British production for Hollywood studio M-G-M resulted in *A Yank At Oxford* (1938) but his most lasting contributions to British cinema came as head of production at Ealing Studios, a position he held from 1938 until the eventual demise of the company in 1959.

Creating a supportive family atmosphere and continuity of employment for the performers and filmmakers working under his auspices, he was able to influence the creation of a body of work that reflected and illuminated the national character in contemporary dramas and, most famously, in popular comedies. The string of Ealing classics includes *Dead of Night* (1945), ▷ *Whisky Galore* (1948), *The Blue Lamp* (1949), ▷ *Kind Hearts and Coronets* (1949), *The Lavender Hill Mob* (1951), *Mandy* (1952) and *The Cruel Sea* (1953).

Later Chairman of British Lion (1964–8) and the Bryanston Company (1959–75), he was associated with the success of films like ▷ *Saturday Night and Sunday Morning* (1960) and *Tom Jones* (1963). Knighted in 1948, he also served as a director of Border television and a Governor of the British Film Institute, publishing the lively memoir *A Lifetime of Films* (1969).

BALL, Lucille Desiree

Born 6 August 1911, Celoron, near Lake Chautauqua, New York, USA. *Died* 26 April 1989, Los Angeles, California, USA.

A stagestruck youngster who performed in high-school plays and community theatre, Ball was a model and showgirl before her selection as the Chesterfield Cigarette Girl and a brief appearance in the film *Broadway Thru A Keyhole* (1933) convinced her to move to Hollywood where she was featured as one of the twelve Goldwyn Girls in *Roman Scandals* (1933).

Numerous bit parts and B-pictures followed and she was under contract to R-K-O from 1935 to 1942 where her more substantial appearances included *Stage Door* (1937), *The Affairs of Annabel* (1938), *Dance, Girl Dance* (1940) and *The Big Street* (1942) (in a rare dramatic role as a crippled nightclub singer).

Later, she partnered ▷ Bob Hope in the comedy films *Sorrowful Jones* (1949) and *Fancy Pants* (1950) but achieved lasting popularity on television as the star of *I Love Lucy* (1951–7). An effervescent redhead with a rasping voice and impeccable comic timing, her antics as a scatterbrained, starstruck housewife endeared her to the American public and the show accumulated over 200 awards, including 5 Emmys. The process of filming each episode in sequence with a three-camera technique was also influential in the general movement from live to pre-recorded television.

Her subsequent domestic comedies comprised *The Lucy Show* (1962–68), *Here's Lucy* (1968–74) and the short-lived *Life With Lucy* (1986), and her many television credits include numerous specials and the dramatic story of homelessness *Stone Pillow* (1985).

A star of radio and Broadway, her later films include the popular comedies *The Facts of Life* (1960) and *Yours, Mine and Ours* (1968) and the disastrously-received musical *Mame* (1974). A shrewd business executive, she ran Desilu Productions with her first husband and initial television co-star Desi Arnaz (1917–86), producing such series as *Our Miss Brooks* (1952–6) and *The Untouchables* (1959–62). In 1984, she was inducted into the Television Academy Hall of Fame.

BARDOT, Brigitte

Born 28 September 1934, Paris, France.

A ballet school student, Bardot's beauty won her employment as a fashion model and her appearance on the cover of *Elle* eventually led to a film debut in *Le Trou Normand* (1952).

Cast in small roles for her decorative value, she was seen internationally in such productions as *Act of Love* (1954), *Helen of Troy* (1955) and *Doctor at Sea* (1955). Married to director Roger Vadim

Brigitte Bardot and Jean-Louis Trintignant in *Et Dieu Créa La Femme* (1956)

(1927–) from 1952 to 1957, she attained stardom as the central attraction in his melodrama *Et Dieu Créa La Femme* (And God Created Woman) (1956) one of the first foreign-language films to gain a major release in Britain.

Adopted universally as a symbol of a new female permissiveness, her roles exploited an image of petulant sexuality that was reinforced by a much-publicised off-camera love life.

Her role as the murderess in *La Vérite* (The Truth) (1960) was an attempt to gain credibility as a serious actress and her more challenging work includes the autobiographical *Vie Privée* (A Very Private Affair) (1962), in which a young woman struggles to cope with the pressures of stardom, *Le Mépris* (Contempt) (1963), and *Viva Maria* (1965) which teamed her with ▷ Jeanne Moreau.

She made a brief appearance in the American *Dear Brigitte* (1965) and was seen in the English-language western *Shalako* (1968) but her remaining roles were an unimpressive group of comedies and romps that traded on her still potent 'sex kitten' image. She retired from the screen after *L'Histoire Très Bonne et Très Joyeuse de Colinot Trousse-Chemise* (1973) and has resisted all offers to return including those for *Someone Is Killing the Great Chefs of Europe* (1978) and *Three Men and A Little Lady* (1990).

More recently, she has devoted herself to campaigning for animal rights, hosting several hard-hitting television documentaries and forming the Foundation for the Protection of Distressed Animals in 1976.

BARKER, Ronnie
Full Name Ronald William George Barker.
Born 25 September 1929, Bedford, UK.

An amateur performer, Barker made his professional debut at Aylesbury Repertory Theatre in *Quality Street* (1948). His London debut came in *Mourning Becomes Electra* (1955) and subsequent theatrical appearances include *Camino Real* (1957), *Irma La Douce* (1958) and *A Midsummer Night's Dream* (1962).

An affable, avuncular figure his comic skills became nationally recognised as a contributor to the television series *The Frost Report* (1966–67) and his ability to create acutely observed and precisely detailed comic characterisations was evident as the prison lag Fletcher in *Porridge* (1974–77) and the stuttering, romantically thwarted Lancashire shopkeeper in *Open All Hours* (1976, 1981–85). His flair for tongue-twisting lyrics and saucily suggestive banter was well to the fore in the long-running *The Two Ronnies* (1971–87) in which he was partnered by Ronnie Corbett (1930–).

His rare film roles include *Wonderful Things* (1958), *Robin and Marian* (1976) and *Porridge* (1979). A noted collector of Victoriana, his lighthearted books on the subject include *Book of Boudoir Beauties* (1975) and *Ooh-la-la* (1983). He announced his retirement in 1988.

BARRYMORE, John
Real Name John Sidney Blythe
Born 15 February 1882, Philadelphia, USA.
Died 29 May 1942, Los Angeles, California, USA.

Born into a distinguished family of actors, Barrymore initially resisted the call of the stage to pursue a career as a cartoonist and illustrator. However, in 1903 sister Ethel (1879–1959) secured him a role in the play *Captain Jinks of the Horse Marines*, and later that year he made his New York debut.

A successful light comedian, he appeared in a film of *An American Citizen* (1913) and signed a contract with Famous Players-Lasky to make a series of comedies. A sternly handsome man, known as 'The Great Profile' and 'The Great Lover' he brought a lithe grace and jaunty manner to swashbuckling roles and relished the odd bravura character study, winning acclaim for such films as *Raffles, the Amateur Cracksman* (1917), *Dr Jekyll and Mr Hyde* (1920), *Sherlock Holmes* (1922), *Beau Brumell* (1924) and *Don Juan* (1926).

A popular matinee idol whose stage *Hamlet* (1922 & 1924) broke box-office records, he easily weathered the transition to sound and his resonant voice enhanced such characterisations as the arch-criminal *Arsene Lupin* (1932), the impoverished jewel thief in *Grand Hotel* (1932), the father in *A Bill of Divorcement* (1932) and the title character in *Counsellor-at-Law* (1933). *Rasputin and the Empress* (1932) marked the only joint appearance of Ethel, John and Lionel (1878–1954) and John's biting comic prowess was expertly deployed as the fading actor in *Dinner at Eight* (1933) and as the egomaniac producer in the frantic screwball farce *Twentieth Century* (1934).

A notorious heavy drinker and bon viveur, his perpetual indulgence in pleasure finally took its toll on his looks and professionalism although there was intermittent evidence of his thespian skill in *Marie Antoinette* (1938), as the reformed drunk in *The Great Man Votes* (1939) and as the devious husband in *Midnight* (1939). However, his final years were largely divided between ripely overdone hamminess and sad self-caricature on radio, on stage and in such unworthy films as *The Great Profile* (1940), *The Invisible Woman* (1941) and *Playmates* (1941).

He published an autobiography *Confessions Of An Actor* (1926) and his children Diana (1921–60) and John Jnr. (1932–) also appeared in films. The much-troubled dynasty continues with Drew Barrymore (1975–) who featured in ▷ *E.T.* (1982) and subsequently underwent treatment for drug abuse and alcohol addiction.

BATMAN
USA 1966–8
Executive producer William Dozier
Adam West, Burt Ward

Launched with a disco cocktail party attended by the likes of ▷ Andy Warhol, *Batman* aimed to solve a ratings problem by adding camp touches and adult appeal to Bob Kane's comic adventure story. The casting of then-unknown Adam West (1929–) as the Caped Crusader aroused criticism – his less than athletic physique tested credulity and the elasticity of his body-stocking – but in retrospect it was an inspired move, with much of the show's appeal resting on his absurdity as a superhero. Sidekick Burt Ward (1945–) – the Boy Wonder, Robin – was also a newcomer, but most of the show's regulars, Commissioner Gordon (Neil Hamilton (1899–1984)), Chief O'Hara (Stafford Repp), butler Alfred (Alan Napier (1903–88)) were seasoned campaigners.

The show was strictly formulaic but was executed with considerable visual flair and clever use of colour. Flummoxed by a threat to the security of Gotham City the police would summon Batman. Alerted by a call on the flashing Batphone, millionaire Bruce Wayne and his reluctant charge Dick Grayson would slide down poles to emerge in the subterranean Batcave as curiously attired crimefighters and bound into their improbably well-equipped car, the Batmobile. Then came the fight scene, each punch greeted by a screen-filling 'Zap!', 'Pow!' and much tilting of the camera.

Batman's task was made no easier by the villains of Gotham whose comic devilment took evil to some very strange places. Among them were the Riddler (Frank Gorshin (1935–), John Astin (1930–)), the Penguin (Burgess Meredith (1908–)), the Joker (Cesar Romero (1908–)), Catwoman (Julie Newmar, (1930–), Eartha Kitt (1928–)). ▷ Joan Crawford, Vincent Price (1911–), Shelley Winters (1922–), Liberace (1919–87) and Milton Berle (1908–) also guested.

The show was an immediate success, spawning a hit record with Neil Hefti's theme tune, a dance craze (the batusi), a full length feature film (complete with splendid rubber shark fight scene) and some ruthless Bat-merchandising. By mid-1967 its popularity was beginning to wane, and for the third series, producer Howie Horwitz introduced Batgirl, played by former ballerina Yvonne Craig (1941–). She was a hit, but the show was axed in 1968. Countless repeats, the re-examination of the Batman character in comic books of the 1980s, and the success of the subsequent 1989 movie starring Michael Keaton (1951–), show the enduring appeal of the character.

THE BATTLE OF ALGIERS (La Battaglia Di Algeri)

Gillo Pontecorvo (Italy/Algeria 1966)
Yacef Saadi, Brahim Haggiag, Jean Martin, Tommaso Neri.
123 mins. b/w.
Venice Film Festival *Golden Lion*

▷ Gillo Pontecorvo's *The Battle of Algiers* is a meticulously detailed recreation of the historical events surrounding the successful rebellion against the French colonial regime in Algeria between 1954 and 1962. Shooting on the actual locations involved with a cast mainly drawn from local non-professionals, Pontecorvo's grainy monochrome images resonate with the authenticity of the newsreel. Yet a statement in the opening titles reminds us that no actuality footage at all was used – thus the film is not only a committed political testament but a commentary on the very techniques of the documentary itself. As former partisan fighter and Youth Secretary of the Italian Communist Party (he left after the 1956 Hungarian invasion) Pontecorvo's sympathies are never in doubt, but while the studiously harrowing scenes of torture at the hands of the French military construct a compelling case against such human rights violations the film admirably refuses to draw its characterisation in the easy ideological shorthand. The colonial Colonel is prepared to admit to the inevitability of the historical process, for example, and a sequence leading up to a rebel bomb attack pulls no punches on the civilian cost to be paid for violent political struggle. By eschewing the caricatures of villainy and heroism and admitting the fictiveness of his filmmaking strategies, the film's willingness to admit the flexibility of political and aesthetic values strengthens rather than weakens its impact.

BATTLESHIP POTEMKIN (Bronenosets Potemkin)

Sergei Esisenstein (USSR 1925)
Alexander Antonov, Grigori

Alexandrov, Vladimir Barsky, Sailors of the Red Navy, Citizens of Odessa, Members of Moscow's Prolekult Theatre.
86 mins. b/w. silent. Reissued 1956 in sound version with new score.

The worldwide success of ▷ Sergei Eisenstein's *Battleship Potemkin* immediately focused international attention on the new Soviet cinema, fulfilling to some extent the Bolshevik authorities' hopes that their state-funded programme of filmmaking for explicitly propagandist purposes would not only consolidate the Revolution at home but promote the notion of class consciousness abroad. Commissioned to mark the 20th anniversary of the 1905 Revolution, Eisenstein chose to concentrate in particular on the naval mutiny and subsequent Tsarist massacre of civilians in the seaport of Odessa. The film embodies the idea of the collective hero he'd absorbed from his days creating a new revolutionary art in the theatre under Meyerhold, using the idea of typage to cast real sailors and actual citizens for greater authenticity of performance. Most significantly however, *Battleship Potemkin* was a textbook demonstration of Eisenstein's theoretical and practical approach to ▷ montage. From Marx's dialectical materialism, Pavlovian work on stimuli and response, Freudian psychology and the post-Revolution Soviet wave of Constructivist art, Eisenstein had worked out a highly mathematical concept of montage according to which a film's meaning was created from the series of synthetic collisions between image and subsequent image. Although few of today's films adopt Eisenstein's doctrinaire approach to the technique, his basic thesis added immeasurably to the widening of film grammar and *Battleship Potemkin*'s typically precise sequence of slaughter on the Odessa steps became one of the cinema's best-known moments.

BAXTER, Stanley

Born 24 May 1926, Glasgow, UK.

Baxter began his acting career in radio and he had a hundred broadcasts to his credit before he was called up towards the end of World War II. After his demob, he gravitated towards the legitimate theatre, making his professional debut as Correction's Varlet at the Edinburgh Festival production of *The Thrie Estates* in 1948.

An appearance in the fantasy pantomime *The Tintock Cup* (1949) led to a radio contract and a long association with the art of pantomime of which he is a master.

In London from 1959, his many stage appearances include *The Amorous Prawn* (1959), *Chase Me, Comrade* (1965), *What The Butler Saw* (1969) and *Phil the Fluter* (1969).

He made his film debut in *Geordie* (1955) and his disappointingly modest list of credits include such comedies as *Very Important Person* (1961), *The Fast Lady* (1962) and *Joey Boy* (1965) (his last to date).

He made his television debut in *Shop Window* (1951) and subsequently starred for the BBC in such critically and popularly acclaimed series as *On the Bright Side* (1959), *Stanley Baxter On...* (1960–4) and *The Stanley Baxter Show* (1967–71).

Moving to London Weekend Television in 1972 for *The Stanley Baxter Picture Show*, his talent for trenchant, rubber-faced mimicry and ability to assume all the guises in technically dazzling, lavish recreations of his favourite Hollywood musicals were given free rein in award-winning specials like *The Stanley Baxter Big Picture Show* (1973), *The Stanley Baxter Moving Picture Show* (1974) and *Stanley Baxter's Christmas Box* (1976).

Rising costs, and his refusal to compromise on standards, spelt an end to such glittering comic extravaganzas, but he returned to the BBC for *Stanley Baxter's Christmas Hamper* (1985) and found fresh popularity with a younger audience as the magician *Mr Majeika* (1987–).

Long-associated with the highest quality of traditional pantomimes, his most recent stage appearance was in *Cinderella* (1990–1) at Edinburgh and his is the voice behind a thousand diverse television commercials.

BEATTY, Warren

Real Name Henry Warren Beaty
Born 30 March 1937, Richmond, Virginia, USA.

The younger brother of actress ▷ Shirley

MacLaine, Beatty was an accomplished high-school football star who rejected numerous offers of sports scholarships to pursue an acting career in New York.

He first attracted attention with a supporting role in the television series *The Many Loves of Dobie Gillis* (1959) and as the indiscreet gas station attendant in *A Loss of Roses* (1959), his only Broadway appearance. He made his film debut in *Splendor in the Grass* (1961) and the pattern of his career was soon established with challenging roles like the gigolo in *The Roman Spring of Mrs Stone* (1961) and the nightclub entertainer in *Mickey One* (1965), interspersed with more commercially-minded escapism that exploited his boyish good looks, sullen rebelliousness and potent sex appeal.

He won new respect as the Oscar-nominated producer and star of ▷ *Bonnie and Clyde* (1967), a seminal work in American cinema for its approach to violence and attempt to illuminate the current state of the nation through a story from its past.

Highly selective in his choice of material, he is said to have turned down *Butch Cassidy and the Sundance Kid* (1969), ▷ *The Godfather* (1972) and ▷ *Last Tango in Paris* (1972) among many others, but gave two of his best performances as the roguish, western pioneer in *McCabe and Mrs Miller* (1971) and the dogged investigative journalist in *The Parallax View* (1974).

Increasingly restricting his appearances to projects of his own creation, he has proved himself an all-round filmmaker of intelligence and ambition, winning an Oscar nomination as the co-writer of *Shampoo* (1975) – a sexually explicit political satire detailing the exploits of a randy hairdresser during the 1968 Presidential election. He was nominated as Best Actor, Producer, Co-writer and Co-Director of the popular, light-hearted fantasy *Heaven Can Wait* (1978) and finally won a Best Director Oscar for *Reds* (1981) a sweeping, romantic epic on the life and love of journalist John Reed told in the manner of ▷ David Lean.

After a lengthy absence, he faltered badly with the witless comic flop *Ishtar* (1987) but restored some of his box-office lustre with the technically well-made (but dramatically somewhat dull) *Dick Tracy* (1990).

Politically active in a number of causes, his well-publicised reputation as a latterday Casanova has frequently diverted attention from his considerable achievements as a creative filmmaker. He will next be seen in the gangster melodrama tentatively entitled *Bugsy* (1991).

BECKY SHARP

Rouben Mamoulian (US 1935)
Miriam Hopkins, Cedric Hardwicke, Frances Dee, Billie Burke.
83 mins. col.

Throughout the Twenties the ▷ Technicolor company had been offering a system which cemented together two simultaneously-filtered (through red and green) 'recordings' to produce an effect (see the ▷ Douglas Fairbanks 1926 swashbuckler *The Black Pirate*) far superior to earlier efforts at tinting or drawing on the film by hand. The major breakthrough however came in 1928 when a new printing method was developed to produce a single final print that combined the twin filter strips without having to glue them together. This method would also of course permit printing from three different filtered images (through cyan, magenta and yellow) so creating the first truly modern colour process. ▷ Walt Disney became the first to try out the new technology with the 1932 *Flowers and Trees* entry in his ongoing *Silly Symphonies* series of cartoons, and the subsequent live-action short in 1934's *La Cucaracha*. Originally begun by Lowell Sherman (1885–1934) who died during shooting, the first three-colour Technicolor feature, RKO's *Becky Sharp*, was started again from scratch by his more adventurous replacement ▷ Rouben Mamoulian, who tried to alter the colour to reflect the emotional tonality of the story (an adaptation of Thackeray's *Vanity Fair* largely confined to the controllable environment of the studio set). Unfortunately the garish reds of the British army uniforms tended to overwhelm the image, and it was only with later offerings like the exquisitely composed *Blood and Sand* (1941) that the so-called Mamoulian palette would reach its maturity.

BEINEIX, Jean-Jacques

Born 8 October 1946, Paris, France.

A medical student until the cataclysmic

events of May 1968 diverted his attention to the cinema as a means of self-expression, Beineix then spent a number of years as an assistant-director on such films as *Les Saintes Cheries* (1969) by Jean Becker (1933–), *La Maison Sous Les Arbres* (The Deadly Trap) (1971) by Rene Clément (1913–), *The Day The Clown Cried* (1972) by ▷Jerry Lewis, *L'Animal* (1977) by Claude Zidi (1934–) and *French Postcards* (1979) by Willard Huyck.

Jean-Hugues Anglade and Béatrice Dalle in *Betty Blue* (1986)

Concluding that he was suffering 'all the inconveniences of the art of directing with none of the advantages', he made his directorial debut with the short film *Le Chien de Monsieur Michel* (1977). His first feature, *Diva* (1981) became a French *cause célèbre* and international success. A labyrinthine story of a postal messenger drawn into murderous mayhem when he illicitly records an opera singer, it displayed a flashy surface style oozing aesthetic seductiveness and an exhilarating use of offbeat camera angles. A trendsetting, visually sophisticated 'designer movie' for the 1980s it inevitably attracted dissenting voices bemused by its narrative incoherence.

The debate between the paramouncy of style or substance grew more acute with his second feature *La Lune Dans Le Caniveau* (The Moon in the Gutter) (1983) a commercially disastrous thriller and the hugely popular *37° 2' Le Matin* (Betty Blue) (1986) an almost delirious emotional rollercoaster of amour fou that sharply divided audiences into those infuriated by its pretensions and the hysterical central character and those besotted by the emotional intensity of a love affair conveyed with directorial bravura.

A frequent director of commercials, he returned to feature films with *Roselyne et Les Lions* (Roselyn and the Lions) (1989) another visually accomplished ode to its female star that lacked the narrative depth or credibility to sustain a lengthy running time.

BEN-HUR

William Wyler (US 1959)
Charlton Heston, Stephen Boyd, Jack Hawkins, Haya Harareet, Hugh Griffith.
217 mins. col.
Academy Award: *Best Picture*; *Best Director*; *Best Actor* Charlton Heston; *Best Supporting Actor* Hugh Griffith; *Best Music* Miklos Rosza; *Best Cinematography* Robert L. Surtees; *Best Editing* Ralph E. Winters, John D. Dunning; *Best Art Direction* and *Set Direction* William A. Horning, Edward Cartagno, Hugh Hunt; *Best Costume Design* Elizabeth Haffenden; *Award for Special Effects* Robert MacDonald, A. Arnold Gillespie, Milo Lory.

The Fifties advent of television as a truly popular entertainment medium initiated the decline of the Hollywood film industry's lucrative primacy. The studios' response was technological innovation designed to emphasise the movie screen's sheer dimensional scale over the flickering cathode ray image on the box at

Charlton Heston in *Ben-Hur* (1959)

home, with the widened screen ratio of ▷ Cinemascope soon in common usage. Fox's lavish biblical spectacular *The Robe* was the first film shot in their new screen format, its success demonstrating that the expansive frame needed expansive action to fill it and thus provoking a slew of similarly vast superproductions. In terms of prestige, the record Oscar tally of eleven statuettes for MGM's $15 million remake of *Ben-Hur* – the studio's silent version of General Lew Wallace's platitudinous novel, starring Latin idol Ramon Navarro (1899–1968), cost a then unprecedented $4 million in 1925 – represented the genre at what passed for an artistic peak. Predictably popular, the ingredients of the formula ran as usual: Heston's physical and moral fortitude as the hero, an educated Jew who survives galley slavery, Roman adoption, and chariot racing stardom before witnessing the Crucifixion; memorable action set-pieces like the battle at sea and the famous chariot clash; a reliably efficient director (in this case veteran ▷ William Wyler) to keep control over the towering sets and multitude of extras; plus healthy doses of Christian religiosity and a lengthy running-time to convince everyone of the importance of it all.

BENNETT, Alan

Born 9 May 1934, Leeds, UK.

An open scholar in history at Exeter College, Oxford, Bennett made his stage debut in the Oxford Theatre Group's revue *Better Late* at the Edinburgh Festival of 1959. He gained wider prominence as a writer and performer in the irreverent satirical show *Beyond the Fringe* which was staged in Edinburgh in 1960, London in 1961 and New York in 1962.

He subsequently wrote the television series *On the Margin* (1966) and his first dramatic play *Forty Years On* (1968).

A prolific writer for the theatre and television, he has managed to transform an acute understanding and observation of everyday lives and incidents into some of the small screen's most highly regarded moments of personal drama, revealing facets of family life, rueful regret over past decisions and the difficulties of personal communications with wit, candour and precision-engineered use of language and speech patterns. His many television plays include *A Day Out* (1972), *Sunset Across The Bay* (1975), *Intensive Care* (1979), *An Englishman Abroad* (1983) and *Talking Heads* (1987), a series of monologues.

His rare work for the cinema comprises *A Private Function* (1984), a hilarious tale of social aspiration in ration-era Britain, and *Prick Up Your Ears* (1987) a boisterous and scandalously witty story of the private life of Joe Orton (1933–67) that skillfully evoked the exuberant qualities evident in Orton's own plays.

His owlish features, schoolmasterish demeanour and inimitable north country diction have also made him a familiar figure as a television actor in work ranging from *Famous Gossips* (1965) to *Fortunes of War* (1987). His stage appearances include *A Cuckoo In The Nest* (1964), *Habeas Corpus* (1974) and *Single Spies* (1989).

His most recent television play is *102 Boulevard Haussmann* (1990), a reunion with actor Alan Bates (1934–) on a story of musical and human obsession culled from incidents in the life of Proust (1871–1922).

BENNY, Jack

Real Name Benjamin Kubelsky
Born 14 February 1894, Waukegan, Illinois, USA.
Died 26 December 1974, Beverly Hills, California, USA.

A child prodigy violinist, Benny performed as part of a vaudeville double-act, 'Salisbury and Benny' and also appeared as 'Ben Benny, the Fiddlin' Kid'.

After navy service during World War I, he returned to the stage and toured extensively before making his first film appearances in the short *Bright Moments* (1928) and as the MC of *The Hollywood Revue of 1929* (1929).

Following his Broadway success in *The Earl Carroll Vanities* (1930) and his radio debut in *The Ed Sullivan Show* (1932), he earned his own radio series which, combined with its subsequent television incarnation, *The Jack Benny Show* (1950–65), won him the loyalty and warm affection of a mass audience.

A gentle, bemused, self-effacing figure, his humour lacked malice, relying for its effect on his mastery of timing and an act based on his ineptitude as a fiddler, perennial claim to youthfulness and an unfounded reputation as the world's meanest man.

A sporadic film career was also the butt of much self-deprecation but after contributing to such all-star entertainments as *Broadway Melody of 1936* (1935) and *The Big Broadcast of 1937* (1936), it developed quite promisingly to encompass a brief string of starring roles in *Buck Benny Rides Again* (1940), *Charley's Aunt* (1941) and the classic black comedy *To Be Or Not To Be* (1942).

He virtually abandoned the cinema after the 1940s, content, apparently, to bask in his dominance of the small screen. Later guest appearances include *A Guide for The Married Man* (1967) and *The Man* (1972). He continued to appear in regular television specials and was planning to star in *The Sunshine Boys* at the time of his death from cancer. The role was inherited by his longstanding friend George Burns (1896–).

BERGMAN, Ingmar Ernst

Born 14 July 1918, Uppsala, Sweden.

The son of a Lutheran clergyman, Bergman escaped from the unhappiness of a strictly-disciplined childhood with a weekly visit to a film screening and, as a youth, his prized possessions included a magic lantern and a puppet theatre.

A student of art history and literature at the University of Stockholm, he later worked at the Royal Opera House in that city before accepting a job in the script department at Svensk Filmindustri and working on *Hets* (Frenzy) (1944) which made a star of ▷ Mai Zetterling.

He made his directorial debut with *Kris* (Crisis) (1945) and spent a number of years cinematically exploring relationships between the sexes before dawning international recognition for such works as *Sommaren Med Monika* (Summer With Monika) (1952) and *Gycklarnas Afton* (Sawdust and Tinsel) (1953).

Appreciation of his talent grew with *Sommarnattens Leende* (Smiles Of A Summer Night) (1955) a charming roundelay of romantic encounters at a country mansion that also served as a critique of bourgeois mores.

One of the richest periods in his career began with *Det Sjunde Inseglet* (The Seventh Seal) (1957), a game of chess between a medieval knight and Death that used luminous imagery and allegory to comment on mortality and man's relation to God. Equally memorable was *Smultronstallet* (Wild Strawberries) (1957) in which an aged professor's journey to accept an honorary doctorate became an expert fusion of past and present, dream and reality as he addressed his own mortality and sense of isolation from the warmth of humanity.

Preoccupied with death, solitude, spirituality and suffering, he has probed the darker side of the human condition with a sensitivity and rigour that make him one of the outstanding ▷ auteurs of world cinema. A selection of his most distinguished work includes *Jungfrukallan* (The Virgin Spring) (1960) (Best Foreign Film Oscar) a further exploration into the cruelty of medieval Sweden, *Sasom I En Spegel* (Through A Glass Darkly) (1961) (Best Foreign Film Oscar) about the torments of an afflicted family, and ▷ *Persona* (1966) in which two women develop a physical and emotional understanding that leads to an intermingling of their personalities.

Working with cinematographers Gunnar Fischer (1911–) and later ▷ Sven Nykvist, he has employed a regular troupe of actors, among them ▷ Liv Ullmann, ▷ Max Von Sydow, Erland Josephson (1923–) and Harriet Andersson (1932–), to create a series of austere masterpieces renowned for their photographic artistry, haunting imagery and subtle exploration of facial characteristics. *Vis Kingar Och Rop* (Cries and Whispers) (1972) proved a harrowing study of a woman's painful death from cancer and the corrosive effect it has on her loved ones. His versatility was also displayed with a joyous version of *Trollflojten* (The Magic Flute) (1974), *Hostsonaten* (Autumn Sonata) (1978) a study of the guilt-ridden relationship between a mother and daughter (superbly played by ▷ Ingrid Bergman and Ullmann) and his final film *Fanny Och Alexander* (Fanny and Alexander) (1982) (Best Foreign Film Oscar) an unexpectedly life-affirming evocation of autobiographical elements from his Dickensian childhood.

Prone to ill-health and nervous disorders, he announced his retirement from the cinema but remains active with the Royal Dramatic Theatre in Stockholm. His autobiography *Laterna Magica* (The Magic Lantern) was published in 1987.

BERGMAN, Ingrid

Born 29 August 1915, Stockholm, Sweden.
Died 29 August 1982, London, UK.

A student at the Royal Dramatic Theatre, Bergman signed a contract with Svenskfilmindustri and made her film debut as a maid in *Munkbrogreven* (1934).

Roles in *Branninger* (1935) and *Valkborgsmassoaften* (1936) established her as a star in her native land and brought her to the attention of ▷ David O. Selznick who signed her to appear in an English-language remake of the popular romantic melodrama *Intermezzo* (1939).

A fresh, unaffected performer with a naturalness and gaiety that won her a devoted following, she was soon the toast of Hollywood with a string of box-office successes that included ▷ *Casablanca* (1942), *For Whom The Bell Tolls* (1943), *The Bells of St. Mary* (1945) and *Notorious* (1946). Most adept at conveying wholesomeness and virtue in an appealing light, she was less convincing when cast unsympathetically, but attempted a wider range as the guttersnipe in *Dr Jekyll and Mr Hyde* (1941) and as the prostitute in *Arch of Triumph* (1948).

In 1950, she gave birth to the illegitimate child of director ▷ Roberto Rossellini and the ensuing scandal led to her ostracisation from the American film industry and a condemnation of her moral turpitude from the floor of the American Senate. She continued to work in Europe on such underrated films as *Stromboli* (1950), *Viaggio In Italia* (Journey to Italy) (1954) and *Elena Et Les Hommes* (1956) and was warmly welcomed back to the Hollywood community in 1956.

She enjoyed further successes with *Indiscreet* (1957), *Inn of the Sixth Happiness* (1958) and *Cactus Flower* (1969) but found that the cinema had few worthy roles to offer an actress of her maturity and stature. Instead, she appeared with growing regularity on stage and television, notably in the mini-series *A Woman Called Golda* (1982).

Her last film appearance was in *Hostsonaten* (Autumn Sonata) (1978), a deeply-felt exploration of a mother-daughter relationship that contained some of her finest emotional acting. Nominated seven times for an ▷ Academy Award, she won ▷ Oscars for *Gaslight* (1944), *Anastasia* (1956) and *Murder on the Orient Express* (1974).

Her autobiography, *Ingrid Bergman: My Story*, was published in 1980.

BERKELEY, Busby

Real Name William Berkeley Enos
Born 29 November 1895, Los Angeles, California, USA.
Died 14 March 1976, Palm Springs, California, USA.

Educated at a New York military academy, Berkeley first displayed evidence of his future skills as an arranger of marching drills. After service in the US Army, he pursued a career in the theatre, working as an actor, stage manager and dance director.

He directed his first Broadway show, *A Night in Venice*, in 1928 and was subsequently hired by ▷ Sam Goldwyn to devise the musical numbers for the film *Whoopee* (1930). He stayed in Hollywood to become one of the cinema's most innovative choreographers, noted for his mobile camerawork and dazzling kaleidoscopic routines involving spectacular multitudes of chorus girls and much sexual innuendo. His work enhanced such Warner Brothers musicals as *42nd Street* (1933), ▷ *Gold Diggers of 1933* and *Dames* (1934).

Working at M-G-M from 1939 as a dance advisor and director, he made such wartime entertainments as *Babes In Arms* (1939), *Strike Up The Band* (1940) and *For Me and My Girl* (1942).

In later years ill-health, bouts of alcoholism and a vexed private life curtailed his professional activities. *Take Me Out To the Ball Game* (1948) was the last film he directed but he did contribute flashes of imaginative choreography to *Small Town Girl* (1953) and *Jumbo* (1962) and made a cameo appearance in *The Phynx* (1970) before enjoying a late Broadway triumph as the supervising producer of the 1971 revival of *No, No, Nanette*.

BERTOLUCCI, Bernardo

Born 16 March 1940, Parma, Italy.

An amateur filmmaker and poet, Bertolucci became an assistant to ▷ Pier Pasolini on *Accatone* (1961). His collection of poetry *In Cerca Del Mistero* (1962) won the Premio Viareggio Prize and he made his directorial debut the same year with Pasolini's script of *La Commare Seca* (The Grim Reaper), a derivative effort depicting the differing stories of the relevant witnesses to the murder of a prostitute.

Working in television documentaries during the 1960s, he also contributed to the scripts of *Ballata De Un Milliardo* (1966) and *C'Era Una Volta Il West* (Once Upon A Time in the West) (1968) before resuming his directorial career with *La Strategia Del Ragno* (Spider's Strategem) (1969) the story of a young man who discovers the truth about his father's apparently heroic death at the hands of the Fascists in 1936.

A member of the Italian Communist Party, his films depict the tension between conventionality and rebellion, and often use some form of dramatic journey (whether exterior or interior) to explore the complex relationship between politics, sex and violence.

Working in frequent collaboration with gifted cinematographer Vittorio Storaro (1946–), he has brought a visual élan to his best work and enjoyed international success with *Il Conformista* (The Conformist) (1970), the story of a man so desperate to prove his normality that he will kill on behalf of the Fascists, and the controversial *Ultimo Tango A Parigi* (Last Tango in Paris) (1972) an intense study of alienation and malaise unravelled through a series of anonymous sexual encounters.

Less widely praised were his sweeping Marxist epic *Novecento* (1900) (1976) covering Italian history from the turn of the century to 1945, the stylish Oedipal saga *La Luna* (1979) and the muddled *La Tragedia Di Un Uomo Ridicolo* (Tragedy of a Ridiculous Man) (1981) in which a father only begins to comprehend the true nature of his son when the latter is held for ransom.

After a number of unrealised projects during the 1980s, he returned to the cameras with a lavish if emotionally reserved life of Pu Yi, *The Last Emperor* (1987), which won him an Oscar as Best Director, and a languid and surprisingly shallow version of the novel *The Sheltering Sky* (1990).

BESSON, Luc

Born 18 March 1959, Paris, France.

Leaving school before completing his baccalaureat, Besson worked on a magazine before gaining a diversity of experiences as an unpaid assistant on such international productions as *Moonraker* (1979), *The Nude Bomb* (1980) and the television mini-series *Gauguin the Savage* (1980).

Between 1977 and 1982, he directed a group of short films, including *Amer*, *Les P'tites Sirenes* and *L'Avant Dernier*, made television commercials, and worked as an assistant director on films like *Deux Lions au Soleil* (1980), *Court Circuit* (1980), *Homme Libre, Tu Cheriras la Mer* (1980) and *Les Bidasses aux Grandes Manoeuvres* (1981).

He made his feature-length directorial debut with *Le Dernier Combat* (The Last Battle) (1983), an imaginative, visually striking post-Apocalypse fable that is boldly shot in monochrome ▷ CinemaScope and virtually dispenses with dialogue. He followed this with the highly popular *Subway* (1985) a dazzling thriller set largely in the labyrinthine tunnels of the Paris Métro.

The second unit director on *Le Grand Carnaval* (1983) and writer of *Kamikaze* (1986), he has also made promotional videos for pop records including *Pull Marine* by Isabelle Adjani (1955–).

A lifelong fascination with the ocean manifested itself in the wondrous, globetrotting *The Big Blue* (1988) in which an accomplished underwater diver is torn between a human love affair and his compulsive attraction to the sea. The most successful film of the decade in France, it was brutally re-edited, re-scored, poorly promoted and failed to make an impression in many English-language markets.

He followed this with the explosive, romantic ▷ film noir *Nikita* (1990). His films have often been criticised for offering only a flashy surface stylishness that masks a lack of emotional or dramatic depth. However, their virtues are those of the purest cinema; original, gripping stories told with a remarkable visual flair, technical virtuosity, energy and pace that remain highly appealing.

BEST BOY The technician who serves as the assistant to the chief electrician who is better known as the gaffer.

THE BEST YEARS OF OUR LIVES

William Wyler (US 1946)
Myrna Loy, Fredric March, Dana Andrews, Teresa Wright, Harold Russell.
182 mins. b/w.
Academy Award: *Best Picture*; *Best Director*; *Best Actor* Fredric March; *Best Supporting Actor* Harold Russell; *Best Screenplay* Robert Sherwood; *Best Editing* Daniel Mandell; *Best Music* Hugo Friedhofer; Special Award to Harold Russell for 'bringing hope and courage to his fellow veterans'.

One of the Hollywood establishment's most admired filmmakers, three-time Oscar winner ▷ William Wyler (the other statuettes were for 1942's *Mrs Miniver* and ▷ *Ben-Hur* in 1959) favoured 'important' subject matter, exercised his skills on a broad range of generic material, and earned the nickname '90-Take Wyler' for his legendary on-set perfectionism. While the seeming absence of any recurring thematic consistency places him outside the auteurist consensus lionising Ford or Welles, Wyler could however be relied on to handle the prestige project with unerring competence. Having just returned to the field himself – where his undemonstrative sensibility made him a fine war documentarist (*The Memphis Belle*, 1944) – he was obviously just the right man for *The Best Year of Our Lives*, the most significant fictional account of America's post World War II homecoming. Collaborating with regular cameraman Gregg Toland (1904–48) to frame the landscape of small town middle-class middle America in immaculately composed detail, the film sympathetically records the pains of readjustment to civilian society as ▷ Fredric March's former sergeant turns to the bottle, air force major Dana Andrews (1909–) finds himself back stacking shelves, and seaman Harold Russell (1914–) (a non-professional actor virtually playing himself) faces home and family having lost both arms in battle. Avoiding overt sentimentality, its responsible and accurate treatment of issues affecting most of the population made it a wide commercial success and perhaps the finest representation of Wyler's consummate, if impersonal, craftsmanship.

BICYCLE THIEVES (Ladri Di Biciclette)

Vittorio De Sica (Italy 1948)
Lamberto Maggiorani, Enzo Staiola, Lianella Carell.
90 mins. b w.
Academy Award: Special Oscar *Most Outstanding Foreign Film*

Former Italian screen idol turned director ▷ Vittorio De Sica's *Bicycle Thieves* (Ladri Di Biciclette) remains one of the best-known films from the Italian ▷ neo-realist cycle which flourished in the mid-Forties in response to the glossily anodyne state-sponsored 'white telephone' movies of the previous Fascist period. As developed by its major theorists, including screenwriter Cesare Zavattini (1902–89) and director ▷ Roberto Rossellini, neo-realism embodied a moral (and hence an aesthetic) position aiming to confront audiences with 'reality', so encouraging the viewer to question the 'real' world instead of merely soaking up the lavishly-prepared images of the screen entertainment industry. Thus De Sica's *Bicycle Thieves* is a cry of despair from the Roman underclass, following the misfortunes of a struggling bill-sticker and his son as they try to recover the stolen bike essential to the father retaining his job and so putting food on the family table. Following the familiar pattern set by Rossellini's ▷ *Rome, Open City*, De Sica and his excellent cast of non-professionals filmed in the capital's most downbeat corners, so providing an accurate picture of struggling ordinary folk. Yet despite such trappings and their accompanying neo-realist claims to truthful objectivity, the film garners its emotional impact from the deliberate sentimental styling of the narrative. Antonio, of course, finds himself surrounded by bicycles just after his own machine is stolen ... lo, bells ring on the soundtrack; it's all part of the contrivance that is film storytelling. The trick is though – and De Sica manages it with moving panache – that you don't notice the contrivance.

THE BIRTH OF A NATION

D. W. Griffith (US 1915)
Henry B. Walthall, Lillian Gish, Miriam Cooper, George Siegman, Mae Marsh.
Approx. 180 mins. censored to approx. 165 mins. b/w. silent.

▷ David Wark Griffith's *The Birth of a Nation* is one of the most problematic milestones in all of cinema history: artistically, it remains a prime achievement in the development of film narrative; ideologically, it is one of the most explicitly racist films ever made. In his myriad shorts from 1908 onwards Giffith had virtually invented the grammar of film: from the close-up and long-shot to the idea of cutting between points of interest to create tension, from the first use of titles to the creation of the flashback, Giffith did it first. Yet inspired by the international success of Giovanni Pastrone's (1883–1959) Italian epic *Cabiria* (1914), he sought to create an American work of art on a similar scale, choosing as his source the Rev Thomas F. Dixon's melodrama *The Clansman.* Himself a Southerner raised in the values of the Old South, Griffith's film is set during the Reconstruction, chronicling the tangled fortunes of Union and Confederate families and at length exhorting both sides to unite against their common enemy, the unruly and sexually avaricious Negro, through the organisation of the Ku Klux Klan ('the saviour of white civilisation' runs one intertitle). Playing on the most blatant racial stereotypes (the coloured characters are played by whites in black faces) and hysterical fears of miscegenation, the merest synopsis is today enough to horrify, but with its expansive battle scenes and detailed plotting *The Birth of a Nation* attracted unprecedented audiences wherever it was shown. While Griffith's admirers still regard him as a giant among American filmmakers, others prove less willing to confer lasting greatness on an artist whose work suffers such grave flaws of conception.

BLACKMAIL

Alfred Hitchcock (UK 1929)
Anny Ondra, Sara Allgood, John Longdon, Charles Paton.
96 mins. b/w.

One of the first sound films to be shot in Britain, *Blackmail* is doubly notable as a progenitor of much of ▷ Alfred Hitchcock's later thematic and stylistic manner, his very particular style. Assigned to shoot the film silent, with the innovation of sound featured only in the final reel, Hitchcock managed however to 'dub' synchronised dialogue (Polish star Anny Ondra (1903–87) mimed her lines while another actress spoke them just off-screen) on to the earlier passages and release *Blackmail* as a talkie. As in most early sound efforts, some of the dialogue stretches are dreadfully slow, but with one striking sequence Hitchcock demonstrates the artistic potential of the new medium: having fended off a stranger's attempted rape by stabbing him to death, Ondra breakfasts the next morning with her family but the soundtrack mutes the entire conversation excepting the word 'knife', each repetition a piercing reminder of her crime. The central figure of the guilty woman was indeed to recur in Hitchcock's work (see *Notorious* (1946), *Psycho* (1960), *Marnie* (1964) for instance), as was the notion of the ordinary protagonist plunged into chaos (*The Thirty-Nine Steps* (1935) and *North By Northwest* (1959) among many others), but what might be most important about *Blackmail* is that it sets the precedent of subjugating the actual meaning of the plotting to the string of darkly comic suspense games that may be exacted from it along the way. Hitch cameo spotters might also like to keep an eye open for a certain scene on the underground.

BLIER, Bertrand

Born 14 March 1939, Boulogne-Billancourt, Paris, France.

The son of French character actor Bernard Blier (1916–89), Blier has become recognised as a master of what he terms 'slapstick, nose-thumbing farces' which provocatively explore the war between men and women.

Deciding to become a director after passing his baccalaureat, he worked extensively as an assistant before making his debut with the documentary *Hitler ... Connais Pas!* (Hitler ... Don't Know Him) (1962). Struggling to establish himself, he made the short film *La Grimace* (1966) and the unsuccessful espionage thriller *Si J'Etais Un Espion* (If I Were A Spy) (1967).

He then spent a long while at work on numerous unrealised projects, with the screenplay for *Laisser Aller, C'Est Une Valse* (1971) his only credit during a bleak and frustrating period. However, he then wrote a bestselling novel which he transformed into *Les Valseuses* (1973) (French slang for testicles), a deliberately offensive story of two boorish, self-centred louts who can neither comprehend nor please the woman in their life and find solace in the comfort of male camaraderie.

He pursued similar themes with rigorous iconoclasm in *Calmos* (Femmes Fatales) (1976), *Preparez Vos Mouchoirs* (Get Out Your Handkerchiefs) (1978), which received an Oscar as Best Foreign Film, and *Beau-Pere* (Stepfather) (1981) which, despite being collectively judged as misogynistic, painted men as emotionally immature, insecure and pathetically juvenile failures who can neither escape nor satisfy the sexual demands of liberated women.

More straightforward work like *Buffet Froid* (1979), a blackly humorous account of the obsessive relationship between a police inspector and a layabout, and *Notre Histoire* (Our Story) (1984) about a strange encounter on a train, tended to be muddled and pretentious.

However, he was back on top form with the outrageous *ménage-à-trois* comedy *Tenue de Soirée* (Evening Dress) (1986) and revealed the subtler shades of his pallette with *Trop Belle Pour Toi* (1989) which homed in on the pain and confusion of three figures caught in a complicated emotional triangle.

He followed this with *Merci, La Vie* (1991).

BLIMP A soundproof cover that fits over a camera during shooting to absorb the noise of the camera and thus allow simultaneous sound recording without interference. Made of either aluminium or magnesium and insulated with rubber and plastic foam, the blimp allows the camera to remain fully operational. A 'barney' also offers sound-proofing whilst protecting the camera from extremes of temperature and weather.

BLOOP When any form of join or splice in a film soundtrack is passed through a sound system the resulting dull, thudding sound is called a bloop. The noise can be reduced or eradicated by covering the offending break with blooping tape or blooping ink on the positive optical soundtrack or by placing a perforation on the negative optical soundtrack.

THE BLUE ANGEL (Der Blaue Engel)
Josef Von Sternberg (Germany 1930)
Emil Jannings, Marlene Dietrich, Rosa Valetti, Hans Albers.
90 mins. b/w. Simultaneously shot in English and German versions.

Although ▷ Marlene Dietrich has claimed that director ▷ Josef Von Sternberg 'discovered' her (casting her in the film was to make both their reputations) the truth is, as ever, a little more prosaic. In 1930 Marlene was already the screen veteran of some seventeen unremarkable movies, while Von Sternberg had in fact returned from a stalled career in Hollywood – invited back to Germany to work with actor ▷ Emil Jannings (1884–1950), whom he'd guided to the first Best Actor Oscar for *The Last Command* (1928). Indeed, although *The Blue Angel* started out as a vehicle for Jannings, the role of a stern professor

Marlene Dietrich in *The Blue Angel* (1930)

destroyed by his infatuation with a night-club singer specially written for him, the eventual signing of fourth-choice Dietrich for the role of chanteuse Lola-Lola was to initiate one of the great director-actress partnerships. The film's basic outline of the weak and masochistic male trapped by an indifferent and enigmatic temptress held an undoubted autobiographical charge for Von Sternberg and he was to use it time and again in his Thirties series of perversely stylish Dietrich showcases for Paramount (*Morocco* (1930), *Shanghai Express* (1932), *The Scarlet Empress* (1934) and others). Always highly aware of her screen image, Marlene for her part conspired in Von Sternberg's complete absorption in filmic style. She was a sculpted presence to be lit and dressed as his often fetishistic impulses demanded, the result a pessimistic and highly personal vision of an existence ruled remorselessly by the unfathomable demands of pleasure. A disastrous 1959 remake, directed by Edward Dmytryk (1909–) cast Curt Jurgens (1912–82) and May Britt (1933–) in the roles of the Professor and Lola-Lola.

BLUE PETER
UK 1958– BBC

An island of Reithian broadcasting under the watchful eye of editor Biddy Baxter (until 1988), *Blue Peter* has encouraged successive generations of children to fill their spare time by making interesting things from discarded household items and 'sticky back plastic' (so called because of the show's pathological aversion to brand names). Created by John Hunter Blair, the magazine's original presenters were actor Christopher Trace and former Miss Great Britain Leila Williams. Though others may canvass for Lesley Judd, Sarah Greene, Caron Keating or John Leslie, the show's golden era came when the reins were held by presenters John Noakes, Valerie Singleton and Peter Purves, along with dogs Shep and Petra, and Jason the cat. On her death Petra was honoured with a bronze bust at the entrance of the BBC. Central to the programme's efforts to encourage do-gooding is the end-of-year charity appeal, when viewers recycle household refuse – typically stamps or bottle tops. ITV's rival show *Magpie* tried to borrow the idea, but simply asked for money. That *Magpie* is long-forgotten tells its own story. Despite the jibes, *Blue Peter* remains, though continuing efforts to revamp the format run the risk of succeeding where the vandals in Percy Thrower's garden failed – rejecting innocence in the name of progress.

BLUE SCREEN PROCESS PHOTOGRAPHY A ▷ travelling-matte method of combining images in which one or more components of the foreground action is shot against a uniform bright blue backing and superimposed on a background scene recorded by another camera. In motion pictures this involves subsequent complex printing at the laboratory to avoid halos showing and the joins being seen.

In television it can be done continuously at the time of shooting by colour keying ('chromakey'): each area of blue is replaced by the corresponding portion of the background scene.

The technique enables an extremely wide range of effects to be achieved: typically studio actors appearing in exotic locations, children playing within a story-book illustration or the fantastic flights of *Superman* (1978).

BOGARDE, Dirk
Real Name Derek Niven Van den Bogaerde.
Born 29 March 1921, Hampstead, UK.

An art student at Chelsea Polytechnic, Bogarde began acting in repertory theatre and worked as an ▷ extra on the film *Come On George* (1940).

After the war, he returned to the theatre in *Power Without Glory* (1947) and had a small role in the film *Dancing With Crime* (1947). Offered a long-term contract by the Rank Organisation, he appeared in a variable collection of titles under their auspices achieving matinee idol status as a personable light comedian and perennially heroic member of His Majesty's Forces. He also gained some critical respect for his portrayal of a young spiv in the now antiquated *The Blue Lamp* (1949).

Among his more noteworthy performances were the pursued murderer in *Hunted* (1952), the breezy Simon

Sparrow in *Doctor in the House* (1953) (the first of a long-running series), the devious villain of *Cast A Dark Shadow* (1955), the title character in *The Spanish Gardener* (1956) and a dignified Sydney Carlton in an adequate version of *A Tale of Two Cities* (1958).

Always a likeable, hard-working talent, he showed evidence of greater depth as the worthless artist Louis Dubedat in *The Doctor's Dilemma* (1959) and took the then brave step of portraying a blackmailed homosexual in ▷ *Victim* (1961).

However, the subtlety and range of his screen technique was only fully explored when he renewed a collaboration with director ▷ Joseph Losey that resulted in his sinisterly manipulative valet in *The Servant* (1963), conscience-stricken defending officer in *King and Country* (1964), high-camp villain in *Modesty Blaise* (1966) and enigmatic, emotionally troubled Oxford don in *Accident* (1967).

He then moved to France, working with some of Europe's most distinguished directors on performances of beautifully modulated sensitivity and power. Notable triumphs include *La Caduta Degli Dei* (The Damned) (1969), *Il Portiere Di Notte* (The Night Porter) (1973), *Providence* (1977) and his finest performance, exquisitely conveying the demons that haunt a dying composer in *Death in Venice* (1971).

More recently, he has developed a reputation as a writer of some perspicacity with several volumes of autobiography and novels to his credit. His acting has been restricted to television appearances in *The Patricia Neal Story* (1981), *May We Borrow Your Husband?* (1986) and *The Vision* (1987) but he returned to the cinema screens after a twelve-year absence in *Daddy Nostalgie* (These Foolish Things) (1990).

BOGART, Humphrey

Full Name Humphrey De Forest Bogart
Born 23 January 1899, New York City, New York, USA.
Died 14 January 1957, Los Angeles, California, USA.

The son of a doctor, Bogart served in the US Navy towards the end of World War I and commenced his theatrical career as a stage manager and walk-on actor, before graduating to 'anyone-for-tennis' juvenile leads in such plays as *Meet the Wife* (1923) and *Cradle Snatchers* (1925).

He made his film debut in the short *Broadway's Like That* (1930) and was signed to a Hollywood contract but made little impression with the modest opportunities afforded him. Back in New York, he astounded audiences with his electrifying performance as hoodlum Duke Mantee in *The Petrified Forest* (1935) and repeated the characterisation on film the following year.

Now employed by Warner Brothers, he was confined to secondary roles in support of their more highly considered stars and served his time as crooked tough guys and gun-wielding gangsters in the likes of *Bullets or Ballots* (1936), *Angels With Dirty Faces* (1938) and *The Roaring Twenties* (1939).

His star potential finally became apparent as the aging gangster in *High Sierra* (1941) and cynical private eye Sam Spade in ▷ *The Maltese Falcon* (1941). Now a fully-fledged leading man, he created an indelible and enduring screen persona of lone wolf; sardonic, anti-authoritarian and heroic, abrasive, romantic and stubbornly true to his own code of ethics. His greatest successes include ▷ *Casablanca* (1942), *To Have and Have Not* (1944) and *The Big Sleep* (1946).

His considerable acting prowess was also displayed as the dishevelled and paranoid prospector in *The Treasure of the Sierra Madre* (1947), the psychotic screenwriter in *In A Lonely Place* (1950) and the psychopathic captain in *The Caine Mutiny* (1954). He won a Best Actor Oscar as the gin-sodden boatman in ▷ *The African Queen* (1951) and made his final appearance in *The Harder They Fall* (1956).

He was married to actress Lauren Bacall (1924–) from 1945 until his death from cancer.

BOGDANOVICH, Peter

Born 30 July 1939, Kingston, New York, USA.

A knowledgeable film buff and durable director, Bogdanovich once commented 'I was born and then I liked movies'.

A student of acting at Stella Adler's Theatre Studio in New York, he worked in the theatre from the late 1950s before gaining renown as a journalist writing criticism and entertaining celebrity profiles for such publications as *Esquire*, *The New York Times* and *Cahiers du Cinema*. His many books include *The Cinema of Orson Welles* (1961), *Fritz Lang in America* (1967) and *Allan Dwan: The Last Pioneer* (1971).

In 1966, ▷ Roger Corman hired him to dub and re-edit the Russian science-fiction film *Planeta Burg* for American consumption and the result was *Voyage to the Planet of Prehistoric Women* which appeared under the pseudonym of Derek Thomas. He also worked in various capacities on Corman's *The Wild Angels* (1966) and it was Corman who backed his legitimate directorial debut *Targets* (1967), a taut thriller melding the stories of an aging horror film star and a psychopathic Vietnam veteran.

The Last Picture Snow (1971), filmed in black and white by Robert Surtees (1906–85), was a masterly evocation of smalltown life in a dusty Texas town in the 1950s and won him Oscar nominations as director and co-screenwriter.

He consolidated his burgeoning reputation as a wunderkind with the lively ▷ screwball comedy *What's Up Doc?* (1972) and the Depression-era con game *Paper Moon* (1973). Thereafter, his nostalgic yearning for the past was ill-served by inappropriate casting and a blind faith in the abilities of his off-screen partner Cybill Shepherd (1949–). The 1930s musical *At Long Last Love* (1975) and *Nickelodeon* (1976), set in the cinema's pioneering days, were notably leaden failures.

He returned to critical favour with *Saint Jack* (1979) an atmospheric adaptation of the novel by Paul Theroux with Ben Gazzarra (1930–) effectively capturing the moral disintegration of a footloose American pimp in Singapore.

An affair with Playmate of the Year Dorothy Stratten (1960–81) ended in tragedy when she was killed by her insanely jealous husband. The events were dramatised in *Star 80* (1983) with Roger Rees (1944–) as Bogdanovich.

They All Laughed (1981) and *Illegally Yours* (1988), further attempts at screwball comedy, passed with little fanfare but he enjoyed his biggest success in years with *Mask* (1985) an emotional study of the relationship between an unconventional mother and her disfigured son.

Texasville (1990), a sequel set many years after *The Last Picture Show*, reunited him with the original cast in a patchwork story of middle-age malaise and disillusionment set against the backdrop of their smalltown's centennial festivities. It received a lukewarm critical reception and his career seemed to be in trouble once more when he subsequently found himself replaced as director of the comedy *Another You* (1991).

An occasional documentarist and actor, he has appeared in such films as *Lion's Love* (1969), *Opening Night* (1977) and Orson Welles's unreleased *The Other Side of the Wind* (1970–75).

BONNIE AND CLYDE

Arthur Penn (US 1967)
Warren Beatty, Faye Dunaway, Gene Hackman, Estelle Parsons, Michael J. Pollard, Gene Wilder.
111 mins. col.
Academy Award: *Best Supporting Actress* Estelle Parsons; *Best Cinematography* Burnet Guffey.

Director ▷ Arthur Penn, rather like his contemporary ▷ Sam Peckinpah, emerged as a creative force in a transitional period for Hollywood. Penn, who had worked on the stage and on television before he made his feature debut with *The Left-Handed Gun* (1958), inherited the conventions of the studio era even if he was too late to truly belong to it, while his consistently thoughtful approach bears the influence of the European screen and looks forward to the idea of a personal American that was to briefly flourish with the first flowering of the

Warren Beatty and Faye Dunaway in *Bonnie and Clyde* (1967)

Movie Brat generation (▷ Martin Scorsese, ▷ Francis Coppola, ▷ Brian De Palma etc) in the early Seventies. With its breakthrough performances for developing performers ▷ Warren Beatty, ▷ Faye Dunaway and ▷ Gene Hackman, brilliantly used banjo score, and insistent screen violence of unprecedented intensity, *Bonnie and Clyde* remains one of those milestones where the diverse ingredients create a perfect chemistry. Yet while the style and trappings of the film are immediately striking it's the amoral tone towards its factually-based chronicle of vigorous criminality which proved most attractive to its young contemporary audiences themselves caught up in the Sixties cultural upsurge of resistance to established authority. Dunaway's beret may have launched new fashion trends, but the deep resonance of the piece lies in its evocation – also present in Penn's earlier *The Chase* (1966) and 1970's native American epic *Little Big Man* – of the conflict between personal impulse and society's constricting, often corrupt, network of control. Explicitly rhyming sexual satisfaction and the breaking of the law, this slice of Thirties Americana might just be the perfect film document of 1968.

BONNAIRE, Sandrine

Born 31 May 1967, Clermont-Ferrand, France.

The seventh of eleven children, Bonnaire left school at 15 with only a vague notion of pursuing hairdressing as a career. When one of her sisters responded to a newspaper advert seeking film extras, she went along to offer moral support and found herself cast by director Maurice Pialat (1925–) as the intensely lovelorn, promiscuous Suzanne in *A Nos Amours* (To Our Loves) (1983) for which she received a French César as most promising actress.

The Best Actress César came her way for *Sans Toi Ni Loi* (Vagabonde) (1985) in which she gave a flawless performance as the tragic, solitary and stubbornly independent itinerant.

Without any formal training as an actress, she has revealed a talent of great honesty and virtuosity that allows her performances to appear without pretence or false technique. Selective in her choice of projects, she has appeared in such controversial and challenging films as *Police* (1985), *Sous Le Soleil de Satan* (Under Satan's Sun) (1987), *Monsieur Hire* (1989) and the hypnotic, almost wordless, *Captive du Désert* (Captive of the Desert) (1990).

The most skilled and versatile French actress of her generation, she recently appeared with ▷ Marcello Mastroianni in *Verso Sera (Towards Evening)* (1990) and was scheduled to make her English-language debut in *Prague* (1991).

BOOM OPERATOR The technician responsible during the shooting of a film for ensuring the most efficacious placement of the long, movable telescopic arm or boom carrying a microphone, camera or light. Keeping the boom out of camera frame, the operator can use its mobility to record synchronous sound by following characters through a scene, photograph a scene from a difficult or humanly inaccessible angle or place extra light in a particular area of the set.

BOORMAN, John

Born 18 January 1933, Epsom, Surrey, UK.

Leaving school at sixteen to run a laundry business, Boorman remained there for a year before pursuing a journalistic career as a critic and broadcaster for *The Guardian* and the BBC. After his National Service, he joined ITN to train as an assistant film editor and gained further experience on current affairs and news programmes at Southern TV.

Employed by the BBC from 1962, he headed a documentary film unit that created such work as the acclaimed series *Citizen 63*, the documentary-drama *The Newcomers* (1964) and *The Great Dictator* (1966), a profile of ▷ D.W. Griffith.

He made his cinema debut as the director of *Catch Us If You Can* (1965), a vehicle for the Dave Clark Five pop group, but made more of an impact with his American debut *Point Blank* (1967), a stylish saga of gangland revenge that used eyecatching directorial technique to emphasise the bleakness and violence at the heart of urban America.

The tensions between man's inherently bestial nature and his pursuit of civilised aspirations was further explored in the raw anti-war statement *Hell in the Pacific* (1968) and *Deliverance* (1972) which pitted sophistication against primitivism as a party of middle-class suburbanites find their mettle tested on a canoeing expedition in the backwoods of Georgia. The latter earned him a Best Director Oscar nomination.

Less successful ventures from this period include the social fable *Leo the Last* (1970), the muddled science-fiction allegory *Zardoz* (1973) and the disastrous sequel *Exorcist II: The Heretic* (1977) which he later described as a 'huge mistake'. However, he redeemed himself with the vigorously executed and popular *Excalibur* (1981) which represented the culmination of his life-long fascination with Celtic mythology, the Arthurian legend and the quest for the Holy Grail.

His interest in the conflict between man and nature surfaced again in the stolid ecological drama *The Emerald Forest* (1985) but he was on form again with *Hope and Glory* (1987), a lovingly recalled autobiographical portrait of his adventurous wartime childhood.

The latter seemed to herald an interest in more mellow and intimate projects and was followed by *Where the Heart Is* (1990), a commercially unsuccessful attempt to revive elements of the ▷ screwball comedy in a contemporary setting.

An executive director of Ireland's Ardmore Studios from 1975 to 1982, he was particularly encouraging to the directorial aspirations of ▷ Neil Jordan and has been a pungent chronicler of his own filmmaking experiences in the volumes *Money Into Light* (1985) and *Hope and Glory* (1987).

BOYER, Charles

Born 28 August 1897, Pigeac, south-west France.
Died 26 August 1978, Phoenix, Arizona, USA.

Educated at the Sorbonne, Boyer pursued a childhood ambition to act by studying drama at the Paris Conservatoire. He made his stage debut in *Les Jardin de Murcie* (1920) and a first film appearance the same year in *L'Homme du Large*. Further supporting roles followed and his stage career blossomed as he developed a following as a matinée idol.

He travelled to Hollywood to appear in French versions of American hits and was subsequently cast in English-language productions like *The Man From Yesterday* (1930) and *Red-Headed Woman* (1932). However, the studios seemed uncertain of his appeal or abilities and he returned to France.

Invited back to Hollywood in 1934, he enjoyed moderate success with *Private Worlds* (1935). However, a succession of strong romantic roles in films like *Mayerling* (1936), *The Garden of Allah* (1936) and *Algiers* (1938) established his stardom. Gravely handsome with an attractively accented bass voice and richly expressive eyes, he came to represent an international ideal of the French lover and was equally adept at comedy or drama.

His performance as Napoleon to Garbo's Marie Walewska in *Conquest* (1937) earned him the first of four Oscar nominations. He brought roguish charm to *History Is Made at Night* (1937), an impeccable romantic presence to *Love Affair* (1939) and a sharp malevolence to the role of the conniving husband in *Gaslight* (1944).

After the War, he worked continuously on both sides of the Atlantic displaying scene-stealing flair as a character actor in films like *Fanny* (1961), *Barefoot in the Park* (1967) and *Stavisky* (1974).

In 1952 with ▷ Ida Lupino, ▷ David Niven and Dick Powell (1904–1963), he founded Four Star Productions and acted in many of their television shows including the series *The Rogues* (1964–65). Later stage performances included *Kind Sir* (1953) and *The Marriage Go-Round* (1958) on Broadway and *Man and Boy* (1964) in London. He received a special Academy Award in 1943 for his efforts in promoting Franco-American cultural relations.

BOYS FROM THE BLACKSTUFF

UK 1982
Writer Alan Bleasdale; *Director* Philip Saville
Bernard Hill, Michael Angelis, Julie Walters

A series of five linked plays showing the impact of unemployment in early 1980s Liverpool, *Boys From The Blackstuff* struck a strong, humanitarian chord, and made a folk hero of one of its characters, Yosser Hughes (powerfully played by Bernard Hill (1945–)). Originally shown on BBC2, *Blackstuff* was repeated within eight weeks on BBC1, and achieved ratings from 5 to 7.9 million – exceptionally high for 'serious' drama.

A development of Bleasdale's 1978 TV play *The Black Stuff* – which chronicled the exploits of a Liverpool tarmac gang in Middlesbrough – the series was originally planned as a seven-parter. Six outlines were commissioned and one play, *The Muscle Market* (about the gang's employer) was shown in a one-off slot in January 1981. The common theme was of a claustrophobic battle against bureaucracy, but the tragedy came iced with black Scouse wit. On paper much of the imagery seems overloaded, particularly the conclusion, where George, the wheelchair-bound representative of honest toil is pushed round his old workplaces by nephew Chrissie (Angelis), only to die at the ghostly Albert Docks. The funeral over, Chrissie visits a pub where much redundancy money is being supped and, on leaving, mutters 'Beam me up Scotty.' As the sunset approaches Yosser steals the last line with his catchphrase of despair, 'Gizza job'.

The punchiness of the drama was due the strong characters and understated acting. The cinematic feel was aided by the decision to shoot with a new, lightweight, outside broadcast unit. Despite Yosser's popularity, Bleasdale considered Chrissie to be the most important character – amid the gloom, he was the only one to talk of fighting back.

BRANDAUER, Klaus Maria

Born 22 June 1944, Alt Aussee, Austria.

Raised in a small Alpine village, Brandauer moved with his family to West Germany in 1954 and was encouraged in his keen appreciation of opera and the theatre. In 1962, he enrolled at the Stuttgart Academy for Music and Performing Arts but left abruptly to join a repertory company in Tubingen, making his professional debut in *Measure for Measure* (1963).

Developing a far-ranging reputation for the power and versatility of his stage performances, he made his Vienna debut in *Poor Richard* (1968) and scored a personal triumph in *Emilia Galotti* (1970). Long associated with the Vienna Burgtheater, Austria's national theatre, he made an inauspicious film debut in the humdrum spy thriller *The Salzburg Connection* (1972). Apart from appearances in several European television mini-series, notably *Jean-Christophe* (1978), he remained essentially a stage performer until 1980.

▷ Istvan Szabo's film of *Mephisto* (1980) won a Best Foreign Film Oscar and brought Brandauer's skills to an international audience as he sympathetically portrayed the ambition, magnetism and theatrical vanity of an actor who abandons all that he holds dear for the recognition of his greatness in Hitler's Third Reich.

Choosing carefully from the many movie offers he made a witty nemesis as Maximilian Largo to ▷ Sean Connery's James Bond in *Never Say Never Again* (1983) and received an Oscar nomination for the charm and ambiguity he invested in the role of the rakish, womanising Baron Bror Blixen in the very popular *Out of Africa* (1985).

Reunited with Szabo he gave an electrifying performance in *Colonel Redl* (1985) as a careerist soldier of the Austro-Hungarian empire destroyed by his own compromises, and the partnership explored similar territory in *Hanussen* (1988).

Established as an international force, Brandauer continued to appear at the Burgtheater (essaying *Hamlet* in 1986) and has calculatedly balanced his film work between leading roles in European productions of distinction and scene-stealing supporting stints in widely-seen Hollywood entertainments. In 1989, he made his directorial debut with *Seven Minutes* telling of a German worker's assassination attempt on Hitler in 1938.

BRANDO, Marlon

Born 3 April 1924, Omaha, Nebraska, USA.

Expelled from a military academy, Brando moved to New York and studied acting with Stella Adler (1895–), making his public debut in *Bobino* (1943) and his Broadway bow in *I Remember Mama* (1944). He was voted Broadway's Most Promising Actor for his performance in *Truckline Café* (1946) and confirmed that potential with his excoriating interpretation of Stanley Kowalski in *A Streetcar Named Desire* (1947).

He made his film debut as the embittered paraplegic in *The Men* (1950) and earned his first Best Actor Oscar nomination for the celluloid reprise of his brutish, inarticulate Kowalski in ▷ *A Streetcar Named Desire* (1951).

A follower of the acting principles advocated by Stanislavsky (1865–1938) and practitioner of 'the Method', he was the pre-eminent figure in the jeans and torn t-shirt generation of film actors who sought a total identification with the characters they portrayed to obtain a psychological and emotional truth that was manifest in naturalistic behaviour and speech patterns. A prodigious talent, he underlined his versatility as the revolutionary leader in *Viva Zapata!* (1952) (Oscar nomination), an articulate Mark Anthony in *Julius Caesar* (1953) (Oscar nomination), the rebellious motorcycle gang leader in *The Wild One* (1953), the misfit longshoreman in ▷ *On The Waterfront* (1954) (Oscar) and singing gambler Sky Masterson in *Guys and Dolls* (1955).

Subsequent choices have proved less felicitous, although he impressed as the Nazi Officer in *The Young Lions* (1958) and made a striking directorial debut with the western *One-Eyed Jacks* (1961).

An anti-establishment figure who was vocal in his disdain for the acting process, he found himself out of favour after a string of box-office failures including *The Appaloosa* (1966), *A Countess From Hong Kong* (1967) and *Candy* (1968).

However, he returned in style as Don Corleone in ▷ *The Godfather* (1972) which brought him a second Best Actor Oscar, which he refused publicly in protest at the film industry's treatment of American Indians.

The renaissance continued with his agonisingly accurate study of a man in mid-life crisis in the controversial ▷ *Ultimo Tango A Parigi* (Last Tango in Paris) (1972) (Oscar nomination) but his subsequent appearances have been erratic and often inspired by financial gain; his contribution to *Superman* (1978) was said to have earned him the highest salary ever paid to a performer. However, his talent has been most effective when allied to a cause close to his heart and he won an Emmy as a neo-Nazi in *Roots: The Next Generation* (1979) and a Best Supporting Actor Oscar nomination for his compelling, worldweary Civil Rights lawyer in *A Dry White Season* (1989).

He delightfully parodied his Don Corleone characterisation in the comedy *The Freshman* (1990), his first substantial role for many years.

BREATHLESS (A Bout De Souffle)
Jean-Luc Godard (France 1959)
Jean-Paul Belmondo, Jean Seberg, Daniel Boulanger, Jean-Pierre Melville.
Berlin Film Festival Award: *Best Director*
89 mins. b/w.

▷ Jean-Luc Godard's *Breathless* (A Bout De Souffle) was an explosive salvo alerting the rest of the film community that the French ▷ *nouvelle vague*, or New Wave, had well and truly arrived. As one of the young critics (including later directors ▷ Francois Truffaut, ▷ Claude Chabrol and ▷ Eric Rohmer), who from the mid-Fifties had been expounding the influential *politique des auteurs* (Americanised by Andrew Sarris as ▷ 'the auteur theory') in the influential pages of film magazine *Cahiers Du Cinema*, Godard and his peers gave vent to a radical polemic that insisted on the director as the author of his film and rejected the discipline of admired French art cinema to acclaim the profundity of hard-boiled American directors like ▷ Sam Fuller and ▷ Nicholas Ray. An upturn in modestly-budgeted French independent production was to allow Godard to give celluloid expression to these ideas. Co-written by Truffaut and with Chabrol as technical adviser, *Breathless* makes no aspiration to social realism or the adaptation of literary values. The film's then-shocking formal play of ▷ jump cuts, scenes extended beyond any narrative function, and unashamed 'quotations' from other movies emphasises cinema-as-cinema, revealing the mechanism of signification so long hidden by the polish of classical narrative styles. Modernist film par excellence it might be, but these days the film's B-Movie plotting, its nods to Bogart, and the laconically iconic presence of Jean-Paul Belmondo (1933–) and Jean Seberg (1938–79), seem to align it more closely to the doomed romanticism of the Hollywood love-on-the-run sub-genre – which is precisely how American director Jim McBride

(1941–) remade it in 1983 with ▷ Richard Gere.

BRESSON, Robert

Born 27 September 1907, Bromont-Lamothe (Puy-de-Dome), France.

After pursuing a career as a painter, Bresson began working in the cinema, contributing dialogue to *C'Etait un Musicien* (1933) and making his directorial debut with *Les Affaires Publiques* (1934) a comedy of which no copies are known to exist.

After spending a year as a prisoner of war in Germany, he directed *Les Anges du Pêche* (1943), a melodramatic tale of a young novice who achieves the redemption of a murderess through the sacrifice of her own life. Uncharacteristic for this director in its uncomplicated style and use of professional actors, its fascination with a quest for spiritual grace still carries his recognisable imprimatur.

Les Dames du Bois de Boulogne (1946), his last film to use professional actors, is based on a script he co-wrote with ▷ Jean Cocteau about the ethics of a woman bent on revenge.

However, his distinctive style emerged in the three features he made during the 1950s – *Journal D'Un Curé de Campagne* (Diary of A Country Priest) (1950), *Un Condamné A Mort S'Est Echappé* (A Man Escaped) (1956) and ▷ *Pickpocket* (1959). A strongly Catholic sensibility dominates these stories of individual anguish on the road to a form of redemption or salvation that often arrives through death or, more rarely, through the power of love. Stripping his work of any flamboyant or conventionally manipulative narrative devices, he creates an austere purity that is a product of non-professional actors, natural sound, restrained emotions and a fastidious camerawork that invokes intensity through its obsessive focus on the tiniest of details and everyday objects.

In the 1960s, he sought to explore concepts of saintliness in a rigorously unsentimental manner, finding unlikely embodiments of its most unadulterated qualities in a much-defiled donkey in *Au Hasard Balthazar* (Balthazar) (1966) and the loveless, abused teenage peasant girl in *Mouchette* (1967).

Une Femme Douce (A Gentle Creature) (1969), in which a woman commits suicide to escape an untenable marriage, was his first film in colour and one of several, including *Quatre Nuits D'Un Reveur* (Four Nights Of A Dreamer) (1971) to contemplate the worth of suicide, although the emotional blankness of the characters and an acute lack of motivation made audience identification more troublesome than ever.

His three most recent films range from *Lancelot du Lac* (Lancelot of the Lake) (1974) a blood-drenched saga of the loss of spirituality among the Knights of the Round Table, to *Le Diable Probablement* (The Devil, Probably) (1977) in which a young man rejects the evils of the world by paying for his own murder, and *L'Argent* (1983) in which the perambulations of a forged bank note form an economical and characteristically elliptical parable about man's greed.

THE BRIDE OF FRANKENSTEIN

James Whale (US 1935)
Boris Karloff, Colin Clive, Elsa Lanchester, Ernest Thesiger.
76 mins. b/w.

Although his directorial career encompassed a broad range of material from the anti-war statement *Journey's End*, (1930) to the musical *Showboat* (1936), the distinctive English talent ▷ James Whale remains best known for the quartet of classic horror films he made at Universal during the Thirties. His 1931 version of *Frankenstein*, introduced by a studio representative warning the audience of the terrors they were about to witness, adapted the visual extremes of the German ▷ Expressionists to a loose adaptation of Mary Shelley's novel, and the director's sympathies for the monster combined with ▷ Boris Karloff's immensely touching performance to create a work of resonance and subversion that stands the test of time. Following this milestone of the macabre with the blood-curdling boisterousness of *The Old Dark House* (1932) and the effects extravaganza *The Invisible Man* (1933), Whale was to revive the Baron and his creation in 1935's dazzlingly eccentric sequel *The Bride of Frankenstein*. A gleefully satirising historical interlude with the author and her literary cohorts sets the note of dark wit that is to be observed throughout, for although Whale again pictures the undead ghoul as noble savage amidst a heartlessly cruel society, much of the film – including the monster's celebrated encounter with a blind hermit and the hissing apparition that is Elsa Lanchester's (1902–86) bride – is played for laughter. Most entertaining is Ernest Thesiger's (1879–1961) Doctor Pretorious who ushers the creature into the wonderful world of gin. 'It's my only weakness' he sighs.

BRIEF ENCOUNTER

David Lean (UK 1945)
Celia Johnson, Trevor Howard, Cyril Raymond, Joyce Carey.
85 mins. b/w.

Written and produced by celebrated playwright/actor/celebrity Noël Coward (1899–1973), ▷David Lean's *Brief Encounter* forever preserves on screen a particularly English, especially middle class, sort of love story. Set against the backdrop of a grimy city railway station, a series of chance meetings pitch suburban housewife Celia Johnson (1908–82) and good-natured GP ▷Trevor Howard into forbidden romance – suddenly confronting them with passionate, illicit emotions that are eventually dampened by the weight of moral conformity. Presented mainly in flashback as Johnson recalls the events of the previous few weeks, the placing of the action in the context of her melancholic reverie licenses the film's Rachmaninoff-accompanied tone of romantic delirium, a heightening of the senses that makes even the most ordinary visit to the park or a restaurant chime with heartfelt intensity. Furthermore, the couple's visit to the cinema where they watch a trailer for a forthcoming epic *The Flames of Passion* should alert us to the element of filmic fantasy working its magic here too, for although we often remark upon *Brief Encounter*'s so-called realism, the artful treatment of the settings (dimly-lit interiors and gloomy noir-ish streets contrast with invigorating interludes in natural surroundings) shows Lean finding a pictorial expression for the theme of emotional containment. His responsiveness throughout to the pair's mingled feelings of elation and fatalism seems much less calculating and rather more moving than the monumental vacuousness of later superproductions *Doctor Zhivago* (1965) and *Ryan's Daughter* (1970).

BRITISH ACADEMY OF FILM AND TELEVISION ARTS (BAFTA) Previously the Society of Film and Television Arts, this was formed in 1959 by an amalgamation of the British Film Academy (founded in September 1948) and the Guild of Television Producers and Directors (formed in 1954). Membership is open to senior creative workers in both industries.

A further reorganization in 1975 resulted in the title change to BAFTA, and Academy branches exist throughout the world to raise standards in all aspects of film-making and television, through screenings, seminars, education campaigns, lectures and other initiatives. Each year it presents a series of awards, although the time-lag between British and American cinema releases has tended to make the film winners dawdle behind some other international awards.

In a recent initiative with Shell-UK it has aimed to improve the international profile of British cinema and increase cinema admissions on the home territory. More recently it has emulated the ▷American Film Institute in presenting a life-achievement award for a distinguished career in the industry. Recipients have been; 1988 – ▷Dirk Bogarde, 1989 – ▷Julie Andrews and 1990 – ▷Sean Connery. In 1990 BAFTA Los Angeles presented their first lifetime achievement award to ▷Michael Caine.

BRITISH FILM INSTITUTE (BFI) Founded in 1933 to support all aspects of film culture, the British Film Institute has steadily expanded over the years to include interests in production, distribution and publishing and currently has a membership in excess of 42,000.

Pursuing a vigorous programme of preservation through the National Film Archive, it also holds the responsibility for running the Museum of the Moving Image and the National Film Theatre, and for supporting a network of 33 regional film theatres throughout the United Kingdom.

The headquarters at Stephen Street in London house an extensive library of books and periodicals whilst the Institute also holds a vast collection of stills and posters. It has financed a series of 'New Directors' short films including *Flames of Passion* (1990) and contributed to the budgets of such British films as *Caravaggio* (1985), ▷ *Distant Voices, Still Lives* (1988), *The Silent Scream* (1990) and *Young Soul Rebels* (1991).

In recent years, it has also hosted an annual awards presentation and bestowed a number of fellowships. The current chairman is ▷ Sir Richard Attenborough.

BROCKA, Lino Ortiz

Born 3 April 1940, San Jose, Nuevo Ecija, Philippines.
Died 22 May 1991, Manila, Philippines.

The son of a fisherman and schoolteacher, Brocka was the most influential and outspoken Filipino filmmaker to have gained international recognition.

After reading science at university he renounced his Catholic upbringing and briefly converted to Mormonism, which led to an 18-month posting to a leper colony on Molokai, a small island in the state of Hawaii. Returning to Manila, he joined PETA (Philippines Educational Theatre Association) and later worked in the film industry on such visiting foreign productions as *Flight to Fury* (1966).

He made his directorial debut with *Wanted: Perfect Mother* (1970) but later abandoned filmmaking when president Ferdinand Marcos declared martial law in 1972.

In 1974, he co-founded the Cinemanilia Corporation and enjoyed critical and commercial success with *Tinimbang Ke Nguni't Kulang* (Ye Have Been Weighed in the Balance And Found Wanting) (1974).

Alternating work in the country's commercial cinema with more personal statements, his films were highly distinctive from the annual output of 200 or so soap operas and kung-fu adventures produced by the local industry. A critic of the social conditions and political repression under the Marcos regime, he won particular praise for *Maynila Sa Mga Kuko ng Liwanag* (Manila In the Claws of Neon) (1975) which tells of a young fisherman who journeys to the capital in search of a sweetheart who has been sold to a brothel keeper. In *In Insiang* (1976) a girl is raped by her mother's gigolo and manipulates events to provoke his murder. *Jaguar* (1979) and *Bona* (1980) complete a trilogy on the subservient role of women in his society and the repression that may lead to alienation and violence.

Also working in television, the theatre and as a cinema historian, he was elected President of the Directors Guild of the Philippines in 1980 and was heavily involved in the Free the Artists movement which campaigned against Marcos's censorship laws and profligate spending on the Manila Film Festival.

Bayan Ko-Kapit Sa Patalim (Bayan Ko: My Own Country) (1984) was a powerful melodrama of a worker destroyed by economic restraints and the conflict between his instinctive urge for working-class solidarity and a system that ruthlessly punishes attempts at collective action and opposition. Likened to ▷ Fritz Lang's *You Only Live Once* (1937) it combined a strong social message with a compelling dramatic plight and led to Brock's incarceration on charges of sedition.

Pressure from Amnesty International and others led to his release. Despite the fall of Marcos and the rise of Cory Aquino, he continued to discover injustices and iniquities in his home land that fuelled the anger in such films as *Macho Dancer* (1988), a dramatic condemnation of police corruption and violence told through the romance of two male prostitutes, and *Fight for Us* (1989) in which a former priest uncovers a ruthless cult of anti-dissident vigilantes who have been allowed to torture and kill innocent villagers.

BRONENOSETS POTEMKIN *see* BATTLESHIP POTEMKIN

BROOKS, Louise

Born 14 November 1906, Cherryvale, Kansas, USA.
Died 8 August 1985, Rochester, New York, USA.

A member of the Denishawn Dancers troupe from 1922, Brooks moved to New York and subsequently performed in the chorus of *George White's Scandals of 1924* and joined the casts of *Louie the*

14th (1925) and the *Ziegfeld Follies* (1925–26).

A striking, bob-haired beauty, she made her film debut in *Street of Forgotten Men* (1925) and signed a contract with Paramount that led to roles in such comedies as *Social Celebrity* (1926) and *It's The Old Army Game* (1926) with ▷ W.C. Fields.

Displaying a naturalness and ease before the cameras that was rare for the time, she became a luminous embodiment of the flapper age in spirited shenanigans like *Just Another Blonde* (1927) and *Rolled Stockings* (1927). Her performances as the alluring circus high diver in *A Girl in Every Port* (1928) and the hobo fugitive in *Beggars of Life* (1928) made more challenging demands of her.

Leaving Paramount, she journeyed to Germany for ▷ G.W. Pabst's *Die Büchse von Pandora* (Pandora's Box) (1928) and her most famous role as Lulu, 'the personification of primitive sexuality'. She brought a vivacity and child-like innocence to a woman who guilelessly provokes and is consumed by intense sexual desire. The purity of her erotic appeal was underlined by Pabst's *Das Tagebüch einer Verlorenen* (Diary of a Lost Girl) (1929) and *Prix de Beauté* (Beauty Prize) (1930) which she made in France.

Against Pabst's advice, she returned to Hollywood and the promise of a contract with Columbia which never materialised. She played only supporting roles in *It Pays to Advertise* (1931) and *God's Gift to Women* (1931) and retired after playing opposite ▷ John Wayne in the B western *Overland Raiders* (1938).

She ran a dance studio in Wichita and worked in radio soap operas before disappearing into virtual obscurity. However, her work was featured in a 1955 Paris exhibition entitled 'Sixty Years of Cinema' and subsequent revivals in Europe and New York brought appreciation from new generations. Prompted by this interest, she became a writer on film and published an autobiography, *Lulu in Hollywood* (1982).

BROOKS, Mel

Real Name Melvin Kaminsky
Born 28 June 1926, Brooklyn, New York, USA.

A talented drummer who performed after school and during summer holidays, Brooks pursued a showbusiness career following his war service as a combat engineer with the U.S. Army. Performing at nightclubs and resorts on the Borscht circuit, he also turned to comedy and secured some radio engagements.

In 1949, his friend Sid Caesar (1922–) hired him to write material for his television series *Broadway Revue*. Their partnership continued through *Your Show of Shows* (1950–54), *Caesar's Hour* (1954–57) and *Caesar Invites* (1958).

He also wrote the books of such Broadway musicals as *Shinbone Alley* (1957) and *All American* (1962), contributed to numerous television specials and issued a recording of *2,000 Years with Carl Reiner and Mel Brooks*, a convulsive collection of memories and quips from a resolutely unimpressionable 2,000 year-old Yiddish sage.

He won an Oscar for the short cartoon *The Critic* (1963) and devised the popular secret agent comedy series *Get Smart* (1965–70). He made his feature-length cinema debut as the writer-director of *The Producers* (1966) a brash, boisterous comedy about an unscrupulous Broadway producer who attempts a get-rich-quick scheme of overselling shares in a certain flop. The show, including the legendary 'Springtime for Hitler' musical number, was an unexpected smash hit.

Brooks won an Oscar for his screenplay and hit his stride with a series of affectionate spoofs including *Blazing Saddles* (1974), *Young Frankenstein* (1974) and *High Anxiety* (1977) that allowed free rein to his style of tasteless humour, zany repartee, offensive barbs and crude vulgarity.

Apart from a guest appearance in *The Muppet Movie* (1979), he has acted exclusively in his own films although the onerous demands of writing, performing and directing may have overstretched his abilities and recent efforts like *History of the World, Part 1* (1981) and *Space Balls* (1987) have betrayed a flagging inspiration. The forthcoming *Life Stinks* (1991) may restore his former sparkle.

His production company Brooksfilms has supported such adventurous films as *The Elephant Man* (1980), *Frances* (1982) and *The Fly* (1986). He has been married to actress Anne Bancroft (1931–) since 1964.

BROOKSIDE

UK 1982–
Executive Producer Phil Redmond

The news that Channel 4, with its commitment to alternative programming, was including a soap opera in its opening schedules drew much critical flak, but Redmond (also behind BBC's realistic school serial *Grange Hill*) aimed to address social issues with a show set on a suburban close in Liverpool.

Scheduled after ITV's *Coronation Street*, the series has consistently been the top-rated programme on Channel 4, and its harder edge has been mirrored in the BBC's *Eastenders*. Its task has not been aided by the setting, a cul-de-sac on a new estate which lacks a communal gathering place. The Close postbox, the scene of numerous chance encounters, is no substitute for the Rovers Return.

But *Brookside* has broken new ground, particularly with its stress on young characters, who are more recognisably of their time than in any other TV soap. The plots have not shirked difficult issues, though cumulatively this has meant that the central characters have been forced to lead exceptionally full lives. Sheila Grant, for example, (played by Sue Johnston (1943–) until a redeemable exit in 1990) endured rape, a middle-aged pregnancy with attendant Catholic guilt, the death of a teenage son (tough not before he sired her a grandchild), the involvement of husband Bobby in a financially-draining strike, separation, and finally an on-off relationship, then marriage, to lovable rogue Billy Corkhill. In quiet episodes she studied for an Open University degree and worked at a deaf school.

Other plotlines have tackled heroin abuse, senile dementia, anti-gay prejudice and the fear of Aids, sexual harassment, picket-line dilemmas, environmental pollution, dyslexia and homelessness along with soap staples like marital infidelity. Curiously, *Brookside*'s best moments have come when the writers have been allowed to forget about their bleeding hearts and simply entertain.

BUÑUEL, Luis

Born 22 February 1900, Calanda, Spain.
Died 19 July 1983, Mexico City, Mexico.

Educated at Jesuit schools in Zaragoza and at the Residencia De Estudiantes in Madrid, Buñuel went to Paris where he worked as an assistant to director Jean Epstein (1897–1953) on such films as *Mauprat* (1926) and *La Chute De La Maison Usher* (The Fall of the House of Usher) (1928).

Influenced by his involvement in the ▷ Surrealist movement, he collaborated with Salvador Dali (1904–89) on *Un Chien Andalou* (1928) and ▷ *L'Age D'Or* (1930), scandalous works of dream-like intensity and shocking imagery – most famously the slicing of a woman's eyeball with a razor.

He then made the documentary *Las Hurdes* (Land Without Bread) (1932) which contrasted the poverty of the peasantry with the wealth of the Catholic Church.

There then followed a long period adrift in a kind of artistic wilderness as he worked as a journeyman supervisor on Spanish musicals and comedies, and made several abortive trips to America.

Resident in Mexico from 1946, he reasserted his individualistic talents with *Los Olvidados* (The Young and The Damned) (1950) a harsh, dispassionate account of the juvenile delinquents living in the slums of Mexico City.

The recipient of the Best Director Prize at the Cannes Film Festival, he then embarked on a prolific period that includes the black comedy *El* (1952), *Ensayo De Un Crimen* (The Criminal Life of Archibaldo De La Cruz) (1955) about the thwarted efforts of a potentially murderous psychopath, and *Nazarin* (1958) in which a humble priest is despised for his sincere efforts to live a purely Christian life.

In 1961, he was invited to return home by the Spanish government and made *Viridiana* (1961) a scabrously imaginative reworking of The Last Supper that savagely attacked the tenets of Catholicism, won the Palme D'Or at Cannes and was promptly banned in Spain.

Subsequently working mostly in France and Italy, he created a rich body of work that used an unassuming technique to convey the most provocative and disturbing of material, consistently addressing his abhorrence of the Catholic church and middle-class morality with biting, black humour, sly satire and the most alarming of surrealistic imagery. His many notable titles include *Belle De Jour* (1967), *Tristana* (1970), *Le Charme Discrèt De La Bourgeoisie* (The Discreet Charm of the Bourgeoisie) (1972) which won a Best Foreign Film Oscar, *Le Fantôme De La Liberté* (The Phantom of the Liberty) (1974) and his last *Cet Obscur Objet Du Désir* (That Obscure Object of Desire) (1977).

BURTON, Richard

Real Name Richard Walter Jenkins
Born 10 November 1925, Pontrhydfen, UK.
Died 5 August 1984, Geneva, Switzerland.

The twelfth son of a Welsh coalminer, Burton developed a love of language and literature from a schoolteacher and mentor of that name and gained a scholarship to Exeter College, Oxford. He first appeared on stage in *Druid's Rest* (1943) at the Royal Court Theatre in Liverpool and, after national service in the RAF, returned to the stage in 1948, the same year as his film debut in *The Last Days of Dolwyn*.

He made his stage reputation in *The Lady's Not For Burning* (1949). A triumphant season at Stratford in 1951 preceded his Hollywood debut in *My Cousin Rachel* (1952) for which he received an Oscar nomination as Best Supporting Actor.

In between seasons at the Old Vic, he pursued an international film career that required little of him beyond a virile presence, brooding looks and the sound of his richly resonant voice. His few qualified successes of the period include *The Robe* (1953) (Oscar nomination), *Alexander the Great* (1956) and *Look Back in Anger* (1959) in which he played Jimmy Porter.

Broadway success in *Camelot* (1960) was followed by the filming of *Cleopatra* (1963) during which his adulterous romance with ▷ Elizabeth Taylor made him a household name. Married in 1964, divorced in 1974 and briefly remarried in 1975, they became the embodiment of showbusiness excess, pursuing a lavish lifestyle in the jetset corners of the globe. Most memorable among their numerous co-starring appearances were *Who's Afraid of Virginia Woolf?* (1966) (Oscar nomination) and *The Taming of the Shrew* (1967).

Adept at conveying disillusion, dissipation and self-loathing, he was also noteworthy as the title character in *Beckett* (1964) (Oscar nomination), the unfrocked priest in *Night of the Iguana* (1964) and the dour secret agent in *The Spy Who Came In From the Cold* (1965) (Oscar nomination).

He enjoyed his last major box-office success with *Where Eagles Dare* (1969) and spent his remaining years living up to a hellraising image and adding little lustre to a far from inspiring assortment of international ventures, although he received a seventh and final Oscar nomination for *Equus* (1977). The like of *The Wild Geese* (1978) and *1984* (1984) enjoyed some popularity in Britain.

He co-directed a version of *Doctor Faustus* (1967) and showed some skill as a writer with the novella *A Christmas Story* (1964). Theatre work included *Hamlet* (1964), *Equus* (1976) and *Private Lives* (1983) in which he again co-starred with Taylor.

THE CABINET OF DR CALIGARI (Das Kabinett Des Dr Caligari)

Robert Wiene (Germany 1919)
Werner Krauss, Conrad Veidt, Friedrich Feher, Lil Dagover.
Approx. 90 mins. b/w. (original release tinted) silent.

Generally recognised as the first major film in the German ▷ Expressionist manner that was to inspire the likes of ▷ F. W. Murnau's *Nosferatu* (1922) and, arguably, G. W. Pabst's *Dir Buchse der Pandora (Pandora's Box)* (1929), and subsequently influence the stylistic development of the horror genre in Hollywood, on its 1922 British release Robert Wiene's (1881–1938) *The Cabinet of Dr Caligari* was billed as 'Europe's greatest contribution to motion picture art'. Certainly, the 'art' was there for everyone to see, for with its disturbing storyline and decors, all misshapen angles and false perspectives, painted shadows and weirdly overemphatic performances, the film made great play of its association with the artists of the

Expressionist movement, their distorted anti-naturalist figures indicative of a fragmenting social fabric. Any creative work is of course a product of the ideological moment, and commentators have fruitfully defined the way in which *The Cabinet of Dr Caligari* is indicative of the mood of Post World War I Germany. While the original screenplay, in which sinister showman Dr Caligari has somnambulist Cesare murder on his behalf, clearly seems an anti-authoritarian statement, the final version (as amended by the producer), in which the horrifying central narrative is revealed as a madman's delusions and the Caligari figure his benevolent psychiatrist, has been read as an expression of the German psyche's fear that individual freedom encourages chaos and must hence be contained by the harshest of leadership.

CAGNEY, James

Full Name James Francis Cagney Junior
Born 17 July 1899, New York City, New York, USA.
Died 30 March 1986, Stanfordville, New York, USA.

Graduating from vaudeville to musicals and the legitimate Broadway theatre, Cagney appeared in *Pitter-Patter* (1920), *Outside Looking In* (1925) and several other shows before his work in *Penny Arcade* (1929) brought him to the attention of Warner Brothers with whom he signed a contract, making his debut in a film version of his theatrical success, now titled *Sinner's Holiday* (1930).

Cast in a secondary role in ▷ *Public Enemy* (1931), he switched places with star Edward Woods after three days of shooting and established himself as the quintessential screen hoodlum: dynamic, cocky, invincible and a fast-talking whirlwind of power and persuasion. He remained at his best on the wrong side of the law in such popular, fast-paced fare as *Lady Killer* (1933), *Angels With Dirty Faces* (1983) (Oscar nomination) and *The Roaring Twenties* (1939).

He was allowed to reveal different facets of his talent as a hustling song and dance man in *Footlight Parade* (1933) and as Bottom in *A Midsummer Night's Dream* (1935), and his affectionate portrayal of patriotic showman and entertainer George M. Cohan (1878–1942) in *Yankee Doodle Dandy* (1942) earned him a popular Best Actor Oscar.

Later, he offered incisive psychological portraits of the hoodlum in ▷ *White Heat* (1949) and *Love Me Or Leave Me* (1955) (Oscar nomination), directed an unremarkable remake of *This Gun For Hire* (1942) entitled *Short Cut To Hell* (1957) and displayed his virtuoso comic skills in *Mister Roberts* (1955) and *One, Two, Three* (1961).

He retired in 1961, returning on doctor's orders for typically pugnacious roles in *Ragtime* (1981) and the television film *Terrible Joe Moran* (1984).

A farmer, a painter and poet, he wrote an autobiography, *Cagney on Cagney*, in 1976 and was the second recipient of the ▷ American Film Institute Life Achievement Award in 1974.

CAINE, Michael

Real Name Maurice Joseph Micklewhite
Born 14 March 1933, London, UK.

The son of a Billingsgate fish porter, Caine survived several dead-end jobs and National Service in the British Army before beginning years of struggle as a small-part actor visible, but only just, in all media.

Taking his new surname from *The Caine Mutiny* (1954), he made his film debut in the war film *A Hill in Korea* (1956) and played many bit parts in the standard fare of comedies and war films produced by the British film industry at the time. Success in the West End play *Next Time I'll Sing To You* (1963) brought him a role as the aristocratic young officer in the hit *Zulu* (1963) and his belated stardom was consolidated with roles as down-at-heel spy Harry Palmer in *The Ipcress File* (1965) and its two sequels, and as archetypal Cockney romeo *Alfie* (1966) for which he received his first Best Actor Oscar nomination.

The cinema's first bespectacled star since the days of ▷ Harold Lloyd, this tall, fair-haired actor with a much-mimicked Cockney drawl was soon one of Britain's few truly international performers.

A staggeringly prolific workaholic, his reputation for consummate professionalism has withstood many inferior films and enhanced superior material. He has repeatedly proved his versatility:

James Cagney in *Yankee Doodle Dandy* (1942)

notably as the 17th century mercenary in *The Last Valley* (1971), a vicious English thug in *Get Carter* (1971) and a surprisingly dashing Scots swashbuckler in *Kidnapped* (1971). He more than held his own against ▷ Laurence Olivier in the devious two-hander *Sleuth* (1972) (Oscar nomination).

Resident in America from 1978, he has remained critically indestructible despite the disasters of *The Swarm* (1978), *Ashanti* (1979) and other egregious adventures but enjoyed huge box-office success with the controversial shocker *Dressed to Kill* (1980) and has increasingly proved his comic skill in the likes of *California Suite* (1978), *Sweet Liberty* (1985) and *Surrender* (1987).

Apparently willing to try his hand at any and every challenging role, he received a further Oscar nomination for his acutely judged performance as the dissolute English professor in *Educating Rita* (1983) and finally won a Best Supporting Actor Oscar for the delicate understanding and sensitivity he brought to the role of a philandering husband in *Hannah and Her Sisters* (1986).

Returning to England in 1986, he contributed a supporting performance of reptilian evil in *Mona Lisa* (1986) and has remained as busy as ever, accepting his first television work since *Hamlet* (1963) with the highly-rated mini-series *Jack the Ripper* (1988) and maintaining a work-rate of at least two major features per annum. His most recent films include the popular comedy *Dirty Rotten Scoundrels* (1988), the black comedy *A Shock to the System* (1990) and the fantasy *Mr. Destiny* (1990).

A gourmet cook, dedicated charity worker, restaurant owner and inimitable purveyor of mind-numbing trivia, he spent most of 1990 writing an eagerly-anticipated autobiography.

CAMERON, (Mark) James (Walter)

Born 17 June 1911, Battersea, London, UK.
Died 26 January 1985, London, UK.

Cameron's distinguished journalistic career began as an office boy for the *Weekly News* (1935) and progressed, via Dundee and Glasgow, to Fleet Street in 1940.

Rejected for military service, he worked as a sub-editor on *The Daily*

Express (1940–45) and subsequently returned to reporting. Covering the atom bomb experiments at Bikini (1946) formed his anti-authoritarian views and convinced him to become a member of CND. He resigned from the *Daily Express* in 1950 and *Picture Post* in 1951 over points of principle before settling with the *New Chronicle* (1952–60) as a roving reporter on war, poverty and injustice.

Renowned for his integrity, dry wit and concise summation of a situation, he painted literary pictures of some of the great events in world affairs, from the Vietnam War to ill-treatment of the underprivileged in India.

Familiar as a writer and presenter of many probing and intelligent television programmes, including a *Western Eyewitness* report from Vietnam in 1965, *Men Of Our Time* (1963), the intermittent *Cameron Country* and the autobiographical *Once Upon A Time* (1984). His radio play *The Pump* (1973) won the Prix Italia and was dramatised for television in 1980.

His books include *Witness In Vietnam* (1966), the autobiography *Point of Departure* (1967) and *An Indian Summer* (1974).

CALLAN

UK 1967–73
Creator James Mitchell
Edward Woodward, Russell Hunter

Bleak and bareboned spy serial which made stars of its two central performers. Woodward (1930–) was nervily clinical in the title role as the disaffected secret agent, and Hunter (1925–) snivelled faultlessly as petty thief Lonely, his long-suffering, and less than fragrant acquaintance.

Created by James Mitchell, *Callan* offered a stark riposte to the spoofish humour of the James Bond movies, and the character was arguably closer to Ian Fleming's original vision. When Woodward dropped to the ground, apparently dead, at the close of the 1969 series Thames were besieged by calls enquiring as to his well-being. Such popularity ensured his return a year later, with a new partner, Cross, played by Patrick Mower (1940–).

Woodward and Hunter reprised their roles in the feature film *Callan* (1973) and the one-off television drama *Wet Job* (1981).

CAMPION, Jane

Born 1955, Wellington, New Zealand.

A graduate of the Australian Film, Television and Radio School in Sydney, Campion made an immediate impression on world audiences with her nine-minute debut effort *Peel* (1982) which won the Cannes Palme D'Or for the best short film.

Other short films like *Passionless Moments* (1983) and *A Girl's Own Story* (1983) also proved to be award-winners and she progressed to a feature-length piece with *Two Friends* (1986) for Australian television.

Interested in dysfunctional relationships, the bonds within families, the thin line between rationality and insanity and 'what most films leave out', her feature film debut *Sweetie* (1989) was a beautifully composed character study of a roly-poly schizophrenic and the disruptive effect she has on the lives of her nearest and dearest, not the least her eccentrically superstitious sister. Unsettling in tone but vastly accomplished in technique, it competed at the Cannes Film Festival.

This was followed by the cinema release of a three-part television series *An Angel At My Table* (1990) a compassionate dramatisation of the autobiographies of novelist Janet Frame (1924–) whose battles against chronic shyness, social inadequacy and a mistaken diagnosis of schizophrenia made compelling viewing.

The recipient of seven separate awards at the Venice Film Festival of 1990, it confirmed her position as one of the international cinema's most distinctive new talents.

CAPRA, Frank

Born 18 May 1897, Bisaquino, Sicily.
Died 3 September 1991, La Quinta, California, USA.

Emigrating to Los Angeles in 1903, Capra later studied chemical engineering at the California Institute of Technology, taught ballistics to the US Army, worked as a lab assistant and eventually made his way into the film industry as a prop man and gag writer for the likes of ▷ Hal Roach and ▷ Mack Sennett.

His first credited film as director is *Fultah Fisher's Boarding House* (1922) and he worked with babyfaced comedian Harry Langdon (1884–1944) on such popular silent comedies as *Tramp, Tramp, Tramp* (1926), *The Strong Man* (1926) and *Long Pants* (1927). When Langdon unwisely decided to go his separate way, Capra proved his versatility at the helm of such technological adventures as *Submarine* (1929) and *Dirigible* (1931) as well as the incisive study of evangelism *The Miracle Woman* (1931) and the exotic romance of *The Bitter Tea of General Yarn* (1933).

He won a Best Director Oscar for the fast-paced comedy romance *It Happened One Night* (1934), a feat he repeated with *Mr. Deeds Goes To Town* (1936) and *You Can't Take It With You* (1938). His most renowned work utilised archetypal all-American actors like ▷ Gary Cooper and ▷ James Stewart to celebrate the decency and integrity of the common man as he combats corruption and malfeasance in high places, triumphing on behalf of truth, justice and the American Way. Criticised for their naive politics and extreme sentimentality (known as 'Capracorn'), films like ▷ *Mr. Smith Goes To Washington* (1939) (Oscar nomination), *Meet John Doe* (1941) and the whimsical ▷ *It's A wonderful Life* (1946) (Oscar nomination) have also proved potent and enduring cinematic fables.

He was also at the helm of the luxurious version of *Lost Horizon* (1937) and found a warm wartime welcome for the black comedy *Arsenic and Old Lace* (1944).

A Major in the Signal Corps during the War, he was responsible for the Why We Fight series of patriotic documentaries.

In a post-war cinema filled with documentary-like thrillers and the doom-laden qualities of ▷ film noir, his talents were less valued and he directed only a handful of further films including the political satire *State of the Union* (1948), a series of science documentaries and his last *Pocketful of Miracles* (1961), a long-winded remake of his own *Lady for A Day* (1933) (Oscar nomination).

He retired in 1964 although retaining his links with the industry. He published an autobiography *The Name Above the Title* in 1971 and was the 1982 recipient of the ▷ American Film Institute Life Achievement Award.

CARNE, Marcel

Born 18 August 1909, Paris, France.

An insurance clerk, Carné served his filmmaking apprenticeship as an assistant to such noted directors as Jacques Feyder (1885–1948) and René Clair, with whom he worked on *Sous Les Toits De Paris* (1929).

A film critic and editor-in-chief of *Hebdo-Films*, he directed the short documentary *Nogent, Eldorado du Dimanche* (1929) and made his fictional debut with *Jenny* (1936) which began his decade long collaboration with poet and screenwriter Jacques Prévert (1900–77).

The farce *Drôle De Drame* (Bizarre, Bizarre) (1937) employed the talents of ▷ art director Alexandre Trauner (1906–) and cameraman Eugen Schüfften (1893–1977) for its evocation of London. This creative ensemble would unite again to make a group of classic romantic melodramas that expressively captured the fatalistic mood of the immediate pre-War years.

Quai Des Brumes (1938) starred ▷ Jean Gabin as an army deserter pursuing a doomed love affair in a melancholy, fog-enshrouded Le Havre. Gabin also starred in *Le Jour Se Lève* (1939) as the perpetrator of a crime passionel trapped by the foregone conclusion of his actions and the literal confinement of a dark tenement room besieged by police and curious crowds.

In a change of pace, Carné then made *Les Visiteurs Du Soir* (1942), a whimsical medieval fairytale, and ▷ *Les Enfants Du Paradis* (1944). Made during the German occupation, it recalled the romance of a theatrical past with wit and sensitivity and is regarded as his masterpiece.

After the War, he made *Les Portes De La Nuit* (Gates of the Night) (1946), a sombre return to earlier themes of doomed lovers that was not successful. When *Le Fleur De L'Age* was abandoned in 1947 and the partnership with Prévert dissolved, his career suffered.

Traces of his former mastery are evident in the Breton melodrama *Le Marie Du Port* (1949) and a version of *Thérèse Raquin* (1953) but his willingness to inexpertly explore French youth culture and much criticism from the directors of the ▷ *nouvelle vague* made him an unfashionable figure.

His last works were *La Merveilleuse Visite* (1973) and the feature-length television documentary *Le Bible* (1976).

CASABLANCA

Michael Curtiz (US 1942)
Humphrey Bogart, Ingrid Bergman, Paul Henreid, Claude Rains.
Academy Award: *Best Film*; *Best Director*; *Best Screenplay* Julius G. and Philip J. Epstein, Howard Koch.

The ever-increasing popularity of an unremarkable 1942 formula Warners production provides continuing evidence of the power of cinematic nostalgia. No better, no worse than any number of similar programmes from the period, *Casablanca* has since become a phenomenon, perhaps the best-loved and most frequently screened of all old Hollywood movies. Based on a play that had never been staged, offered to three other directors before ▷ Michael Curtiz took the assignment, and at one time planned as a vehicle for Ronald Reagan (1911–) and Ann Sheridan (1915–67), the film's origins hardly inspire, yet for today's audiences it exemplifies the kind of snappily-crafted studio entertainment they don't make 'em like anymore. We revel in the cherishable array of supporting actors (Sidney Greenstreet (1879–1954), Peter Lorre (1904–64) and Claude Rains (1889–1967) each in characteristic form), in the tart dialogue exchanges and the film's memorable song *As Time Goes By*, while the presence of ▷ Humphrey Bogart in a role that established the worldweary anti-hero of his future screen persona makes us forget the implausibilities of the plotting and the dullness of fugitive couple ▷ Ingrid Bergman and Paul Henreid (1908–) we're supposed to be rooting for. Certainly it satisfies the current appetite for retro-chic, but the film's enduring appeal may lie in the way its attitudes have somehow remained contemporary – the notion of doomed romance in a world on the brink of chaos still affects us, after all – so enabling a functional product to transcend its original source and pass into the realm of folklore.

CASSAVETES, John

Born 9 December 1929, New York, USA.
Died 3 February 1989, Los Angeles, USA.

A graduate of the New York Academy of Dramatic Arts, Cassavetes acted in Rhode Island stock companies before making his film debut with a brief role in *14 Hours* (1951) which wound up on the cutting-room floor.

Seen in *Taxi* (1953), he appeared as representatives of an angry, alienated generation in potent contemporary dramas like *Crime in the Streets* (1956) and *Edge of the City* (1957) before enjoying great popularity in the television series *Johnny Staccato* (1959–60). His earnings from that helped finance his directorial debut ▷ *Shadows* (1959), an influential experimental piece that broke new ground in its use of improvisational acting and ▷ cinema verité techniques to achieve a new form of emotional integrity and a greater sense of fly-on-the-wall verisimilitude in the depiction of human relationships.

Much praised, it led him to direct within the Hollywood studio system on two acrimonious and unhappy productions: *Too Late Blues* (1961) and *A Child Is Waiting* (1962).

He concentrated on acting for a while, receiving a Best Supporting Actor Oscar nomination for *The Dirty Dozen* (1967) and starring in the mammoth box-office success *Rosemary's Baby* (1968). He resumed his directorial career with *Faces* (1968) a brilliantly acted portrait of a marriage in crisis that features actress Gena Rowlands (1934–) his wife from 1954 until his death.

Throughout the 1970s, in the company of such regular actors as Rowlands, Ben Gazzara (1930–) and Peter Falk (1927–), he refined his techniques on a series of excoriating extemporised dramatic pieces that laid bare the torments of his characters with an unflinching honesty. His best known work includes *Husbands* (1970), *Minnie and Moskowitz* (1971) and *A Woman Under the Influence* (1974) in which Rowlands' painfully believable portrayal of a woman at the end of her tether earned a Best Actress Oscar nomination.

She was nominated again for her husband's *Gloria* (1980), a more tightly scripted, conventional yarn of a pistol-packin' gangster's moll that rattles along with great verve.

Still occasionally seen as an actor, his later appearance included *Two Minute Warning* (1976), *Whose Life Is It Anyway?* (1981) and *The Third Day Comes* (1986). He directed, co-wrote and co-starred with Rowlands in *Love Streams* (1984) about the intense bond between a brother and sister and subsequently took over the reins of the beleagured comedy *Big Trouble* (1985), his final film as director.

His son Nick (1959–) has appeared in such films as *Black Moon Rising* (1986) and *The Wraith* (1986).

CATHY COME HOME

UK 1966

Writer Jeremy Sandford; *director* ▷ Ken Loach

Sandford's play about a young girl's drift into homelessness against a backdrop of bureaucratic apathy became a touchstone for TV drama, and has been cited as a factor in the establishment of the charity organisation Shelter.

The story of how Cathy (Carol White (1941–)) and husband Reg (Ray Brooks (1939–)) slid to the foot of the housing ladder, losing each other and much else along the way, made its mark largely because of the way the message was relayed. The play broke new ground by mimicking a drama-documentary, and was filmed on location in the style of a newsreel – a break with studio-bound tradition. These innovations raised fears about the blurring of news and drama, and by viewing the world through Cathy's eyes, Sandford made plain where his sympathies lay – his aim was to influence public opinion. In succeeding, *Cathy* set the tone for countless, inferior dramas, none of which could repeat its impact.

The author of the documentary books *Down and Out In Britain*, *Gypsies*, and *Prostitutes*, Sandford described himself as a campaigning journalist rather than a playwright. He endured two and a half years of rejection before *Cathy* was produced, and also wrote *Edna The Inebriate Woman*, a 1971 *Play For Today*. Curiously, Loach's 1990 film, *Hidden Agenda*, about political corruption in Northern Ireland, revived many of the questions about the convergence of fact and fiction.

CEL Derived from the word celluloid, a cel is the transparent sheet used by animators to trace and then paint part of a scene for use in the process of cel animation. When the various elements of the scene on individual cels are superimposed upon each other there is a depth to the image created and the process also eradicates the need to make repeated drawings of elements from the scene that will remain stationary.

CHABROL, Claude

Born 24 June 1930, Paris, France.

A student of pharmacy and law at the University of Paris, the young Chabrol was a dedicated cinephile who became a distinguished film critic for *Cahiers du Cinema* before co-writing an influential book-length study of ▷ Hitchcock (1958) with ▷ Eric Rohmer.

His directorial debut *Le Beau Serge* (Bitter Reunion) (1958), a grey view of smalltown life, was credited with inaugurating the ▷ nouvelle vague but the follow-up *Les Cousins* (The Cousins) (1959) was more typical in its study of decadent Paris through the eyes of a naive country cousin who goes to study at the Sorbonne. Other notable early works include the glossy thriller *A Double Tour* (Web of Passion) (1959), his first in colour, and *L'Oeil du Malin* (The Third Lover) (1962) in which a journalist insinuates his way into a couple's life and destroys their happiness.

A dispassionate observer of relationships and moral quandries, he then turned to a series of purely commercially-minded assignments before establishing his skill as a chronicler of bourgeois mores. Using a murder, or the intrusion of an outsider into an established relationship, he was able to explore the motivations behind acts of violence, reveal the emotions that seethe beneath the most respectable of middle-class veneers and apportion guilt with a God-like assurance. Frequently compared to ▷ Hitchcock, his best films from this period include: *Les Biches* (The Does) (1968) an elegant thriller in which an architect disrupts a lesbian liaison, *La Femme Infidèle* (The Unfaithful Wife)

(1968) the first of several films in which marital infidelity results in murder, and *Le Boucher* (The Butcher) (1969) a singularly well-observed story in which a shy smalltown butcher is revealed to be a sex murderer.

In the 1970s when he turned his attention to television drama and international co-productions, the results were less satisfying and his prolific body of work is remarkably variable in quality. However, he found critical favour with *Violette Noziere* (1977) a well-crafted re-enactment of a true story from the 1930s when a teenage girl's hedonistic lifestyle culminated in her attempt to poison her parents. He has enjoyed box-office successes with the sardonic detective thrillers *Poulet au Vinaigre* (Cop Au Vin) (1984) and its sequel *Inspector Lavardin* (1986) and was very much at home with *Une Affair des Femmes* (1988) a sympathetic account of the last French woman to be guillotined.

However, the pendulum has swung again with recent excursions into international co-productions like *Quiet Days in Clichy* (1990) and *Dr. M* (1990) proving little short of disastrous. A new version of *Madame Bovary* (1991) starring ▷ Isabelle Huppert (1955–) holds more promise.

Since 1964 he has been married to actress Stephane Audran (1932–) whose feline grace and aloof manner have been effectively deployed in a number of his better films.

CHANEY, Lon

Real Name Alonzo Chaney
Born 1 April 1883, Colorado Springs, Colorado, USA.
Died 26 August 1930, Los Angeles, California, USA.

The son of deaf-mute parents, Chaney acquired an all-round showbusiness education as an actor, comic and song-and-dance man before making his first credited screen appearance in *Poor Jake's Demise* (1913).

Scores of films followed as he became an all-purpose character actor under contract to Universal and he also directed several films including *The Oyster Dredger* (1915) and *The Chimney's Secret* (1915).

Gaining attention as the Barbary Coast politician in *Hell Morgan's Girl* (1917) and for his dyed-in-the-wool villainy in the western *Riddle Gawne* (1918) and in *The Devil Bateese* (1918), his greatest popularity came in the wake of *The Miracle Man* (1919) in which he portrayed a con-man posing as a cripple.

Throughout the 1920s, often in collaboration with director Tod Browning (1880–1962), he developed a reputation as 'The Man of a Thousand Faces', creating a gallery of screen grotesques through the use of elaborate disguises and the painful contortion of his body. Beneath the make-up and contortionist's art lay power and pathos in classic examples of the horror genre like *The Hunchback of Notre Dame* (1923), *The Phantom of the Opera* (1925) and *London After Midnight* (1927).

He died of throat cancer shortly after completing his sound debut in a remake of *The Unholy Three* (1930). He was impersonated by ▷ James Cagney in the film biography *The Man Of A Thousand Faces* (1957). His son Creighton Tull Chaney (1906–1973) followed in his footsteps as the actor Lon Chaney Jnr and worked primarily in horror films. Familiar as *The Wolf Man* (1941), he gave his best performance as Lennie in *Of Mice and Men* (1939).

CHARIOTS OF FIRE

Hugh Hudson (UK 1981)
Ben Cross, Ian Charleson, Nigel Havers, Ian Holm.
123 mins. col.
Academy Award: *Best Picture*; *Best Original Screenplay* Colin Welland; *Best Music* Vangelis Papathanassiou; *Best Costume Design* Milena Canonero.

Hoisting his winner's statuette high in the air, writer Colin Welland (1934–) gave his famous warning to the members of the American Academy that 'The British are Coming!'. Even if this fervently expressed threat was not fulfilled in the succeeding years, *Chariots of Fire* secured its place in British screen history for (at least momentarily) restoring the faith in the home industry after a decade which had witnessed an apparently inexorable decline in production statistics and creative standards. Having opened the New York Film Festival some months before its enthusiastic Oscar night endorsement, the film served to indicate that British producers need not opt for the risky and often unsatisfying compromise of mid-Atlantic product, but could indeed win success with indigenous subject matter, providing the budget was modest and the treatment effective. Skilfully performed, with its hummable theme tune and attractive visual polish courtesy of a top adman turned debutant director, *Chariots of Fire* offered inspiration and comfort to an international audience obviously hungry for it. Based on the victorious exploits of two British athletes at the 1924 Paris Olympics – Jewish Harold Abrahams, whose obsessive drive towards accomplishment in the face of anti-semitic prejudice takes him to gold in the 100 yards sprint; and Scot Eric Liddell whose religious beliefs prevent him from competing against his countryman in the Sunday heats, but spur him on to first place in the 400 yards – the film looked back to a time when patriotism and a sense of honour were still valuable currency, doing so with such obvious yearning that some critics saw in it the overblown and self-congratulatory jingoism close to the heart of many a Thatcherite zealot.

CHAPLIN, Charles Spencer (Sir)
Born 15 April 1889, London, UK.
Died 25 December 1977, Corsier-sur-Vevey, Switzerland.

The son of theatrical parents, Chaplin's childhood was one of extreme poverty and hardship. Abandoned by an alcoholic father and left with a mentally unstable mother who was unable to support him, he struggled through life in the poor house and on the streets.

A newsboy and glass-blower, he harboured showbusiness aspirations, participating in a team of clog dancers, acting and performing in music hall. He learnt much of his timing and technique in the employment of impresario Fred Karno (1866–1941) whose troupe he left during an American tour in 1913.

Offered a contract by Keystone films, his tramp character was first glimpsed in *Kid Auto Races at Venices* (1914). A splay-footed, bowler-hatted, down-and-out with a moustache and walking cane, he railed against authority and evil, endured romantic disappointments and endeared himself to a vast global audience as an elementary embodiment of the common man.

In scores of silent shorts, he established the grammar and ground rules of screen comedy using his physical dexterity and pantomime skills to create expertly choreographed, visually humorous entertainments that mixed irreverence, romance, chases and pathos. Among the more notable titles are *The Pawn Shop* (1916), *The Vagabond* (1916), *Easy Street* (1917) and *The Immigrant* (1917).

He progressed to lengthier work with *Shoulder Arms* (1918), *The Kid* (1920), ▷ *The Gold Rush* (1924) and *The Circus* (1928) which earned him a special Oscar for 'versatility and genius in writing, acting, directing and producing'.

He also directed but did not star in the serious drama *A Woman of Paris* (1923).

Resistant to the arrival of sound and the use of dialogue, he created the wordless ▷ *City Lights* (1931) a film that illustrates the difficulties for modern audiences in appreciating his work. Filled with moments of high comedy and athletic grace, it is also soaked in sentimentality and mawkishness as the little tramp falls in love with a blind flower-girl.

Increasingly painstaking in his filmmaking methods and more prone to use his character for social or naive political comment, he made *Modern Times* (1936) (a satire on mechanisation) and the well-intentioned anti-Hitler diatribe *The Great Dictator* (1940) (Best Actor Oscar nomination).

After a lengthy absence, he appeared as a dapper mass-murderer in the black comedy *Monsieur Verdoux* (1947) and wallowed in the lachrymose final days of a fading vaudevillian in *Limelight* (1952) (Best Musical Score Oscar), in which he was outshone by ▷ Buster Keaton.

His left-wing sympathies caused him to fall foul of the rabid anti-Communist factions in post-war America and he emigrated to Switzerland, making only two further features; the embittered anti-American satire *A King In New York* (1957) and the arthritic romantic comedy *A Countess From Hong Kong* (1967).

There was a very public reconciliation with the American establishment in April 1972 when he received an honorary Oscar for 'the incalculable effect he has had in making motion pictures the art form of this century'. He published *My Autobiography* in 1964 and was knighted in 1975.

A writer, performer, director, composer and icon, he was a vital figure in the development of screen comedy. Changing public tastes and critical attitudes will not deprive him of this historical significance.

CHAYEFSKY, Paddy

Real Name Sidney Chayefsky.
Born 29 January 1923, Bronx, New York, USA.
Died 1 August 1981, New York City, New York, USA.

A graduate of the City College of New York, Chayefsky served in the United States Army during the Second World War and began his writing career whilst recuperating from war wounds.

A writer of short stories, radio and television drama, his first film credit was the minor comedy *As Young As You Feel* (1951). He enjoyed television success with *Marty* and won an Oscar when he re-wrote it for the screen in 1955. The story of a Bronx butcher and his love for a plain girl, it was a surprise commercial hit winning the year's Best Picture Oscar and encouraging Hollywood in its attempts to depict more realistic characters.

Chayefsky developed a reputation for his sensitive and affecting stories of ordinary people and provided big screen rewrites of such television plays as *The Catered Affair* (1956) about the wedding of a Bronx taxi driver's daughter, *The Bachelor Party* (1957) about the emotional impact on his friends of a bridegroom's stag night, and *Middle of the Night* (1959) a May–December romance.

With only two screenplays to his credit in the 1960s, (*The Americanization of Emily* (1964) and *Paint Your Wagon* (1969)) it was some time before his work made a similar impact but he won further Oscars for *The Hospital* (1971) and *Network* (1976), savagely satirical attacks on contemporary American society that bristled with indignant rage and bile-ridden rhetoric as he railed against a general loss of humanity and the repression of individuality.

Criticised for a quality of writing that bordered on the hysterical, he remained a distinctive and recognisable authorial voice even if he expressed his ire at the fate of his last screenplay for *Altered States* (1980) by insisting on the pseudonym of Aaron Sidney.

CHEN KAIGE

Born 12 August 1952, Beijing, China

The son of a film director, Chen spent three years of his youth working in rural China during the Cultural Revolution. After army service, he returned to Beijing to work in a film laboratory and, in 1978, was accepted as one of a first year intake of 153 students to the directing course at China's only Film Academy, which had reopened after twelve years.

The graduates of 1982 became known as China's 'Fifth Generation' of Filmmakers. Chen worked as an assistant for two years before making his directorial debut on *Huang Tudi* (Yellow Earth) (1984). Set in the spring of 1939, it tells of a soldier who arrives in a remote northern province to research folk songs and finds himself unable to alter the community's staunch adherence to traditional ways and entrenched practices. A simple story told with controlled, unobtrusive camerawork that evokes the dusty, arid beauty of the province, its strengths lie in the ability to convey human emotions and its critical stance towards feudal China and the cruel treatment of women in that society.

He followed this with *Da Yuebing* (The Big Parade) (1986), a strikingly shot and well-acted dramatisation of the intensive training experienced by 400 army recruits as they compete for the honour of a place in the National Day Parade in Peking's Tiananmen Square. Falling foul of official disapproval, the film was taken from his hands in 1985 and only allowed a release after certain changes in its content.

His third feature, *Hai Zi Wang* (King of the Children) (1988) reflected elements of his own life with an attractively photographed story of a labourer dispatched to the country during the Cultural Revolution who becomes an unorthodox, iconoclastic teacher to the peasant community.

He also accepted a small role as the Captain of the Imperial Guard in the epic *The Last Emperor* (1987). His latest film is *Life on a String* (1991).

THE CHILDREN OF PARADISE *see* **LES ENFANTS DU PARADIS**

CIMINO, Michael
Born 1940, New York, New York, USA.

The oldest son of a musical publisher, Cimino majored in Fine Arts at Yale University before studying acting and ballet and learning about filmmaking with a small company that specialised in documentaries and industrial commissions.

In 1971, he moved to Hollywood and entered the mainstream industry with his contributions to the scripts of *Silent Running* (1971) and *Magnum Force* (1973) before making his directorial debut on *Thunderbolt and Lightfoot* (1974) which made effective use of its star ▷ Clint Eastwood in an entertaining buddy yarn about a gang of bank robbers.

He followed this with *The Deerhunter* (1978), the story of three steelworkers who leave the strong bonds of their community to fight in Vietnam. Despite charges of racism and gross disregard for historical truth, it remains a compelling attempt to convey the trauma that the Vietnam experience represented for ordinary Americans and won Oscars for Best Picture and Best Director.

He subsequently received complete freedom to make ▷ *Heaven's Gate* (1980), an epic account of the Johnson County Wars that bankrupted its studio and inspired the book *Final Cut* (1985). A sumptuous if elliptical vision of the conflict between the poor European-born farmers and the avaricious Wyoming cattle barons, its socialist sympathies did not endear it to American viewers and Cimino was punished for his time-consuming perfectionism and profligacy when the film was swingeingly edited and performed miserably at the US box-office. However, there remains much to be admired in its striking photography, superbly orchestrated set pieces and grandiose ambition.

Considered unemployable by traditionalists in the film industry, he returned with the racist thriller *Year of the Dragon* (1985), then a visually fetching if ludicrously inaccurate biography of peasant hero Salvatore Guiliano, *The Sicilian* (1987), and an ill-advised remake of *The Desperate Hours* (1990).

CINEMASCOPE Invented by Henri Chrétien (1879–1956) in 1927, this system of wide-screen cinematography was adopted by 20th Century Fox in 1953 and first used in the Biblical epic *The Robe* (1953), starring ▷ Richard Burton. An ▷ anamorphic lens on the camera produces a laterally compressed image on 35mm film, which is expanded on projection by a similar optical system. A squeeze factor of 2:1 horizontally is used, resulting in a screened picture of aspect ratio 2.55:1 with stereophonic magnetic sound, and 2.35:1 with optical sound. Similar systems were adopted by other studios but the difficulties of composing images and finding suitably epic stories for the wide-screen did not endear it to many of the directors of the day with ▷ George Stevens declaring 'It's fine if you want a system that shows a boa constrictor to better advantage than a man.'

CINEMATOGRAPHER Also known as a lighting cameraman or director of photography, the cinematographer is the individual responsible for composing the look and ambience of a film through its images, collaborating with the ▷ director to create a mood, atmosphere or dramatic point through a combination of light and shadow, the judicious use of a rich palette of colours or the creation of a striking visual metaphor.

From the silent era, when cinematographer Billy Bitzer (1874–1944) worked with ▷ D.W. Griffith on such films as ▷ *Birth Of A Nation* (1915) and ▷ *Intolerance* (1916), there are numerous examples of the creative and technical innovations to have resulted from the collaboration between the cinematographer and the director. The deep-focus cinematography and expressionistic lighting of Gregg Toland (1904–48) added enormously to the impact of ▷ *Citizen Kane* (1941), whilst, more recently, ▷ Sven Nykvist developed a rewarding partnership with ▷ Ingmar Bergman that resulted in his Oscar win for the staggering use of a sumptuous array of red hues in *Cries and Whispers* (1972).

Currently among the most celebrated cinematographers in world cinema are Nestor Almendros (1930–) an Oscar-winner for *Days of Heaven* (1978), Gordon Willis (1931–), who has visualised ▷ Woody Allen's view of New York in ▷ *Annie Hall* (1977) and *Manhattan* (1979), and Vittorio Storaro (1946–), an Oscar-winner for ▷ *Apocalypse Now* (1979), *Reds* (1981) and *The Last Emperor* (1987), whose most recent work includes *Dick Tracy* (1990) and *The Sheltering Sky* (1990).

CINEMA-VERITE Now a general purpose description of a realistic documentary-like approach to drama or fact-based cinema and television, the term first gained wide critical currency in the 1960s when it was applied to films like *Chronique D'un Eté* (Chronicle of A Summer) (1960) and *The Chair* (1962), and work by David (1931–) and Albert Maysles (1933–), including *Showman* (1963) and *Salesman* (1969). The approach prefers non-professional actors and minimal script and rehearsal, using the mobile viewpoint of a hand-held camera and natural sound to allow the observation of an approximation to spontaneous, uninhibited human behaviour.

CINERAMA In the 1950s, as television began to make in-roads into the loyal and multitudinous American audience, Hollywood struck back with a number of gimmicks and innovations, from ▷ 3-D to wider screens, all designed to make the cinema-going experience bigger and better. *This Is Cinerama* (1952) was the first feature film to use a system pioneered by Fred Waller (1886–1954) for the World's Fair of 1939 and originally known as Vitarama. Shot with three adjacent cameras, the resulting images are presented on a large curved screen via three synchronised projectors to give a continuous picture. By covering the viewer's peripheral vision, there is a greater illusion of the viewer actually participating in events and the first Cinerama film included an exhilarating roller-coaster ride that was the closest cinematic equivalent to being there. The difficulty of co-ordinating the three separate images did lead to problems of visible joins but the system was later used for such expansive feature films as *The Wonderful World of the Brothers Grimm* (1962) and *How The West Was Won* (1962). Eventually the expense and technical drawbacks of adapting cinemas proved prohibitive, and Cinerama subsequently became a single-film process.

CITIZEN KANE
Orson Welles (US 1941)
Orson Welles, Joseph Cotten, Dorothy Comingore, Everett Sloane.
120 mins. b/w.
Academy Award: *Best Original Screenplay* Herman J. Mankiewicz, Orson Welles.

▷ Francois Truffaut's oft-quoted statement that 'Everything that matters in cinema since 1940 has been influenced by *Citizen Kane*' meets with little disagreement from the critical consensus, for ▷ Orson Welles's miraculous debut feature has twice topped the major poll of international film writers taken every ten years by cinema magazine *Sight and Sound.* A former child prodigy whose acclaimed work with his own Mercury Theatre Company on stage and radio while barely out of his teens seemed to genuinely merit the word 'genius', Welles was put on contract at RKO in 1939. In the subsequent two years of unfulfilled projects he obviously absorbed a great deal about film technique, for *Citizen Kane*, a triumph for its 25-year-old writer/producer/actor/director – to be followed almost instantaneously by years of decline – was a milestone in the developing art of the cinema. Welles's story has reporter Joseph Cotten (1905–) piecing together the biographical details of media magnate Charles Foster Kane, played by Welles himself, from his dying word 'Rosebud'. Though apparently based on real-life mogul William Randolph Hearst (whose newspapers unanimously slammed the film) the widely-admired structure deploying overlapping narratives from various witnesses seemed to embody the concept of relativism central to the modernist impulse. Examining the complex relationship between individual power, personal choice and the context of society that Welles found fascinating in Shakespeare and was to return to in his own later work, *Citizen Kane* was also a stylistic landmark. Eschewing the conventional mise-en-scene of edited long-shot/medium-shot/close-up in favour of ▷ deep-focus composition within the frame, Welles reveals meaning through the way in which the characters are positioned in their surroundings (the dying Kane, for instance, lost in the expansive void of his mansion). The greatest film ever made? Bring on the competition ...

CITY LIGHTS

Charles Chaplin (US 1931)
Charles Chaplin, Virginia Cherrill, Florence Lee, Harry Myers.
86 mins. b/w. synchronised music and sound effects.

In 1928 Charles Chaplin temporarily halted production on his latest feature. The realisation that the cinema's recently found ability to talk was no flash in the pan left him pondering the status of his newest self-financed silent movie. The completed *City Lights* was to eventually open in January 1931, with an added musical score (by Chaplin himself) and occasional sound effects, but, to all intents and purposes, a silent film released in the fifth year of the sound era. A risk it certainly was, but from its initial welcome by a public already nostalgic for the glories of the wordless past *City Lights* now stands as one of the director's most universally loved films. Never an influential formalist in the manner of, say, ▷ D. W. Griffith or ▷ F. W. Murnau, Chaplin's almost total concentration here on the pantomimic expertise and appeal of his emblematic little tramp confirmed that the core of his art lay in the performance before the camera and so transcended the day's technical innovation behind it. A simple narrative – tramp falls for blind flower seller and after a series of misadventures involving a drunken millionaire and a spell in prison is able to pay for the operation that restores her sight – offers plenty of scope for the transition from broad comedy to tear-strewn tragedy, with the sophistication of expression that Chaplin brought to his role very probably at its peak. In his 1949 essay *Comedy's Greatest Era*, critic and screenwriter James Agee (1909–1955) was moved to write of the tramp and the girl's moving climactic reunion 'It is enough to shrivel the heart to see, and it is the greatest piece of acting and the highest moment in movies'.

CLAIR, René

Real Name René Lucien Chomette
Born 11 November 1898, Paris, France.
Died 15 March 1981, Neuilly, France.

A member of the Ambulance Corps during the First World War, Clair wrote for the Paris newspaper *L'Intransigéant* and was a poet and actor in such film serials as *Les Deux Gamines* (1920) and *Parisette* (1921) before beginning his career behind the camera.

He made his directorial debut with the comic fantasy *Paris Qui Dort* (The Crazy Ray) (1923) and his close involvement with the then burgeoning ▷ avant-garde movement was reflected in the playful, ▷ surrealistic nonsense of *Entr'acte* (1924). He enjoyed success with the more conventional farce *Un Chapeau de Paille d'Italie* (The Italian Straw Hat) (1927) but came into his own with the advent of sound, when he was able to orchestrate the bustle of the streets and exuberance of his characters into fanfares for the common man.

Films like *Sous Les Toits De Paris* (1929), *Le Million* (1931), *A Nous La Liberté* (1931) and *Quatorze Juillet* (1932) reveal his pioneering use of music and natural sounds, fluid camerawork and mastery of cinematic comedy. Richly atmospheric and inventive, they present the most sparkling of whimsy that salutes the vicissitudes of everyday life and the camaraderie experienced by the denizens of a never-never land evocation of Paris.

Absent from France for many years, he directed the elegant fantasy *The Ghost Goes West* (1935) in Britain and found his light touch and fondness for droll whimsy much in demand during a wartime sojourn in Hollywood where he made such lighthearted concoctions as *I Married A Witch* (1942) and *It Happened Tomorrow* (1944).

He resumed his French career with *Le Silence Est D'Or* (Silence Is Golden) (1947) a bittersweet elegy for the silent cinema era, and enjoyed some measure of success with the time-hopping fantasy *Les Belles De Nuit* (Night Beauties) (1952) and *Les Grandes Manoeuvres* (Summer Manoeuvres) (1955). An ironic romantic comedy, it displayed unusually serious overtones in its story of a dragoon whose Don Juan reputation sours his genuine love for a haughty divorcee.

The rise of the ▷ *nouvelle vague* saw his reputation suffer and the elegant artifice and gossamer-light gaiety of his genuinely innovative early work dismissed as old-fashioned and superficial. He made only a few more films and retired from the cinema after the laboured farce *Les Fêtes Galantes* (1965), although he continued to work in the theatre during the 1970s.

CLAPPERBOARD A slate with a pair of boards hinged together that is photographed at the beginning of each shot and then 'clapped' to mark a synchronisation point in the separate picture and sound records. The slate will contain such identifying information as the name of the production, cameraman, scene and take numbers. Recent technical advances mean that picture and sound recording can now be synchronised electronically.

CLARK, Kenneth Mackenzie (Baron)

Born 13 July 1903, London, UK.
Died 21 May 1983, Hythe, Kent, UK.

Educated in modern history at Trinity College, Oxford, Clark embarked on a career of lifelong devotion to the arts as a historian, gallery administrator, writer and, ultimately, television proselytiser.

He was director of the National Gallery in London (1934–45), surveyor of the King's Pictures (1934–44) and a formative influence in the creation of the Arts Council, which he chaired from 1953 to 1960.

Appointed the first chairman of the Independent Television Authority (1954–7), he was instrumental in defending its impartiality by insisting that a news service remained under direct control of the authority and not be placed in the hands of contractors.

A dapper figure of aquiline profile and imperturbable urbanity, he became a respected specialist writer, radio broadcaster, lecturer and television presenter. He therefore brought a wealth of experience to bear on the 13-part BBC television series *Civilisation* (1969), a vast, panoramic history of the world's cultural riches which he popularised with a presentation style that combined erudition with enthusiasm.

Created a life peer in 1969, he later wrote two volumes of autobiography *Another Part of the Wood* (1974) and *The Other Half* (1977).

CLEESE, John *see* MONTY PYTHON'S FLYING CIRCUS.

CLIFT, Montgomery

Full Name Edward Montgomery Clift.
Born 17 October 1920, Omaha, Nebraska, USA.
Died 23 July 1966, New York City, USA.

After working in summer stock as a teenager, Clift moved to New York and for ten years remained exclusively a stage actor with notable performances in *There Shall Be No Light* (1940), *The Skin Of Our Teeth* (1942) and *You Touched Me!* (1945).

Finally succumbing to one of many film offers, he appeared opposite ▷ John Wayne in *Red River* (1948) and earned his first Best Actor Oscar nomination as the sensitive young soldier in *The Search* (1948).

Broodingly handsome, he was a slight, intense figure particularly adept at conveying the introspective turmoil of society's drifters and outsiders. Briefly considered the most promising of post-war cinema actors, he was daring and non-conformist in his choice of roles, rejecting such prestigious projects as ▷ *Sunset Boulevard* (1950), ▷ *On the Waterfront* (1954) and *East of Eden*

Montgomery Clift

(1955) but proving his versatility and range as the caddish fortune hunter in *The Heiress* (1949), the murderously smitten drifter in *A Place in the Sun* (1951) (Oscar nomination) and the pacifist soldier in *From Here to Eternity* (1953) (Oscar nomination). All of his work revealed an emotional and psychological complexity and an entirely convincing search for the truth of each individual character.

A car accident during the making of *Raintree County* (1957) left him permanently scarred. Troubled by his homosexuality and beset by poor health, his subsequent career never fulfilled its early promise, although his sincerity and conviction remained evident, particularly as the traumatised concentration camp survivor in *Judgement At Nuremberg* (1961) (Best Supporting Actor Oscar nomination) and in his last major role as *Freud* (1962).

His longstanding friend and supporter ▷ Elizabeth Taylor had insisted on him for a leading role in *Reflections In A Golden Eye* (1967) at the time of his death from a heart attack. His final film *L'Espion* (The Defector) (1966) was released posthumously.

CLOSE, Glenn

Born 19 March 1947, Greenwich, Connecticut, USA.

A boarding-school thespian and talented lyric soprano, Close travelled throughout Europe as a singing member of Up With People before returning to America and majoring in drama at William and Mary College in Virginia.

Joining the New Phoenix Repertory Company, she understudied Mary Ure (1933–75) in *Love for Love* and when that actress was dismissed from the production stepped forward to make her Broadway debut in November 1974. She continued to work in repertory and regional theatre, making her television debut in *Too Far To Go* (1979) and building a Broadway reputation as a villainess in the Sherlock Holmes mystery *Crucifer of Blood* (1978) and as the feisty wife of showman *Barnum* (1980–81) for which she received a Tony nomination.

Blonde, with expressive gray-blue eyes and a well-scrubbed patrician beauty, she made her film debut in *The World According to Garp* (1982) (Best Supporting Actress Oscar nomination) and gave intelligent interpretations of goodness and virtue in 'Earth Mother' roles in *The Big Chill* (1983) and *The Natural* (1984) (both Oscar nominated).

Continuing her multi-media career, she won a Tony Award for the Broadway production of *The Real Thing* (1984) and an Emmy nomination for *Something About Amelia* (1984), an incest drama seen by over 60 million American viewers.

A radical change of image as the psychotic mistress in the box-office sensation *Fatal Attraction* (1987) (Best Actress Oscar nomination) enhanced her international renown and she brought a chilling malice and vulnerable edge to her portrayal of the Machiavellian manipulator behind *Dangerous Liaisons* (1988) (Oscar nomination).

She served as the executive producer of the documentary *Do You Mean There Are Still Cowboys?* (1987) and returned to television for *Stones for Ibarra* (1988).

Recent film performances include the comatose Sunny Von Bulow in *Reversal of Fortune* (1990), Gertrude to ▷ Mel Gibson's *Hamlet* (1990) and a temperamental opera diva in *Meeting Venus* (1991).

CLOUZOT, Henri-Georges

Born 20 November 1907, Niort, France.
Died 12 January 1977, Paris, France.

Educated at the naval school in Brest, a student of law and political science and a journalist before entering the film industry, Clouzot began by creating French-language versions of German films before ill health forced him to retreat to a variety of hospitals and sanatoriums between 1934 and 1938.

He returned as the scriptwriter of such films as *Le Duel* (1938) and *Le Dernier des 6* (1941), making his directorial debut with the witty thriller *L'Assassin Habite à 21* (1942). His second feature *Le Corbeau* (1943) attracted more attention; charting the effect of poison pen letters on a small French town, its sour view of provincial life was deemed as being anti-French whilst Clouzot was branded a collaborator and banned from filmmaking for two years.

Quai des Orfèvres (1947), a murder mystery, emphasised his jaundiced view of life and he gained an international reputation with the desperately tense thriller *Le Salaire de La Peur* (The Wages of Fear) (1953) in which four greedy reprobates agree to transport truckloads of nitroglycerin over a treacherous three hundred mile journey. The equally popular *Les Diaboliques* (1955) an influential murder story with a vanishing corpse, confirmed the director as a ▷ Hitchcock-like master of mystery whose technical expertise with the mechanics of suspense was allied to a distinctively pessimistic view of human morality.

Thereafter, chronic ill health restricted his filmmaking career. His last features were *La Vérite* (1960) with Brigite Bardot at the centre of a crime of passion and the similarly disappointing *La Prisonnière* (1968).

COCTEAU, Jean

Born 5 July 1889, Maisons-Lafitte, near Paris, France.
Died 11 October 1963, Milly La Flôret, France.

A poet, actor, novelist, dramatist and librettist, Cocteau's involvement with the cinema was spasmodic but highly influential as he explored a poetic vision of the fragile, dream-like potential of film as a medium of self-expression.

He made his directorial debut with *Le Sang D'Un Poète* (Blood of The Poet) (1930), a reflection of his involvement with the ▷ Surrealist movement that follows a young poet on a journey through the surface of a mirror and into a world of living statues, strange tamperings with mortality and a fatal snowball fight.

He worked as a screenwriter on such films as *L'Eternel Retour* (1942) and *Les Dames Du Bois De Boulogne* (1946) and directed again with the magical fairytale *La Belle et La Bête* (Beauty and the Beast) (1946) and ▷ *Orphée* (Orpheus) (1950), a haunting immersion in his recurring preoccupation with the dichotomy between the real and the imaginary as Orpheus follows Heurtebise into the underworld to secure the return of his wife Eurydice.

In the 1950s, various films were adapted from his works including *Le Bel Indifferent* (1957) and *Charlotte at Son Jules* (1959) and he directed his final feature *Le Testament D'Orphée* (The Testament of Orpheus) (1959), a form of self-portrait in which he appeared as a poet taking stock of his work and images of his life in the company of friends and colleagues.

COHN, Harry

Born 23 July 1891, New York, USA.
Died 27 February 1958, Phoenix, Arizona, USA.

A singer in the show *The Fatal Wedding* (1905), Cohn was a performer, a vaudevillian, cavalryman and songplugger before entering the film industry as a secretary to Carl Laemmle (1867–1939) at Universal Studios.

In 1920 he joined with his brother Jack (1889–1956) to form the CBC Sales Corporation which expanded into Columbia Pictures in 1924 allegedly on the strength of the profits from a film entitled *Traffic in Souls* (1923), which is estimated to have cost £1,900 to produce and earned £117,000 at the box-office.

President of the company from 1932, he also acted as head of production and fulfilled the stereotype of an old-style mogul in his autocratic behaviour, boorishness and acute business acumen. Building the studio from a poverty-row operation into one of Hollywood's major players, he gave an uncharacteristic level of artistic freedom to director ▷ Frank Capra who responded by bringing Columbia prestige and healthy box-office returns from a string of Oscar-winning successes including *It Happened One Night* (1934), *Mr Deed Goes To Town* (1936) and ▷ *Mr Smith Goes To Washington* (1939).

Among the outstanding successes of his reign are *The Jolson Story* (1946), *All the King's Men* (1949), *Born Yesterday* (1950), *From Here To Eternity* (1953) and ▷ *On the Waterfront* (1954) and he also acted as a starmaker to the careers of ▷ Rita Hayworth and Kim Novak (1933–).

The biography *King Cohn* (1967) painted a vivid picture of his thick-skinned vulgarity and pugnacious personality.

COMPLETION GUARANTEE
The agreement, usually between a financier and a filmmaker, that a production will be completed within a certain time frame and on a fixed budget. Other considerations may be included within the fine print and there is usually a contingency fund of around 10 per cent of the total budget to allow for overruns, illnesses, inclement weather or acts of God that might result in delays and extra expenditure.

CONNERY, Sean
Real Name Thomas Connery
Born 25 August 1930, Edinburgh, UK.

After a succession of jobs from milkman to coffin-polisher, lifeguard, artist's model and Mr Universe contestant, Connery's powerful physique won him a position in the chorus line of the London production of *South Pacific* (1951–2).

Sporadic film work followed, with early roles in *Lilacs in the Spring* (1954) and *No Road Back* (1956) but there were more significant opportunities in television drama, notably *Requiem for a Heavyweight* (1956) and a production of *Anna Karenina* (1957) in which he played Vronsky.

Seen in such international ventures as *Tarzan's Greatest Adventure* (1959), ▷ Walt Disney's *Darby O'Gill and the Little People* (1959) and *The Longest Day* (1962), the course of his career was altered dramatically when he was cast as secret agent James Bond in *Dr No* (1962).

Chosen over such fancied contenders as ▷ Richard Burton and Peter Finch (1916–77), he was not, perhaps, the most obvious candidate for the role but quickly made it his own. A virile figure who brought a rough-edged masculinity to the character's suave nature, he proved equally at home risking life and limb to save the world, amorously pursuing some international beauty or dispensing a sardonically witty one-liner. The film was an enormous worldwide success, bestowing on him instant celebrity. He would play the character on seven occasions, most memorably in *From Russia With Love* (1963), *Goldfinger* (1964), *Diamonds Are Forever* (1971) and finally

Andy Garcia, Sean Connery, Kevin Costner and Charles Martin Smith in *The Untouchables* (1987)

Never Say Never Again (1983).

Determined to avoid typecasting, he starred for ▷ Hitchcock in *Marnie* (1964), impressed as the rebellious soldier in *The Hill* (1965) and showed an adventurous nature as a poet in *A Fine Madness* (1966).

Appreciation of his considerable talent has often lagged behind his achievements but he has brought a charisma and conviction to bear on all his work and is particularly drawn to the portrayal of the heroic or noble. His best performances include the 19th-century union leader in *The Molly Maguires* (1969), a disturbed police officer in *The Offence* (1972) and a roistering adventurer destroyed by folie de grandeur in *The Man Who Would Be King* (1975).

His star seemed to dim in the late '70s and early '80s, but he returned with increased authority and ease, winning a British Academy Award for his depiction of a medieval sleuth in *Der Name Der Rose* (The Name of the Rose) (1986) and a popular Best Supporting Actor Oscar for his aging Irish cop with true grit in *The Untouchables* (1987).

He followed this with a number of box-office successes, including *Indiana Jones and the Last Crusade* (1989) and *The Hunt for Red October* (1990), and is currently one of the most sought-after of international stars. A highly competitive golfer and professional of unimpeachable integrity, he also seemed to derive much amusement from a recent magazine selection of him as the world's sexiest man.

CONTINUITY The essential narrative structure of a film or television production from beginning to end, initially established by the scriptwriter, given form by the ▷ director during shooting, and finally assembled in physical form by the ▷ editor.

A detailed shooting script is prepared for the day-to-day filming of a production and it is the responsibility of the continuity person on a crew to keep precise and detailed records of each scene to ensure that there are no apparent alterations of setting or action during any completed sequence. Sharp-eyed cinemagoers may have noticed a recent continuity error in *Days of Thunder* (1990) when an injury to ▷ Tom Cruise results in a red ring round the iris of his eye which switches from left to right during the course of the film.

COOPER, Gary

Real Name Frank James Cooper.
Born 7 May 1901, Helena, Montana, USA.
Died 13 May 1961, Los Angeles, California, USA.

Originally a political cartoonist, Cooper began working in motion pictures as an extra and stunt rider in silent westerns like *The Thundering Herd* (1925) and *Wild Horse Mesa* (1925).

He graduated to small roles in hits of the day like *The Winning of Barbara Worth* (1926), *It* (1927) and *Wings* (1927) before his work as the laconic cowboy in *The Virginian* (1929) made him a star.

A tall, sensitively handsome man, initially cast as naive suitors or gauche, gangly country boys, his screen image developed into that of the decent, peace-loving and courageously determined all-American citizen and whilst most fondly recalled as a cowboy star, he also proved adept at light, frothy comedy and rousing high adventure. His many popular successes include *The Lives Of A Bengal Lancer* (1935), *Mr Deeds Goes To Town* (1936) (Best Actor Oscar nomination), *Ball of Fire* (1941), *The Pride of the Yankees* (1942) (Oscar nomination) and *For Whom The Bell Tolls* (1943) (Oscar nomination).

He won Best Actor Oscars for his performances as the World War I Quaker hero *Sergeant York* (1941) and as the sheriff who stood alone in *High Noon* (1952), which marked a career resurgence after a string of box-office disappointments.

He displayed a tougher side to his determined westerner in such subsequent successes as *Vera Cruz* (1954), *Man of the West* (1958) and *The Hanging Tree* (1959) and made his final appearance in the thriller *The Naked Edge* (1961).

He was awarded a special Oscar in April 1961 for his 'many memorable screen performances and the international recognition he, as an individual, has gained for the motion picture industry'.

COPPOLA, Francis Ford

Born 7 April 1939, Detroit, Michigan, USA.

Raised in New York, Coppola was an avid creator of home movies as a child. Educated in theatre arts at Hofstra University, Long Island, he later studied film at the University of California, Los Angeles.

Determined to make his way into the existing film establishment, he worked tirelessly as a scriptwriter, dialogue director and sound man on such ▷ Roger Corman productions as *The Tower of London* (1962) and *The Young Racers* (1963) and was rewarded for his tenacity by being allowed to direct the low-budget, axe-murder shocker *Dementia 13* (1963).

In demand as a screenplay writer, he contributed to the scenarios of such films as *Is Paris Burning?* (1966) and *Reflections In A Golden Eye* (1967) and later won an Oscar for his screenplay of *Patton* (1970).

He returned to direction with the fresh, kaleidoscopic youth comedy *You're A Big Boy Now* (1966) and followed this with the old-fashioned musical *Finian's Rainbow* (1968) and *The Rain People* (1969), detailing a pregnant woman's odyssey of self-discovery.

He arrived at the forefront of his profession with ▷ *The Godfather* (1972), a magnetic study of power and corruption in a Mafia family that brought him a Best Director Oscar nomination and an Oscar as co-writer of the Screenplay.

He has tended to intersperse epic statements on power and patriarchy with more intimate portraits of personal failure and alienation. Thus a psychological study of a wire-tapper's crisis of conscience in *The Conversation* (1974) was followed by the richer and thematically more complex continuation of the Corleone saga in *The Godfather Part II* (1974) which brought him Oscars for direction, screenplay and as co-producer of the year's Best Picture.

The next few years were spent in the arduous creation of ▷ *Apocalypse Now* (1979) (Oscar nomination), a stunningly visualised journey into the hell of Vietnam and the barbarity of man's worst nature.

Supportive of other filmmakers, his American Zoetrope Company had made such films as *THX-1138* (1971) and now he attempted to run his own studio. The commercial failure of his vastly expensive, hi-tech romance *One From the Heart* (1982) put paid to such ambitious dreams and left him with crippling financial burdens.

Whilst paying the bills as a director for hire on such projects as *Peggy Sue Got Married* (1986) and *Gardens of Stone* (1987), he has created a diverse body of work that includes the existential youth drama *Rumble Fish* (1983) and *Tucker* (1988), a biography of the maverick automobile designer Preston Tucker (1903–57) that suggested a strong affinity with the director's own quest for independence and innovation.

A film promoter and wine grower, he re-captured the elusive combination of critical acclaim and commercial success with *The Godfather Part III* (1990), a grandiose meditation on themes from *King Lear* which earned a total of seven Oscar nominations.

CORMAN, Roger

Born 5 April 1926, Detroit, Michigan, USA.

Affectionately known as the Pope of Pop Cinema and the King of Z Movies, Corman's career as a maverick low-budget filmmaker has almost been eclipsed by his latterday position as mentor to an influential generation of Hollywood talents and as the American promoter of prestigious European art films.

A graduate in Industrial Engineering at Stanford, he began his film career in 1948 as a messenger at Twentieth Century-Fox and then became a story analyst. Taking advantage of the G.I. Bill to study English at Oxford, he travelled in Europe before returning to America and working in television and attempting to sell scripts.

Financing his own early efforts, he produced *The Monster from the Ocean Floor* (1954) and turned director with the Civil War western *Five Guns West* (1955). Over the next few years, he directed dozens of exploitation subjects (sex, violence, drugs) that illustrated his philosophy of speed, economy and resourcefulness and also reflected their age in tales of juvenile delinquency and mutant threats from outer space. Notable titles include *Attack of the Crab Monsters* (1957), *Sorority Girl* (1957) and *Teenage Caveman* (1958).

On *The Viking and the Sea Serpent* (1957) he achieved the breathtaking feat of 77 camera set-ups in one working day and his growing technical dexterity and increasing budgets were soon allied to more sophisticated storytelling. He mixed black humour and horror with considerable panache in *Bucket of Blood* (1960) and *Little Shop of Horrors* (1960) which was shot over two days.

Fall of the House of Usher (1960) began a series of flamboyant and inventive adaptations that used Freud's psychological theories to interpret the work of Edgar Allan Poe (1809–1849). Utilising the acting talents of Vincent Price (1911–) and the colourful camerawork of Floyd Crosby (1899–1986), the eight titles include such considerable achievements as *The Raven* (1963), *The Masque of the Red Death* (1964) and *The Tomb of Ligeia* (1965).

The Intruder (1961), an incisive account of a racist rabble-rouser, was his most explicitly political work and a rare financial flop.

Attracted to stories about outcasts and antiheroes, his work in the 1960s explored the drug subculture and biker counterculture in such titles as *The Wild Angels* (1966) and *The Trip* (1967).

Attempts to work within the conventional studio system seemed to frustrate him and he abandoned a directorial career after the zestful gangster yarn *Bloody Mama* (1970) and the World War I adventure *Von Richthofen and Brown* (1971).

He subsequently formed New World Films which churned out lucrative, opportunistic titles with underlying vestiges of political commentary or brash satire, and he seemed to relish the irony in a company that could produce *Night Call Nurses* and release Bergman's ▷ *Cries and Whispers* in the same year (1972).

Over the years, he has provided countless opportunities for aspiring film-makers, showing initiative and a prudent sense of financial restraint. ▷ Francis Coppola, ▷ Peter Bogdanovich, ▷ Jack Nicholson, ▷ Martin Scorsese and ▷ Jonathan Demme are among the many to have benefited from his largesse, and he acted in Coppola's *The Godfather Part II* (1974).

In 1989, he returned to direction in Milan with the large budget *Roger Corman's Frankenstein Unbound*. His autobiography *How I Made A Hundred Movies in Hollywood and Never Lost A Dime*, written with Jim Jerome, was published in 1990.

CORONATION STREET

UK 1960–

Creator Tony Warren

In Autumn 1960, disaffected 23-year-old writer Tony Warren overcame the boredom of adapting a script for Biggles by conceiving an idea for a serial to be set in North-West England. His idea, designed to exploit the temporary fashion for Northern drama in films and theatre, was to be called *Florizel Street*, though when the first episode went out live on 9 December 1960, the title had been changed. Apparently, Agnes, the Granada tea lady, felt Florizel sounded like a disinfectant.

Initial reluctance to advertise on the twice weekly serial disappeared after it was successfully networked in Spring 1961, and the show has remained top of the ratings for most of its lifetime. (The BBC's ▷ *Eastenders* and Australian ▷ *Neighbours* challenged this supremacy briefly, by adding viewing figures of repeat showings.) Early scripts following the class-conscious doings of Annie Walker (Doris Speed (1901–)), Ena Sharples (Violet Carson (1905–83)), Elsie Tanner (Pat Phoenix (1924–86)) and the Ogdens, Hilda and Stan (Jean Alexander (1926–), Bernard Youens (1914–86)) were closer to the kitchen sink ideal, and reluctance to change a winning formula led to the Street slipping out of synch with a fast-changing society. The stasis in Weatherfield (the mythical Manchester setting) means that though the women characters are exceptionally strong (to appeal to the traditional viewer-base of soap operas), black faces are strangely rare.

Realism long abandoned, the Street has sugared its agonies with a winning brand of gentle humour, a formula which is hard to mimic since its success depends on familiarity. Some effort at updating was made when the serial began to be shown thrice-weekly in 1989, with new characters moving into the houses built on the former site of Mike Baldwin's factory. But attempts to turn the 'Rovers Return' into an American theme bar were defeated. The Street has survived the departure of all its original cast except Ken Barlow (William Roache (1932–)), whose marital strife remains a rich dramatic seam.

COSBY, Bill

Full Name William Henry Cosby Junior
Born 12 July 1937, Germantown, Pennsylvania, USA.

After service in the US Navy (1956–60), Cosby enrolled at Temple University, Philadelphia on a track and field scholarship. Supplementing his income as a nightclub comic, he eventually abandoned his studies to pursue this career full time.

An appearance on *The Tonight Show* in 1965 led to him being cast as secret agent Alexander Scott in the television series ▷ *I Spy* (1965–8) which won him three consecutive ▷ Emmy Awards and broke new ground in the small-screen's belated acceptance of black leading-men. He won a fourth Emmy for *The Bill Cosby Special* (1969) and subsequent series include *The Bill Cosby Show* (1969–71), *The New Bill Cosby Show* (1972–3) and the moralistic cartoon series *Fat Albert and the Cosby* (1972–84) which he hosted.

A congenial figure, his gentle, wholesome humour is based on quirky observations of the world around him and offbeat anecdotes based on personal and domestic experience. Involved with children's educational television for many years, and a frequent personality in commercials, his series *The Cosby Show* (1984–), an inoffensive middle-class domestic comedy, has consistently topped the ratings and made him one of the wealthiest men in showbusiness.

Strangely, he has been unable to convert his television popularity into any kind of consistent cinema success. He made his film debut in *Hickey and Boggs* (1971) and enjoyed some box-office reward for his appearances in *Uptown Saturday Night* (1974) and *California Suite* (1978) but recent lamentable efforts like *The Devil and Max Devlin* (1981), *Leonard, Part VI* (1987) and *Ghost Dad* (1990) have failed miserably.

He has recorded more than twenty albums and won 8 Grammy Awards and more recently gained a reputation as a wry guru of family life with bestselling humorous manuals like *Parenthood* (1986).

COSTA-GAVRAS, Constantin

Real Name Konstantinos Gavras.
Born 12 February 1933, Athens, Greece.

The son of a Greek bureaucrat, Costa-Gavras was a promising ballet dancer before his father's politically outspoken status forced the family to relocate to Paris where he studied comparative literature at the Sorbonne and film-making at the IDHEC (Institut Des Hautes Etudes Cinématographiques).

An assistant to Yves Allegrèt (1907–) on *L'Ambiteuse* (1959), he also worked in this capacity for a number of directors including ▷ René Clair, René Clement (1913–) and Jacques Demy (1931–90), before making his directorial debut with *Compartiment Tueurs* (The Sleeping Car Murders) (1965), an agreeably old-fashioned mystery concerning a murder on the overnight express from Marseilles to Paris.

Another thriller *Un Homme de Trop* (Shock Troops) (1967) was followed by ▷ *Z* (1968), a passionate dramatisation of a true incident in which the 'accidental' death of an opposition leader in a Mediterranean country is revealed to be the murderous act of a covert government organisation. A gripping mixture of excitingly directed investigation and a veiled attack on the Greek dictatorship of the 1960s, it won an Oscar as Best Foreign Film and established its director as a master of the political thriller, dedicated to exposing the denial of human rights and practise of atrocities in totalitarian regimes of all persuasions.

Able to deliver his polemics in a popular entertainment form, his controversial work in the 1970s includes *L'Aveu (The Confession)* (1970), an attack on the Stalinist purges in 1950s Czechoslovakia, *Etat de Siège* (State of Siege) (1973), an indictment of clandestine American involvement in Latin America, and *Séction Speciale* (Special Section) (1975), in which six Frenchmen are shot in a placatory act of political expediency by the Vichy Government in wartime France.

After the sombre romance *Clair de Femme* (1979), he made his Hollywood debut with *Missing* (1981) which earned him an Oscar as co-writer of a script that used true events to tell a powerful, human tale of the American Government's duplicitous involvement in covering up the deaths of American citizens caught up in a CIA-led overthrow of the democratically elected Allende Government.

After failing to convey the complexities of the Palestinian question in the unusual romance *Hanna K* (1983), he revealed a mellower, less excitable approach to potentially explosive material in *Betrayed* (1988), about contemporary American racism, and *The Music Box* (1989), dealing with the trial of a Nazi war criminal.

He has also acted in the film *La Vie Devant Soi* (Madame Rosa) (1977).

COSTNER, Kevin

Born 18 January 1955, Los Angeles, California, USA.

A graduate from California State University with a B.A. in marketing, Costner left a job in marketing after only thirty days to pursue a career in acting. A movie fan and accomplished sportsman, he had hoped to become a professional baseball player but an involvement with the South Coast Actors Co-op fired his acting aspirations.

Moving to Hollywood, he worked as a stage manager for Raleigh Studios (where he now maintains an office) and made his film debut in the undistinguished low-budget *Sizzle Beach U.S.A.* (1981). He followed this with small roles in the likes of *Night Shift* (1982) and *Table for Five* (1983) but found that more substantial opportunities in *Frances* (1982), and as the pivotal character of Alex in *The Big Chill* (1983), were sacrificed during the final editing.

He began to make an impact as the father whose daughter dies of radiation sickness in *Testament* (1983) and furthered his career with starring roles in *Fandango* (1985) and *American Flyers* (1985). He won popular acclaim as a devil-may-care gunslinger in the western *Silverado* (1985) and consolidated his leading man status in *The Untouchables* (1987), capturing the naivety and complexity of incorruptible law enforcer Eliot Ness.

A tall, ruggedly handsome figure with piercing blue eyes, he was able to project the all-American decency and virility that previous generations had admired in ▷ Gary Cooper or ▷ James Stewart. His sexual charisma and athleticism were well-deployed in the thriller *No Way Out* (1987) and he enjoyed further major successes with the baseball romance *Bull Durham* (1988) and the heartwarming fantasy *Field of Dreams* (1989).

The violent, style-conscious thriller *Revenge* (1990) proved to be his first major failure but he triumphed again as star (Oscar nomination) and director (Oscar award) of the western ▷ *Dances With Wolves* (1990) which reflected his love of that genre (he has cited *How The West Was Won* (1963) as his favourite film) and his partial descent from Cherokee Indian stock. Nominated for 12 Oscars, it was the first commercially successful western in many years.

Highly selective in the roles he chooses, he is said to have rejected *Big* (1988), *Mississippi Burning* (1988) and *War of the Roses* (1989) among many others. In 1991 he was seen in heroic mould as *Robin Hood: Prince of Thieves* and was subsequently signed to appear as New Orleans district attorney Jim Garrison in *JFK* (1991), an examination of the Kennedy assassination.

COSTUME-DESIGNER In collaboration with the ▷ director and possibly the ▷ art director and ▷ cameraman, the costume designer will select the colours and fabrics to be used in the creation of a wardrobe for a particular character and co-ordinate their activities with the over-all visual concept of a film, whether it is a period piece or contemporary scenario. Among the more famous of Hollywood studio costume designers were Adrian (1903–59) of M-G-M, Orry-Kelly (1897–1964), who won Oscars for ▷ *An American in Paris* (1951) and ▷ *Some Like It Hot* (1959), and the doyenne of the profession Edith Head (1907–1981) who received 34 Oscar nominations and won Oscars for such films as ▷ *All About Eve* (1950), *Sabrina* (1954) and *The Sting* (1973).

The links between the fashion world

and costume design have always been strong with, for instance, ▷ Audrey Hepburn sporting the creations of Givenchy (1927–) or Cecil Beaton (1904–80) in *My Fair Lady* (1964) and, more recently, Jean-Paul Gaultier (1952–) working on *The Cook, the Thief, His Wife and Her Lover* (1989).

The growing commercial and merchandising spin-off opportunities from film productions have also resulted in well-publicised attempts to popularise certain 'looks': a return to the 1920s after *The Great Gatsby* (1974) or the pursuit of Diane Keaton's casual inelegance after ▷ *Annie Hall* (1977).

COX, Paul
Born 16 April 1940, Venlo, Netherlands.

A student at the Dutch School of Professional Photography, Cox first went to Australia in 1963 and became resident there two years later, initially as a part-time student at Melbourne University. Setting up a small photographic studio, he also moved into filmmaking with the short *Matuta* (1965).

Working on numerous short films and documentaries, including *Skindeep* (1968), *Symphony* (1969) and *Calcutta* (1970), he also taught at the Prahan Institute of Advanced Education before making his feature-length debut with *Illuminations* (1975).

Displaying a distinctly European sensibility, he enjoyed a modicum of success with *Kostas* (1978), the story of a love affair between a Greek immigrant taxi driver and an Australian divorcee. This was followed by greater international acclaim for *Lonely Hearts* (1981), a computer-dating romance between a painfully shy bank teller and a middle-aged piano tuner.

Concerned with the emotional minutiae of often ill-starred relationships, his films rigorously eschew melodrama in their use of painterly compositions, naturalistic performances and a sympathetic but unsentimental view of his lonely, repressed or eccentric suitors.

Subsequent work, including *Man of Flowers* (1983) and *My First Wife* (1984) offered telling examinations of jealousy, insecurity and the fragility of relationships, a theme continued in *Cactus* (1986) starring ▷ Isabelle Huppert (1955–) as a woman rendered semi-blind in a car accident who finds strength and hope in her love of a man blind since birth.

He then directed *Vincent: The Life and Death of Vincent Van Gogh* (1987) an unconventional documentary biography of Van Gogh (1853–90) featuring selected readings from his letters illustrated by scenes of the locales he visited, sketches and paintings he made and reconstructions of the settings that inspired his artistry.

Recent work includes *Island* (1989) and the more characteristically intimate personal drama *The Golden Braid* (1990), the story of an introverted clock repairer who becomes besotted with a braid of hair, much to the chagrin of his current lover.

He has maintained a parallel career in documentaries and television with a wide range of credits, including *For A Child Called Michael* (1979), about childbirth, *Handle With Care* (1985) on breast cancer, and such children's dramas as *The Paper Boy* (1985) and *The Gift* (1987).

CRAB SHOT A movement of the camera sideways, at right angles to its optical axis.

CRANE SHOT Designed to provide the filmmaker with greater flexibility and movement of the camera, a crane is a large trolley with a long projected arm that extends much like fire-fighting equipment, and ends in a platform space large enough to accommodate a camera, camera operator, assistant and ▷ director. The high-angle camera can then be moved up, down or laterally while mounted on the travelling crane. A crane shot embodying this movement can be used to give an aerial view of a scene, pass over a tumultuous crowd or follow a character from a distance to a near point. A crane is credited with first being employed on the musical *Broadway* (1929) and one of the most spectacular uses of the device can be seen over the opening titles of ▷ Orson Welles's *Touch of Evil* (1958) when a bomb-rigged car is followed through a small town until its eventual explosion.

CRAWFORD, Joan
Real Name Lucille Fay Le Sueur
Born 23 March 1904, San Antonio, Texas, USA.
Died 10 May 1977, New York, New York, USA.

A dancer and Broadway chorus girl, Crawford received her new name as the result of a movie magazine competition and grew to become the epitome of the glamorous Hollywood Movie Queen.

Joan Crawford and Ann Blyth in *Mildred Pierce* (1945)

An extra and bit-part player at M-G-M from 1925, she determinedly worked her way to more prominent roles, embodying the zest of the jazz age flapper in *Our Dancing Daughters* (1928) and *Our Blushing Bride* (1930) and creating a star niche for herself in a succession of formulaic melodramas in which she played working-class girls with their sights set on wealth and sophistication. Her greatest successes include *Grand Hotel* (1932), *Dancing Lady* (1933), *Forsaking All Others* (1935) and *The Gorgeous Hussy* (1936).

She was declared 'box-office poison' in 1938, but returned as the wickedly-witty husband stealer in *The Women* (1939) and gave some of her finest performances in *Strange Cargo* (1940) and as the scarred beauty in *A Woman's Face* (1941).

Later, she portrayed a succession of repressed older women, beset by emotional problems and suffering in ermine and pearls, winning a Best Actress Oscar for her performance as the cafe waitress turned restaurant owner *Mildred Pierce* (1945) and further nominations for *Possessed* (1947) and *Sudden Fear* (1952).

Whatever Happened to Baby Jane? (1962), a stylish exercise in gothic horror, inaugurated the final phase of her career when she appeared in several low-budget shockers. Her final film was *Trog* (1970).

An autobiography, *A Portrait of Joan*, appeared in 1962 and her adopted daughter Christina wrote a scathing attack on her domestic tyranny in *Mommie Dearest* (1978), which was filmed in 1981 with ▷ Faye Dunaway offering an uncanny impersonation or malicious defamation, depending on one's perspective.

CRONENBERG, David
Born 15 May 1943, Toronto, Canada.

The son of a musician-mother and a father who frequently wrote for Canadian pulp fiction magazines, Cronenberg began making short films whilst still at the University of Toronto. Early efforts like *Transfer* (1966) and *From the Drain* (1967) have been described as technically rough, and surreal in tone.

He worked for French and Canadian television before securing the funding for *The Parasite Murders* (1974), released in Britain as *Shivers*. The story of parasites running rampant in an apartment block established a blend of cerebral and visceral horror that the director was to make his forte. Special-effects, which border on the disgusting, serve stories that offer food for thought about society at risk from bizarre, uncontrollable and often sexually-driven forces in films like *Rabid* (1976), *The Brood* (1978), *Scanners* (1980) and *Videodrome* (1982). At one time, Cronenberg declared, 'It's a small field, Venereal Horror, but at least I'm king of it.'

The Dead Zone (1983) marked a step away from the world of gore with a mystery about second sight that used suspense and good acting to tell its story. *The Fly* (1986) was an inventive and superbly acted reinterpretation of the 1959 classic in which the plight of the afflicted scientist evoked potent parallels with the AIDS virus. This more refined and imaginative approach to the horror genre continued with *Dead Ringers* (1988) which examined the emotional travails of twin gynaecologists.

He has also acted in *Into the Night* (1985) and *Nightbreed* (1990) in which he played the unhinged psychiatrist.

In 1991, he was preparing to film *The Naked Lunch* by William Burroughs (1914–).

CROPPING The process of reducing an image either by masking it on film or using a smaller projector aperture, cropping has a number of purposes most commonly associated with wide-screen films where it has been used to lower the height of an image or delete the sides of a wide-screen image for general screening.

CROSBY, Bing

Real Name Harry Lillis Crosby.
Born 2 May 1901, Tacoma, Washington, USA.
Died 14 October 1977, Madrid, Spain.

A contender for the title of the century's most popular showbusiness figure, Crosby had a lengthy career that flourished in all media from radio to television, concert halls to record sales. His three decades of cinema success may be only one component of that career but it is not an insignificant one.

A singer with Paul Whiteman's Rhythm Boys, he appeared in the short *Two Plus Fours* (1930) and made his feature-length debut in *King of Jazz* (1930). In characteristically unassuming fashion, he was prone to disclaim the possession of any significant talent, but his crooning, deftness with light comedy and relaxed, easygoing manner made him a screen natural and beloved all-rounder.

His popularity grew in musicals like *We're Not Dressing* (1934), *Anything Goes* (1936) and *Pennies From Heaven* (1937) and he was soon established as one of the industry's most dependable box-office draws. When Fred MacMurray (1907–) and George Burns (1896–) proved unavailable for the minor comedy *The Road to Singapore* (1940), he was enlisted as a co-star for ▷ Bob Hope and thus begin an innovative screen partnership that brought new standards of spontaneity and irreverence to screen comedy as the two men ad-libbed, bantered, indulged in in-jokes and traded affectionate insults. With Dorothy Lamour (1914–) in tow, they made seven Road films between *Singapore* and *The Road to Hong Kong* (1961).

At his peak in the 1940s, other notable films include *Holiday Inn* (1942) and *Blue Skies* (1946), in which he appeared with ▷ Fred Astaire, and *Going My Way* (1944) for which his interpretation of singing priest Father O'Malley received a Best Actor Oscar. He received further nominations for the sequel *The Bells of St Mary's* (1945) and for *The Country Girl* (1954), a rare dramatic role in which he proved highly effective as the alcoholic has-been.

Poor material dented his popularity in the 1950s but top musicals like *White Christmas* (1954) and *High Society* (1956) showed his stature undiminished in the right setting.

He later contributed scene-stealing cameos to *Robin and the Seven Hoods* (1964) and *Stagecoach* (1966) and made his last film appearance as one of the celebrity narrators in *That's Entertainment* (1974).

One of the most enduring stars of radio, he worked extensively in television on annual holiday specials with his second family, in the series *The Bing Crosby Show* (1964–5) and the film *Dr Cook's Garden* (1970). His records are estimated to have sold over 350 million copies, with *White Christmas* alone accounting for 30 million of those. Still giving sell-out concerts up to his sudden demise, he died after participating in his favourite pastime – a round of golf.

CROSS-CUTTING The process of cutting between two or more separate sequences to establish a relationship between the events being depicted. A dramatic device that can create suspense, add commentary or counterpoint to an event, its use had been explored as early as ▷ *The Great Train Robbery* (1903) and there was a more sophisticated illustration of its effectiveness in the climax of ▷ *Intolerance* (1916) where the editing establishes the connections between four interlocking stories and also contributes to a sense of pace and gathering momentum. A frequently cited example from more recent filmmaking is ▷ *The Godfather* (1972) when the increasingly murderous actions of ▷ Al Pacino's Michael Corleone are contrasted, via cross-cutting, with the baptism of his godson.

CRUISE, Tom
Full Name Tom Cruise Mapother IV.
Born 3 July 1962, Syracuse, New Jersey, USA.

Compensating for his difficulties with dyslexia by excelling on the sports field, Cruise appeared in such high school productions as *Guys and Dolls* and *Godspell*, before moving to New York in 1980 and allowing himself a decade to become established as an actor. In between odd jobs, he landed a small role in the film *Endless Love* (1981) and found his supporting assignment in *Taps* (1981) gradually increasing in size and significance as the makers responded to his charisma and application.

After joining the ensemble of ▷ Francis Coppola's *The Outsiders* (1982), he enjoyed a major box-office hit with the teenage comedy *Risky Business* (1983) in which he attained stardom on the basis of a superior script and a scene in which he mimicked a rock'n'roll number whilst dancing in his underwear.

Following the disappointment of the lavish but vacuous fairytale *Legend* (1985), he re-asserted his commercial clout as the gung-ho pilot in *Top Gun* (1986).

A handsome actor with a winning smile, personal magnetism and a relentless focused concentration and intensity in his performances, he began to gain wider respect for his acting as the impetuous pool player jousting with old pro ▷ Paul Newman in *The Color of Money* (1986).

The classic Cruise character has become the hotheaded, ambitious young man in a hurry who learns lessons about life from an experienced older man. This stereotype has surfaced in such unadulterated star vehicles as *Cocktail* (1988) and *Days of Thunder* (1990) in which he indulged a passion for motor-racing. However, he has alternated such calculating crowd-pleasers with more challenging material like *Rain Man* (1988) in which he portrayed the opportunistic brother of autistic savant ▷ Dustin Hoffman and, more especially, *Born on the 4th of July* (1989) in which his compelling and moving performance as a wheelchair-bound Vietnam veteran earned him a Best Actor Oscar nomination.

Tom Cruise in *Cocktail* (1988)

CUKOR, George Dewey
Born 7 July 1899, New York, USA.
Died 24 January 1983, Hollywood, California, USA.

Involved with the theatre from an early age, Cukor was the assistant stage manager of *The Better 'Ole* (1918) and worked extensively in this capacity before making his Broadway directorial debut with *Antonia* (1925).

He moved to Hollywood at a key moment in the transition to sound and served as a dialogue director on *River of Romance* (1929) and ▷ *All Quiet on the Western Front* (1930). He began his directorial career collaborating on such films as *Grumpy* (1930) and *The Royal Family of Broadway* (1930) and made his mark as the solo director of *What Price Hollywood?* (1932) and *A Bill Of Divorcement* (1932), the first in a ten film partnership with actress ▷ Katharine Hepburn that lasted until his death.

A literate man of the theatre, drawn to adaptations of novels and plays, he betrayed no 'staginess' in his direction which served the narrative with fluid elegance, allowing a viewer to savour the richness of dialogue and brightness of performance without losing sight of cinematic virtues. His many notable successes from the zenith of the studio system era include *Little Women* (1933) (Best Director Oscar nomination), *David Copperfield* (1935), *Camille* (1936) and *The Women* (1939).

Renowned for his sensitive handling of major stars, he was the director associated with such Oscar-winning performances as those of ▷ James Stewart in the sparkling sophisticated comedy ▷ *The Philadelphia Story* (1940) (Oscar nomination), ▷ Ingrid Bergman in *Gas-*

light (1944), Ronald Colman (1891–1958) in *A Double Life* (1947) (Oscar nomination) and Judy Holliday (1922–65) in *Born Yesterday* (1950) (Oscar nomination).

His range also extended to archetypal battle-of-the-sexes comedies like *Adam's Rib* (1949) and dramatic musicals of style and scope like ▷ *A Star Is Born* (1954) and *My Fair Lady* (1964) for which he won a Best Director Oscar.

Latterly active in television, he won a Best Director Emmy for *Love Among the Ruins* (1975) and one of the longest Hollywood careers ended unworthily with *Rich and Famous* (1981), an uncharacteristically coarse re-make of *Old Acquaintance* (1943).

CURTIZ, Michael

Real Name Mihaly Kertesz.
Born 24 December 1888, Budapest, Hungary.
Died 11 April 1962, Los Angeles, California, USA.

A stage actor in his youth, Curtiz is frequently credited with directing the first Hungarian feature film *At Utolso Bohem* (The Last Bohemian) (1912). He worked in Denmark between 1912 and 1914, served in the Hungarian infantry during World War I and was the managing director of the Phoenix Studios when he fled Hungary in 1918.

He continued his prolific directorial career in various European locales before he was invited to Hollywood where he made his American debut with *The 3rd Degree* (1926).

The Biblical epic *Noah's Ark* (1929) displayed his mastery of large-scale productions and tumultuous crowd scenes and he also made his mark in the horror genre with *Doctor X* (1932) and *The Mystery of the Wax Museum* (1933), a macabre classic filmed in the early two-banded ▷ Technicolor process.

A contract director at Warner Brothers noted for his dictatorial control of a film set and only passing acquaintance with the intricacies of the English language, he directed some of the most popular and enduring entertainments of Hollywood's Golden Age. He brought a sense of crisp narrative drive, pace and professionalism to every conceivable genre from peerless swashbucklers like *The Adventures of Robin Hood* (1938) and *The Sea Hawk* (1940), to the gangster story *Angels With Dirty Faces* (1938), the patriotic musical *Yankee Doodle Dandy* (1942) and the film noir melodrama *Mildred Pierce* (1945). He won his only Best Director Oscar for ▷ *Casablanca* (1942).

He left Warners in 1953 but his string of box-office successes continued with *White Christmas* (1954) and *King Creole* (1958). Even if his work as a freelancer lacks the distinctive edge of his studio assignments, he maintained his grasp of what would appeal to the general public right up to his final film, the typically punchy western *The Comancheros* (1961).

DAD'S ARMY

UK 1968–77
Creator Jimmy Perry, *Producer/Director* David Croft

Jimmy Perry used his own experiences in local defence corps in Barnes and Watford as the basis for this nostalgic, quintessentially English comedy. Clearly derived from the music hall, *Dad's Army* follows the bumbling efforts of a Home Guard platoon in the fictional town of Walmington-on-Sea. The idea was originally rejected by BBC programme controller Paul Fox (1925–   ), who was fearful about the impact of a show debunking wartime heroes, but rich characters and inspired casting ensured its popularity. The humour was always gentle, and great play was made on the problems encountered when people from different classes and backgrounds were pushed together by war. The central joke was the upset order which placed blustering bank manager Captain Mainwaring (▷ *Coronation Street* regular, and former sergeant-major, ▷ Arthur Lowe) above the ineffectual Sergeant Wilson, (former member of the Royal Armoured Corps John Le Mesurier (1912–83)). Catchphrases were used – 'Do you think that's wise sir?' was Wilson's frequent challenge to Mainwaring's authority. Butcher Mr Jones (Clive Dunn (1923–   )) and undertaker Private Frazer (a memorable creation by John Laurie (1897–1980)) took turns to herald the crisis which invariably befell the platoon with their respective cries of 'Don't Panic, don't panic,' and 'We're doomed.'

Helping set the mood was Perry's Ivor Novello Award winning theme tune 'Who Do You Think You Are Kidding

Mr Hitler?', sung by Bud Flanagan (1896–1968). Producer/director Croft, who had worked on *Beggar My Neighbour* and *This Is Your Life*, went on to become a comedy stalwart with *Hi-de-Hi*, *'Allo 'Allo*, and *Are You Being Served?* Perry adapted his later experience in the Royal Artillery into *It Ain't 'alf 'ot Mum*. A full-length *Dad's Army* feature film was made in 1970, and premiered the following year.

DALLAS

USA 1978–91
Creator David Jacobs

The show which heralded and defined the 1980s fashion for the glossy, high-budget serials, started as a five hour mini-series detailing the rivalries between two Texas oil clans. Jacobs wrote the drama without ever visiting Dallas. The original story followed the romance of Bobby Ewing (Patrick Duffy (1949–)) and Pamela Barnes (Victoria Principal (1950–)), and its impact on their feuding families. Their on-off relationship was a constant of the subsequent series, as was the Barnes/Ewing rivalry, but the central figure on Southfork Ranch was Bobby's ruthless brother JR, who would certainly have worn a black hat had he been cast in a Western. Played by Larry Hagman (1930–), JR was called a 'human oilslick' by *Time* magazine, and the public interest at the close of the second series proper (1980) rivalled the Kennedy assassination. In the US 76% of the TV audience tuned in, in the UK it scored 27 million viewers. The answer to the much asked the question 'Who shot JR?' was Kristin Shepard, a pregnant ex-lover, a disappointment to those who wanted revenge from his much-maligned, alcoholic wife, Sue-Ellen (Linda Gray (1942–)).

Dallas was hugely successful worldwide, going to 90 countries, partly because of high production values. An average episode was budgeted at $700,000, seven times more than a traditional US daytime soap. Many agonised over the meaning of the phenomenon, wondering if it celebrated or satirised American capitalism. Perhaps unknowingly, it did both. More importantly, it was a tale of heightened human passions in a luxurious setting. The show began to falter in the mid-1980s, as arguments over contract renewals infected the storyline. Bobby was killed in a car accident in 1985, though not before uttering the immortal last words 'Be a family'. A full season later Pam had re-married and was on her honeymoon when the formerly-deceased Ewing emerged from the shower asking for a towel. The public wanted Patrick Duffy back, and the happenings of the previous series were, not unreasonably, passed off as a bad dream. The series outlived imitators like *Dynasty*, but has lost its impact. Often mistakenly referred to as soap opera, *Dallas* was characteristic of Hollywood TV in the 1980s.

Dallas: Linda Gray, Larry Hagman, Barbara Bel Geddes, Patrick Duffy, Victoria Principal.

DANCES WITH WOLVES

Kevin Costner (USA 1990)
Kevin Costner, Mary McDonnell, Graham Greene.
Academy Awards: *Best Picture*; *Best Director*; *Best Score*; (John Barry); *Best Cinematography* (Dean Semler); *Best Adapted Screenplay* (Michael Blake); *Best Sound* (Russell Williams, Jeffrey Perkins, Bill Benton); *Best Editing* (Neil Travis)
180 mins. col.

Cited as evidence of the western's enduring popularity and the malleability of the genre in supporting the dramatic demands of successive generations, *Dances With Wolves* proved to be an unexpected international box-office sensation and the recipient of twelve Oscar nominations (thus tying with *Reds* (1981) as the most nominated film since *Who's Afraid of Virginia Woolf* (1966) which received thirteen nominations).

An epic saga of personal fulfilment, a paean to the simpler values of a vanished age, a rousing adventure yarn and a statement on America's guilty legacy of racism and genocide, the film is the kind of reckless gamble that Hollywood is no longer supposed to make. Running for three hours and using Lakota Sioux dialect throughout, it is an astonishingly accomplished début from ▷ Costner the ▷ director who shows an eye for landscape, an astute judgement on pacing a story and a gentle and engaging humanism that treats the audience with intelligence.

Costner plays Lieutenant John Dunbar, a distinguished Civil War veteran who chooses a posting on the western frontier of the Dakota Plains and finds his life transformed by his affinity with the Sioux Indian, his immersion in their culture and a marriage to Stands With A Fist, a white orphan woman who has been raised by the Indians. Through knowledge comes understanding and respect between the races and a sense of harmony that is inevitably shattered by the advance of his ignorant and fearful army colleagues who now regard him as a traitor and the Indians as mere savages.

Standing alongside *Broken Arrow* (1950) and *Cheyenne Autumn* (1964) as a rare voice on behalf of the native American, *Dances With Wolves* also seemed to strike a very contemporary chord in its advocacy of the community, family, environmentally sound action and an almost utopian mode of existence. It also seems to have inspired a crop of productions dealing with the problems of native Indians in American culture from both historical and contemporary perspectives.

DAVIS, Bette

Real Name Ruth Elizabeth Davis
Born 5 April 1908, Lowell, Massachusetts, USA.
Died 6 October 1989, Neuilly-sur-Seine, France.

After studying at the John Murray Anderson School, Davis worked with repertory and summer stock companies; making her Broadway bow in *Broken Dishes* (1929). A screen test brought her a contract with Universal and she arrived in Hollywood in December 1930, making her film debut in *Bad Sister* (1931).

A highly dedicated actress who did not fit the conventional mould of screen stardom, she was determined not to languish in supporting roles as the hero's girl or some other simpering representative of feminine decoration. Dropped by Universal, she signed a long-term contract with Warner Brothers and commanded attention as the slatternly tramp in *Cabin in the Cotton* (1932) and as the crippled Mildred in *Of Human Bondage* (1934). Following that with a Best Actress Oscar for *Dangerous* (1935) she was encouraged to fight for juicier scripts and better conditions.

Saucer-eyed, with a declamatory style that could be electrifying or mannered, she puffed expressively on an omnipresent cigarette as she illuminated a vast gallery of fully-rounded women, bringing an emotional honesty to the most unprepossessing of melodramas.

A prime box-office attraction between 1937 and 1946, she won a second Best Actress Oscar for *Jezebel* (1938) and her greatest successes include *Dark Victory* (1939) (Oscar nomination), *The Letter* (1940) (Oscar nomination), *The Little Foxes* (1941) (Oscar nomination) *Now Voyager* (1942) (Oscar nomination) and *Mr. Skeffington* (1944) (Oscar nomination).

Leaving Warner Brothers in 1949, she freelanced with erratic success; vividly memorable as the flamboyant but vulnerable theatrical grande dame Margot Channing in ▷ *All About Eve* (1950) (Oscar nomination) and as the grotesquely deranged Jane in *Whatever Happened to Baby Jane?* (1962) (Oscar nomination).

An indomitable workhorse, her later years were marked by many unworthy ventures and courageous battles with ill-health but she did manage to secure some substantial television roles, winning an Emmy for *Strangers: The Story Of A Mother and Daughter* (1979) and critical acclaim for *A Piano for Mrs. Cimino* (1982).

The first woman to receive the ▷ American Film Institute Life Achievement Award (1977) her last cinema appearances were *The Whales of August* (1987) and *The Wicked Stepmother* (1989).

Bette Davis in *The Private Lives of Elizabeth and Essex* (1939)

Her several volumes of autobiography include *The Lonely Life* (1962), *Mother Goddam* (1975, with Whitney Stine) and *This' N' That* (1987, with Michael Herskowitz).

DAY FOR NIGHT A convenient process that allows the filmmaker to shoot during the daylight but use filters, underexposure or printing techniques to create the illusion that the events are taking place under cover of night. The term was also adopted as the English-language title for ▷ Francois Truffaut's 1973 film *La Nuit Americaine*.

DAY, Doris

Real Name Doris Von Kappelhoff
Born 3 April 1924, Cincinnati, Ohio, USA.

An accomplished, popular vocalist with several big bands, and a radio favourite in the 1940s, Day made her film debut in *Romance on the High Seas* (1948).

Her sunny personality, singing talent and girl-next-door image made her an asset to many standard Warner Brothers musicals like *On Moonlight Bay* (1951) and *April in Paris* (1953) but more satisfying material followed with her energetic performance in *Calamity Jane* (1953), *Young at Heart* (1954) opposite ▷ Frank Sinatra and the film version of the Broadway show *The Pajama Game* (1957).

A top-selling recording artist, she was also able to prove her dramatic worth in *Storm Warning* (1950) and *Love Me or Leave Me* (1955).

The public affection for the lightweight sex comedy *Pillow Talk* (1959) and her sure comic touch earned her a Best Actress Oscar nomination, revitalised her career and saddled her with an image as the perennially virginal career girl in a series of frothy farces where she was often partnered by Rock Hudson (1925–85).

During her twenty year film career, she was a constant feature in listings of the most potent box-office attractions, but even her loyal following could not endure the inanities of *Caprice* (1967), *The Ballad of Josie* (1967) or *With Six You Get Egg Roll* (1968) (her last film to date).

She retired from the cinema screen in 1968 but has appeared occasionally on television specials and the series *The Doris Day Show* (1968–73) and *Doris Day's Best Friends* (1985). More recently, her name has been associated with the

Doris Day Animal League, an organisation dedicated to the protection of animal rights.

Her autobiography, *Doris Day, Her Own Story* (1976), revealed much of the personal heartache and turmoil beneath her apparently carefree vivacity.

DAY, Robin (Sir)

Born 24 October 1923, London, UK.

After army service in the Royal Artillery (1943–47), Day studied law at St Edmund Hall, Oxford (1947–51) and was called to the Bar in 1952.

Employed with the British Council in Washington, he became a freelance broadcaster in 1954, working at ITN from 1955 to 1959 and leaving to join the BBC where he later presented their flagship current affairs series *Panorama* (1967–72).

His combative style was in stark contrast to the more reverent pre-Macmillan (1896–1986) era tradition of questioning politicians and world figures, and he was influential in modernising interview techniques and setting standards of incisive, informed and determined television presentation. His credibility was enhanced by a vast theoretical and practical understanding of the political sphere and, in 1959, he stood as an unsuccessful Parliamentary candidate for the Liberal Party.

He is rarely seen without his trademark bow-tie; the gurgles and wheezes of his much-mimicked speech pattern and a refreshingly acerbic approach have made him one of the small screen's most recognisable personalities.

His extensive radio work includes *It's Your Line* (1970–76) and *The World At One* (1979–87) whilst he proved a firm but fair chairman of television's *Question Time* (1979–89).

In 1990, he was briefly associated with the ill-starred satellite station British Satellite Broadcasting and has since been seen in many an unlikely setting from a chat show to a light entertainment song 'n' dance act with singer Des O'Connor (1932–).

A dedicated campaigner for the televising of Parliament, his books include *Television—A Personal Report* (1961), *Day By Day* (1975) and the aptly-titled autobiography *The Grand Inquisitor* (1989).

DAY-LEWIS, Daniel

Born 20 April 1958, London, UK.

The son of poet laureate Cecil Day-Lewis (1904–72) and grandson of the film producer ▷ Sir Michael Balcon, Day-Lewis was attracted to acting from an early age, appearing in nursery school Nativity plays, boarding school theatricals and making an early film debut with a non-speaking part in *Sunday, Bloody Sunday* (1971).

Active on stage with the National Youth Theatre and Bristol Old Vic Theatre School, he made a strong impression as Guy Bennett in the West End production of *Another Country* (1982–3). His television appearances include such plays as *How Many Miles to Babylon* (1982) and *The Insurance Man* (1985) and he can be seen in small roles in such epic international films as *Gandhi* (1982) and *The Bounty* (1984). However, the vividness and range of his talent was established in two sharply contrasting film performances.

In *My Beautiful Laundrette* (1985) he brought passion and conviction to the role of a homosexual punk who is torn between his ties to a white, working-class culture and his love for a young Pakistani entrepreneur. At the same time, he was

Daniel Day-Lewis in *My Left Foot* (1989)

equally affecting and credible as a prissy, mannered Edwardian suitor in *A Room With A View* (1985).

Tall and gaunt with an intense Byronic demeanour and a desire to protect his privacy that left him ill-at-ease with the celebrity status that has ensued, he won rave reviews as radical poet Mayakovsky (1894–1930) in *The Futurists* (1986) at the National Theatre and began to accept leading roles for the cinema.

Miscast as the womanising neurosurgeon in the soporific *Unbearable Lightness of Being* (1988) and as the comic Englishman in *Stars and Bars* (1988), he received decidedly mixed notices for a National Theatre production of *Hamlet* (1989) that he was forced to leave when diagnosed as suffering from nervous exhaustion.

However, there was unanimity about the excellence of the astounding verisimilitude he brought to the screen as handicapped Irish writer Christy Brown in *My Left Foot* (1989), a performance that seemed to go beyond acting in its subtle display of a complex gamut of emotions, physical gestures and its testimony to a prickly but unquenchable human spirit. The film won him innumerable awards, including a popular Best Actor Oscar.

DAYBREAK *see* **LE JOUR SE LEVE**

DEAN, James
Full Name James Byron Dean
Born 8 February 1931, Fairmont, Indiana, USA.
Died 30 September 1955, intersection of highways 466 and 41, Southern California, USA.

Dean turned to acting at UCLA where he joined a drama group run by James Whitmore (1921–). He made his film debut with a bit part in *Fixed Bayonets* (1951) and also secured small assignments in *Sailor Beware* (1952) and *Has Anybody Seen My Gal?* (1952).

Moving to New York, he studied at the Actor's Studio and gradually learned his craft in commercials, television drama and on stage in the likes of *See The Jaguar* (1952) and *The Immoralist* (1953).

His first major film role came in *East of Eden* (1955) (Best Actor Oscar nomination) as the much troubled, misunderstood farmer's son Cal. He refined this characterisation memorably as Jim in ▷ *Rebel Without A Cause* (1955). He had just completed a more expansive role, ageing from surly cowhand to cynical oil baron as Jett Rink in ▷ *Giant* (1956) (Oscar nomination), when he was killed in a car crash.

James Dean in *Rebel Without A Cause* (1955)

An intense, waif-like figure, moodily handsome, inarticulate and petulant, he had become, in those three films, the personification of American youth – restlessly rebellious, self-assertive and deprived of security or direction.

Following his death, he was embraced as a cult figure and twentieth-century icon whose potent, angst-ridden vulnerability and sexual magnetism have not dimmed. Amidst the numerous reminisces, picture books and biographies revealing a wildly hedonistic lifestyle, there have been two film biographies; ▷ Robert Altman's *The James Dean Story* (1957) and *James Dean: The First American Teenager* (1975).

DEEP FOCUS The standard photography of a film or television series allows the viewer to observe events on only one plane whilst other action in the foreground or background appears to be blurred. Deep focus is a style of photography that brings all levels (foreground, middleground and background) into equal focus and sharp clarity. The stylistic advantages in the technique are the heightened reality it can bring to a scene and its use as a less manipulative means of involving the viewer in all aspects of the action. Rather than demanding attention by close-ups or editing the ▷ director allows the viewer to see all perspectives and select their own choice of germane material for absorption.

Gregg Toland (1904–48) is credited as the first ▷ cinematographer to make expressive use of the technique in the filming of ▷ *Citizen Kane* (1941).

DE HAVILLAND, Olivia Mary

Born 1 July 1916, Tokyo, Japan.

Raised in California with her sister, the actress Joan Fontaine (1917–), De Havilland made her stage debut with the Saratoga Community Players in *Alice in Wonderland* (1933). Her subsequent appearance as Puck in *A Midsummer Night's Dream* (1934) brought her to the attention of Max Reinhardt (1873–1943) who cast her in his Los Angeles production of the play and his 1935 film version in which she portrayed Hermia.

Signed to a long-term contract with Warner Brothers, she was cast opposite ▷ Errol Flynn in the swashbuckler *Captain Blood* (1935). Popular as a romantic couple – her demure charm matched his reckless bravado – they appeared in seven films together including *The Charge of the Light Brigade* (1936), *The Adventures of Robin Hood* (1938), in which she played Maid Marian, and *They Died With Their Boots On* (1941).

A warm and likeable performer in all genres, she established her claim to greater consideration as an actress with an intelligent performance as Melanie in ▷ *Gone With The Wind* (1939). Through her delicacy, sensitivity and sincerity a shy, timid character whose nobility and virtue might otherwise seem too good to be true is rendered credible and moving. Her performance received a Best Supporting Actress Oscar nomination.

Further opportunities came as the gentle, understanding wife in *The Strawberry Blonde* (1941) and as the shy schoolteacher in *Hold Back the Dawn* (1941) which earned her a Best Actress Oscar nomination.

Her Warner Brothers contract ended in 1942. She had continuously fought with the studio over the calibre of her roles and she sued them when they added an extra six months to her period of employment, to compensate for her time on suspension. Supported by the Screen Actor's Guild she won a landmark victory that limited all film contracts to a seven year maximum.

She returned to the screen and began one of the richest periods of her career by winning an Oscar as a self-sacrificing unwed mother in *To Each His Own* (1946). *Dark Mirror* (1946) allowed her to play twin sisters of contrasting gentility and malevolence and some of her best work is featured in *The Snake Pit* (1948) as a young woman undergoing archaic treatment in a mental home. The emotional vulnerability of her gauche ugly duckling in *The Heiress* (1949) brought a second Oscar win.

In 1951, she made her Broadway debut in *Romeo and Juliet* and later moved to France when she wed the editor of *Paris Match*. Film appearances grew rarer but her presence was welcome as a pioneer woman in *The Proud Rebel* (1958) and as the deceitful sister in *Hush, Hush, Sweet Charlotte* (1964).

Recent work has often been as part of all-star casts in films like *Airport '77* and *The Swarm* (1978) although television has offered more demanding assignments including *Noon Wine* (1967), *The Screaming Woman* (1972), *Roots: The Next Generation* (1979) and *Anastasia: The Mystery of Anna* (1986) for which she received a Golden Globe as Best Supporting Actress.

She published an autobiography, *Every Frenchman Has One*, in 1963.

DE MILLE, Cecil B(lount)

Born 12 August 1881, Ashfield, Massachusetts, USA.
Died 21 January 1959, Hollywood, California, USA.

The son of an Episcopalian minister, De Mille followed his education at a

Pennsylvania Military Academy (1896–1898) and the American Academy of Dramatic Arts (1898–1900) by pursuing a livelihood in the theatre as an actor, stage manager and undistinguished playwright.

Together with ▷ Sam Goldwyn he travelled to the then unknown village of Hollywood in California to make his directorial debut with *The Squaw Man* (1913) and helped to establish the locale as a filmmaking centre.

Drawn to spectacular yarns of the great outdoors, he was instrumental in advancing the fluidity of cinematic narrative through the pioneering use of such devices as ▷ cross-cutting, and also developed a reputation for sophisticated explorations of bourgeois mores, quite often starring ▷ Gloria Swanson. His most famous silent features include *The Cheat* (1915), *Male and Female* (1919) and *The Affairs of Anatol* (1921).

An extravagant showman who revelled in a stereotypical image of the autocratic filmmaker, he enjoyed enormous success with the Biblical epic *The Ten Commandments* (1923) which he would remake in 1956 in ▷ Cinemascope.

Typecast as a purveyor of gaudy, action-packed spectaculars, he worked in all genres from adventure to comedy, but did achieve his greatest renown and box-office success as a master of the sensationalist epic – lavish productions, often culled from the Bible, that adopted a high moral tone of spurious respectability as the maker wallowed in that which he sought to condemn among the seven deadly sins. His many accomplishments include *King of Kings* (1927), *The Sign of the Cross* (1932), *Reap The Wild Wind* (1942), *Samson and Delilah (1949) and The Greatest Show On Earth* (1952) (Best Director Oscar nomination).

A man of many talents, he also worked extensively in radio, organized Mercury Aviation, the first commercial passenger Airline service in the USA, and refused a nomination to the US Senate in 1938. His last film was *The Buccaneer* (1958) which he produced under the direction of his then son-in-law Anthony Quinn (1915–).

He received a special Oscar in March 1951 for 'thirty-seven years of brilliant showmanship' and published *The Autobiography of Cecil B. De Mille* in 1959.

DEMME, Jonathan

Born 22 February 1944, Baldwin, Long Island, New York, USA.

A movie usher as a youth, Demme studied to become a veterinarian at the University of Miami where he wrote film criticism for the student newspaper, *The Florida Alligator*, and for the bi-weekly shopping guide, *The Coral Gable Times*.

Briefly in the US Air Force, a meeting with producer Joseph E. Levine (1905–87) brought him employment in the New York publicity department of Embassy Pictures. Continuing to write film and rock journalism, he made his first 16mm amateur film entitled *Good Morning, Steve* (1968).

Resident in Britain for a spell from 1969, his involvement with the film industry deepened. He worked as music co-ordinator on *Sudden Terror* (1970) and was hired as a unit publicist on ▷ Roger Corman's *Von Richthofen and Brown* (1971).

Given the opportunity by Corman to work on low-budget exploitation fare, he contributed to such scripts as *Angels Hard As They Come* (1972) and *The Hot Box* (1972) before making his directorial debut with the melodrama *Caged Heat* (1974) set in a woman's prison.

He brought a lively sense of pace to such action-packed fare as the gangster yarn *Crazy Mama* (1975) and the vigilante thriller *Fighting Mad* (1976) whilst *Citizen's Band* (1977) (later titled Handle With Care), showed a gentler side to his talents as he explored the radio maniacs of a small town and created an engaging slice of Americana.

Displaying a feel for the highways and by-ways of his native land and the eccentricities of human nature, he has shown an uncondescending appreciation of blue-collar America and popular culture. His subsequent films include the undervalued ▷ Hitchcockian thriller *Last Embrace* (1979) and the quirkily original *Melvin and Howard* (1980).

Dissatisfied with studio interference in his wartime saga of female solidarity *Swing Shift* (1984), he has enjoyed a modicum of commercial success with the darkly humorous thriller *Something Wild* (1986) and the Runyonesque Mafia comedy *Married to the Mob* (1988). Both of these display his penchant for affectionate characterisation and use of eclectic musical soundtracks.

His unpredictable and diverse career also encompasses the direction of pop videos, documentaries like the exemplary concert movie *Stop Making Sense* (1984) featuring Talking Heads, and *Haiti, Dreams of Democracy* (1987), the film of Spalding Gray's autobiographical monologue *Swimming to Cambodia* (1987) and television work like the Columbo mystery *Murder in Aspic* (1978) and the PBS drama *Who Am I This Time?* (1982).

His most recent film is the chilling psychological thriller *Silence of the Lambs* (1991).

DE NIRO, Robert

Born 17 August 1943, New York City, New York, USA.

A student of acting with Lee Strasberg (1901–82) and Stella Adler (1895–), De Niro worked off-Broadway and in television commercials before making his film debut as an extra in *Trois Chambres A Manhattan* (1965).

Supporting roles followed in such films as *Bloody Mama* (1970) and *The Gang That Couldn't Shoot Straight* (1971) before he gained critical attention as the fatally stricken baseball player in *Bang The Drum Slowly* (1973) and won a Best Supporting Actor Oscar for his performance as the young Vito Corleone in *The Godfather Part II* (1974).

He began a fruitful creative relationship with director ▷ Martin Scorsese on the film *Mean Streets* (1973). In their seven feature films together, they have explored various facets of contemporary American society and mores, relentlessly focusing on the brutality of patriarchal order, the conflicting currents of obsession and repression in aggressive masculinity and the violence of the dispossessed and marginalised members of a bankrupt society. Noted for his chameleon-like versatility and dedicated quest for authenticity in his characterisations, he has excelled as the sociopath in ▷ *Taxi Driver* (1976) (Best Actor Oscar nomination), self-destructive boxer Jake La Motta in ▷ *Raging Bull* (1980) (Oscar), celebrity-seeking creep Rupert Pupkin in *King of Comedy* (1982) and hoodlum Jimmy 'The Gent' Conway in *Good Fellas* (1990).

In his other films, he chose to work with some of the most distinguished of international talents including ▷ Bernardo Betolucci on *Novecento* (1900) (1976), ▷ Michael Cimino on *The Deerhunter* (1978) (Oscar nomination), Elia Kazan on ▷ *The Last Tycoon* (1976) and ▷ Sergio Leone on *Once Upon A Time in America* (1983).

Whilst he has attempted to keep his private self as anonymous as possible, his astonishing range of work has proved his most eloquent testimony as he has employed the tenets of Method Acting to seek the emotional and psychological truth of his characters.

He added a ripe performance as bloated gangster Al Capone (1899–1947) to his gallery in *The Untouchables* (1987) and has edged towards a more mainstream acceptance with lighter and less consequential roles in the popular comedy-thriller *Midnight Run* (1988) and *We're No Angels* (1989).

In 1989 he purchased a converted coffee factory in Manhattan which has become the TriBeCa Film Center housing production offices, screenings facilities, state of the art equipment and a restaurant. His base for future endeavours as a neophyte movie mogul, it is a concrete commitment to his aspirations as a producer, director and all-round filmmaker.

Meanwhile he continues to act, clocking up a breathtaking number of major feature films. His most recent roles include a postencephalitic in *Awakenings* (1990) (Oscar nomination) opposite ▷ Robin Williams, a victim of the 1950s Hollywood blacklist in *Guilty By Suspicion* (1990) and the forthcoming remake of the 1962 thriller *Cape Fear* (1991) which again reunites him with Scorsese.

DE PALMA, Brian

Born 11 September 1940, Newark, New Jersey, USA.

An undergraduate at Columbia University, New York, De Palma began his directorial career with short films like *Icarus* (1960), *660214, The Story of an IBM Card* (1961) and *Wotan's Wake* (1962) which received a number of prizes and earned him a writing fellowship to Sarah Lawrence College.

He made his feature-length debut with *The Wedding Party* (1966) and followed this with a number of rough-edged satirical commentaries on American mores and domestic responses to the Vietnam War including *Greetings* (1968) and its sequel *Hi, Mom!* (1970).

His Hollywood career began with the offbeat comedy *Get To Know Your Rabbit* (1972) which suffered from substantial studio interference and met a dim fate at the box-office. However, he received some critical plaudits for *Sisters* (1973), in which one Siamese twin proves to be a murderer, and later enjoyed a tremendous box-office success with *Carrie* (1976), the story of a teenage girl's use of telekinesis to exact revenge on a heartless world.

Established as a purveyor of pyrotechnic horror stories that offered hearty acknowledgment to ▷ Alfred Hitchcock, he remained with glossy, special-effects driven yarns for *The Fury* (1978).

Often accused of placing his genuinely creative cinematic talents at the disposal of flashy, blood-drenched thrillers that traded in glossy sensationalism, repellent misogyny and derivative storylines, he aroused considerable ire with the stylishly seductive slasher-on-the-loose yarn *Dressed to Kill* (1980) and the even sleazier *Body Double* (1984).

However, his technical virtuosity was more appealingly allied to the less gory *Obsession* (1975) with its overtones of ▷ *Vertigo*, and *Blow-Out* (1981) an excellent, but undervalued political thriller.

Less characteristic work includes the satirical rock musical *The Phantom of Paradise* (1974) and the feeble comedies *Home Movies* (1979) and *Wise Guys* (1986).

Scarface (1983) took an operatic, wildly overemphatic approach to its study of a drug baron's bloody rule but he redeemed himself in many eyes with the rousing, old-fashioned tale of resolute law officers tackling the gloated criminal empire of Al Capone (1899–1947) in *The Untouchables* (1987).

He followed this with *Casualties of War* (1989) a respectable study of morality under fire during the Vietnam War, and a controversial adaptation of *Bonfire of the Vanities* (1990).

DEPARDIEU, Gérard

Born 27 December 1947, Châteauroux, France.

The son of an illiterate sheet-metal worker, Depardieu was, by all accounts, an unruly child who drifted into truancy and delinquency at an early age. He dropped out of school when he was twelve and worked at various jobs from lifeguard to dishwasher and amateur boxer before arriving in Paris and attending some acting classes.

He made his film debut in *Le Beatnik et Le Minet* (1965) and worked at the Théatre National Populaire and in television before securing small cinema roles. His first appearances tended to reflect his own experiences of life and he was cast as a sinister salesman in *Nathalie Granger* (1972), a loutish womaniser in the popular *Les Valseuses* (1973) and a boxer in *Vincent, Francois, Paul et Les Autres* (Vincent, Francois, Paul and The Others) (1974).

Gérard Depardieu in *Jean De Florette* (1986)

Over six feet tall with the hulking physique of a rugby player and the soul of a poet, he has brought an animal magnetism to bear on a vast array of characterisations that combine strength and sensitivity.

Appearing in more than sixty films over the past twenty years, he has worked with some of Europe's finest directors, displaying his versatility, prodigousness and an infectious love of his craft. His skill and the wide international release of his work have won him recognition as France's greatest actor and her most ubiquitous.

As he has refused to impose any restrictions on the type of challenge he will accept, any handful of titles will reflect his adventurousness. In the 1970s, his best work includes the self-emasculating slob in the extraordinary *La Dernière Femme* (The Last Woman) (1975), the peasant Olmo in ▷ Bertolucci's *Novencento* (1900) (1976) and the inadequate husband in *Préparez Vos Mouchoirs* (Get Out Your Handkerchiefs) (1978).

He won a César, the French equivalent of the Oscar, for *Le Dernier Métro* (The Last Metro) (1980), brought a persuasive earthiness to *Le Retour De Martin Guerre* (The Return of Martin Guerre) (1981) and, in *Danton* (1982) approached the title character not as a distant historical cypher but as a richly human, tolerant and fallible individual.

He directed himself in an adaptation of *Tartuffe* (1984) and, more recently, has received wave after wave of acclaim for such diverse performances as the racist, passionate cop in *Police* (1985), the child-like bi-sexual thief in the black comedy *Tenue De Soirée* (Evening Dress) (1986), the heartrending, indomitable hunchback in *Jean De Florette* (1986), a country priest in *Sous Le Soleil De Satan* (Under Satan's Sun) (1987), sculptor Rodin (1840–1917) in *Camille Claudel* (1988), the tragically besotted husband in *Trop Belle Pour Toi!* (Too Beautiful For You!) (1989) and as *Cyrano De Bergerac* (1990) (Best Actor Oscar nomination), a performance that captured both the boisterous swagger and pathos of the lovestruck poet-swordsman.

Among his more recent roles are the alcoholic bar owner in *Uranus* (1990) and his English-language debut as a rumbustious French musician embroiled in an American marriage of convenience to gain his *Green Card* (1990).

A bon viveur who cultivates vineyards in the Loire Valley, he acquired the French distribution rights to *Henry V* (1989) and produced ▷ Satyajit Ray's *Shakha Proshakha* (Branches of the Tree) (1990).

DEREN, Maya

Real Name Eleanora Derenkowsky
Born 29 April 1917, Kiev, the Ukraine.
Died 13 October 1961, Queens, New York, USA.

Often referred to as the mother of the American ▷ avant garde, Deren fled to New York with her parents in 1922. A major in journalism and political science at Syracuse University, she was a left-wing activist in the late 1930s and regional organiser of Syracuse Young People's Socialist League.

In 1941, she was employed by choreographer Katherine Dunham (1910–) and began a lifelong fascination with dance and movement. Her interest in the cinema was stimulated by Czech filmmaker Alexander Hammid (1907–) who became her second husband in 1942. Together, they collaborated on *Meshes of the Afternoon* (1943) and *At the Land* (1944) poetic, trance-like films that use slow motion and subjective camerawork to tell stories that dispense with conventional narrative requirements of time and space.

Notions of movement and the camera's intimate participation in the structure and fluidity of a film were explored in dance films like *A Study in Choreography for Camera* (1945), *Ritual in Transfigured Time* (1946) and *Meditation on Violence* (1948) which uses the camera as an active sparring partner for boxer Wu-Tang and creates 'cubism in time' by cinematically evoking the mental act of meditation.

In the 1940s, she worked tirelessly on behalf of independent, experimental cinema, establishing a national network of non-theatrical screenings and writing the pamphlet *An Anagram of Ideas on Art, Form and Film* (1946). She also worked on the uncompleted documentary *Divine Horsemen: The Living Gods of Haiti* (1947–51) a reflection of her interest in Haitian culture and voodoo.

In 1954 she established the Creative Film Foundation. Viewing her work as a 'slow spiral around some central essence', she believed her final film *The Very Eye of Night* (1958) to have completed a spiral of reflections and regarded it as 'the coolest, the most classicist'.

DEROCHMENT, Louis *see* MARCH OF TIME

DE SICA, Vittorio

Born 7 July 1902, Sora, Italy.
Died 17 November 1974, Paris, France.

A graduate in accounting from Rome University, De Sica entered the film industry as an actor with a small role in the film *Il Processo Clémenceau* (1918).

Appearing with a variety of theatre companies, he essayed an occasional film role before his performance in *La Vecchia Signora* (1931) helped establish him as something of a matinee idol.

Screen popularity allowed him to move behind the camera and he made his directorial debut with *Rose Scarletti* (1940) in which he also acted.

His reputation rests on his influential role within the ▷ neo-realist movement of the post-War years when the personal traumas and social ills of a battle-ravaged Italy were conveyed in simplistic but heartrending stories utilising non-professional actors, realistic locations and naturalistic dialogue to achieve a heightened poetic approximation to real life.

Sciuscia (Shoeshine) (1946), which conveyed the futile aspirations and delinquency of shoeshine boys Giuseppe and Pasquale, received an honorary Oscar for 'proof to the world that the creative spirit can triumph over adversity'.

Other notable titles from this period include ▷ *Ladri Di Biciclette* (Bicycle Thieves) (1949) (Best Foreign Film Oscar) in which the theft of a bicycle endangers a man's already precarious livelihood, *Miracolo a Milano* (Miracle in Milan) (1950) an allegorical fable in which the poor establish a make-shift sense of community in their self-created village, and *Umberto D* (1952) a harsh portrait of an aged man's extreme poverty.

In the 1950s, his style and concerns altered to include glossier, frothy entertainments featuring stars like ▷ Sophia Loren and ▷ Marcello Mastroianni and, whilst often entertaining, they lacked the impact and emotion of earlier work. *Ieri, Oggi Domani* (Yesterday, Today And Tomorrow) (1963) however did win another Best Foreign Film Oscar.

Less characteristic of his later years but more compelling were *La Ciociara* (Two Women) (1960) a melodramatic tale of a mother and daughter's ordeals in wartime Italy that won a Best Actress Oscar for Loren, and *Il Giardino Dei Finzi Contini* (The Garden of the Finzi Continis) (1970) (Best Foreign Film Oscar) a nostalgically inclined account of Italy's guilty participation in the Holocaust.

His last films as director were *Una Breve Vacanza* (A Brief Vacation) (1973) a novelettish but winning tale of a woman's self-discovery and *Il Viaggio* (The Voyage) (1974) an unremarkable romance that provided his final collaboration with Loren.

An avuncular character actor, he also played minor roles in scores of international productions including *Madame De* (1952), *A Farewell To Arms* (1958), *The Shoes of the Fisherman* (1968) and ▷ Andy Warhol's *Dracula* (1974) and was seen in the television series *The Four Just Men* (1959).

THE DICK VAN DYKE SHOW

USA 1961–66

Creator Carl Reiner

Without getting the go-ahead, comedy writer Carl Reiner (1922–) scripted 13 episodes of *Head Of The Family*, a semi-autobiographical show in which he would star. The similarities with his own life were many – the central character, Rob Petrie, was a comedy writer on a the Alan Sturdy Show, a fictional sitcom – Sturdy was a corruption of the yiddish *shtarker*, meaning big bruiser, a sideways reference to Reiner's old employer Sid Caesar (1922–). The setting was New York, and the fictional Petries lived on the same street as the real-life Reiner.

The pilot show was less than sensational, and the project seemed doomed until the intervention of producer Sheldon Leonard (1907–), who cast ▷ Van Dyke from a shortlist of two (Johnny Carson (1925–) was the other, more famous contender), subtly changing the show's ethnicity (Alan Sturdy became Alan Brady, and was played by Reiner). The series was indirectly financed by the Kennedy family, and is a living testament to the liberal hopes of the early 1960s, with Brady and wife Laura (▷ Mary Tyler Moore) as US TV's first (and last) un-selfconsciously progressive, white, middle-class family.

This apart, the show broadened the scope of the family-based sitcom by taking it into the workplace, (albeit the glamorous setting of a TV studio). The dialogue was sophisticated and skilfully executed. Van Dyke's career would falter later, but Tyler Moore moved into quality TV, heading a successful production company. Cancellation was ordered by Reiner in 1966. He feared a decline (the show had slipped to 16th place in the ratings), and doubted the wisdom of converting to colour, in line with a CBS edict covering all their prime time shows, for 1966–67. The series lived on in countless repeats, its timelessness aided by the strict ban on ephemeral slang in the scripts.

DIETRICH, Marlene

Real Name Maria Magdalena Von Losch.
Born 27 December 1901, Berlin, Germany.

Accepted into the Deutsche Theaterschule, Dietrich made her film debut as a maid in *Der Kleine Napoleon* (a.k.a. So Sind Die Manner) (1922) and appeared in a number of roles, including the lead in *Prinzessin Olala* (Princess Olala) (1928), before her performance as the temptress Lola-Lola in ▷ *Der Blaue Engel* (The Blue Angel) (1930) (Germany's first sound feature) brought her international renown and a contract to film in Hollywood.

Under the direction of ▷ Josef Von Sternberg she created an indelible image of enigmatic sexual allure in a succession of exotic starring vehicles like *Morocco* (1930) (Best Actress Oscar nomination), *Blonde Venus* (1932), *The Scarlet Empress* (1934) and *The Devil Is A Woman* (1935) which served as stunningly pictorial paeans to her artfully created mystique, aloof glamour and magnificently photographed features.

Marlene Dietrich

Labelled 'box-office poison' in 1938, she returned in triumph as brawling saloon singer Frenchie in *Destry Rides Again* (1939). Later film work tended to rely on the camera's love affair with a screen goddess and little of her subsequent screen work could be termed challenging acting; she was merely required to lend her now legendary presence. However, she was effective as the cabaret entertainer in *A Foreign Affair* (1948), the western outlaw in the outré *Rancho Notorious* (1952) and as the more homely widow in *Judgment At Nuremberg* (1961), a role said to be close to her off-screen persona as a contented hausfrau.

After extensive tours to entertain troops during World War II she developed a further career as an international chanteuse and cabaret star that took her to the most exclusive venues in the world and brought adoring audiences to their feet at her husky-voiced rendition of such songs as 'Falling in Love Again' and 'Where Have All The Flowers Gone'.

After a long absence, she made her final film appearance in *Just A Gigolo* (1978).

Increasingly reclusive in her Parisian residence, she refused to be photographed for the 1984 documentary *Marlene*, directed by Maximilian Schell (1930–) but did contribute a cantankerous vocal commentary and the result won an Oscar nomination as Best Documentary.

Her books include *Marlene Dietrich's ABC* (1961) and *My Life Story* (1979).

DIMBLEBY, Richard

Born 25 May 1913, Richmond-on-Thames, UK.
Died 22 December 1965, London, UK.

Dimbleby began his distinguished journalistic career as a teenager in the composing room of the family newspaper *The Richmond and Twickenham Times* and later worked on *The Southern Daily Echo* and *The Advertisers' Weekly* (1935–36) before joining the Topical Talks Department of the BBC in 1936.

The Corporation's first 'news observer', he did much to set standards of gravitas, integrity and objectivity in his pursuit of a fair if reverential reporting of the facts, gaining experience and respect for his broadcasts from the Spanish Civil War. As the BBC's first war correspondent from September 1939, he reported from France, the Middle East, the Mediterranean and Germany, accompanying over twenty bombing raids and providing the first vivid testimony of the horrors of Belsen.

After the War he established a similar position of authority on television with the BBC's monopoly contributing to his status as the voice of the nation. An anchorman of unflappability, and a doughty political analyst as host of *Panorama* (1957–63), he was also a sonorous, sometimes pompous commentator on state occasions, general elections, budget days and noted events like the Coronation of 1953 and the funeral of President Kennedy in 1963.

He also remained actively involved in the family newspaper and loyal to his radio roots as chairman of *Twenty Questions* and presenter of *Down Your Way*.

Ill for the final five years of his life, his last great commentary was on the state funeral of Winston Churchill (1874–1965). A fellowship of cancer research was established in his name at St Thomas's Hospital in London and two of his sons, David (1938–) and Jonathan (1944–), have successfully followed in his footsteps.

His writings include *The Frontiers Are Green* (1943), *Storm At the Hook* (1948) and *Elizabeth Our Queen* (1953).

DIRECTOR The individual primarily responsible for all the creative aspects entailed in the making of a motion picture or television production.

A director may have participated in the origination or writing of the screenplay and will exert approval over the casting of all the dramatic roles and the selection of technical crew members. The director's visualisation is the basis for the work of the ▷ art director and ▷ costume designer and will guide the contribution of the ▷ cinematographer. The director will rehearse and help mould the actors' performances and it is his/her concept that the ▷ editor must realise in assembling picture and sound.

Television directors have similar responsibilities but often work during shooting from the studio control suite, giving direct instructions to the floor manager and camera operators in the studio.

Directors may also double as their own ▷ producers and many also function in dual or multiple capacities: ▷ Woody Allen is a screenwriter and performer, ▷ Charles Chaplin wrote his own musical scores and Peter Hyams (1943–), director of *The Presidio* (1988) and *Narrow Margin* (1990) often serves as his own ▷ cinematographer.

To crudely differentiate the responsibilities of the producer from that of the director, it has been stated that the producer is the one who raises the money to make a film, whilst a director is the one who spends it.

See also **Auteur** and **Second-unit director**.

DIRTY HARRY

Don Siegel (US 1971)
Clint Eastwood, Harry Guardino, John Vernon, Andy Robinson.
101 mins. col.

The hard-hitting story of a San Francisco cop who uses the most extreme methods to track down a crazed serial killer before 'executing' him, Don Siegel's (1912–91) *Dirty Harry* was vociferously attacked as a fascistic statement endorsing any means necessary if the end was to rid the streets of crime. Certainly in the wake of the film's increasingly routine and meretricious sequels and the disturbing urban vigilante cycle – exemplified by Phil Karlson's (1908–) *Walking Tall* (1973) and Michael Winner's (1935–) *Death Wish* (1974) – that its commercial success undoubtedly spawned, it is easy to make this judgement, but such a position may possibly ignore the work's obvious sense of ambiguity. 'Dirty' Harry Callahan has earned his nickname for his willingness to take on the most dangerous assignments, but the moniker also signifies his refusal to shirk from his own violent nature in carrying out his duties. A plethora of Christian images (churches, crosses, Jesus signs) suggests the viewer question the moral and

Clint Eastwood in *Dirty Harry* (1971)

psychological justification for this avenging angel, while Harry's brutal relish, in symbiosis with the maniac's psychotic carnage, perhaps acts out the violent fantasies suppressed by the rest of society. In this light, Harry's climactic disposal of his police badge is evidence that his particular brand of individualistic endeavour has no place within the proper restraints of today's law enforcement agencies, and as such it's hardly a call to arms for the would-be vigilante. Subsequent sequels, all starring ▷ Eastwood, comprise *Magnum Force* (1973), *The Enforcer* (1976), *Sudden Impact* (1983), and *The Dead Pool* (1988).

DISNEY, Walt

Full Name Walter Elias Disney
Born 5 December 1901, Chicago, Illinois, USA.
Died 15 December 1966, Los Angeles, California, USA.

A student at the Kansas City Art Institute, Disney subsequently worked as a commercial artist in the local advertising business producing, directing and animating a series of *Laugh-O-Gram* cartoons in 1920.

In Hollywood from 1923, he worked on the animated series *Alice in Cartoonland* (1923–7) and *Oswald the Lucky Rabbit* (1927–8). However, his greatest early success came with the release of *Steamboat Willie* (1928). This was the first cartoon to use synchronised sound and the third to feature his creation of Mickey Mouse (originally to be called Mortimer Mouse).

A lively rodent with large ears, a bulbous nose, a loveable nature and a squeaky voice that Disney himself initially provided, Mickey Mouse captivated successive generations and was soon joined by the irascible Donald Duck, slow-witted Goofy, Pluto the dog and a gallery of other characters in the Silly Symphonies series of cartoons.

Flowers and Trees (1932) was the first cartoon in ▷ Technicolor and the first to win an Oscar in the new Best Short Subject Cartoon category. The same year Disney was awarded a special Oscar for his creation of Mickey Mouse and his other Oscar-winning cartoons include *The Three Little Pigs* (1933), *The Tortoise and the Hare* (1934), *Ferdinand the Bull* (1938), *Toot, Whistle, Plunk and Boom* (1953) and many others.

In 1934, work began on ▷ *Snow White and the Seven Dwarfs* (1937); the first feature-length cartoon. Its richly detailed animation, strong narrative and endearing characters made it an instant classic. The Disney studios received a special Oscar for their innovative use of ▷ multiplane camera techniques in the film and Disney received another special Oscar from ▷ Shirley Temple that took the form of one Oscar representing Snow White and seven dwarfs descending in a row.

Never directly responsible for the direction or animation of the films that bear his name, Disney was the guiding figure at his studio – a conservative, all-American patriarch whose work reflects these qualities. After the forward-looking attempt to interpret classical music with animated images in ▷ *Fantasia* (1940), his future work lacked the same edge of ambition. Instead, his aim was to entertain the young at heart with anthropomorphic yarns that stimulated the senses and delighted the eye. He succeeded admirably with *Pinocchio* (1940),

Dumbo (1941), *Bambi* (1942), *Lady and the Tramp* (1955), *One Hundred and One Dalmatians* (1959) and a host of others.

He gradually built the Walt Disney name into a global business empire that included the Disneyland theme park in California (1955) and numerous television series. The studio's output also diverged into a True Life Adventure series that includes *Seal Island* (1948) (Oscar), *Men Against The Arctic* (1955) (Oscar) and *White Wilderness* (1958) (Oscar) and such live-action family adventures and comedies as *Treasure Island* (1950), *20,000 Leagues Under the Sea* (1954) and *The Absent-Minded Professor* (1960).

He had combined animation and live-action in *The Three Caballeros* (1944) but brought new refinements to the technique with the highly popular *Mary Poppins* (1964). The last animated feature that he personally supervised was *The Jungle Book* (1967).

After his death, the studio struggled to maintain its reputation but the earnings from Walt Disney World (1971) in Florida and other interests kept them afloat until a diversification into more adult fare under the Touchstone label brought the box-office bonanza of films like *Splash* (1984), *Three Men and A Baby* (1987), *Good Morning, Vietnam* (1987), *Honey, I Shrunk the Kids* (1989) and *Pretty Woman* (1990). A return to higher standards of classical animation resulted in their collaboration with ▷ Steven Spielberg on *Who Framed Roger Rabbit* (1988) and the recent *The Little Mermaid*, the most commercially successful animated feature of all-time.

DISSOLVE A transition effect in which the whole image of a scene gradually ▷ fades out and disappears as it is replaced by the following scene as it, in turn, fades in. It is also termed a lap dissolve (as the pictures are overlapped), or a mix. Most frequently utilised during the heyday of Hollywood studio production it was effective in conveying the passage of time or to suggest a flashback or dream sequence, although overuse has rendered it a cinematic cliché.

DISTANT VOICES, STILL LIVES

Terence Davies (UK 1988)
Peter Postelthwaite, Freda Dowie, Angela Walsh, Dean Williams.
Cannes Film Festival *International Critics Prize*
84 mins. col.

Terence Davies' (1945–) autobiographical diptych *Distant Voices, Still Lives*, with its admiring reception at the 1988 Cannes Film Festival the prelude to a virtual apothoeosis upon the film's later domestic release, justifiably remains the most lauded British film of the late Eighties. As in his earlier *Trilogy* (1974–83), wherein a terminally-ill cancer patient contemplates the tension between Catholicism and homosexuality which has marked his entire experience, a concentrated cinematic technique elevates the fearless revelation of deeply personal scars towards the dignity of a true work of art. Seamlessly fusing two short films – *Distant Voices* was shot in autumn 1985, *Still Lives* two years later – he chronicles the turbulent fortunes of his own working-class Liverpool family through the Forties and Fifties, the first half concentrating on the domineering violence of his autocratic father, the second documenting the marital discord of the household's next generation. Isolating key moments into a mosaic of short scenes (weddings, funerals, christenings, wartime) Davies' elliptical approach cumulatively coheres into a moving picture of alienated male authority tyrannically suppressing feminine domesticity. Yet despite the film's insistence on the courageous silent suffering of the women, Davies avoids utter bleakness by stressing the period's close-knit sense of community and celebrating the cathartic potency of popular culture – the pub singalongs, movie musicals and songs on the radio teem with words of hope and romance, the distance between lyric and image, art and reality, ironically opposing our capacity for love with its sparing manifestation in family life's ongoing daily grind.

DISTRIBUTOR The company responsible for selling films to ▷ exhibitors, the distributor will rent them a film on the producing company's behalf for a set sum or percentage of the profits. A distributor will also be responsible for the marketing of specific titles under their auspices – arranging publicity campaigns, seeking the cooperation of stars and ▷ directors in promotional interview activity, identifying the potential audience, devising possibilities for merchandising spin-offs and exploring all avenues that will raise the public profile of their product and create as wide a potential audience as possible.

In America, before the Supreme Court Paramount decision of 1948, the major studios ran integrated business empires with responsibility for the streamlined production, distribution and exhibition of their wares. When this was declared illegal, many sold off their chains of cinemas and theatres and now must act as distributors selling to the highest bidder among exhibitors.

Traditionally, a distributor may demand up to 90% of a film's box-office receipts during its first week of release and a decreasing percentage for each successive week of its run. The figures vary enormously according to the desirability of any particular title and whether a distributor is well-placed to demand favourable terms. Taking into account the costs of publicity, advertising and prints, it is reckoned that a film will need to earn 2.5 times its production costs before it can be considered to have broken even. Creative book-keeping has meant that some very popular films, *Coming to America* (1988) for instance, appear never to have made a profit, a source of concern for those creative talents who had worked on the basis of receiving a share of the profits.

One of the biggest British distributors is UIP (United International Pictures) who handle the release of film production from Universal, Paramount and MGM/UA, whilst independent distributor Palace have enjoyed great success with their shrewd purchase of British rights to an eclectic mixture of international productions. Their range of 1990 box-office hits in this country include *When Harry Met Sally* (1989), *Nuns On The Run* (1990), *Family Business* (1989), *Wild at Heart* (1990) and *Cinema Paradiso* (1989).

DIXON OF DOCK GREEN

UK 1955–76
Creator Ted Willis
Jack Warner

In the light of what has followed, the idea behind the first British TV police drama seems quaint. In reviving the character of Constable George Dixon from the 1950 movie *The Blue Lamp* Willis (1918–) hoped to show that the lives of ordinary people could be quite extraordinary. He spent over six weeks at Paddington police station researching the lead character, based on a real policeman from Leman Street in London's East End. Willis decided that though the show would nod to the then-fashionable documentary style, it would not concentrate on violence, preferring marriages to murder (ironically, since Dixon is killed early in the original film, by ▷ Dirk Bogarde).

Warner (1895–1981) was popular as the friendly beat bobby who opened each show with a cosy 'Evenin' all.' No crime was so great that it could not be summed up with a shrug and an assurance that just desserts had been administered. By the 1970s (despite a promotion to Sergeant) Dixon looked tired and tame in comparison with ▷ *Z-Cars* and tougher shows like *The Sweeney* (1974–8). When Warner died in 1981 his coffin was borne by policemen, and the show's theme tune, *An Ordinary Copper* was played over the PA. As well as Warner, the police must have mourned the passing of unquestioned public goodwill.

LA DOLCE VITA (The Sweet Life)

Federico Fellini (Italy/France 1960)
Marcello Mastroianni, Yvonne Furneaux, Anouk Aimée, Anita Ekberg.
Academy Award: *Best Foreign Film*; Cannes Film Festival *Palme D'Or*
180 mins. b/w.

A widespread *succès de scandale* for a supposed sexual candour mild by today's standards, *La Dolce Vita* fully established ▷ Federico Fellini as an internationally acclaimed ▷ auteur, the movie's cachet indeed being such that its title and the name given to one of its characters – the sleazy press photographer Paparazzo – have since passed into the language. A sort of rake's progress following magazine hack ▷ Marcello Mastroianni through the vigorously decadent urban landscape of contemporary Rome, the film charts a week of variously caricatured activity: from the protagonist's infidelities to the arrival of a statue of Christ by helicopter; from a vision of the Madonna that causes a riot to an extended bout of orgiastic revelry that ends with the discovery of a strange sea creature. It has often been said of the film that Fellini pretends to expose the moral collapse of an entire society while all the time gleefully enjoying the extremities of its lifestyle, but the Italian is hardly an ideas man in the first place. The religious imagery and the suicide of the film's token intellectual matter much less than the showmanlike display (rather than shoot on Rome's famous Via Veneto Fellini built his own version in the city's Cinecitta studio) of the director's varied peccadilloes and the tangible emotional effects evoked by his bizarre tableaux. It's arguable that no other major filmmaker has so little to say but such a striking way of saying it – a position apparently addressed by Fellini himself in his subsequent masterly extravaganza *8½* (1963).

DOCTOR WHO
UK 1963–

The BBC's longest-running drama series was born in the era when the Corporation's ideas about educative broadcasting were first under attack. Though conceived as a children's programme, it came from the drama department under new head Sydney Newman (1915–), who had been brought in from populist ITV. Previously the BBC's children's programming had consisted mainly of worthy literary adaptations.

The everyday intergalactic adventures of a mystical Time Lord able to travel through space and time, even the title had inbuilt mystery. 'Doctor Who?' was a pertinent question given the initially sketchy genealogy of the show's central figure. Over the years the myth has grown. The first Doctor, played by William Hartnell (1908–75), was styled after an Edwardian grandad, but as ratings fell – to the point where cancellation was seriously considered – he was reinvented as a cosmic nomad (Patrick Troughton (1920–87)). Jon Pertwee (1919-) took the character into a dandified action phase, and Tom Baker (1936–) was memorable for his trailing scarf and one-liners, though the show was beginning to trade in self-parody. The turn to comedy that came was a response to the success of ▷ *Star Wars*, and persistent criticism of the show's alleged sex and violence (as perceived by campaigners like Mary Whitehouse (1910–)).

In a 1983, twentieth-anniversary special entitled *The Five Doctors*, all five of the actors were involved in an epic saga, with the deceased Hartnell replaced by Richard Hurndall (1910–1984).

Vital to the show's success was its inventive, cost-effective hardware. The Doctor's time machine, the Tardis (Time and Relative Dimensions In Space) is a curiously roomy police box designed with the original intention of being able to change shape to accommodate the environment in which it landed. Budgetary restrictions meant that it has remained a police box.

Notable too are such recurring adversaries as the monsters-on-wheels Daleks (crackly catchphrase 'Exterminate! Exterminate!'), the Cybermen, Yeti, and evil, renegade Time Lord The Master. The musical input of the BBC's Radiophonic workshop was vital to the futuristic feel of the early series.

There was an attempt to return to the original concept with Peter Davison (1951–) (also of the veterinary drama *All Creatures Great and Small* (1977–)) but in playing safe, the show was frustrating its audiences, young and old. So diminished is its impact, that subsequent Doctors Colin Baker (1943–) and Sylvester McCoy (1943–) are less well-known than their predecessors from decades before.

In the 1960s, there were two feature films starring Peter Cushing (1913–) as the Doctor: *Doctor Who and the Daleks* (1965) and *Daleks—Invasion Earth 2150* AD (1966). Recently there has been much speculation surrounding both the possible demise of the television series and a large-budget cinema incarnation with names like Dudley Moore (1935–), Donald Sutherland (1935–) and ▷ John Cleese mooted as possible interpreters of the role.

DOLLY A mobile platform on wheels for a film or video camera and its operational crew, allowing forward and sideways movement of the point of view during the action of the scene. The camera rests on an arm which can be raised or lowered, panned or tilted and the device allows for noiseless movement. A system of tracks is also frequently employed to assist in the smooth flow of the apparatus and the muscled individual pushing the equipment is termed a dolly grip or dolly pusher.

A dolly shot is a camera movement using this scope of mobility to give a different perspective on events or merely focus attention on a character or crucial detail.

A crab dolly allows the further versatility of sideways movement and other variations include the spider dolly which provides improved access to awkward spaces via its adjustable legs.

DONAT, Robert

Born 18 March 1905, Withington, near Manchester, UK.
Died 9 June 1958, London, UK.

Of Polish descent, Donat took elocution lessons to cure him of a boyhood stammer and this led to his involvement with the theatre and, eventually, to his stage debut with a small part in *Julius Caesar* (1921).

He spent several years touring with the Benson Company before joining Liverpool Rep in 1928. His London debut in *Knave and Queen* (1930) was followed by successes in *Saint Joan* (1931) and *Precious Bane* (1932) which brought him a contract with ▷ Alexander Korda.

He made his film debut in *Men of Tomorrow* (1931) and his gracefulness, melodious voice and a gentle disposition which bordered on the ethereal made him one of the most popular stars of 1930s British cinema.

After a supporting role as Culpepper in *The Private Life of Henry VIII* (1932) he made the only trip of his career to Hollywood and cut a dashing blade as *The Count of Monte Cristo* (1934). Back in Britain, his versatility was well illustrated with diverse roles as a romantic Richard Hannay in ▷ Hitchcock's *The Thirty-Nine Steps* (1935), a Scottish hero and his ghostly double in the delightful whimsy *The Ghost Goes West* (1936) and the idealistic young doctor in *The Citadel* (1938) which earned him a Best Actor Oscar nomination.

Against the strongest competition from ▷ Clark Gable's Rhett Butler to ▷ Olivier's Heathcliff and ▷ James Stewart's Mr Smith, he won a richly deserved Best Actor Oscar for *Goodbye Mr. Chips* (1939) conveying, with rare sensitivity, the life, love and career of a gentle schoolmaster.

A chronic asthmatic and insecure performer, his later career was blighted by ill-health and indecision. He appeared in few leading roles, but did contribute a beautifully modulated performance as the defence counsel in *The Winslow Boy* (1948) and also directed the poor north country comedy *The Cure for Love* (1949).

He died shortly after completing a poignant performance as the Chinese mandarin in *Inn of the Sixth Happiness* (1958), uttering one of the cinema's most apt and heart-rending final lines 'We shall not see each other again, I think. Farewell.'

Always sought after in Hollywood and Britain, the many projects that might have been born the imprimatur of his talent had the circumstances of his health and personality been otherwise, include *Captain Blood*, *Beau Geste*, *Oliver Twist*, *A Double Life*, *Hobson's Choice* and, literally, dozens of others.

His stage work included *The Devil's Disciple* (1940), *Heartbreak House* (1943) and a much-lauded *Murder in the Cathedral* (1953) for the Old Vic.

DONSKOI, Mark Semenovich

Born 6 March 1901, Odessa, Russia.
Died 24 March 1981, Moscow, USSR.

Educated in medicine and music, Donskoi studied law at the University of Simferopol, graduating in 1925 after having served with the Red Army and endured a spell in jail which he chronicled in the book *The Prisoners* (1925).

He attended the State Film school, studying under ▷ Eisenstein, and gradually involved himself in the film industry as an editor, scriptwriter, actor (in, for instance, *Prostitutka* (The Prostitute) (1926)) and co-director of the short *Zihzn* (Life) (1927) and the feature *V Bolshom Gorode* (In the Big City) (1928).

He made his solo debut with *Pizhon* (The Fop) (1929) and attracted wider attention with *Pesnya O Shchastye* (Song of Happiness) (1934). However, he gained a lasting international reputation when his friendship with Maxim Gorky

(1868–1936) resulted in a vivid trilogy culled from his autobiographical works *Detstvo Gorkovo* (The Childhood of Maxim Gorky) (1938), *Vlyudyakh* (My Apprenticeship) (1939) and *Moi Universiteti* (My Universities) (1940). A loving and deeply felt recreation of the writer's youth and growing politicisation, the films sensitively evoke the people and landscapes of 19th century rural Russia and patriotically convey the struggle of ordinary Russians against Tsarist oppression with warmth and humanity.

Kak Zakalyalas Stal (How the Steel Was Tempered) (1942) used a story of 1918 Ukranian resistance to German aggression for its obvious parallels with World War Two, and he remained a director at the disposal of the state, saluting Russian courage and fortitude in films like *Raduga* (The Rainbow) (1944) and *Nepokorenniye* (Unvanquished) (1945).

Later work generally failed to emulate the poetic social realism of the Gorky trilogy and he incurred official displeasure with his laudatory view of Lenin (1870–1924) in *Alitet Ukhodit V Gory* (Alitet Leaves for the Hills) (1948) which was banned. He subsequently returned to the inspiration of Gorky's writing for *Mat* (Mother) (1956) and *Foma Gordeyev* (1959) and went some way to creating the equal of the trilogy with the diptych *Serdtse Materi* (A Mother's Heart) (1966) and *Vernost Materi* (A Mother's Devotion) (1967) on the early life of Lenin and the influence of his mother.

Rewarded with many prizes, including the Stalin Prize (1941, 1946, 1948) and the Order of Lenin (1944 & 1971) his last films were a portrait of the singer *Chaliapin* (1972) and the little-seen *Nadezhda* (1973).

DOUBLE EXPOSURE A widely used film technique, the process involves the intentional combination of two or more images separately exposed on a single photographic record either in the camera or by subsequent printing.

The images may be superimposed so that one is seen through the other, or appear without any overlapping by the use of masks or ▷ mattes to reserve specific areas.

The uses are many: in ▷ dissolves to convey the passage of time; to create a ghostly spectre in supernatural tales; or when two characters played by the same actor have to meet within one scene as in *Dark Streets* (1929), *The Prisoner of Zenda* (1937) or *A Stolen Life* (1946).

DOUBLE-HEADED PRINT A print in which the soundtrack and pictures are on separate reels of film. This can occur during the late editing stages of a production before approval has been given for a final, combined print to be struck. Equipment able to provide double-head projection allows the synchronisation of the separate reels to screen the film as a work in progress or to check on specific sequences.

DOUBLE INDEMNITY
Billy Wilder (USA 1944)
Fred MacMurray, Barbara Stanwyck, Edward G. Robinson, Porter Hall.
107 mins. b/w.

Adapted from James M. Cain's novel by fellow crime fiction doyen Raymond Chandler (1888–1959) and notably acerbic director ▷ Billy Wilder, it's hardly surprising that *Double Indemnity* epitomises the narrative concerns of the ▷ film noir. Although their working relationship was by all accounts a strained one, Chandler and Wilder did succeed in bringing from page to screen the familiar fatalistic package of stinging-dialogue, cold-blooded slaying and sexual duplicity that marks out the genre. Cast against his usual jocular type, Fred MacMurray (1907–91) excels as the insurance man ensnared by ▷ Barbara Stanwyck's paradigmatic femme fatale. Together they hatch a plot to sell her husband a hefty new insurance policy, murder him and collect on the return, but they fail to account for the close attentions of company claims investigator ▷ Edward G. Robinson. The confident male protagonist who soon becomes the confused victim of the predatory female – attracted by the twin goals of financial gain and sensual reward, MacMurray believes himself the prime mover but the opposite is, of course, the case – is a recurrent notion in film noir plotting, which frequently investigates the disruption of patriarchal structures (here the insurance company's network of trust, often a male business partnership) by the sexual manipulation of the dangerous but desirable woman. Told in flashback as his life is ebbing away, MacMurray's narration, measuring what he knows as the story unfolds against his perception as he now retells it to the audience, underlines that the die is cast. He never had a chance.

DOUGLAS, Bill

Born 17 April 1937, Newcraighall, near Edinburgh, UK
Died 18 June 1991, Barnstaple, Devon.

The illegitimate son of a Midlothian miner, Douglas's childhood of almost unbelievable misery and penury would later form the basis of his famed autobiographical trilogy. Frequent visits to the cinema provided escapist relief from the unrelenting gloom of his existence and he told one journalist 'I would rather my life had been an M-G-M musical'.

After his National Service, he pursued a career as an actor and writer for the theatre, working as an assistant to Joan Littlewood (1914–) and ultimately attending film school before making his directorial debut with *My Childhood* (1972).

Together with *My Ain Folk* (1973) and *My Way Home* (1977), it formed an intense, dramatic interpretation of his childhood and adolescence, following a boy from life with his grandmother in a mining village to National Service in the Middle East.

Austere and rigorously unsentimental, their spartan style of lingering monochrome images, sparse dialogue and naturalistic performance brought a bleak poetry to bear on the moving story of the human spirit confronted by a harsh and loveless world.

A painstaking director to whom compromise was an alien concept, he spent a number of years at work on *Comrades* (1986), an epic account of the Tolpuddle Martyrs that simultaneously celebrates the struggle of working-men and conveys a pre-history of the cinema with its magical delight in the lantern shows, camera obscura and diorama of its narrator – an itinerant lanternist.

He spent the years since then preparing a screen version of the macabre 19th century novel *Private Memoirs and Confessions Of A Justified Sinner* by James Hogg (1770–1835). In 1990, he served as a Carnegie Visiting Fellow at Strathclyde University and worked with pop group Fine Young Cannibals on the video for the AIDS fund-raiser *Red, Hot and Blue*.

DOUGLAS, Michael Kirk

Born 25 September 1944, New Brunswick, New Jersey, USA.

The eldest son of veteran star Kirk Douglas (1916–), Douglas has had to work hard to win respect and recognition in his own right.

His first experience of the film industry came as a second assistant director on his father's film *Cast A Giant Shadow* (1966). A drama major at the University of California in Santa Barbara, and was named the school's best actor for *Candida* (1967) and best director for his staging of the one-act social satire *Muzeeka* (1968).

After graduating in 1968, he moved to New York, working in the theatre and on television in *The Experiment* (1969). He made his film debut as a hippie pacifist in *Hail, Hero!* (1969) but subsequent roles in films like *Adam at 6 A.M.* (1970) and Walt Disney's *Napoleon and Samantha* (1972) failed to establish him as a star.

A frequent performer on television, he was cast as assistant to Karl Malden (1913–) in the top-rated police series *The Streets of San Francisco* (1972–5). Interested in producing films, he left the series to co-produce ▷ *One Flew Over the Cuckoo's Nest* (1975) which won five Oscars including one for Best Picture that he shared.

Alternating between acting and producing, he was often cast as heroic or moral protagonist in films like *Coma* (1978) and *The Star Chamber* (1983). His interest in liberal causes and political issues manifested itself most obviously in *The China Syndrome* (1979) a prescient thriller about the attempted cover-up of an accident at a nuclear power plant.

His dual careers were successfully combined on *Romancing the Stone* (1984) a breathtaking high adventure in which his own performance as rugged adventurer Jack Colton was much admired. An inferior sequel *Jewel of the Nile* (1985) proved almost as popular.

Now a bankable star, his likable authority garnered sympathy for the philandering husband in the enormously successful *Fatal Attraction* (1987) and he won a Best Actor Oscar as the embodiment of 1980s materialism and corporate greed in *Wall Street* (1987).

Clearly intent on underlining his versatility, he then portrayed a tough, racist cop in *Black Rain* (1989) and brought a brittle edge to a black comedy of martial discord, *The War of the Roses* (1989) directed by his friend of twenty years Danny De Vito (1944–).

He is still an active producer, most notably on *Starman* (1984) and *Flatliners* (1990).

DRAGNET

USA 1951–8, 1967–70
Jack Webb

'Ladies and gentlemen, the story you are about to see is true. Only the names have been changed to protect the innocent.' Thus began every stylish episode of producer/director/actor Jack Webb's (1920–82) landmark police series, adapted from his earlier radio thriller of the same name.

Webb was Joe Friday, a police sergeant of few words. 'This is the city,' he would say, 'Los Angeles, California. I work here. I carry a badge.' The show, the first American drama to be shown on British TV (starting on ITV in 1955) was filmed in a similarly sparse semi-documentary style, with frequent use of close-ups. Webb based the stories on cases handled by the LA police – as research, he rode in their cars for two years. He prided himself on his knowledge of procedures, and the minutiae of investigations ('just the facts, ma'am,' was another catchphrase).

Dragnet was tougher than its British near-contemporary ▷ *Dixon Of Dock Green* in its depiction of crime as an incurable urban reality, but the scriptwriters were only allowed one gunshot for every four episodes. Over a 300 episode run, Friday was promoted to lieutenant. When he returned in colour for a second run he was a sergeant again, and was partnered by Officer Bill Gannon played by Harry Morgan (1915–) (later *M*A*S*H*'s Colonel Sherman Potter).

DREYER, Carl Theodor

Born 3 February 1889, Copenhagen, Denmark.
Died 20 March 1968, Copenhagen, Denmark.

Raised in a strict Lutheran family, Dreyer's early career as a reporter and feature writer brought him into contact with the developing film industry and he began writing scripts with *Bryggerens Datter* (The Brewer's Daughter) (1912).

He made his debut as a director with *Praesidenten* (The President) (1919), and followed this with the comedy *Prastankan* (The Parson's Widow) (1920).

Blade Of Satans Bog (Leaves from Satan's Book) (1920) reflected the influence of ▷ D. W. Griffith's ▷ *Intolerance* (1916) as it recounts four stories of Satan's nefarious activities through the ages and shows, in embryonic form, the director's enthusiasm for technical innovation in editing and composition and his attraction to spiritual themes and the pernicious influence of evil on the human soul.

Du Skal Aere Din Hustru (Master of the House) (1925) brought international attention for his sophisticated approach to affairs of the heart and he then created ▷ *La Passion de Jeanne D'Arc* (The Passion of Joan Of Arc) (1928), an intense focus on the trial and execution of Joan of Arc (1412–31) that conveyed her torture and martyrdom through an extensive use of lingering close-ups that scrutinised the exquisitely agonised features of actress Falconetti (1901–46) in her only screen role.

He first worked in sound with *Vampyr* (1932), a luminously photographed and chillingly atmospheric excursion into the macabre that stands as one of the benchmarks in the cinematic supernatural.

An exacting perfectionist who elevated cinema as a means of artistic self-expression, his work rarely found commercial favour and he returned to journalism in the 1930s, subsequently concentrating on documentaries like *Vandet Pa Landet* (Water from the Land) and *Kampen Mod Kraeften* (The Struggle Against Cancer) (1948).

His few remaining fictional works continued to refine his explorations of spirituality and deliverance through death and include *Vredens Dag* (Day of Wrath) (1943), an austere study of a 17th century pastor and his wife caught in the repercussions of a witch's curse; *Ordet* (The Word) (1955), a powerful tale of religious divisions and resurrection in a remote farming community; and *Gertrud* (1964) which conveys in unrelentingly hypnotic fashion a woman's realisations that her ideal lover may not exist.

Until his death he continued to work on a long-cherished project to film the life of Christ.

DUNAWAY, Dorothy Faye
Born 14 January 1941, Bascom, Florida, USA.

The daughter of an army career man, Dunaway's childhood was spent on a variety of Army camps in America and Germany. A student at the University of Florida, she also attended the Boston University School of Fine and Applied Arts, making her Broadway debut in *A Man for All Seasons* (1962) and appearing with the Lincoln Center Repertory Company in such productions as *After the Fall* (1964), *But For Whom Charlie* (1964) and *Tartuffe* (1965).

An off-Broadway success in the play *Hogan's Goat* (1965), led to her television debut in an episode of *The Trials of O'Brien* (1966) and she was signed to a personal contract with ▷ Otto Preminger, making her film debut in *The Happening* (1966) and appearing in his version of *Hurry Sundown* (1967).

However, she quickly achieved star status, winning a Best Actress Oscar nomination for her electrifying performance in the trendsetting ▷ *Bonnie and Clyde* (1967), superbly capturing the hurt, longing and naivety of a small town girl with rapacious and tragic aspirations.

Working at a feverish pitch she maintained a fair average of hits with co-starring appearances in *The Thomas Crown Affair* (1968), *The Arrangement* (1969) and *Little Big Man* (1970) and gave one of her more underrated performances as the desperately unhappy fashion model in *Puzzle of a Downfall Child* (1971).

Her star appeared to be fading, but she enjoyed popular successes in the all-star casts of *The Three Musketeers* (1973), *The Four Musketeers* (1974) and *The Towering Inferno* (1974) and then impressed with a run of excellent characterisations, subtly conveying the neuroses of the enigmatic victim at the centre of *Chinatown* (1974) (Oscar nomination), bringing flesh-and-blood conviction to the cliched role of the romantically entangled innocent bystander in *Three Days of the Condor* (1975) and winning a well-deserved Oscar for the mixture of aggression and vulnerability she conveyed as the monstrously ambitious television executive in *Network* (1976).

Absent from the screen for two years, she found little work of any substance until the controversial and much misunderstood *Mommie Dearest* (1981). Achieving an uncanny physical resemblance to ▷ Joan Crawford, she also transcended the sensationalistic script with a sympathetic and perceptive tour-de-force as a deeply troubled woman trapped by her past and the demands of stardom.

Outraged reaction to the film and a lengthy domicile in Britain undoubtedly harmed her career and appearances in the likes of *The Wicked Lady* (1983) and *Supergirl* (1984) merely traded on the theatricality inherent in her sometimes mannered style.

Returning to Hollywood, she received critical plaudits for her work as the fragile alcoholic in *Barfly* (1987) and has remained intensely active on a number of, generally unworthy, international ventures.

She has returned to the stage in such productions as *Old Times* (1972), *A Streetcar Named Desire* (1973) and *Circe and Bravo* (1986) and television has offered both employment and some attractive roles in the likes of *The Disappearance of Aimee* (1976), *Ellis Island* (1984) and *The Cold Sassy Tree* (1989).

DUPE An abbreviation for the term duplicate negative, this is a copy of a film prepared from the original camera negative. Made for the purpose of peace of mind and protection of the original it can also be a first step in the creation of many release prints, and the process allows for the duping of individual scenes to correct errors or incorporate visual effects not originally photographed.

DURBIN, Deanna
Real Name Edna Mae Durbin
Born 4 December 1921, Winnipeg, Manitoba, Canada.

Moving with her family to California,

Durbin's singing attracted the attention of Hollywood talent scouts and she appeared with ▷ Judy Garland in the short film *Every Sunday* (1936).

Signed to a contract with Universal and already popular on the radio, she became an immediate star with the release of *Three Smart Girls* (1936). A high-spirited youngster with a lifting soprano and a charmingly fresh personality, she captivated audiences with a succession of lighthearted folderols that includes *One Hundred Men and A Girl* (1937), *Mad About Music* (1938) and *Three Smart Girls Grow Up* (1939). A purveyor of welcome good cheer through the Depression and war years, she received a Special Oscar in February 1939 with Mickey Rooney (1920–) for 'their significant contribution in bringing to the screen the spirit and personification of youth, and as juvenile players setting a high standard of ability and achievement'.

She weathered the transition to adulthood with her flair for comedy evident in *It Started With Eve* (1941), and she gave creditable dramatic performances in *Christmas Holiday* (1944) and the enjoyable *Lady On A Train* (1945). However, mismanagement and her own lack of ambition resulted in the swift termination of her career after such lacklustre vehicles as *Up In Central Park* (1948) and her last *For The Love Of Mary* (1949).

Although sought-after for many roles, including the original production of *My Fair Lady*, she has enjoyed a long and contented retirement in Paris where she has been resident with her third husband for the past forty years.

Frequent revivals of her films on British television have been in response to viewer's requests that confirm her enduring popularity as a joyful symbol of the cinema's less worldly past.

E.T. – The Extraterrestrial

Steven Spielberg (USA 1982)
Henry Thomas, Dee Wallace, Peter Coyote, Robert MacNaughton, Drew Barrymore.
115 mins. col.

▷ Steven Spielberg's *E.T.* seems to transcend categorisation as mere movie, but stands instead as a fully-fledged cultural phenomenon. It is estimated that worldwide more than 240 million people have seen the film, earning it in excess of $700 million in box office revenues, while the associated merchandising jamboree turned it into one of the first multi-media movie 'events' now ubiquitously assailing the consumer with numbing regularity (cf. Tim Burton's (1958–) *Batman* (1989)). The highly lucrative source of all this public and critical adoration – though the film was determinedly ignored by the American Academy – is the sentimental tale of a lonely suburban youngster and his one true friend, an alien being left behind by a recent mission to Earth, whom he helps escape from the clutches of government scientists. Redolent of the boy-and-his-dog school of weepie and described by US trade magazine Variety as 'the best Disney movie Disney never made', the protagonist's deep and abiding companionship with his sexless, ageless extraterrestrial pal offers a vision of profound and reciprocated affection in a world where the institutions of home (much play is made of the absent father) and organised religion no longer offer such much-needed succour. The film's overwhelming emotional appeal lies in the opportunity it affords the audience to participate in this genuinely childlike love, an experience heightened by the drama of the creature's 'death' and resurrection. The alien Other of paranoid Fifties sci-fi has thus become bestest buddy and Christ figure rolled into one, the feel-good Spielbergian theology of space as heaven – evinced by *Close Encounters of the Third Kind*'s (1977) pseudo-spiritual uplift – sending us out of the cinema reassured by its notion of the great nuclear family in the sky.

EASTENDERS

UK 1985–

The BBC, saddled with its mission to inform, had never been happy with the notion of ▷ soap operas, trying half-heartedly with shows like *Compact* and *The Newcomers*. But with the introduction of Channel 4 leaving the two public service channels just 40% of the TV audience, urgent action was required. In February 1985, actor Leslie Grantham, soon to be known to all as

Dennis 'Dirty Den' Watts, kicked down a door in Walford E20 (a fictional borough on a purpose-built set at Elstree Studios), and *EastEnders* was born. Seventeen million viewers tuned in.

The show was devised by producer Julia Smith and script editor Tony Holland, whose team track record included *Z-Cars* and *Angels*. The setting was a market square, with the Queen Vic pub at its centre, run by battling rough diamonds Den and Angie Watts (Anita Dobson (1949–)). The other core families were the Beales and the Fowlers, loosely based on Holland's relatives.

Despite initial interest and press controversy (Grantham was revealed as a convicted murderer) ratings slipped to a low of 8.2 million in May. The show stayed afloat thanks to the innovation of BBC chief Michael Grade (1943–), whose idea of a Sunday omnibus kept viewing figures respectable even in decline. Eventually the serial took root, and on Christmas day 1986, when two episodes were shown, 30.15 million viewers tuned in to see Den and Angie go their separate ways. It was the highest viewing figure ever recorded by BARB, and the beginning of the end for the show's central story-line.

Since then, *EastEnders* has struggled. Flirtations with gangsters and an absurd prison subplot subsided with the departure of Den, whose charisma was not wholly replaced by the cock sparrer charm of Frank Butcher (comedian Mike Read). Plots have never shirked difficult subjects. The blighted borough of Walford has seen rape, murder, mental illness, AIDS, death from drug overdoses, prostitution, homelessness and a rash of single parenting – often in the same episode. Frequently the grimness is unrelenting, and *EastEnders* seems reluctant to entertain. Ironically, its main contribution may have been the rejuvenation of its rival, ▷ *Coronation Street*.

EASTWOOD, Clint

Full Name Clinton Eastwood Junior
Born 31 May 1930, San Francisco, California, USA.

An athletic youngster who swam competitively and played high school basketball, Eastwood worked as a piano player, lumberjack, forest firefighter and steelworker before being drafted into the army. On his discharge, he enrolled in business administration at Los Angeles City College but left when offered a standard contract by Universal.

He made his film debut in *Revenge of the Creature* (1955) and can be seen in several unprepossessing, minor roles in films like *Francis in the Navy* (1955) and *Tarantula!* (1955).

Clint Eastwood in *The Good, the Bad and the Ugly* (1966)

Eventually the studio dispensed with his services and he had almost decided to leave the acting profession, when he was cast as cowboy Rowdy Yates in the long-running television western *Rawhide* (1959–66).

During one summer vacation from the series he travelled to Spain to make the ▷ spaghetti western ▷ *Per Un Pugno Di Dollar* (A Fistful of Dollars) (1964) which belatedly catapulted him to international stardom.

A tall, laconic figure, whose Mount Rushmore-like features squinted expressively into the sunlight, he was ideal casting for the taciturn, cheroot-chomping gunslinger known as The Man With No Name and returned for the further spaghetti westerns *Per Qualche Dollari In Pui* (For A Few Dollars More) (1965) and *Il Buono, Il Brutto, Il Cattivo* (The Good, the Bad and the Ugly) (1966).

His newfound cinematic renown spread to America with the 1967 release there of all three films and he began to make regular appearances in the list of top ten box-office draws thanks to his roles in such popular action fare as *Hang 'em High* (1967) and *Where Eagles Dare* (1969).

Coogan's Bluff (1968) in which he played an Arizona lawman adjusting to the urban jungle of New York, began an association with director Don Siegel (1912–91) that flourished in the Civil War grand guignol tale *The Beguiled* (1970) and, most famously, ▷ *Dirty Harry* (1971) in which he starred as a ruthless policeman whose actions are only slightly less reprehensible than the criminals he seeks to bring to justice.

He made his directorial debut with the ▷ Hitchcock-style thriller *Play Misty For Me* (1971) and has taken firm control of his career, alternating commercial projects with more personal statements or challenging acting assignments.

His most popular films include the broad asinine comedies *Every Which Way But Loose* (1978), *Any Which Way You Can* (1980) and further Dirty Harry thrillers like *Magnum Force* (1973) and *Sudden Impact* (1983), although he has shown a willingness to explore the darker side of his screen image in films like the psychological thriller *Tightrope* (1984).

As a director, he has placed his signature on such interesting works as the Gothic Civil War vengeance saga *The Outlaw Josey Wales* (1976), the light-hearted slice of Americana *Bronco Billy* (1980) and the stylish western *Pale Rider* (1985).

Recently, his seemingly indestructible box-office lustre appears to be crumbling with disappointing results netted by *The Dead Pool* (1988), *Pink Cadillac* (1989) and *The Rookie* (1990) but his artistic credibility has soared as the director of *Bird* (1988), an accomplished biography of jazz musician Charlie Parker (1922–55), and *White Hunter, Black Heart* (1990), a study in obsession which allowed him to display a more garrulous acting range in an affectionate portrayal of a ▷ John Huston-style filmmaker.

From 1986 to 1988, he served as mayor of the California community Carmel where he resides.

EASY RIDER

Dennis Hopper (USA 1969)
Peter Fonda, Dennis Hopper, Jack Nicholson, Phil Spector.
94 mins. col.
Cannes Film Festival: *Best First Film.*

Although ▷ Dennis Hopper's *Easy Rider* was not the first film to examine the mood and aspirations of late Sixties America's burgeoning youth counterculture – the ever-enterprising American International Pictures had already turned out exploitation items like ▷ Roger Corman's *The Trip* (1967) – it was perhaps the first to do so from within. Made on a tiny budget by a crew shooting fast and travelling light like the film's protagonists, *Easy Rider* was a triumph for independent production at a time when the studios were facing ruin, the prototype road movie's unexpected commercial success highlighting just how out of touch the majors were from the day's predominantly young moviegoing audiences. The odyssey of biker-cum-outlaws (actor/producer Peter Fonda (1939–) and actor/director Hopper) journeying on low-slung Harley-Davison choppers across country to Mardi Gras in New Orleans penetrates the heart of middle America, where the degree to which their whole ethos of drugs, rock music and anti-establishment attitudes is disaffected from the mainstream becomes radically apparent; such is the paranoia of this deeply conservative society its bigots can only respond with bullets. However, if Hopper and Fonda's lyrical cruise through the big country represents the only true freedom left, the film's various visions of social alternatives (an unconvincing hippy commune, a hard-up rancher, a bad acid trip) prove unanimously limited. Sincere, portentous, rather over-directed, *Easy Rider* has dated badly but remains both eulogy and elegy for a crucial moment in cultural history, and gateway to the new Hollywood of the Seventies.

EDISON, Thomas Alva

Born 11 February 1847, Milan, Ohio, USA.
Died 18 October 1931, West Orange, New Jersey, USA.

Regarded as the most prolific inventor

Dennis Hopper and Peter Fonda in *Easy Rider* (1969)

the world has ever known, or at least the most commercially astute, Edison was expelled from school for being retarded, became a railroad newsboy on the Grand Trunk Railway, and soon printed and published his own newspaper on the train, *The Grand Trunk Herald.*

During the Civil War he worked as a telegraph operator in various cities, and invented an electric vote-recording machine. In 1871, he invented the paper ticker-tape automatic repeater for stock exchange prices, which he then sold in order to establish an industrial research laboratory at Newark, New Jersey, which moved in 1876 to Menlo, New Jersey, and finally to West Orange, New Jersey in 1887. He was now able to give full scope to his astonishing inventive genius.

He took out more than 1000 patents in all, including the gramophone (1877), the incandescent light bulb (1879) and the electric valve (1883).

In 1888, his company was responsible for the kinetograph, a motion picture camera that created short strips of film. The following year, using a durable film stock created by George Eastman (1854–1932), a kinetoscope was devised to permit the viewing of what had been filmed on the kinetograph. The machine was privately demonstrated in 1891 and received its public bow in 1893 when short, unedited sequences, including *Sneeze* (1891) were shown. The commercial potential of the device was soon exploited in peep-shows and ▷ nickelodeons throughout the land.

The short film strips were made at the Black Maria, commonly regarded as the world's first film studio, and further developments included the Kinetophone which combined the peephole machine and the phonograph to provide non-synchronous musical accompaniment and sound effects for the films. He continued to experiment in matching images with sound until 1916.

Purchasing the rights to the ▷ Vitascope, he gave a public projection of motion pictures on 23 April 1896 in New York and is recognised as the first major pioneer of the American industry responsible for initiating such experimental notions as the narrative film ▷ *The Great Train Robbery* (1903).

Between 1909 and 1917, he joined with other companies in establishing the monopolistic Motion Pictures Patents Company which was designed to control all aspects of film manufacture and distribution. Its influence was eventually broken by independent opposition and the Edison Company ceased its filmmaking activities and sold the studio in 1918.

EDITOR The individual who works with the director to achieve the final assembly of a film, he is responsible for integrating and modulating sound, selecting and arranging, in sequence, the individual filmed scenes in accordance with the director's interpretation of the script. The potential creative impact of editing on a final film is enormous given its effect on pace, timing, atmosphere, performance and the creation of an overall ambience that can flow smoothly or jar depending on the skill and sympathy of the editor.

▷ Hitchcock once illustrated the fundamental importance of the cutting together of scenes with a simple but effective example. A filmmaker may have two shots of actor ▷ James Stewart. In one he is fairly impassive. In the other his face lights up with evident joy. Cut those together with a shot of a gurgling baby in between and the impression is created of Stewart as a kindly old gentleman. Cut those together with a shot of a nubile woman undressing and Stewart is a dirty old man.

There are many examples of editors who went on to become directors; ▷ David Lean edited *Pygmalion* (1938) among others, Robert Wise (1914–) edited ▷ *Citizen Kane* (1941) and ▷ *The Magnificent Ambersons* (1942), and Hal Ashby (1932–88) won an Oscar for his editing of *In The Heat of the Night* (1967).

One of the industry's most enduring editors is Margaret Booth (1902–) whose career from 1921 included the position of supervising film editor at M-G-M from 1939 to 1968. Since then she has edited such films as *Fat City* (1972) and *The Way We Were* (1973) and received an honorary Oscar in March 1978 for her 'exceptional contribution to the art of film editing'.

Significant collaborations between ▷ directors and editors are legion and more recent ones of note include that between ▷ Martin Scorsese and Thelma Schoonmaker (1945–) that resulted in her Oscar win for ▷ *Raging Bull* (1980); ▷ Mike Nichols and Sam O'Steen (1923–) on ▷ *The Graduate* (1967), *Carnal Knowledge* (1971), *Silkwood* (1983) (Oscar nomination) etc; ▷ Arthur Penn and Dede Allen (1924–) on ▷ *Bonnie and Clyde* (1967), *Little Big Man* (1970) and many others.

EISENSTEIN, Sergei Mikhailovich

Born 23 January 1898, Riga, Latvia.
Died 11 February 1948, Moscow, USSR.

The son of an architectural engineer, Eisenstein served with the Red Army before working in the theatre as a scenic artist and innovative director.

He made his first use of film as an addition to one of his stage productions and made his feature-length directorial debut with *Stachka* (Strike) (1924), a dynamic recreation of a conflict between workers and the state in 1912 that made use of ▷ montage to emphasise such dramatic points as the senseless slaughter of the strikers by state cavalrymen.

Drawn to the celebration of heroic collective struggles, he followed this with ▷ *Bronenosets Potemkin* (Battleship Potemkin) (1925), a further excursion into polemical historical drama, set during the sailors mutiny of 1905 and featuring the famous 'Odessa Steps' sequence.

Oktyabr (October) (1928) and *Generalnaya Linya* (The General Line) (1929) were further fusions of art and propaganda that put his advanced use of editing and montage techniques at the disposal of stories saluting the events of 1917 and the establishment of a collective farm.

With *Que Viva Mexico* (1932) unfinished, *Bezhin Meadow* abandoned in 1937, and the director felled by a nervous breakdown, it would be a decade before his next film ▷ *Alexander Nevsky* (1938), a more conventional narrative epic closely allying music by Prokofiev (1891–1953) with the visual images, and famed for the Battle of the Ice sequence.

His final work, ▷ *Ivan Groznyi* (Ivan The Terrible) (1942–46) was planned as a monumental triptych on the epic struggles of 17th century Tsar Ivan IV. In the end, only two parts were completed, with the rapid editing and dynamism of the director's earliest work superseded by a more measured pace and ornate visual style. Arousing the ire of Stalin (1879–1953) with its unflattering portrait of the secret police, Part II was suppressed by the authorities until 1958 and the director died of a heart attack with Part III barely begun.

A great craftsman and impassioned propagandist whose influence has extended over all future generations, his films tend to betray the hand of an academic and theoretician rather than the beat of a natural storyteller's heart.

ELLIOTT, Denholm Mitchell
Born 31 May 1922, London, UK.

A student at RADA, Elliott served in the RAF during the Second World War and spent a number of years as a p.o.w. in Germany (1942–45). After his return to Britain he made his professional stage debut in *The Drunkard* (1945) at Amersham and his London debut in *The Guinea Pig* (1946).

A prolific stage career followed with highlights including *Venus Observed* (1950), *Ring Round the Moon* (1950), *Camino Real* (1957), *Write Me A Murder* (1961), *Come As You Are* (1970) and *Hedda Gabler* (1972).

He made his film debut in *Dear Mr. Prohack* (1949) and saw service as breezy juveniles, gallant officers and gentlemen in such wartime heroics as *The Cruel Sea* (1953) and *They Who Dare* (1954).

A thoroughly reliable asset to any cast, he came into his own in middle-age as an incisive character actor illuminating a gallery of rogues, shabby ne'er-do-wells and outright bounders. Notable roles include the uppercrust layabout in *Nothing But the Best* (1964), the seedy abortionist in *Alfie* (1966) and the burnt-out bookkeeper in *Saint Jack* (1979).

A versatile talent at ease with high farce, intense drama, leading roles and scene-stealing secondary parts, his more recent international career has included a sardonic butler in *Trading Places* (1983), the snobbish Dr Swaby in *A Private Function* (1984), a dissolute journalist in *Defence of the Realm* (1985), the uncouth Mr Emerson in *Room With A View* (1985), a kindly doctor in *September* (1987) and a recurring role as Brody in *Raiders of the Lost Ark* (1981) and *Indiana Jones and the Last Crusade* (1989).

A familiar face on British television, his work ranges from such ▷ Dennis Potter plays as *Follow the Yellow Brick Road* (1972) and *Blade on the Feather* (1980) to the series *Bleak House* (1985) and the American television film *Mrs. Delafield Wants to Marry* (1986).

EMMY Presented annually, the Emmy is the television world's equivalent of the Oscar. Established by the American Academy of Television Arts and Sciences it recognises 'The advancement of the arts and sciences of television and achievements broadcast nationally in primetime'. Awards are voted upon by Academy members in a bewildering number of catgories from comedy, drama, and variety series, to technical achievement, best actress in a lead role in a mini-series etc. Since 1983, the Academy has also inducted individuals into a Television Hall of Fame in recognition of their outstanding contributions to 'the electronic medium'. The 1984 recipient, for instance, was ▷ Lucille Ball.

Emmy award (Dick Van Dyke and Mary Tyler Moore)

LES ENFANTS DU PARADIS
(The Children of Paradise)
Marcel Carné (France 1944)
Jean Louis Barrault, Arletty, Pierre Brasseur, Marcel Herrand.
Originally two parts, running time 195 mins; later combined and cut to 188 mins. b/w.

It's not too fatuous to suggest ▷ Marcel Carné's *Les Enfants Du Paradis* as a sort of French equivalent to ▷ *Gone With The Wind*; the culmination of pre-war studio production expertise at its most lavish, most perfectly finished, most appealing to the emotions. The troubled gestation of ▷ Selznick's sweeping vision has its match too in the absurd folly of Pathé mounting France's most expensive ever project at a time when the country was languishing under Nazi occupation. The romantic pessimism of Carné's celebrated collaborations with poet and screenwriter Jacques Prévert (1900–1977), *Les Quai Des Brumes* (Port of Shadows, 1938) and *Le Jour Se Lève* (Daybreak, 1939), saw them banned under the Germans – forcing the pair towards the less successful medieval allegory of *Les Visiteurs Du Soir* (The Devil's Envoys, 1942). The 18th century setting of *Les Enfants Du Paradis*, a swirling fresco of life and love on the French 'Theatre Street', would enable it to escape the Nazi censor, even though the film was intended as a paean to the enduring vitality of the French spirit.

Arletty and Marcel Herrand in *Les Enfants du Paradis*

Shooting for eighteen months on a huge set off the famous Boulevard Du Crime, Carné stalled his progress in the hope, eventually fulfilled, that the movie's premiere would take place after the liberation. Its sumptuous trappings making the historical tableaux tangibly immediate, Carné's masterpiece charts the ebb and flow of an ingenue actress's contesting affections for three very different men, bringing the most delicate of emotions into crystalline focus. As the determinedly independent Garance, ▷ Arletty's central performance places her alongside ▷ Vivien Leigh's Scarlett O'Hara in the pantheon of the screen's great romantic heroines.

EXHIBITOR From peephole machines in penny arcades to ▷ nickelodeons, the grand cathedrals of the Depression-era and the vast multiplexes of the modern age, the exhibitor has been the cinema proprietor or owner who rents films from a ▷ distributor and screens them to a paying audience.

The rental system will be based on some percentage of the exhibitors' box-office takings (less their expenses) which can be as much as 90% in the first week, decreasing by degrees in subsequent weeks. The figure will depend on the desirability of the film, whether it is offered exclusively and whether it is a first-run.

The exhibitor will share advertising and promotional costs with the distributor and may be the owner of an individual screen or an entire chain. In Britain the Rank Organisation is the owner of numerous Odeon cinemas, the British Film Institute provides support for a network of 33 regional non-commercial film theatres and there are independent exhibitors running single cinemas. The Cinematograph Exhibitors Association (CEA) is the exhibitors' official organisation.

THE EXORCIST
William Friedkin (USA 1973)
Ellen Burstyn, Max Von Sydow, Jason Miller, Linda Blair.
122 mins. col.
Academy Award: *Best Screenplay* William Peter Blatty; *Best Sound* Robert Knudson and Chris Newman

A major current in Seventies cinema was the commercial recognition of the previously marginalised horror and science fiction genres, with both ▷ Steven Spielberg's *Jaws* (1975) and George Lucas's ▷ *Star Wars* (1977) in their day the most successful films ever released. Similarly phenomenal in its ability to draw in a broad spectrum of the mainstream audience, William Friedkin's (1939–) *The Exorcist*, an expensive diabolical thriller with a highly respectable cast, drew on the talk show currency of the 'Is God dead?' debate to wrap its graphically exploitive study of the demonic possession in a 12-year-old girl in an aura of thematic credibility. Thrill-seekers could thus ostensibly ponder society's crisis of faith at the same time as they (guiltily?) revelled in Friedkin's gloating spectacle of profanity, levitation, pea-soup vomit, swivelling heads and – the pièce de resistance – masturbation with a crucifix. Still, the notion of the monster within emphasised the degree of anxiety created by the nuclear family in crisis; the same ruptured domesticity had surfaced earlier in ▷ Roman Polanski's *Rosemary's Baby* and ▷ George A. Romero's *Night of The Living Dead* (both 1968), while Richard Donner's (1930–) *The Omen* (1976) and Larry Cohen's (1947–) *It's Alive* (1973), both with sequels, concentrated firmly on the demonic offspring factor. Necessarily, *The Exorcist*'s stric ken daughter is 'cured' by a surge of faith courtesy of the Catholic church, though the box office imperative meant the treatment was not successful enough to prevent a couple of rather wayward sequels in ▷ John Boorman's *Exorcist II: The Heretic* (1977) and *Exorcist III: Legion* (1990), directed by William Peter Blatty (1928–), the source novelist and writer/producer of the original installment.

EXPRESSIONISM A movement that flourished in all areas of German artistic endeavour from the turn of the century until the establishment of Nazi rule in 1933, expressionism in the cinema began with ▷ *Das Kabinett Des Caligari* (The Cabinet of Dr. Caligari) (1919) directed by Robert Wiene (1881–1938).

A reaction against impressionism in which the artist attempts to faithfully convey the immediate impression a scene makes upon a viewer, expressionism turns from the outer world to the inner life, distorting and exaggerating that scene until it becomes an outward projection, or 'expression', of inner turmoil, sensitivity or feeling.

This was accomplished through use of shadows, distorted settings, unusual perspectives in both shape and composition, unnatural make-up and highly-stylised performances.

Other examples from the period include the gothic sets and chiaroscuro lighting in *Der Golem: Wie Er in Die Welt Kam* (The Golem) (1920), the doom-laden shadows in ▷ *Nosferatu, Eine Symphonie Des Grauens* (Nosferatu) (1921) and the magnitude of the futuristic sets in ▷ *Metropolis* (1926).

The flood of German emigrants to America during the Nazi era meant that the influence of expressionism spread and is most obviously visible in the work of directors like ▷ Josef Von Sternberg, and in both the horror and gangster classics of the early 1930s like *Dracula* (1930), *Frankenstein* (1931) and ▷ *Public Enemy* (1931).

EXTRA A non-speaking actor temporarily engaged for an incidental role in a production, usually as part of a crowd or some background action. Although it is a profession in its own right, the individual extra will receive no credit for his or her work. However, it can lead to more illustrious careers. ▷ David Niven began his Hollywood career as a Central Casting extra in *Mutiny on The Bounty* (1935), ▷ Dirk Bogarde's first film work was as an extra in *Come On George* (1940) and ▷ Robert De Niro made his screen debut as an extra in *Trois Chambres à Manhattan* (1965).

FADE The gradual disappearance of a picture to uniform black is a fade-out; the opposite sensation of an image coming into focus from blackness is a fade-in. Achieved by the opening or closing of a camera aperture, it can also be obtained in an optical printer when the intensity of the exposure light is increased or decreased. Effective in marking the passage of time or a change in locale, it has served as a blunt form of dramatic punctuation.

FAIRBANKS, Douglas (Senior)
Real Name Douglas Elton Ulman
Born 23 May 1883, Denver, Colorado, USA.
Died 12 December 1939, Los Angeles, California, USA.

Treading the boards from the age of twelve, Fairbanks studied briefly at Harvard and worked on Wall Street before returning to the stage and making his Broadway debut in *Her Lord and Master* (1902).

His most significant stage work includes *A Case of Frenzied Finance* (1905), *The Man of the Hour* (1906) and *He Comes Up Smiling* (1913).

Offered a three-year screen contract, he made his film debut in *The Lamb* (1915). Success came swiftly as personable black sheep and ne'er-do-wells discovering their basic American decency in comedies like *His Picture in the Papers* (1916), *The Americano* (1916) and *His Majesty, the American* (1919) and in 1919 with ▷ Charles Chaplin, ▷ D.W. Griffith and his future wife ▷ Mary Pickford, he formed United Artists, a company designed to offer creative freedom and financial reward to the individuals concerned.

Enjoying great success in *The Mark of Zorro* (1920), his image became that of the zestful swashbuckler, displaying his acrobatic, physical prowess, virility and devil-may-care zest in a series of expertly choreographed and increasingly spectacular epics of adventure and derring-do like *The Three Musketeers* (1921), *Robin Hood* (1922), *The Thief of Bagdad* (1924) and *The Black Pirate* (1926).

Silent films spread his renown throughout the civilised world and his heroic endeavours remain one of the most potent images from the cinema world of the 1920s.

However, even he could not stem the aging process and after co-starring with Pickford in a poorly-received version of *The Taming of The Shrew* (1929) his career dwindled away and his last appearances were in the dreary *Mr Robinson Crusoe* (1932) and *The Private Life of Don Juan* (1934).

A posthumous Oscar was given at the March 1940 ceremony, honouring his 'unique and outstanding contribution to the international development of the motion picture'.

His son Douglas Fairbanks Junior (1909–) has pursued a diverse showbusiness career but remains most familiar to cinemagoers as a similarly dashing swashbuckler in such films as *The Prisoner of Zenda* (1937), *Gunga Din* (1939), *The Corsican Brothers* (1941) and *Sinbad The Sailor* (1947).

FANNY AND ALEXANDER (Fanny och Alexander)
Ingmar Bergman (Sweden 1982)
Pernilla Allwin, Bertil Guve, Ewa Frohling, Jan Malmsjo, Erland Josephson, Stina Ekblad, Alan Edwall.
300 mins television version reduced to 189 mins for theatrical release. col.

Self-consciously intended as a farewell to the cinema, ▷ Ingmar Bergman's serio-comic domestic tableau *Fanny and Alexander* is an intriguing and enjoyable index to a momentous career. Set in the provincial Sweden of 1907, the film opens in decorative good humour as the Ekdahl family, who run the local theatre, get together for Christmas, yet their festivities are soon to be tempered with the news of the father's sudden fatal collapse during rehearsals for a production of *Hamlet*. Swiftly remarried to the stern Bishop Vergerus (Malmsjo), the widow Ekdahl (Frohling) takes her two children, Fanny (Allwin) and Alexander (Guve), to live under his disciplinarian rule. The youngsters are rescued by avuncular Jew Isak Jakobi (Josephson), who hides them in his magical antiques shop until the Bishop's residence burns

Bertil Guve in *Fanny and Alexander* (1982)

down – killing the hated cleric in the blaze, but sparing the pair's mother to deliver them a baby sister. To some degree autobiographical, Bergman's film foregrounds the magic lantern apparatus as a metaphor for the cinema itself, while the theatrical setting and references to Shakespeare and Strindberg deliberately underscore the director's characteristic examination of psychological/metaphysical tensions within the dreamplay of filmic representation (most notably in ▷*Persona*, 1966). Here the family unit as the locus of distress (cf *Cries and Whispers*, (Viskningar och rop), (1972) and the lingering spiritual doubts (cf *Winter Light*, (Nattvardsgasterna), (1961) reveal how the obsessions of the man are created in the travails of the young boy.

FANTASIA

(Production Supervision) Ben Sharpsteen (USA 1940)
Narrated by Deems Taylor. Music performed by the Philadelphia Orchestra, conducted by Leopold Stokowski.
124 mins. Col. 1942 cut to 81 mins. 1946 120 mins 'popular version' released. Reissued 1982 with soundtrack re-recorded by Irwin Kostal and orchestra in digital stereo. 1990 print restored and digitally remastered Stokowski score reinstated for 50th anniversary reissue.
Academy Awards: *Special Award for Advancement of Sound in Motion Pictures*; *Special Award* (Leopold Stokowski) *for his Achievement in the Creation of a New Form of Visualised Music*; *Irving G. Thalberg Award* Walt Disney.

In the course of the Thirties the ▷ Walt Disney operation had grown from a small animation production company to a substantial studio complex capable of mounting the hitherto unimagined resources necessary to produce the first cartoon feature, 1937's ▷ *Snow White and the Seven Dwarfs*. The same year however, Disney was also looking for a new vehicle to re-establish the pre-eminence of his favourite Mickey Mouse character – latterly eclipsed by the burgeoning popularity of one Donald Duck Esq – and approached noted conductor Leopold Stokowski (1869–1946) with the notion of featuring America's foremost rodent in an animated setting of French composer Paul Dukas's (1865–1935) orchestrated fairy tale *The Sorcerer's Apprentice*. As work progressed into 1938, Disney began to conceive of a complete concert feature visualising a number of classical favourites in this way and so *Fantasia* was born. While vulgarisation of the music is not entirely avoided, the film's moves away from anthropomorphic pictorialism towards a free abstraction (as in the J.S. Bach (1685–1750) *Toccata and Fugue*) proved stylistically influential, though some sections certainly work better than others – the frolicsome bestiary of Ponchielli's (1834–86) *Dance of the Hours*, for instance, is rather more winning than the greeting card religiosity of Schubert's (1797–1828) *Ave Maria*. Still, Disney's commitment to broadening the scope of the medium is self-evident, and the prohibitively-expensive 'Fantasound' multi-speaker system developed to offer directional sound proved an early precursor of stereo recording, even if the full impact as Stokowski envisaged it has only been fully realised by the expansive digital Dolby of the film's restored 50th anniversary reissue.

FASSBINDER, Rainer Werner

Born 31 May 1946, Bad Wörishofen, West Germany.
Died 10 June 1982, Munich, West Germany.

A student of acting at the Fridl-Leonhard Studio in Munich, Fassbinder worked with fringe theatre companies in the 1960s before founding his own anti-theater group working within a contemporary interpretation of Brechtian principles.

Moving into films, he wrote and directed *Der Stadtsreicher* (The City Tramp) (1965) and *Das Kleine Chaos* (The Little Chaos) (1966) before gaining a wider renown with adaptations of his stage pieces *Liebe is Kalter Als Der Tod* (Love Is Colder Than Death) (1969) and *Katzelmacher* (1969).

Looking to the accessible narrative style of American cinema for inspiration, he was also greatly influenced in terms of content by ▷ Jean-Luc Godard as he attempted to create works that were not about manipulating an audience's emotions, or merely entertaining, but that served a greater social purpose as well.

His first film to gain international recognition was *Warum läuft Herr R. Amok* (Why Does Herr R. Run Amok) (1979) and this was followed by a staggeringly prodigious succession of films, often as many as four a year.

They were usually politically committed criticisms of contemporary Germany contrasting personal failure and frustration with the perceived boon of the country's economic miracle; they illustrated the misuse of power, his observations of widespread social oppression and fears of a latent fascism in the German character.

Among the most notable titles to place him at the forefront of the 'New German Cinema' of the 1970s were *Die Bitteren Tränen Der Petra Von Kant* (The Bitter Tears of Petra Von Kant) (1972), detailing the power struggles within a sado-masochistic lesbian relationship; *Fontane Effi Briest* (Effi Briest) (1974), a damning indictment of the hypocrisies of bourgeois morality; *Angst Essen Seele Auf* (Fear Eats the Soul) (1974), on the plight of immigrant Arab workers in Germany and the myriad bigotries a special relationship encounters; *Faustrecht Der Freiheit* (Fox) (1975), which revealed class conflict through a homosexual relationship and *Die Ehe Der Maria Braun* (The Marriage of Maria Braun) (1978), which details the rise and rise of indomitable, opportunistic wife Hanna Schygulla (1943–) and serves as an allegory for post-War Germany.

Brad Davis in *Querelle* (1982) directed by Rainer Werner Fassbinder

Drawn to the Hollywood work of his countryman ▷ Douglas Sirk, he worked on a series of stylised, colour-drenched melodramas that dealt with recent German history through the eyes of women. Titles include *Lili Marleen* (1980), *Lola* (1981) and *Veronica Voss* (1982).

Staggeringly prolific, he also worked for radio and television and frequently acted in his own films. He had completed a highly mannered adaptation of the homoerotic *Querelle* (1982) shortly before his premature death from a suspected overdose of sleeping pills and cocaine.

FELLINI, Federico

Born 20 January 1920, Rimini, Italy.

A comic-strip artist for such publications as *420* and *Avventuroso*, Fellini became involved in the film industry as a writer of comic material and jokes for films like *Lo Vedi Come Soi . . . Lo Vedi Come Sei?* (1939) and *Non Me Lo Dire!* (1940) before graduating to the role of co-scenarist on some of the key films in the Italian ▷ neo-realist movement like ▷ *Roma, Citta Aperta* (Rome – Open City) (1945) and *Paisa* (1946).

He gradually moved towards direction, collaborating on *Luci Del Varieta* (1950) and made his solo directorial debut with *Lo Sceicco Bianco* (The White Sheik) (1951), an ironic comment on fantasy and reality as a honeymooning bride spends most of her time in the company of a womanising star of romantic picture stories.

An international reputation grew with the sentimental *La Strada* (1954) (Best Foreign Film Oscar) starring his wife Guilette Masina (1921–) as a pitiable waif tragically enamoured of circus strongman Anthony Quinn (1915–), and *Le Notte Di Cabiria* (Nights of Cabiria) (1956) (Oscar) with Masina as a prostitute cheerfully resolute as she survives every vicissitude and humiliation that life can offer.

He enjoyed his greatest combination of critical and commercial success with ▷ *La Dolce Vita* (1960) (Oscar), a strikingly shot, witty illumination of decadence and a society that has lost sight of

Federico Fellini

values and purpose; ▷ Marcello Mastroianni stars as a gossip columnist adrift among *la jeunesse dorée*, trying to find both himself and the vestiges of a sane and orderly world.

He followed this with the blatantly autobiographical *Otto E Mezzo* ($8\frac{1}{2}$) (1963) in which a famous film director escapes his creative block by retreating into ruminations on his past recalled in elaborate dreams and fantastical memories.

Since then, his work has tended towards the repetitive, overblown and self-indulgent, preoccupied with the director's somewhat hermetic creative universe of grotesques and ever more elaborate and inaccessible dream landscapes.

The most notable of his subsequent works include *Giulietta Degli Spiriti* (Juliet of the Spirits) (1965), which catapulted Masina into her dream-world as a middle-age housewife mentally escaping the tedium of her marriage; *Amarcord* (1973) (Oscar), a more entertaining and accessible portrait of a family's life during the Fascist years in Rimini, and the long-winded but touching *Ginger E Fred* (Ginger and Fred) (1986) a satire on the banalities of the contemporary entertainment media with Mastroianni and Masina as a former song 'n' dance act who are reunited for a garish and grotesque television spectacular of showbusiness nostalgia.

His most recent film is the chaotic and bewildering *La Voce Della Luna* (Voices of the Moon) (1990).

A brilliantly imaginative visual stylist and the creator of some of European cinema's most memorable screen images, he received the second Lifetime Achievement award of the European Cinema Society in 1989 but, perversely, his remains an idosyncractic talent seen at its best when constrained within a more conventional or warmhearted narrative.

FIELDS, W.C.

Real Name William Claude Dukenfield
Born 10 February 1879, Philadelphia, USA.
Died 25 December 1946, Pasadena, California, USA.

The victim of an unhappy home life,

Fields ran away at a tender age and spent his itinerant teenage years as a carnival juggler. He later progressed to the burgeoning medium of vaudeville and toured the world as a stellar attraction before settling in as a regular crowd puller at the *Ziegfeld Follies* between 1915 and 1921.

He made his film debut in *Pool Sharks* (1915) and later secured more substantial roles in such silent features as *Sally of the Sawdust* (1925) and *It's The Old Army Game* (1926).

A bulbous nose and gravelly-voiced rasp enhanced his creation of a bibulous, child-hating misanthrope, an image said to be not unlike his own personality; his eccentric comic style came into its own with the advent of sound.

Skilled in the art of verbal insult and visual wit, he used such outlandish pseudonyms as Mahatma Kane Jeeves and Otis J. Criblecoblis to devise the screenplays of such original, abrasive and surreal comedy classics as *It's A Gift* (1934), *The Bank Dick* (1940), *My Little Chickadee* (1940) and *Never Give A Sucker An Even Break* (1941). Often dispensing with the niceties of plot, pacing or logic he created a unique universe in which his disreputable conniver was continually and comically at odds with the rest of the planet and all of its inhabitants, whether human or inanimate.

In more conventional films he also appeared as Humpty Dumpty in *Alice in Wonderland* (1933) and was singularly effective as Micawber in *David Copperfield* (1935).

Illness and a legendary reputation for being cantankerous restricted later creative opportunities and his final appearance was as a guest artist in *Sensations of 1945*.

He was portrayed by Rod Steiger (1925–) in an ill-conceived biography *W.C. Fields and Me* (1975).

FILM NOIR A French term meaning 'dark film', the antecedents of film noir can be discerned in the German ▷ expressionist films of the 1920s and the hardboiled detective fiction of Dashiell Hammett (1894–1961) and Raymond Chandler (1888–1959). Derived from the term *roman noir* ('dark novel') which describes English gothic fiction of the 19th century, it has come to describe a series of disparate Hollywood films from the late 1940s and early 1950s that reflected the gloomy, post-War mood of the nation in their explorations of corruption and betrayal with characters that stoically faced an inescapable, doom-laden fate seen to reflect the worry and confusion of those now living in the nuclear age.

Set in the rain-drenched, shadow-enshrouded mean streets of an urban jungle, the films used expressive contrasts of light and shade, distorted perspectives and new levels of screen violence to achieve their atmospheric results. Examples include ▷ *Double Indemnity* (1944), *The Killers* (1946) and *Out of the Past* (1947).

With time, the brutality and pessimism darkened even further, the mood grew ever more nihilistic and the films focused on the twisted psychology and neuroses of such characters as the mother-fixated gangster Cody Jarrett in ▷ *White Heat* (1949) and the psychotic screenwriter Dixon Steele in *In A Lonely Place* (1950).

A combination of factors from the onset of the Eisenhower era to the cinema's experimentation with wide-screen and colour processes signalled an end to the development of this style of filmmaking but there have been recent attempts to emulate the form and ambience of the films, most notably *Chinatown* (1974) and the steamy *Body Heat* (1981).

FINAL CUT A self-explanatory term, final cut is the last edited version of a film and the one that will be seen by audiences in cinemas. The right to determine this ultimate version is much sought after within the creative echelons of the filmmaking community with ▷ directors of high repute or a strong commercial track record insisting on it as theirs whilst others must bow to the wishes of the producer, financier or studio depending on the form of their contract. It also served as the title of Steven Bach's 1985 book detailing the making of ▷ *Heaven's Gate* (1980).

FISHER, Terence

Born 23 February 1904, London, UK.
Died 18 June 1980, London, UK.

Coming from a naval background, Fisher

worked as a runner and a clapperboard boy before graduating to become an assistant editor on *Brown On Resolution* (1935). A full editor by the time of *Tudor Rose* (1936), he worked in this capacity on such films as *On The Night Of The Fire* (1939), *The Flying Fortress* (1942) and *The Wicked Lady* (1945).

He made his directorial debut with *A Song for Tomorrow* (1948) and then co-directed a number of distinguished Rank releases including *The Astonished Heart* (1949) and *So Long At the Fair* (1950).

In the 1950s he directed numerous episodes of television drama and a slew of undistinguished 'B' movies often featuring American stars well past the peak of their popularity. When Hammer films embarked on a plan to produce colour remakes of the Universal horror classics from the 1930s, Fisher was entrusted with the direction of *The Curse of Frankenstein* (1956) thus beginning a long and fruitful association with the macabre and the fantastical.

He helped foster a distinctively Gothic in-house Hammer style that consisted of richly sensual colours, dripping blood, jangling music and low budgets. Leading characters were strongly etched in the classical manner but there was a new, contemporary emphasis on the previously implicit erotic appeal of horror. His most striking accomplishment was a stylish and chilling version of *Dracula* (1958) that introduced Christopher Lee's (1922–) suave, aristocratic interpretation of the vampire. Other successes include *The Mummy* (1959), *Brides of Dracula* (1960), *The Curse of the Werewolf* (1961) and *Dracula – Prince of Darkness* (1965).

Apart from the odd excursion into equally gruesome subject matter, he continued to ring variations on the Dracula and Frankenstein series until his final film *Frankenstein and The Monster from Hell* (1973).

A FISTFUL OF DOLLARS (Per un Pugno di Dollari)

Sergio Leone (Italy/West Germany/ Spain 1964)
Clint Eastwood, Gian Maria Volonte, Marianne Koch, Pepe Calvo.
100 mins. col.

By the early Sixties, the Western genre was no longer the filmic wellspring of pioneer values it had once seemed: the sturdy myths of old had already been questioned by the psychologically probing Fifties films of director ▷ Anthony Mann (*Man of the West*, 1958), while television's ubiquitous horse operas (▷ *Gunsmoke*, *Wagon Train*, etc) nightly played out the same situations to the point of redundancy. At roughly the same time, the Italian film industry was mounting its own challenge to the dominant Hollywood model, and with the international success of ▷ Sergio Leone's Spanish-shot *A Fistful of Dollars* (1964) the new sub-genre which came to be known as the 'spaghetti western' had well and truly arrived. Playing on the hitherto unsuspected screen charisma of its somewhat second division American star ▷ Clint Eastwood (previously a supporting player on TV's *Rawhide*), Leone's film was initially treated with critical suspicion – just another badly dubbed, cheapjack Italian rip-off – yet closer investigation revealed a sardonic and stylistically distinctive revision of the old West. Against the backdrop of ▷ Ennio Morricone's half-romantic, half-crazy score, Leone's storyline (borrowed from ▷ Akira Kurosawa's *Yojimbo*, 1961) established the taciturn archetype of Eastwood's future anti-heroic screen persona, playing up the rampant greed of its universally sleazy cast and exaggerating the conventions of gunplay into a ritual dance of death. Audiences loved the baroque mayhem enough for Leone to expand the film into a trilogy of increasing aplomb – *Per Qualche Dollari in Piu*, (For a Few Dollars More) (1965) and *Il Buono, il Brutto, il Cattivo* (The Good, The Bad and The Ugly) (1966) – and to tempt a host of lesser talents to turn out a slew of derivative, occasionally arresting imitations.

FLAHERTY, Robert John

Born 16 February 1884, Iron Mountain, Michigan, USA.
Died 23 July 1951, Black Mountain, Vermont, USA.

Acknowledged as the father of modern documentary, Flaherty's body of work stands as a visually arresting, acutely observed and poetic attempt to convey the struggles of cultures and races on the brink of fundamental changes that would soon render their way of life obsolete. Too subjective and poetic to be judged an honest ethnographer, his pioneering status, influence and pure artistry are less disputatious.

The son of a miner and prospector, Flaherty worked as an explorer and surveyor for the Canadian Grand Trunk Railroad and developed an interest in Eskimo culture whilst employed to search for iron ore along the Hudson Bay. He filmed their activities and endeavours in 1913, although the footage was subsequently destroyed in a fire.

Returning with more advanced equipment and secure funding, he directed *Nanook of the North* (1922), a stark, simple and expressively photographed examination of the conflict between man and nature as the Eskimos seek food and shelter in the Arctic wastes.

Acclaimed for his work, he received funding to making *Moana* (1926), a similar attempt to convey the South Seas lifestyle through the initiation into manhood of a Polynesian native.

He worked on a number of abortive dramatic projects and contributed to *Tabu* (1931), before travelling to Britain and immersing himself in the lives of the Aran islanders. The result was *Man of Aran* (1934) a documentary again focusing on the hardiness of a remote community and their titanic struggles to eke a living from the bleak and unyielding forces of nature.

His later work includes supervising the location sequences for *The Elephant Boy* (1937), and *The Land* (1942), which focused on America's migrant farmers.

Involved with many abortive projects over the years, his last successfully realised venture was *Louisiana Story* (1948) which melded his highly manipulative documentary techniques and apolitical stance with a fictional story to capture the visual beauties of the bayous and the benign arrival of the oil industry.

His final films were the short *Green Mountain Land* (1950) which he produced and *St Matthew's Passion* (1951) which he edited and narrated.

FLEISCHER, Max

Born 17 July 1883, Vienna, Austria.
Died November 1972, Motion Picture County Home, California, USA.

A key pioneer in the animator's art, Fleischer was resident in America from the age of four. An errand boy for *The Brooklyn Daily Eagle*, he later joined the staff as an artist and during his tenure as art editor of *Popular Science Monthly* (1914) he invented the rotoscope, a device still used for transferring live action film into animated cartoons via frame-by-frame tracing.

In 1915, with his brother Dave (1894–1979), he began work on the *Out of the Inkwell* series of cartoons which combined live action with animation and featured a cartoon clown called Ko-Ko.

The series continued when they formed their own company in 1921. Among their innovations were the 'bouncing-ball' singalong cartoons which were silent but synchronised to the cinema orchestras. The Ko-Ko films lasted until 1929, by which time the brothers had been experimenting with sound cartoons in a *Talkartoons* series that led to stardom for the character of Betty Boop in shorts like *Betty Co-Ed* (1931), *Boop-Oop-A-Doop* (1932) and *Red Hot Mama* (1934).

They followed this with their most famous series giving animated life to Popeye the Sailor Man from 1933 in films like *Strong to the Finich* (1934), *The Hyp-Nut-Ist* (1934) and the lengthier *Popeye the Sailor Meets Sinbad the Sailor* (1936), *Popeye the Sailor Meets Ali Baba and the Forty Thieves* (1937) and *Aladdin and His Wonderful Lamp* (1939) which made early use of ▷ Technicolor and another Fleischer invention the Turntable Camera, which created three-dimensional effects by photographing animation ▷cels (sheets of celluloid on which the cartoons have been drawn) in front of revolving miniature sets.

Encouraged by the success of ▷ Disney's *Snow White and the Seven Dwarfs* (1937), they made a feature-length début with a version of *Gulliver's Travels* (1939) and followed this with *Mr Hoppity Goes to Town* (1941), but the financial failure of the latter led to Paramount foreclosing on their loans and the demise of the company.

Dave became head of the Columbia cartoon department and later worked at Universal until his retirement in 1967. Max worked on industrial projects and later made animation for television.

Max's son is the director Richard Fleischer (1916–) who has such films as *20,000 Leagues Under the Sea* (1954), *The Vikings* (1958) and *The Boston Strangler* (1968) to his credit.

FLYNN, Errol
Born 20 June 1909, Hobart, Tasmania.
Died 14 October 1959, Vancouver,

The son of a distinguished zoologist and marine biologist, Flynn's early background is as colourfully romantic and adventurous as one of the screen characters he would later portray and includes diamond-smuggling and a charge of murder on which he was acquitted. He worked briefly as a shipping-clerk, tobacco-plantation manager and journalist before sailing to New Guinea in search of gold.

The documentary-travelogue *Dr H. Erben's New Guinea Expedition* (1932) contains his first screen work and, after playing the role of Fletcher Christian in the Australian-made film *In The Wake of The Bounty* (1933), he came to England to gain acting experience with the Northampton Repertory Company. An appearance in *Murder at Monte Carlo* (1934), made by the London branch of Warner Brothers, brought him an offer of a Hollywood contract.

When ▷ Robert Donat declined the role, *Captain Blood* (1935) established him as a virile swashbuckling star and, along with ▷ Douglas Fairbanks Senior, he remains the quintessential historical hero of Hollywood fantasy, dashing, athletic, handsome, humorous and totally at ease with the physical demands of the genre. His greatest successes include *The Charge of the Light Brigade* (1936), *The Adventures of Robin Hood* (1938) and *The Sea Hawk* (1940).

Equally at home in westerns and war films, he played his part in a succession of wartime thrillers and patriotic adventure yarns; with further popular titles including *They Died With Their Boots On* (1941), *Gentleman Jim* (1942) and *Edge of Darkness* (1943).

A notoriously larger than life figure off-screen, his excessive indulgence in wine, women and song eventually took its toll on his youthful exuberance and hearty demeanour, with *The Master of Ballantrae* (1953) and *The Dark Avenger* (1955) (US: The Warriors) being rather wan reflections of more cavalier triumphs from the past.

However, he later proved himself a character actor of some promise, using his own hedonistic experiences to illuminate the dissolute and drunken older figures in the likes of *The Sun Also Rises* (1957), *The Roots of Heaven* (1958) and *Too Much Too Soon* (1958), in which he played ▷ John Barrymore.

Active on television in one-off dramas and the series *The Errol Flynn Theatre* (1957–8), he also directed the short films *Cruise of the Zaca* (1952) and *Deep Sea Fishing* (1952) and made his last film appearance in *Cuban Rebel Girls* (1959) a disastrous semi-documentary that he wrote, co-produced and narrated in tribute to Fidel Castro (1927–).

His books include *Beam Ends* (1937), *Showdown* (1946) and the appropriately entitled autobiography *My Wicked, Wicked Ways* (1959).

FOCUS PULLER Also known as the assistant cameraman, the focus puller is responsible for the maintenance and care of the camera equipment, for changing lenses and magazines and for following focus during shooting to ensure that the image is captured with clarity and sharp definition.

FONDA, Henry Jaynes
Born 16 May 1905, Grand Island, Nebraska, USA.
Died 12 August 1982, Los Angeles, California, USA.

A chance involvement with the Omaha Community Playhouse led to Fonda's professional association with the theatre and his New York appearance in such plays as *I Loved You Wednesday* (1932), *Forsaking All Others* (1933) and *New Faces of 1934*.

He made his film debut in *The Farmer Takes A Wife* (1935) and secured a niche in colourful outdoor adventures like *The Trail of the Lonesome Pine* (1936) and *Spawn of the North* (1938) whilst evidencing his dramatic worthiness in *You Only Live Once* (1937) and his status as a capable leading man opposite ▷ Bette Davis in *Jezebel* (1938).

His self-effacing manner, resolute

nature and dry, flat tones were effectively used to portray honesty and decency and his earliest major successes include *Young Mr Lincoln* (1939), the comedy *The Lady Eve* (1941) and ▷ *The Grapes of Wrath* (1940) (Best Actor Oscar nomination) where he eloquently conveyed the dignity and idealism of Tom Joad.

Never totally deserting his commitment to the theatre, he was absent from Hollywood during his wartime service in the Navy and later between 1948 and 1955 when he worked on Broadway. However, notable performances from the period include lawman Wyatt Earp (1848–1928) in *My Darling Clementine* (1946) and the military martinet in *Fort Apache* (1948).

He was lured back to the screen for the popular film version of *Mister Roberts* (1955) and found acclaim as the resolute juror in *Twelve Angry Men* (1957) and as the Presidential candidate in *The Best Man* (1964), in between dozens of westerns and dramas where he brought conviction and presence to a succession of authority figures from generals and admirals to Presidents and policemen.

Cast against type and sporting brown contact lenses, he made a menacing villain in *C'Era Una Volta Il West* (Once Upon A Time in the West) (1968) and seized whatever opportunities that came his way amongst a morass of more mediocre assignments; he was memorable as a backwoodsman in *Sometimes A Great Notion* (1971) and as a semi-literate convict who challenged the American legal system in the television film *Gideon's Trumpet* (1980).

The recipient of the ▷ American Film Institute Life Achievement Award in 1978 and an honorary Oscar in March 1981, he finally won a Best Actor Oscar for his touching performance as a cantankerous octogenarian achieving a belated emotional reconciliation with his daughter in *On Golden Pond* (1981).

A frequent stage performer, he enjoyed long runs with *Mister Roberts* (1948–51), *Two for the Seesaw* (1958) and *Clarence Darrow* (1974–5).

Married five times, his professional footsteps have been followed by daughter ▷ Jane and son Peter (1939–), who directed him in *Wanda Nevada* (1979).

His autobiography, *My Life* as told to Howard Teichman was published in 1981.

Henry Fonda, Jane Fonda and Katharine Hepburn in *On Golden Pond* (1981)

FONDA, Jane Seymour

Born 21 December 1937, New York City, New York, USA.

The daughter of ▷ Henry Fonda, Jane made her acting debut opposite her father in a production of *The Country Girl* (1955) and pursued a dilettanteish existence as an art student and model before she began to take acting seriously as a pupil of Lee Strasberg (1899–1982).

She made her Broadway debut in *There Was A Little Girl* (1959–60) and her first film appearance as the cheerleader in *Tall Story* (1960).

Blue-eyed with chestnut hair and a healthy vigour in her appearance she was seen, in casting terms, as the all-American girl, but found slightly more challenging work as the prostitute in *Walk on the Wild Side* (1962) and as the frigid wife in *The Chapman Report* (1962).

Work in Europe and a brief marriage to director Roger Vadim (1927–), found her labelled as a sex kitten, an image further encouraged by her appearance as naive comic strip heroine *Barbarella* (1968).

In America, she had displayed a delicious sense of the rhythms and timing required for comedy playing in *Cat Ballou* (1965) and *Barefoot in the Park* (1967) and she managed to recapture her dramatic credibility as the hard-bitten Depression-era marathon dancer in *They Shoot Horses, Don't They?* (1969) (Best Actress Oscar nomination).

There was a compelling warmth and complexity to her performance as the threatened prostitute in *Klute* (1971)

(Oscar) and had her outspoken criticisms of America's involvement in Vietnam not aroused such hostile ire (she was dubbed 'Hanoi Jane') her career might have developed more rapidly.

She was seen intermittently over the ensuing years, but returned dramatically with a string of box-office successes including the farce *Fun With Dick and Jane* (1977), the elegant period drama *Julia* (1977) (Oscar nomination) and the brittle comedy *California Suite* (1978), and she won a second Best Actress Oscar for *Coming Home* (1978), as an American wife soured by the realisation of the price America has paid in Vietnam.

Her own company IPC enjoyed a commercial success with films that shrewdly blended entertainment with a strong commitment to the exploration of social and political issues. Notable hits include *The China Syndrome* (1979) (Oscar nomination) and the comedy *Nine to Five* (1980).

On Golden Pond (1981) (Oscar nomination), a tale of reconciliation between an emotionally timid old man and his daughter, provided a fitting swansong for her father and a dramatic catharsis for her.

Devoting herself to political causes, the pace of her screen career slowed but she won an Emmy as an indomitable Kentucky matriarch in *The Dollmaker* (1984) and has been seen recently as a chainsmoking psychiatrist in *Agnes of God* (1985), fading, alcoholic actress in *The Morning After* (1986) (Oscar nomination) and virginal spinstress bewitched by the romance of revolutionary Mexico in *Old Gringo* (1989).

A noted political activist and radical, she was once listed on the 'enemies list' of President Nixon (1913–) and has recently gained a different renown as an energetic proselytiser for physical fitness through bestselling videos, books and a line of exercise clothing.

FORD, Harrison

Born 13 July 1942, Chicago, Illinois, USA.

A performer in student productions, Ford later worked in summer stock where he was discovered by a talent scout from Columbia and signed to a standard contract. Groomed for stardom he made small appearances in *Dead Heat On A Merry-Go-Round* (1966), *Luv* (1967), *Journey to Shiloh* (1968) and other films.

Dissatisfied with the calibre of roles he was receiving in films and television, he drifted away from the profession whilst pursuing a more successful secondary career as a carpenter. However, he began to establish himself with small but key roles in films like *American Graffiti* (1973) and *The Conversation* (1974) before being cast as space age gunslinger Han Solo in ▷ *Star Wars* (1977).

Amidst the hardware and special-effects in this science-fiction blockbuster, he stood out for the way he underscored his heroism with a sardonic humour and a touch of irony that encouraged viewers not to take anything too seriously. The popularity of sequels *The Empire Strikes Back* (1980) and *The Return of the Jedi* (1983) brought him fame and wealth but his stardom was consolidated when he took over as a last minute replacement for Tom Selleck (1945–) in *Raiders of the Lost Ark* (1981).

A whip-totin' academic ophidiophobe archaeologist with a battered fedora and a limitless resourcefulness, his creation of Indiana Jones struck a responsive chord in international audiences and he has reprised the role in *Indiana Jones and the Temple of Doom* (1984) and *Indiana Jones and the Last Crusade* (1989).

Outwith such grandly heroic escapism, his sincere, personable and conscientious acting skills have been seen to effect in his subtle shading of the police officer adrift in the Amish community in *Witness* (1985) (Best Actor Oscar nomination).

Other performances of note include the futuristic cop in *Blade Runner* (1982), the distraught husband in *Frantic* (1988), the dashing corporate Prince Charming in *Working Girl* (1988) and the finely detailed authority in his emotionally repressed attorney destroyed by the torrents of passion in *Presumed Innocent* (1990).

FORD, John

Real Name Sean Aloysius O'Feeney
Born 1 February 1895, Cape Elizabeth, Maine, USA.
Died 31 August 1973, Palm Desert, California, USA.

The son of Irish immigrant parents, Ford

arrived in Hollywood in 1914 to join his brother Francis (1883–1953), who was already active in the embryonic film industry. He began his career as an actor and stuntman in the serial *Lucille Love, the Girl of Mystery* (1914) and such films as ▷ *The Birth Of A Nation* (1915) and *Chicken-Hearted Jim* (1916), before making his directorial debut with the western *The Tornado* (1917).

Forever identified with the western, he worked extensively in the genre throughout the silent era, most notably on *The Iron Horse* (1924), an epic account of transcontinental railway construction.

Throughout the 1930s, he proved his versatility as a master storyteller and unassuming visual stylist in rural comedies like *Dr Bull* (1934), the South Seas adventure *The Hurricane* (1937), the Kipling yarn *Wee Willie Winkie* (1937) and the ▷ Expressionist *The Informer* (1935) for which he won a Best Director Oscar.

However, his greatest renown came as an affectionate, patriotic chronicler of American history in films like *Young Mr Lincoln* (1939), *My Darling Clementine* (1946) and *Fort Apache* (1948), poetic visions of how the west was won, its rugged heroes, pioneering families, sense of values and a male camaraderie generally forged in intoxication or fisticuffs.

Interested in myth and morality, an American landscape rich in rugged beauty and hidden dangers, his body of work, whilst often flawed by maudlin sentimentality and both racist and sexist undertones, is unrivalled in its breadth and consistency.

He won Best Director Oscars for ▷ *The Grapes of Wrath* (1940), *How Green Was My Valley* (1941) and *The Quiet Man* (1952), one of many films resulting from his long collaboration with ▷ John Wayne whose screen image was defined and refined as the Ringo Kid in ▷ *Stagecoach* (1939) (Oscar nomination), the aging cavalry officer Nathan Brittles in *She Wore A Yellow Ribbon* (1949), the unbending Ethan Edwards in ▷ *The Searchers* (1956) and Ranson Stoddard, the man whose gunfighting prowess is behind the legend of politician ▷ James Stewart's bravery in *The Man Who Shot Liberty Valance* (1962).

He made his last feature film *Seven Women* in 1965 and later directed the documentary *Chesty: A Tribute to a Legend* (1970) and supervised *Vietnam! Vietnam!* (1971). During the war he had served as a Lieutenant Commander in the U.S. Navy Corps and directed such documentaries as *The Battle of Midway* (1942) and *We Sail at Midnight* (1943).

Among the many honours he received late in his career were a Golden Lion at the 1971 Venice Film Festival and the first ▷ American Film Institute Life Achievement Award in 1973.

FORMAN, Milos
Born 18 February 1932, Caslav, Czechoslovakia.

Educated in Prague at the Academy of Music and Dramatic Art and later at the Film Academy, Forman's first involvement with the cinema was as the screenwriter of *Nechte to na Mne* (Leave It To Me) (1955) and as assistant-director on *Dedecek Automobil* (Old Man Motorcar) (1955).

After a spell with the multi-media group *Laterna Magika* and two short documentaries, he made his directorial debut with *Cerny Petr* (Peter and Pavla) (1963), a lightly satirical view of a teenage trainee store detective's travails as he copes with life, love and an uncommunicative father. Subsequent films like *Lasky Jedne Plavovlasky* (A Blonde in Love) (1965) and especially *Hori, Ma Panenko* (The Fireman's Ball) (1967) brought fresh vigour to the Czechoslovakian cinema of the period with their documentary-like use of locations, partially improvised dialogue and mixture of professional and non-professional actors contributing to boisterous, affectionately observed portraits of ordinary life with allegorical undertones.

In Paris when the USSR invaded Czechoslovakia in 1968, he moved to New York and began the American phase of his career with *Taking Off* (1971) which cast a caustic eye over American mores and cross-generational conflicts.

He won a Best Director Oscar for ▷ *One Flew Over the Cuckoo's Nest* (1975), a crackling version of the novel about a nonconformist's subversive influence on the inmates of a rigidly-run asylum. His subsequent work has disappointed as he has turned to tasteful and unadventurous adaptations of stage successes and bestsellers like *Hair* (1979) and *Amadeus* (1984) for which he won a second Best Director Oscar. There was more life to his dexterous handling of the many plotlines in the kaleidoscopic slice of Americana *Ragtime* (1981) but *Valmont* (1989), his version of *Les Liaisons Dangereuses*, was poorly released and virtually ignored in the wake of the superior *Dangerous Liaisons* (1988).

He also made an acting appearance in *Heartburn* (1986).

FORSYTH, Bill

Full Name William David Forsyth
Born 29 July 1946, Whiteinch, Glasgow, UK.

The son of a plumber, Forsyth entered the film industry by chance when he answered a newspaper advertisement seeking a young lad for employment in a production company. Within a year, he was dedicated to the notion of making films and gained a wide-ranging technical

Housekeeping (1987) directed by Bill Forsyth

expertise in his profession at work on sponsored films and documentaries.

He later worked at the BBC and was a short-lived student among the first intake to the National Film School in 1971. His own early experiment with film grammar and content include the ▷ avant-garde *Language* (1969) and *Waterloo* (1970).

In a country bereft of any strong film-making tradition, he endured a hand-to-mouth existence before striking out on his own with the extremely low-budget social comedy *That Sinking Feeling* (1979) ('a fable for the workless') which was warmly received at the Edinburgh Film Festival of that year.

The encouraging response brought him the backing for *Gregory's Girl* (1980), an acutely observed comedy of schoolhood days and a gauche, gangly youngster's unrequited infatuation for a female soccer wizard.

He followed this with the popular *Local Hero* (1983), a melancholy tale of how the oil business affects a small Scottish community and bewitches the American executive who is dispatched there to buy their land.

Generally perceived as a purveyor of gently humorous and ineffably charming whimsy, his work is open to many more complex interpretations and is rich in unpredictability and darker overtones.

He has subsequently concentrated on illuminating the fragile lives of those on the periphery of conventional society. *Comfort and Joy* (1984) followed the personal and professional crises in the life of a radio deejay, the beautifully crafted *Housekeeping* (1987) tells of a dangerously eccentric aunt burdened with the care of two young girls, and the less satisfying *Breaking In* (1989) slyly unravels the loneliness of the criminal life.

For television, he has directed the drama *Andrina* (1981) and a commercial for Fosters Lager featuring ▷ Burt Lancaster. More recently, he has been at work on an original script entitled *Being Human*.

FOSTER, Jodie

Real Name Alicia Christian Foster
Born 19 November 1962, Los Angeles, USA.

One of the youngest veterans of the showbusiness world, Foster's career began at the age of three when she was chosen as the bare-bottom child in the Coppertone sun tan commercials.

Appearing in over forty commercials as a child, she made her dramatic debut in a 1969 episode of the television comedy series *Mayberry, R.F.D.* (1968–70) and subsequently appeared in individual episodes of such series as ▷ *Gunsmoke*, *Bonanza* (1959–71) and *Kung Fu* (1971–74) and had a starring role in *Paper Moon* (1974).

She made her film debut in *Napoleon and Samantha* (1972) and tended to play tomboyish characters in family adventures before attracting attention as the streetwise urchin in *Alice Doesn't Live Here Anymore* (1974).

Her precocious talents were well displayed in a trio of films: as the speakeasy vamp Tallulah in the children's musical *Bugsy Malone* (1976), the resourceful orphan in *The Little Girl Who Lives Down The Lane* (1976), and, especially, as the baby-faced hustler exuding a feigned wordliness but underlying vulnerability in ▷ *Taxi Driver* (1976) (Best Supporting Actress Oscar nomination).

Acclaimed as one of the most promising young performers of the day she was seen in a succession of Disney films like *Freaky Friday* (1976) and *Candleshoe* (1977) before tackling more adult roles in *Foxes* (1979) and *Carny* (1980), in which she played a waitress whose dreams of a more exciting life are partially fulfilled as a carnival performer.

She then concentrated on her education, graduating first in her class from the bilingual Lycée Français in Los Angeles and subsequently achieving a B.A. in Literature from Yale.

Re-establishing herself as an adult performer in the cinema proved difficult, but she found substantial opportunities in *The Hotel New Hampshire* (1984) and the offbeat *Five Corners* (1988), which eerily echoed her private life in its story of a man's obsessive infatuation. (In 1981, John W. Hinckley Jr had attempted to kill President Reagan as a mark of his love for her).

However, she made a powerful transition with a harrowing performance as a rape victim in search of justice in *The Accused* (1988), which won her a Best Actress Oscar.

She has capitalised on this success with roles in *Backtrack* (UK: Catchfire) (1989) and ▷ Jonathan Demme's thriller *Silence of the Lambs* (1991) and has recently completed her directorial debut with *Little Man Tate* (1991).

FRANKENHEIMER, John

Born 19 February 1930, Malba, New York, USA.

Few careers display the inconsistency of Frankenheimer's; his films of the 1960s show such an electrifying verve and technical virtuosity that it is difficult to distinguish the same talent behind recent failures like *Dead-Bang* (1989) or *The Fourth War* (1990).

After serving in the US Air Force, he joined CBS-TV in New York and became the director of the series *You Are There* in 1954. Over the next five years he is reckoned to have directed in excess of one hundred television plays including *The Last Tycoon* (1957), *Days of Wine and Roses* (1958) and *For Whom The Bell Tolls* (1959).

He made his cinema debut with *The Young Stranger* (1957) and enjoyed a rewarding partnership with ▷ Burt Lancaster on films like *The Young Savages* (1961), *Birdman of Alcatraz* (1962), *Seven Days in May* (1963) and *The Train* (1964) which vividly illustrate his proficiency with well-paced, action-packed thrillers, lucid approach to convoluted plots and disciplined modulation of human drama.

The Manchurian Candidate (1962), a dazzling mixture of McCarthyite hysteria and conspiracy thriller, was a directorial tour de force and he continued to interest with the rejuvenation drama *Seconds* (1966) and *Grand Prix* (1967). After the desultory comedy *The Extraordinary Seaman* (1968) his career grew more erratic; proof of his flair for explosive set-pieces was still evident in *French Connection II* (1975) and *Black Sunday* (1976) and his adaptation of *The Iceman Cometh* (1973) is beautifully acted by a superb cast. However, the comic book gangster yarn *99 and 44/100% Dead* (1973) (UK: Call Harry Crown) or the ecological horror film *Prophecy* (1979) or the wooden thriller *The Holcroft Covenant* (1985) are lacklustre fare that display little authorial individuality.

As with all his recent ventures, hopes are high that *Year of the Gun* (1991) may mark a return to form.

FREARS, Stephen Arthur

Born 20 June 1941, Leicester, England.

Graduating from Trinity College, Cambridge in 1963 with a degree in law, Frears moved to London and worked at the Royal Court Theatre before learning the craft of filmmaking as an assistant on such productions as *Morgan, A Suitable Case for Treatment* (1966), *If!* (1967) and *Charlie Bubbles* (1968).

He made his directorial debut with *Burning* (1967) a thirty minute drama set in South Africa. His feature-length debut came with *Gumshoe* (1971) a whimsical and affectionate parody of the romantic myths surrounding trenchcoated detectives, with Albert Finney (1936–) as a bingo caller who advertises his services as a private eye.

Over the next decade, he worked extensively for television on such feature-length plays as *A Day out* (1972), *Sunset Across the Bay* (1973), *Me, I'm Afraid of Virginia Woolf!* (1978) and *Walter* (1982), screened on the first night of transmission by Channel 4.

Attracted to acutely-observed stories of everyday lives and of wider social significance, he has worked with such distinguished dramatists as ▷ Alan Bennett, Tom Stoppard (1937–) and Christopher Hampton (1946–). His body of work illustrates a belief in the classical virtues of a well-written script, provocative themes and deceptively self-effacing direction that serves the author's intent.

A television film *Bloody Kids* (1979) received a theatrical release but his full-scale return to the cinema came with the offbeat revenge thriller *The Hit* (1984). He then directed *My Beautiful Launderette* (1985), a script by Hanif Kureishi (1954–) that provided a richly dramatic and humorous account of the racial tensions in modern Britain as a white Englishman and an Anglo-Pakistani entrepreneur fall in love while renovating a launderette.

He collaborated with Kureishi again on *Sammy and Rosie Get Laid* (1987), a more fragmented and excoriating exploration of the inherent racism and class snobbery of contemporary England.

He displayed a mordant wit, visual economy and strong sense of rhythm in *Prick Up Your Ears* (1987), Alan Bennett's wickedly funny Ortonesque dramatisation of playwright Joe Orton's private life.

In 1988 he tackled his first big budget production *Dangerous Liaisons* and rendered a tale of 18th century sexual intrigue vividly alive for contemporary audiences. He then proceeded to his first American film *The Grifters* (1990) (Best Director Oscar Nomination), based on a novel by Jim Thompson (1906–77).

FREE CINEMA Reacting against documentaries focused on institutions or collective actions, and the prevailing tone of British cinema, Free Cinema was a documentary movement of the 1950s determined to depict everyday lives through the commonplace events experienced by those who live them. 'Free' from the restraints of sentimentality or structure, the aim was to create a more caring, personal and poetic approach.

The first three films in the movement, shown in February 1956, were *O Dreamland* (1953) directed by Lindsay Anderson (1923–), *Momma Don't Allow* (1955) a collaboration between Karel Reisz (1926–) and Tony Richardson (1928–), and *Together* (1956) which Anderson edited.

A relatively shortlived movement that also produced *Every Day Except Christmas* (1957), *The March to Aldermaston* (1958) and *We Are the Lambeth Boys* (1959), its principal figures brought some of their techniques and concerns with them as they revolutionised mainstream British cinema with films like *Look Back in Anger* (1959), ▷ *Saturday Night and Sunday Morning* (1960) and *This Sporting Life* (1963).

FREEZE FRAME Achieved in an optical printer by repeating one frame of film many times within a single sequence, the freeze frame process gives the illusion of the action being held stationary or 'frozen'. It serves as a form of dramatic punctuation.

FRIESE-GREENE, William
Real Name William Green
Born 7 September 1855, Bristol, UK.
Died 5 May 1921, London, UK.

A prolific inventor, Friese-Greene was educated in Bristol and later opened a portrait studio in Bath as his interest in photography blossomed into a commercial concern.

He became interested in exploring the

possibilities of moving pictures as a natural progression of his work on movable lantern slides for the 'dissolving lantern'.

In 1889 with patent no 10131, he patented a motion-picture projector and a camera that took photos in rapid succession on a roll of sensitised paper. The paper was replaced by celluloid in 1890 but his device could only photograph ten frames per second, six less than the number required to produce visible, flowing motion.

The results of his film experiments were also too short in duration to qualify as genuine motion-pictures and, despite his lodging of patents for improved mechanisms in 1895 and 1896, his work had been superseded by others and his place in the history of film's evolution is much more marginal than the romantic and inaccurate film biography *The Magic Box* (1951) would suggest.

Among the seventy-nine British patents he registered were a two-colour method of cinematography, rapid printing and processing machines, a three-colour camera, a cigar lighter and a process of inkless printing.

FROST, David Paradine

Born 7 April 1939, Tenterden, Kent, UK.

Educated at Cambridge, Frost participated in the Footlights Revue and edited *Granta* before moving into television in 1961.

Known for his robust, staccato greeting of 'Hello. Good evening. And welcome', he hosted ▹ *That Was The Week That Was* (1962–63) a popular, late-night revue show whose topical satire and irreverent attitude to the establishment were innovations in television light entertainment.

Subsequent shows in Britain and America include *The Frost Report* (1966–67), *Frost on Friday* (1968–70), *The David Frost Show* (1969–72) and many others.

His trans-Atlantic television career divides into two areas: an anchorman and satirist in shows like *The Guinness Book of World Records* (1981–) and *The Spitting Image Movie Awards* (1987), and as a diligent inquisitor of world leaders in such programmes as *The Nixon Interviews* (1976–77) and *The Shah Speaks* (1980).

A businessman with a diverse portfolio of interests, he was a producer of the film *The Slipper and the Rose* (1976) and a co-founder of Britain's TV-AM (1983–) breakfast television service. The only one of the original participants to remain with the channel, he still presents a morning of chat and political interviews on Sundays.

His many international honours include The Golden Rose of Montreux (1967) and the ▹ Emmy (1970 and 1971).

His publications include *How To Live Under Labour* (1964), *I Gave Them A Sword* (1978) and *The World's Shortest Books* (1987).

FULLER, Samuel Michael

Born 12 August 1911, Worcester, Massachusetts, USA.

A copy boy for the New York Journal at the age of 13, Fuller is reputed to have been the youngest crime reporter in New York before becoming a fiction writer and moving to Hollywood.

He wrote such scripts as *Hats Off* (1936), *It Happened in Hollywood* (1937) and *Bowery Boy* (1940) before his war service with the 16th Regiment of the Army 1st Division, when he received the Bronze Star, Silver Star and Purple Heart.

He briefly resumed scriptwriting before making his directorial debut with *I Shot Jesse James* (1948). Gainfully employed at Twentieth Century-Fox between 1951 and 1957, he developed a recognisable style that favoured fast-paced, violent melodramas using rapid editing and ▹ montage to punch across stories torn from the day's headlines. A specialist in war films and westerns, his many titles from the period include *The Steel Helmet* (1950), *The House of Bamboo* (1955) and *40 Guns* (1957). His journalistic past was reflected in *Park Row* (1952), a tale of rival newspaper magnates, whilst his anti-Communist stance found a voice in *Pickup on South Street* (1953) and *Hell and High Water* (1954).

Perceived as having a broad interest in outsiders and the conflicts between questions of personal and national identity, he was less prolific in the 1960s, although *Merrill's Marauders* (1962) and the raw psychological thriller *Shock Corridor* (1963) showed that he lacked none of his former force.

He worked in television, directing an episode of *The Virginian* in 1962 and devising the series *Iron Horse* (1966). He also worked abroad, where his oeuvre had been more readily appreciated, on films like *Dead Pigeon on Beethoven Street* (1972).

He returned to mainstream American cinema with the accomplished autobiographical war drama *The Big Red One* (1980) and the much-admired anti-racist statement *White Dog* (1981) and continues to direct from his European base.

As an actor, his inimitable rasping growl and diminutive cigar-chomping figure have been seen in a variety of roles from *The Last Movie* (1971) to *1941* (1979) and *The State of Things* (1982) although his best known appearance remains *Pierrot Le Fou* (1965) in which he uttered his famous dictum, 'Film is like a battleground; love, hate, action, violence, death ... in one word – emotion.'

GABIN, Jean

Real Name Jean-Alexis Moncorgé.
Born 17 May 1904, Paris, France.
Died 15 November 1976, Neuilly-sur-Seine, Paris, France.

Neither the most expressive or versatile of actors, Gabin's direct, unflamboyant style and potent screen presence made him a star for almost all of the four decades he reigned as a commercial constant in the changing French cinema.

A car mechanic from a showbusiness family, his father's connections secured him a job at the Folies Bergere and he built his career in the theatre and as a cabaret performer. He made his film debut in *Chacun Sa Chance* (1931) and made numerous screen appearances before his screen image was defined by a key role as *Pepé le Moko* (1936), a jewel-thief whose love for a beautiful woman draws him from the sanctuary of the Algerian casbah to inevitable arrest.

He was instrumental in bringing ▷ *La Grande Illusion* (1937) to the screen and his image was crystallised with roles as the army deserter grabbing a futile chance of love in *Quai Des Brumes* (1938) and as the hunted and haunted perpetrator of a crime passionel in ▷ *Le Jour Se Lève* (Daybreak) (1939).

Cast as loners, doomed to the most pessimistic of fates, his performances as fatalistic working-class anti-heroes were seen to eloquently personify the mood of pre-War France.

He fled to America in 1941, later joining the Free French Navy, participating in the Normandy invasion and receiving the Croix de Guerre and Medaille Militaire.

It took time to establish his career in

Jean Gabin in *Le Désordre de la Nuit* (1957)

an altered post-War cinema, but in the 1950s his old dominance was re-asserted in films like *Touchez Pas Au Grisbi* (Grisbi) (1953) and *Archimède Le Clochard* (Archimede The Tramp) (1958) and he brought a new worldliness to his portrayal of a range of mature, patriarchal figures.

Popular in gangster yarns and comedies, his name on a film was a guarantee of success in France and his bankability was further enhanced by his tendency to participate in attractive partnerships with, for instance, Jean-Paul Belmondo 1933–) in *Un Singe en Hiver* (Monkey in Winter) (1962), Alain Delon (1935–)

in *Melodie En Sous-Sol* (The Big Snatch), ▷ Simone Signoret in *Le Chat* (1971) and ▷ Sophia Loren in *Verdict* (1974).

He was characteristically cast as an aging convict attempting to recover a cache of stolen money near the Vatican in his final film *L'Année Sainte* (Holy Year) (1976).

GABLE, Clark

Full Name William Clark Gable
Born 1 February 1901, Cadiz, Ohio, USA.
Died 16 November 1960, Reno, Nevada, USA.

A lumberjack and labourer, Gable began working in the theatre as a handyman. Touring in stock companies, he worked as an extra in such silent films as *Forbidden Paradise* (1924) and *The Merry Widow* (1925). After working on Broadway, his appearance in a Los Angeles production of *The Last Mile* (1930) brought him to the attention of film studio scouts and won him a screen-test.

He signed the first of several contracts with M-G-M where he would work for the next twenty years and a concentrated assault on stardom was made with twelve film appearances in 1931.

A powerfully built, ruggedly masculine individual with outsize ears and a roguish manner, he was initially cast as rough, brutish gangsters and hoodlums, chauvinistic tough guys who treated women with a slap of the hand. His swaggering virility proved popular after the smooth Latin lovers of the silent era and his early successes include *A Free Soul* (1931), *Night Nurse* (1931) and, especially, *Red Dust* (1932).

Soon established as one of the screen's favourite idols, he won a Best Actor Oscar as the newshound in the comedy-romance *It Happened One Night* (1934) and his popularity soared with a succession of manly roles in high adventures like *Mutiny on the Bounty* (1935) (Oscar nomination) and *Test Pilot* (1938) and he also showed an impish sense of humour in lighter fare like *Wife Versus Secretary* (1936) and *Idiot's Delight* (1939).

Voted King of Hollywood in 1937 (▷ Myrna Loy was Queen), he made the definitive Rhett Butler in ▷ *Gone With the Wind* (1939) (Oscar nomination).

A member of the U.S. Army during World War II, he returned to the screen in *Adventure* (1945) and enjoyed box-office success with *The Hucksters* (1947), but subsequent films relied on the authority of his now more sombre screen image and provided little of a challenge in terms of characterisation or settings.

Mogambo (1953), a remake of *Red Dust*, provided a perfect showcase for his still virile appeal and subsequent late career hits include *The Tall Men* (1955) and *Run Silent, Run Deep* (1958).

The elegiac quality of his dignified final performance as cowboy Gay Langland in the posthumously released *The Misfits* (1961) represents his finest hour as an actor.

GAFFER The chief electrician in a film or television production crew, the gaffer is responsible for positioning, operating and maintaining the lighting equipment required by the ▷ cinematographer. The charge-hand electrician working directly under the gaffer is known as the ▷ best boy.

GANCE, Abel

Born 25 October 1889, Paris, France.
Died 10 November 1981, Paris, France.

A law office clerk and stage performer, Gance's first involvement with film was as an actor in comedy shorts and the feature *Molière* (1909). He also wrote scripts and made his directorial debut with *La Digue ou Pour Sauver La Hollande* (1911).

A technical innovator throughout his career, he experimented with distorting lenses on *La Folie du Docteur Tube* (1916) and used split-screen techniques as early as *J'Accuse* (1919), an eight-hour anti-war statement told within a romantic triangle. A sojourn in America after the death of his wife exposed him to a variety of filmmaking styles and this influence was evident in *La Roue* (1923) and his masterpiece ▷ *Napoleon Vu Par Abel Gance* (1927), a rousing celebration of Bonaparte's life that was well ahead of its time in its fluid and restlessly mobile camerawork, use of frenziedly inventive ▷ montage, and climactic eruption into a triple-screen effect.

His first talking picture *La Fin du Monde* (1931) was not a success and his subsequent career was overshadowed by his achievements in the silent era. He developed a stereophonic sound process and frequently returned to an obsessive interest in the Bonaparte era, re-editing *Napoleon* into *Napoleon Bonaparte* (1934) and *Bonaparte et la Revolution* (1971).

Latterly restricting his work to adaptations of plays and popular novels, he enjoyed a commercial hit with *La Tour de Nesle* (1954) and ended his career with the costume dramas *Austerlitz* (1960), *Cyrano et d'Artagnan* (1964) and, for television, *Marie Tudor* (1965). He lived to see *Napoleon* restored to its original glory in 1980 and his work hailed afresh for its pioneering artistry.

GANZ, Bruno

Born 22 March 1941, Zurich, Switzerland.

After his military service, Ganz moved to West Germany to work in the Student Theatre and made his film debut in *Chikita* (1961). One of the co-founders of the Berlin theatre group Schaubuhne, he built a formidable theatrical reputation with leading roles in such productions as *Hamlet* (1967), *Torquato Tasso* (1969) and *Peer Gynt* (1972).

Aside from isolated film appearances in *Es Dach Uberem Chopf* (1962) and *Der Sanfte Lauf* (1967), he remained primarily a stage actor until 1975. He subsequently attracted international attention for his roles in *Die Marquise Von O* (The Marquise of O) (1976) directed by ▷ Eric Rohmer and *Lumière* (1976) by ▷ Jeanne Moreau.

Particularly associated with the directors who came to prominence during the flourishing of German cinema in the 1970s, he was seen to notable effect as the dying hit man in *Der Amerikanische Freund* (The American Friend) (1977), the central protagonist's husband in *Der Linkshändige Frau* (The Left-Handed Woman) (1977), the critically wounded biogeneticist in *Messer im Kopf* (Knife in the Head) (1978) and as Harker in the new version of *Nosferatu: Phantom der Nacht* (Nosferatu, The Vampire) (1979).

A soft-spoken man with intensely expressive eyes, he is adept at portraying laconic, taciturn or somewhat enigmatic characters removed from the mainstream of life and consumed by a search for self-knowledge or inner awareness. Reflective and thoughtful, he can portray a range of emotions from the philosophical to the heartrendingly romantic.

Continuing to associate with the best of European directors throughout the 1980s, his selective appearances include *Die Fälschung* (Circle of Deceit) (1981), *Dans La Villa Blanche* (In The White City) (1983) and *Der Himmel Über Berlin* (Wings of Desire) (1987), in which his grave demeanour and gentle tones brought a graceful note to his portrayal of a benign angel yearning to experience the emotions of mere mortals.

His infrequent appearances in English-language productions include *The Boys from Brazil* (1978) and *Strapless* (1989) and he made his directorial debut as the co-creator of the documentary *Gedachtnis: Ein Film für Curt Bois und Bernhard Minetti* (1982).

GARBO, Greta

Real Name Greta Lovisa Gustafsson
Born 18 September 1905, Stockholm, Sweden.
Died 15 April 1990, New York City, New York, USA.

A shop assistant and model who won a bathing beauty contest at the age of 16, Garbo also studied at Stockholm's Royal Dramatic Theatre (1922–4) and made her film debut as an extra in *En Lyckoriddare* (A Fortune Hunter) (1921).

Small film roles followed until she came under the influence of director Mauritz Stiller (1883–1927) who chose her new name and featured her as an Italian countess in *Gösta Berlings Saga* (The Atonement of Gösta Berling) (1924). She accompanied him to America in 1925 and he co-directed her in *The Temptress* (1926).

A langorous creature on whose sphinx-like face a map of pain and amorous anguish would be carved, the fine beauty of her features and delicacy of her understated acting were displayed in *Flesh and the Devil* (1927), *Love* (1927) and *A Woman of Affairs* (1927).

Publicised as a remote deity in the early stages of her American career, her sound debut in *Anna Christie* (1930) (Best Actress Oscar nomination) was promoted with the line 'Garbo Talks!' as

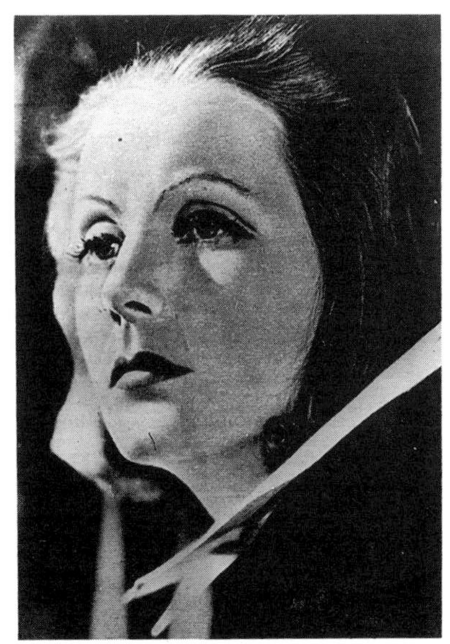

Greta Garbo in *Queen Christina* (1933)

if a miracle had occurred.

She continued to play women of mystery, who suffered the heartaches and misfortunes of impossible love in such exquisitely crafted works as *Queen Christina* (1933), *Anna Karenina* (1935) and *Camille* (1936) (Oscar nomination).

Reputed to be an insecure performer and timorous individual, she was off-screen for two years before returning to make fun of her icy image in the witty romantic comedy *Ninotchka* (1939) (Oscar nomination) ('Garbo Laughs') but was apparently distressed by her scathing reviews for the more unprepossessing farce *Two-Faced Woman* (1941) and retired from the screen shortly after its release.

She became an American citizen in 1951 and a reclusive figure for the rest of her life. The recipient of an honorary Oscar in March 1955 for her 'unforgettable screen performances', the award was accepted on her behalf by Nancy Kelly (1921–).

The numerous attempts to woo her back to the screen range from ▷ Hitchcock's *The Paradine Case* (1947) to *My Cousin Rachel* (1952) and even *Airport '75* (1974) but the closest she appears to have come to a return was a version of *La Duchesse De Langeais* in 1949 to have been directed by ▷ Max Ophuls and co-starring ▷ James Mason. A contract was signed but the project, like so many others, came to naught and the mystique remained untarnished.

GARDNER, Ava Lavinnia

Born 24 December 1922, Grabtown, Smithfield, North Carolina, USA.
Died 25 January 1990, London, UK.

Born on a tenant-farm, Gardner endured a tough Depression-era childhood before a visit to New York and a photographic session with her brother-in-law brought her to the attention of Hollywood and she was signed to a standard long-term contract with M-G-M.

She made her screen debut in the short *Fancy Answers* (1941) and appeared in many small roles before emerging from the ranks of decorative starlets with her magnetic portrayal of a ravishing femme fatale in *The Killers* (1946).

More interesting roles followed as she developed into one of the most glamorous stars of the era. A green-eyed brunette, once voted the world's most beautiful woman, she was given few really first-rate opportunities by her home studio but she impressed as the American socialite in *Pandora and the Flying Dutch-*

Ava Gardner in *The Killers* (1946)

man (1951), as Julie LaVerne in *Show Boat* (1951) and in *The Snows of Kilimanjaro* (1952), and she received her only Best Actress Oscar nomination as the tough, warmhearted showgirl in *Mogambo* (1953).

Over the next few years she brought an earthy combination of sensuality and cynicism to such roles as the Hollywood legend recalled in *The Barefoot Contessa* (1954), Lady Brett in *The Sun Also Rises* (1957) and, in one of her best roles, as the hotel-proprietor in *Night of the Iguana* (1964).

With sympathetic direction and a good role, she could be highly effective as can be seen from her performance as the Anglo-Indian woman whose loyalties are torn in *Bhowani Junction* (1956), directed by ▷ George Cukor. Too often however, she was cast merely for the vision of loveliness that she presented.

Never the most ambitious of performers, she later brought a warm presence and vast experience to character parts and supporting roles in a number of all-star ventures from *The Life and Times of Judge Roy Bean* (1972) to *Earthquake* (1974) and *The Bluebird* (1976) and her last work was in such television miniseries as *The Long Hot Summer* (1985), *Harem* (1986) and the long-running soap opera *Knot's Landing*.

Married to Mickey Rooney (1920–), Artie Shaw (1910–89) and ▷ Frank Sinatra, she gave her account of a tempestuous lifetime in a posthumously published autobiography, *Ava: My Story* (1990).

GARLAND, Judy

Real Name Frances Ethel Gumm
Born 10 June 1922, Grand Rapids, Minnesota, USA.
Died 22 June 1969, London, UK.

In one genuine instance of showbusiness running in the veins, Garland made her stage debut at three and performed as a member of The Gumm Sisters, making her film debut with them in *The Meglin Kiddie Revue* (1929).

Other novelty film appearances followed and she signed a contract with M-G-M that brought her employment in the studios' musicals and as Dorothy in the magical ▷ *The Wizard of Oz* (1939). A replacement for ▷ Shirley Temple, she won a special Oscar for the charm and poignancy of her performance and the song 'Over the Rainbow' became her lifelong trademark.

At a time when the American public seemed besotted by the precocious antics of juvenile stars like Temple, she was a distinctive young performer, full of unaffected high spirits and genuine abilities. Frequently partnered with Mickey Rooney (1920–), she brought youthful exuberance to such escapist entertainments as *Babes In Arms* (1939), *Strike Up the Band* (1940) and *Babes on Broadway* (1941).

Her transition to adult performer was less painful than most, and she went on to star in some of M-G-M's most accomplished and enduring musicals including the warmhearted *Meet Me in St Louis* (1944), *The Pirate* (1948) and *Easter Parade* (1948). She also impressed with a rare serious role in *The Clock* (1944).

Living under the studio's strict regime and beset by emotional difficulties and weight problems, her life grew more problematic and her ability to meet gruelling work demands diminished. After she had failed to fulfil her commitments to such films as *Annie Get Your Gun* (1950), *Showboat* (1951) and *Royal Wedding* (1951), the studio unceremoniously terminated her contract.

Playing concerts at the London Palladium and New York Palace Theatre she triumphed, as the throbbing emotion and power of her vocal talents made her an electrifying live performer. She returned to the screen as Vicki Lester in ▷ *A Star Is Born* (1954) giving a multifaceted tour de force of musical excellence and dramatic depth. When her Best Actress Oscar nomination failed to win her the statuette ▷ Groucho Marx declared it to be the biggest robbery since Brinks.

Gradually worn out by an increasingly turbulent lifestyle, she played only a few more roles, receiving a Best Supporting Actress Oscar nomination as a German housewife in *Judgement at Nuremberg* 1961 and making her last appearance in *I Could Go On Singing* (1963), a thinly veiled autobiographical portrait of a star's public adulation and private anguish. Films that she was announced for but never made include *Harlow* (1964) and *Valley of the Dolls* (1967).

On television, she starred in *The Judy Garland Show* (1963–4) and her daughter Liza Minnelli (1946–) has followed in her footsteps, winning an Oscar for *Cabaret* (1972).

GARSON, Greer

Born 29 September 1908, County Down, Northern Ireland.

A University graduate with aspirations to teach, Garson entered the advertising industry instead and pursued an interest in amateur dramatics, making her professional debut at Birmingham Repertory in *Street Scene* (1932). Moving to London, she enjoyed a successful West End career, playing opposite ▷ Laurence Olivier in *Golden Arrow* (1935); it was her performance in *Old Music* (1938) that led to the offer of a contract with M-G-M.

She made her film debut in *Goodbye Mr Chips* (1939) and although her screen time was brief the impact of her fresh, good-humoured personality earned her the first of seven Best Actress Oscar nominations and a call to Hollywood.

Her arrival at M-G-M coincided with the retirement of both Norma Shearer (1900–83) and ▷ Greta Garbo, allowing her to swiftly inherit the mantle of being the studio's first lady; if not the Queen of the lot she was at least the Duchess. Cast in their most prestigious productions, she was applauded for her regal dignity, luminous complexion and an unruffled manner that was leavened with a hint of mischievous red-headed charm.

Blossoms in the Dust (1941) (Oscar nomination) began a popular partnership with Walter Pidgeon (1897–1984) and the sentimental, propagandistic *Mrs Miniver* (1942), in which her ladylike English mother effortlessly coped with a Nazi intruder and the rigours of the Blitz, earned her an Oscar and was said to have hastened America's greater involvement in World War II. Other notable successes include the deft romance *Random Harvest* (1942) and such well-produced 'women's pictures' as *Madame Curie* (1943) (Oscar nomination) and *Mrs Parkington* (1944) (Oscar nomination).

She found less favour with post-War audiences and her film career declined. Her performance as Eleanor Roosevelt in *Sunrise at Campobello* (1960) brought a final Oscar nomination and her last film appearance was in Walt Disney's *The Happiest Millionaire* (1967).

Later stage performances include *Auntie Mame* (1958) on Broadway and *Captain Brassbound's Conversion* (1968) in Los Angeles. An infrequent television performer, her appearances in the medium include *The Little Foxes* (1956), *Crown Matrimonial* (1974) and *Little Women* (1978).

IL GATTAPARDO

see **THE LEOPARD**

GAUMONT, Leon Ernest

Born 10 May 1864, Paris, France.
Died 11 August 1946, Paris, France.

A manufacturer of photographic equipment, by 1896 Gaumont had progressed to the production and exhibition of films and, two years later, had built a small studio at Buttes-Chaumont.

An entrepreneur and inventor, he left the creative side of filmmaking to such associates as Alice Guy-Blaché (1873–1968), who directed *Esmeralda* (1905) and *La Vie Du Christ* (1906), screenwriter Louis Feuillade (1873–1925), who eventually directed cliffhanging serials like *Fantômas* (1913), and Émile Cohl (1857–1938) animator of *Le Cauchemar de Fantoche* (1908) and many others.

The studio was rebuilt in 1905, becoming the largest film production centre in Europe and production of regular newsreels commenced three years later. His dominance of pre-World War I Europe was accelerated by the establishment of an extensive network of production and distribution branches including a London outfit that began their famous newsreel series in 1910 and became an independent concern in 1927, retitled Gaumont-British.

He synchronised a projected film with a phonograph in 1901 and was responsible for a demonstration of talking pictures at the Academie des Sciences at Paris in 1910 and at the Royal Institute in London in 1912. He also introduced an early form of coloured cinema film using a three-colour separation method with special lenses and projectors.

He retired in the late 1920s and in 1928 the company was partially absorbed into the operations of M-G-M with a production arm continuing as Gaumont-Franco Film-Aubert until it went into liquidation in the early 1930s.

THE GENERAL
Buster Keaton & Clyde Bruckman (USA 1926)
Buster Keaton, Marion Mack, Glen Cavender, Jim Farley.
74 mins. b/w. silent. Reissued 1928 with musical score and sound effects.

The General remains the best known of the masterly silent comedy features directed by and starring stone-faced clown ▷ Buster Keaton, an artist whose ill-deserved but unavoidable decline during the sound era has since been substantially countered by a great resurgence in critical admiration. While the one time vaudevillian fortunately lived long enough to appreciate this belated recognition, *The General* provides ample evidence of his uniquely lugubrious comic persona, spellbinding physical agility and clinical filmmaking control. Loosely based on actual events during the American Civil War, the film has Keaton as a typically inconsequential railway engineer who, when faced by the Union forces' hijacking (with his intended on board) of his much-cherished locomotive 'The General', commandeers another engine to go behind enemy lines and prove his courage by retrieving both the loves of his life. His expression drolly implacable throughout a cavalcade of memorable sight gags – among them the risky spectacle of the hero seated on his engine's moving driving bar and the famous cannon-shot from train to train – all presented in long-shot to reveal the authentically daring stunt work, Keaton's always precise camera placement is often crucial to the comic impact and compares favourably with Chaplin's functional mise-en-scène which is concerned solely with recording the little tramp's pantomimic paces. Not that we really notice the direction or the seamless integration of the various action setpieces into the narrative however, for the keynote of Keaton's art and his deliciously unharried performances is the way their meticulous élan comes across as effortless.

GERE, Richard
Born 31 August 1949, Philadelphia, Pennsylvania, USA.

A talented musician and trumpet player, Gere wrote background music for school productions and eventually made his way on to the stage. A student at the University of Massachusetts, he later worked with the Provincetown Playhouse and Seattle Repertory Company and acquired extensive experience in the theatre that included the London production of *Grease* (1972), an appearance with the Young Vic company in *The Taming of the Shrew* (1973), acclaim for an off-Broadway appearance in *Killer's Head* (1975) by Sam Shepard (1943–) and further success in *Habeas Corpus* (1975) and *Awake and Sing* (1976).

Richard Gere in *American Gigolo* (1980)

He made his screen debut with a small role as a pimp in *Report to the Commissioner* (1975) and other supporting assignments followed before his electrifying impact as the edgily intense Italian stud in the controversial *Looking For Mr Goodbar* (1977).

The emotional intensity of his characterisations and his propensity for unselfconscious nudity in most of his films found him labelled as a smouldering sex object but that press image has tended to distract attention from his genuine dramatic abilities.

Stardom initially beckoned with leading man roles in *Days of Heaven* (1978) and *Yanks* (1979) and was confirmed with his performance as the self-absorbed, emotionally aloof male escort in *American Gigolo* (1980).

His popularity reached a peak as the cocky recruit with a tough exterior and a vulnerable heart in the bellicose *An Officer and A Gentleman* (1982) and he used this box-office potency to test his ambition as an actor; he brilliantly conveyed the vibrant narcissism of doomed smalltime car-thief Jesse Lujack in *Breathless* (1983), but was less convincing as the morally compromised Paraguayan doctor in *The Honorary Consul* (US: Beyond the Limit) (1983) or the cornet-player turned movie idol in *The Cotton Club* (1984).

The commercial failure of the Biblical story *King David* (1985) and *Power* (1985), a well-intentioned study of media manipulation, signalled a downturn in his career and he busied himself with much activity on behalf of Tibetan causes and Buddhism.

His only appearances in a while were in the mindless thriller *No Mercy* (1986) and the underrated rural melodrama *Miles From Home* (1988).

However, grey-haired and exuding a more mature charisma and authority, he re-established his potency with contrasting performances: he portrayed the chillingly malevolent corrupt policeman in *Internal Affairs* (1989) and the besotted business tycoon in the phenomenally popular *Pretty Woman* (1990), in which his relaxed and smoothly underplayed comedy technique subtly counterpointed the vivacity of newcomer ▷ Julia Roberts.

His only stage work in many years was a well-received Broadway production of *Bent* (1979–80). He will next be seen on screen in *Rhapsody in August* (1991).

GIBSON, Mel

Born 3 January 1956, Peekskill, USA.

Having emigrated to Australia with his family in 1968, Gibson expressed an interest in becoming a chef or a journalist before his sister sent an application form on his behalf to the National Institute of Dramatic Art in Sydney. After several stage appearances, he made his film debut in *Summer City* (1977) and showed early promise as the mentally retarded young man who falls in love with an older woman in *Tim* (1978).

Boyishly handsome and well built with a considerable screen presence, he was then cast as the title character in *Mad Max* (1979). His performance as the vengeance-seeking cop in a lawless, post-apocalypse society won international attention and the sensitivity of this brooding, vulnerable vigilante was explored in two action-packed sequels: *Mad Max II – The Road Warrior* (1981) and *Mad Max Beyond Thunderdome* (1985).

Mel Gibson in *Mad Max Beyond Thunderdome* (1985)

He increased his standing as the young soldier in *Gallipoli* (1981) and as the journalist in *The Year of Living Dangerously* (1982) and was well received in a 1983 Australian stage production of *Death of a Salesman*.

Accepting a number of American offers, he worked ceaselessly over the next two years but the results tended to highlight a certain callowness and inexperience on his part and he was often overshadowed by such strong leading ladies as Diane Keaton (1946–) in *Mrs Soffel* (1984) and Sissy Spacek (1949–) in *The River* (1984).

After a two year absence, he returned to the screen in *Lethal Weapon* (1987) where his reckless, near-suicidal cop seemed to reflect his own intensity, love of language and goofy sense of humour. A sequel, *Lethal Weapon 2* (1989), proved even more popular and his personal drawing-power was illustrated by the successes of such lacklustre projects as *Bird On A Wire* (1990) and (to a lesser extent) *Air America* (1990) which traded on his potent combination of romantic allure and near-psychotic penchant for life-endangerment.

His morally compromised drug dealer in *Tequila Sunrise* (1988) showed different facets of his talent and he gave a well-received interpretation of *Hamlet* (1990).

GILLIAM, Terry *see* MONTY PYTHON'S FLYING CIRCUS

GISH, Lillian Diana

Born 14 October 1896, Springfield, Ohio, USA.

An actress for almost ninety years and a screen force for three-quarters of a century, Gish's remarkable career began with her stage debut at the age of five.

She worked in the theatre, touring with sister Dorothy (1898–1968) in *Her First False Step* (1903–4); they made a joint film debut as featured players in *An Unseen Enemy* (1912) directed by ▷ D.W. Griffith.

Dedicated to the craft of acting, she became a regular member of the acting troupe in Griffith's films and rose to become the principal female character in ▷ *Birth Of A Nation* (1915) and to feature as the cradle rocker in ▷ *Intolerance* (1916).

Under Griffith's tutelage she created a gallery of sensitively delineated, waif-like heroines whose surface frailty often concealed an indomitable spirit and emotional intensity. Her greatest performances include *Hearts of the World* (1918), *Broken Blossoms* (1919) and *Orphans of the Storm* (1921).

Her work for other directors confirmed her as perhaps the pre-eminent silent actress and includes *La Bohème* (1926), *The Scarlet Letter* (1926) and *The Wind* (1928) in which she suffered mightily for her art as a genteel Virginian girl enduring the very vividly conveyed brutalities of life in the barren wilderness of Texas.

She made her sound debut in *One Romantic Night* (1930) and also appeared in *His Double Life* (1933) but the Hollywood of that era seemed to have no use for a woman whose acting style was unjustly perceived as archaically Victorian.

She worked extensively on the stage, playing Ophelia to the *Hamlet* of John Gielgud (1904–) in 1936 and appearing in *Crime and Punishment* (1947), *A Passage to India* (1962), *Romeo and Juliet* (1965) and *I Never Sang for My Father* (1967) among many others. She was also active on television and made occasional film appearances, earning a Best Supporting Actress Oscar nomination for her performance as the consumptive wife in *Duel In The Sun* (1946).

Notable among her rare later films were *Night of the Hunter* (1955), *A Wedding* (1978) and *Sweet Liberty* (1985) and she returned to a major screen role, offering an understanding and well-modulated interpretation of the warm-natured woman who has stoically endured the cantankerousness of sister ▷ Bette Davis in *The Whales of August* (1987).

She directed one film, *Remodelling Her Husband* (1920) and has written several volumes of memoirs including *Life and Lillian Gish* (1932) and *The Movies, Mr Griffith and Me* (1969). She received an honorary Oscar in April 1971 for her 'superlative artistry' and was the 1984 recipient of the ▷ American Film Institute Life Achievement Award.

GODARD, Jean-Luc

Born 3 December 1930, Paris, France.

A student at the Sorbonne, Godard met such initially kindred spirits as ▷ Francois Truffaut, ▷ Claude Chabrol and ▷ Eric Rohmer whilst attending screenings at the Cinématheque Française and his involvement with film became total as he produced *Quadrille* (1950), appeared in *Présentation et son Steack* (1951) and wrote criticism for *Gazette du Cinema* and *Cahiers du Cinema*.

He subsequently directed his first short film *Operation Béton* (1954) and worked in a variety of capacities from ▷ editor to actor and producer on the work of his contemporaries. A leading figure in the ▷ nouvelle vague, he was a passionate advocate of a more modern approach to cinema that railed against the stuffy traditions of narrative-driven films and saw more value in the pace, energy and populism of some American counterparts.

He began to replace theory with practice as the director of the feature ▷ *A Bout De Souffle* (Breathless) (1960), the story of a fugitive car thief and his American girlfriend that displayed a youthful vigour and freshness in its use of hand-held cameras, natural locations, and jump cuts to bustle along the story and distinguish it from more conventional approaches.

He continued to experiment with form and content in an attempt to achieve an artistic liberation whereby the technique of telling a story and the acknowledgement of the artifice taking place become almost as central a consideration as the narrative itself.

Films of note from the early part of his career include *Vivre Sa Vie* (It's My Life) (1962), a documentary-like exploration of prostitution and a valentine to the expressive features of his then wife Anna Karina (1940–); *Le Mépris* (Contempt) (1963), a fascinating autobiographical look at international filmmaking, the problems of communication and the failure of a marriage; *Pierrot le Fou* (1965), which makes bold use of colour in its ruminations on the transience of love and prevalence of global violence; and the futuristic *Alphaville* (1965), a part of a growing politicisation of his work in its depiction of a world where individuality and love have been suppressed.

Increasingly involved with left-wing groups and distressed by the adulterating commercial concerns of mainstream cinema, his work grew more radical and experimental as he sought cinematic explanations of the significance of words and images.

Tout Va Bien (1972) starring ▷ Jane Fonda and ▷ Yves Montand was his first film in many years aimed at a wider audience.

In recent years he has dismantled every individual element of the conventional dramatic and technical approach to filmmaking in a bid to revolutionise the medium and comment on the process of what is taking place.

The results have often been more of an obfuscation than illumination of his intentions but amidst a welter of confusion and obscurity the more accessible of his recent works include *Prénom Carmen* (First name: Carmen) (1983) a witty treatise on cinema, B-films and his own pretensions, *Je Vous Salue Marie* (Hail Mary) (1985) a stunningly photographed and inevitably controversial contemporary interpretation of the Nativity, and *Nouvelle Vague* (1990) starring Alain Delon (1935–) as two men of diverse character who allow a woman to discover the true priorities of her life.

THE GODFATHER

Francis Ford Coppola (USA 1972)
Marlon Brando, Al Pacino, James Caan, Robert Duvall, Diane Keaton.
176 mins. col.
Academy Awards: *Best Picture*; *Best Actor* Marlon Brando; *Best Screenplay* Francis Ford Coppola and Mario Puzo.
Followed by two sequels: *The Godfather Part II* (1974); *The Godfather Part III* (1990). Material from the first two films re-edited with previously unseen footage to form *Mario Puzo's The Godfather: The Complete Novel for Television* (1977).

The huge critical and commercial success of ▷ Francis Coppola's *The Godfather* launched its director on a subsequent career of almost operatic highs and lows, at the same time affirming the arrival of a new filmmaking generation within the American mainstream. While the film school credentials of Coppola's contemporaries and much of their work to date – compare ▷ Martin Scorsese's *Who's That Knocking on My Door?* (1969), ▷ George Lucas's *THX 1138* (1970) – had associated them with the European notion of an author's cinema, Coppola was then alone among the so-called 'Movie Brats' in having moved between studio journeyman work (1968 musical spectacular *Finian's Rainbow*), and more personal projects (1969's road movie *The Rain People*). Nonetheless, it was as hired hand that he took on the *Godfather* assignment.

Paramount had already been planning their adaptation of Mario Puzo's mafia chronicle as a major blockbuster, and with its bankable source, established stars and fashionable violence, Coppola's vividly recounted story of the Corleone family was certainly to fit the bill. Combining traditional Hollywood strengths (lavish production resources and narrative clarity) with a genuine cinematic artist's handling of muted colour, densely-texture composition and the nuances of performance, the film cast the crime 'family' as the apotheosis of the American dream; yet, such pseudo-radical irony aside, its primary appeal and achievement is that of the 19th century realist novel. As the expansive footage of two sequels and a television presentation amply demonstrate, here is an alternate society of complex social relations, ruthless machinations and explicit codes of behaviour, where the accumulation of coherent detail in itself becomes a luxuriant and satisfying experience.

GOLD DIGGERS OF 1933

Mervyn LeRoy (USA 1933)
Warren William, Joan Blondell, Ruby Keeler, Dick Powell, Ginger Rogers.
96 mins. b/w.

Hollywood's response to the widespread depression of the early Thirties took various forms. Although the period did produce so-called 'social conscience' movies along the lines of Mervyn LeRoy's (1900–87) *I Was a Fugitive from a Chain Gang* (1932), the industry for the most part concentrated on escapist product to distract the audience from harsh everyday realities, initiating a successful era for screen comedy, the horror film and the resurgent film musical. The annus mirabilis of 1933 saw Warner Brothers' ace choreographer ▷ Busby Berkeley, once regarded as box office poison, entirely revitalise the genre with a trio of classic 'puttin' on a show' extravaganzas; *42nd Street*, *Gold Diggers of 1933* and *Footlight Parade*. Sharing much of the same cast, notably love interest duo Dick Powell (1904–1963) and Ruby Keeler (1909–), and pretty much the same plot (Broadway show is threatened but goes ahead with plucky chorus girl drafted into starring role), the films' rudimentary wish fulfilment narratives would barely be worthy of note were it not for the startling artifice of Berkeley's production numbers. These well-drilled cavalcades of objectified sexuality – flesh at times turned to mere abstraction as the aerial camera peers down on Berkeley's characteristic kaleidoscopic patterning – proved hugely influential in the development of the medium by deviating from the set notion of filmed theatrical hoofing. To some extent, *Gold Diggers'* most famous numbers, the ironic *We're in the Money* and the elegiac *My Forgotten Man* sequences, are the most intriguing of all, because here at least the brilliant dance director's filmic world teasingly, obliquely offers comment on life outside the movie house.

THE GOLD RUSH

Charles Chaplin (USA 1925)
Charles Chaplin, Mack Swain, Tom Murray, Georgia Hale.
82 mins. b/w. silent. Reissued 1942 with music and narration at 74 mins.

▷ Charles Chaplin's critical reputation began to decline somewhere around the late Fifties, a process spurred on by the relative failure of his very late films, *A King in New York* (1957) and *A Countess from Hong Kong* (1967), and the rediscovery around the same time of ▷ Buster Keaton, whose work appears rather more interesting as cinema (cf. ▷ *The General*) and whose understated personality has much more appeal these days than Chaplin's ripe sentimentality. Nonetheless, Chaplin's achievement should not be undervalued: he is after all the one figure who first embodied the cinema's worldwide power of communication, and his finest Twenties feature *The Gold Rush* represents a joyful highpoint in his art before the sound revolution, marital problems and political controversy rendered his subsequent output rather more contentious. Typically, the film contrasts the innate goodness of the outcast little tramp with the rampant unscrupulousness of his fellow prospectors, while all concerned struggle for survival in Alaska's frozen wastes; his roommate's half-starved delirium might very well transform our hero into a human chicken to be pursued around the room, but Chaplin's protagonist still retains his benevolence of spirit and an enviable resourcefulness. Thus, the famous scene which he fends off the thought of cannibalism by tucking into his boots – the hobnails become bones to be sucked dry, the laces succulent strands of spaghetti – and from the most humdrum of props, Chaplin creates one of the screen's most enduring comic moments.

GOLDEN GLOBE AWARDS

Presented annually since 1944, the Golden Globes are the awards of the Hollywood Foreign Press Association (itself formed in 1940), a group of around 80 foreign journalists who cover the entertainment industry in Los Angeles for the delectation of a global readership in excess of 100 million. Given on much the same kind of basis as the ▷ Oscars, the main difference is that the Golden Globes distinguish between work in the categories of Drama and Comedy/Musical which means more awards and more recognition for work in the latter fields that are notoriously under rewarded in Oscar terms. Other distinctive awards have been given for Most Promising Newcomer (now Best Acting Début) and in the category of World Film Favourites. The ▷ Cecil B. De Mille Award salutes a lifetime of achievement within the industry and there are also awards given for television work. The results of the ceremony in late January are now regarded as one of the prime indicators as to who will win the Spring Oscars.

GOLDWYN, Samuel

Real Name Samuel Goldfisch
Born 27 August 1882, Warsaw, Poland.
Died 31 January 1974, Beverly Hills, Los Angeles, California, USA.

An orphan who ran away to England at the age of eleven and worked as a blacksmith's apprentice, Goldwyn arrived in America during the last years of the 19th century and found work as a glove salesman.

Rising to a modest prosperity on his wits and business acumen, he joined with his brother-in-law Jesse Lasky (1880–1958) and ▷ Cecil B. De Mille in a venture to make motion pictures. An early result was *The Squaw Man* (1913), an influential title in establishing the varied terrain and temperate conditions of California as an attractive home for film-making.

He subsequently left their company to form a new company with the Selwyn brothers entitled Goldwyn Pictures (1918) and from this came his new name. By 1922, he had formed his own Samuel Goldwyn Productions and was to remain a stubbornly independent producer for the rest of his career.

His first solo effort was *Potash and Perlmutter* (1923) and he enjoyed successes with *Stella Dallas* (1925), *The Winning of Barbara Worth* (1926), *Bulldog Drummond* (1929) and *Arrowsmith* (1931) (Best Picture Oscar nomination).

Among the several stars under contract to him over the years were Eddie Cantor (1892–1964), Ronald Colman (1891–1958) and Danny Kaye (1913–87), although his exhaustive efforts to foist the talents of Anna Sten (1908–) on an unappreciative American public were fruitless.

Attracted to prestigious literary titles and stage productions of high repute, he brought glossy good taste and high production values to such films as *Dodsworth* (1936) (Oscar nomination), *Dead End* (1937) (Oscar nomination), *Wuthering Heights* (1939) (Oscar nomination), *The Little Foxes* (1941) (Oscar nomination) and ▷ *The Best Years of Our Lives* (1946), which won an Oscar as Best Picture.

The gaudy vulgarity in his character was more apparent in a succession of colourful musicals and comedies, among them *Roman Scandals* (1933), *Up in Arms* (1944), *The Secret Life of Walter Mitty* (1947), *Guys and Dolls* (1955) and his last, *Porgy and Bess* (1959).

A showman with a legendary reputation for malapropisms, the many 'Goldwynisms' attributed to him include 'a verbal contract isn't worth the paper it's written on', 'in two words: impossible' and 'include me out'.

He received the Irving Thalberg Award in 1946 and the Jean Hersholt Humanitarian Award in 1958. His son Samuel Goldwyn Junior (1925–) has produced a number of films including the recent *Stella* (1990).

GONE WITH THE WIND

Victor Fleming (USA 1939)
Vivien Leigh, Clark Gable, Leslie Howard, Olivia De Havilland, Hattie McDaniel.
220 mins. col. 1969 blown-up to widescreen 70mm. 1989 restored print reissued in original academy ratio.
Academy Awards: *Best Picture*; *Best Director*; *Best Actress* Vivien Leigh; *Best Supporting Actress* Hattie McDaniel; *Best Screenplay* Sidney Howard; *Best Colour Cinematography* Ernest Halleer; *Best Editing* Hal C. Kern and James E. Newcom; *Best Interior Decoration* Lyle Wheeler; *Special Awards to* William Cameron Menzies *for Colour Achievement and to* Don Musgrave and Selznick International Pictures *for use of Co-ordinated Equipment.*

If ▷ D.W. Griffith's ▷ *Birth of a Nation* encapsulates the panoply of American film technique up to 1915, then another Civil War epic, producer ▷ David O. Selznick's truly phenomenal screen presentation of Margaret Mitchell's best-seller *Gone with the Wind*, takes us up to 1939. Having secured the rights before publication in 1936 to a source novel whose popularity was to rank second only to the Bible, the ambitious Selznick single-mindedly set out to turn the melodramatic story of southern belle Scarlett O'Hara's wartime romance with dangerously dashing rogue Rhett Butler into the biggest movie event America had yet experienced. Over the next three years, every move Selznick made was calculated to whip up a flurry of excited publicity: he wrested ▷ Clark Gable, the biggest male star of the day, from MGM; he conducted a high-profile search across America to find the ideal Scarlett O'Hara before casting almost unknown English actress ▷ Vivien Leigh; he battled the Production code over the final inclusion of a mild expletive; and in honour of a final product that was the longest and most expensive film ($4.25 million) ever made, he held three days of celebrations before the Atlanta world premiere in December 1939. A slew of writers may have worked on the script and a trio of directors – Victor Fleming (1883–1949), ▷ George Cukor and Sam Wood (1884–1949) – may have been involved, but the flamboyant sweep, charismatic star performances and tearful emotional appeal of this intimate spectacular have continued to win over audiences and preserve its status as the most famous of Hollywood pictures. *Gone with the Wind* remains perhaps a victory for logistics over artistry, but, like ▷ Francis Coppola's ▷ *The Godfather* or the later work of ▷ David Lean, it is a supreme example of a novelistic cinema lavishly-achieved.

Clark Gable and Vivien Leigh in *Gone With The Wind* (1939)

GOOD LIFE

UK 1974–78

Though a strong case could be made for the Terry Scott (1927–)/June Whitfield (1925–) vehicle *Terry and June* (1979–87), it is *The Good Life* which has come to epitomise the latterday British sitcom. A (now) typical tale of ordinary middle class folks coming to terms with the slightly unusual, it centred on the efforts of draughtsman Tom (Richard Briers (1934–)) and wife Barbara (Felicity Kendal (1947?–)) to become self-sufficient in suburban England. The couple were treated sympathetically by writers Bob Larbey and John Esmonde, though their efforts to be ecologically-sound predated the rise in green consciousness by more than a decade.

The humour sprang from the reactions of their neighbours, the snootily dismissive Margo (Penelope Keith (1939–)) and Jerry (Paul Eddington (1927–)), who, as well as being troubled by the noise from the pigs in the back garden, were constantly undermining all efforts to get back to the land.

The series was a great success, and is still a popular repeat. All the principals have gone on to enjoy further successes; Eddington with *Yes, Minister*, Keith in *To The Manor Born*, Kendal as *The Mistress* and Briers in *Ever Decreasing Circles*, coffee commercials and as part of the theatrical endeavours of ▷ Kenneth Branagh (1960–). Writer Larbey scored another hit with the comedy *A Fine Romance* (1981–4).

GRADE, Lew

Real Name Louis Winogradsky
Born 25 December 1906, Odessa, Russia.

The eldest of three brothers who were to dominate British showbusiness management for four decades, Grade arrived in Britain in 1912 with his parents and younger brother Boris, who changed his name to Bernard Delfont (1909–).

The brothers both became dancers (semi-professional at first) winning competitions during the Charleston craze of the 1920s but both gave up to become theatrical agents, booking variety acts into theatres.

Whilst Delfont became a noted theatrical impresario, responsible for many Royal Variety Performances, Grade managed a theatrical agency until 1955 before becoming an early entrant into the world of commercial television, serving as managing director of ATV (Associated Television Corporation) until 1973 and chairman thereafter until 1977.

His style has always been to seek the popular touch with television series designed to appeal to the home audience and assist his attempts to penetrate the lucrative American market. Successes include ▷ *The Adventures of Robin Hood*, *Danger Man* (1959–62), ▷ *The Saint*, and *The Julie Andrews Hour* (1972). His company was also responsible for the long-running soap-opera *Crossroads* which began in 1964.

He later moved into film production with a slate of films designed as international crowd-pleasers that proved uniformly poor in quality and generally failed under any definition of quality entertainment or popular success. Titles include *March or Die* (1977), *Escape to Athena* (1979) and the particularly disastrous *Raise the Titanic* (1980).

For many years his vast business empire appeared to survive on the profitability of his association with ▷ Jim Henson's Muppet characters and such films as *The Muppet Movie* (1979) and *The Great Muppet Caper* (1981).

More recently, he has been responsible for cable versions of romantic bodice-rippers like *The Lady and the Highwayman* (1989).

He was made a life peer (Baron Grade of Elstree) in 1976 and published an autobiography *Still Dancing* in 1987.

His son Michael Grade (1943–) served as Controller of BBC1 (1984–86) and Director of Programmes at the BBC (1986–88) and is currently Chief Executive of Channel Four (1988–).

THE GRADUATE

Mike Nichols (USA 1967)
Dustin Hoffman, Anne Bancroft, Katharine Ross, William Daniels.
108 mins. col.
Academy Award: *Best Director*.

While 1968's upsurge of violent opposition to the status quo in general and the Vietnam war in particular would be succeeded by a slew of would-be radical 'protest' movies as diverse as Richard Rush's (1930–) *Getting Straight* (1970) and ▷ Michelangelo Antonioni's *Zabriskie Point* (1969), Mike Nichols' enjoyable satire was one of a number of earlier films – ▷ Arthur Penn's ▷ *Bonnie and Clyde*, ▷ Peter Bogdanovich's *Targets* (both 1967) among them – in which the young protagonists appeared rather adrift from the constraining values of the older generation and society around them. In his major film debut, ▷ Dustin Hoffman's awkward college graduate Benjamin Braddock is no revolutionary; instead, and perhaps more sympathetically, he is an ordinary junior member of the middle classes, rather isolated from his parents, whose dispiriting comic affair with Anne Bancroft's (1931–) predatory married older woman, leaves him questioning the set of cosy bourgeois values it's assumed he will comfortably adopt. Confidently maintaining an undercurrent of seriousness beneath the film's often farcical encounters, director ▷ Nichols consistently shoots from Benjamin's point of view, often distorting the image to formally italicise the protagonist's gnawing alienation. Stylistic panache aside, the eminently agreeable pop soundtrack and a sympathetic eye for the social traumas of young adulthood were more than enough to make the 18–25 age group take to the film as if it were their very own.

LA GRANDE ILLUSION

Jean Renoir (France 1937)
Erich Von Stroheim, Jean Gabin, Pierre Fresnay, Marcel Dalio.
117 mins. b/w. 1946 reissued with authorised cuts. 1958 restored to original version.

La Grande Illusion is the most popular and accessible work by great French director ▷ Jean Renoir, whose richly varied *oeuvre* spanning some 45 years has seen certain commentators acclaim him as the finest filmmaker of all. Renoir's agility across a broad spectrum of thematic approaches has often indeed tended to befuddle the strictest of auteurist critics, whose championing of 1939's rather mischievous satiric escapade ▷ *La Regle du Jeu* has perhaps threatened to overshadow the fine achievement of *La Grande Illusion.* Although the latter is usually read as a more conventional exercise in anti-war humanism, both pieces offer wise reflection on the flexibility of previously rigid social constructs. Set in a World War I German POW camp, *La Grande Illusion* shows how the camaraderie between officers and soldiers on both French and German sides transcends the barriers of class, creed, rank and even nationality. Thus, when Pierre Fresnay's (1897–1975) upper class French officer allows his opposite number and fellow aristocrat ▷ Erich Von Stroheim to shoot him, (so allowing a pair of French internees – proletarian ▷ Jean Gabin and Jewish nouveau riche Marcel Dalio (1901–) – to escape to freedom), his fatal self-sacrifice is as much an admission of the passing of the old order and the dawn of a new society as it is an act of patriotic heroism. However, in the brave new world outside the two escaped men are still looked on as working class ruffian and hated Jew once more; the great illusion referred to by the title ironically drawing out the contrast between the ideals of brotherhood experienced in confinement and the stubborn prejudices of 'free' society.

GRANT, Cary

Real Name Alexander Archibald Leach
Born 18 January 1904, Bristol, UK.
Died 29 November 1986, Davenport, Idaho, USA.

Leaving school to join a troupe of comedians and acrobats, Grant travelled with them to America and remained there to perform in circuses, at Coney Island and in vaudeville.

Progressing to the legitimate theatre, he appeared in such productions as *Golden Dawn* (1927), *Boom, Boom* (1929) and *Nikki* (1931) before signing a contract with Paramount and making his film debut in the short *Singapore Sue* (1932).

A leading man to such flamboyant stars as ▷ Marlene Dietrich in *Blonde Venus* (1932) and ▷ Mae West in *She Done Him Wrong* (1933) and *I'm No Angel* (1933), he developed in to a polished light comedian and romantic idol.

In comedies that demanded verbal dexterity and physical agility he had the ability to make his every action and gesture appear spontaneous and effortless and he excelled in such fast-talking, outlandish ▷ screwball comedies as *The Awful Truth* (1937), *Bringing Up Baby* (1938), *His Girl Friday* (1940) and *My Favourite Wife* (1940).

His dramatic abilities were evident in *Only Angels Have Wings* (1939) and *Gunga Din* (1939) and he received Best Actor Oscar nominations for the romantic melodrama *Penny Serenade* (1941) and his portrayal of a Cockney waif in *None But the Lonely Heart* (1944).

The epitome of wittily debonair sophistication in ▷ *The Philadelphia Story* (1940), he also delighted as the manic nephew, indulging in a plethora of pratfalls and double-takes in *Arsenic and Old Lace* (1944).

▷ Alfred Hitchcock cast him against type as the sinister-seeming husband in *Suspicion* (1941) and the callous American agent in *Notorious* (1946) and the two men collaborated to memorable effect on the elegant *To Catch A Thief* (1954) and *North By Northwest* (1959), a classic tale of an innocent man embroiled in a web of larcenous intrigue.

Latterly, making films for his own production company he enjoyed a string of successes that stand as a testimony to his sure touch and enduring charm. Notable titles include *Operation Petticoat* (1959) and *Charade* (1963).

After his last acting appearance in *Walk Don't Run* (1966), he announced his retirement from the screen. He received an honorary Oscar in April 1970 for his 'unique mastery of the art of screen acting' and remained resolutely unavailable for work despite many offers, among them *Man of La Mancha* (1972), *That's Entertainment* (1974), *Heaven Can Wait* (1978) and *Gorky Park* (1983).

THE GRAPES OF WRATH

John Ford (USA 1940)
Henry Fonda, Jane Darwell, John Carradine, Charley Grapewin.
128 mins. b/w.
Academy Awards: *Best Director*; *Best Supporting Actress* Jane Darwell.

On August 8th 1939 author John Steinbeck personally approved Nunally Johnson's (1897–1977) screenplay adaptation of his novel *The Grapes of Wrath*, the latest step in producer ▷ Darryl F. Zanuck's drive towards filming the writer's emotive chronicle of Depression America. On both printed page and celluloid, *The Grapes of Wrath* broke new ground with its startlingly direct treatment of the Thirties' mass farmland evictions. The Joad family are left with only the few belongings they can load on to their truck before heading on the great trek west – only to find exploitative enslavement to the fruit harvest when they get there. The narrative follows ▷ Henry Fonda's untutored moral reaction, as the son and ex-con Tom Joad, to the injustice around him before he finally carries on the labour agitation started by John Carradine's (1906–1988) ill-fated preacher Casey in response to the employers' unscrupulous machinations. Such indeed is our belief in the integrity of this imperfect soul's moral conviction that the audience, surprisingly for a Hollywood movie of the period, is actually left empathising with this radical activity. Still, Zanuck's added coda, which has Jane Darwell's (1880–1967) Ma Joad professing her faith in 'the people', probably runs truer to the life-giving communal spirit frequently the keystone of Ford's work. The plight of the Okies here is not so far from that of the townsfolk in *My Darling Clementine* (1946), the cavalry division in the *Fort Apache* (1948) trilogy or even the native Americans in *Cheyenne Autumn* (1964) – the Fordian imperative throughout is the quest by the group to secure a peaceful home they can call their own.

THE GREAT TRAIN ROBBERY

Edwin H. Porter (USA 1903)
George Barnes, A.C. Abadie, Marie Murray, G.M. Anderson.
10–12 mins. b/w. silent.

At the turn of the century the new-fangled ▷ nickleodeon was just one of a number of forms of popular entertainment alongside stage melodrama, circuses, vaudeville and magic lantern shows; the challenge to the new medium was to develop as an artform while at the same time communicating with the audience in a way they could understand. As one of America's first story films, ▷ Edwin S. Porter's *The Great Train Robbery* was a significant point in the process. Billed as 'The sensational and startling "hold up" of the "gold express" by famous western outlaws', the significance of the film was that its twenty shots connected to tell the story of an exciting chase, and as such its complexity and impact dwarfed the mere curiosity value of the single scene subjects so common in the era when the movies were not much more than a fairground attraction. Although each piece of action, from the bandits breaking into a telegraph office to their actual robbery of the train, was presented unedited as a kind of visual tableau, Porter was among the first to generate excitement by cutting between two points of view – as the bandits make off with their booty, he shifts back to the office to show the telegraph operator's daughter freeing her father, who then raises a posse to chase after the desperadoes. After a climactic gunbattle, Porter then pulls off his most extraordinary visual coup, when the film's first close-up has an unruly varmint pointing his rifle straight into the camera; for an audience who had never actually seen a close-up of any kind before it was a moment as thrilling as that first sight of the ▷ Lumière brothers' train pulling into the station.

GREED

Erich Von Stroheim (USA 1924)
Gibson Gowland, Zasu Pitts, Jean Hersholt, Tempe Pigott.
Initial release prints c. 160 mins. b/w. silent. 1986 new print, with added score composed by Carl Davis, runs 132 mins. Many existing prints run 110 mins.

Not one of the films made by the extravagant Austrian expatriate ▷ Erich Von Stroheim today exists as he originally intended it. The greater part of his Twenties output offering a sardonic vision of European decadence, *Foolish Wives* (1921) offered the first evidence that his relentless eye for authenticity would create results too expansive or outrageous for the producers of the day. Adapted from Frank Norris's chronicle of San Franciscan low-life *McTeague*, the subject matter of Von Stroheim's 1924 masterpiece *Greed* made it an exception in his *oeuvre*, but the film's treatment at the hands of production company Metro-Goldwyn still exemplifies the ongoing conflict of creative cinematic artist and intransigent moneymen. Von Stroheim achieved a visual equivalent of Norris's Zolaesque naturalism by shooting everything on location, but this heavily-detailed realist style pushed the running time of the finished result to lengths unprecedented even by Von Stroheim's ever-generous standards. From his first nine-hour assembly he reduced the footage to seven and then four hours, but producer ▷ Irving Thalberg still baulked at the notion of screening the film over two consecutive nights. Metro then carried out their own drastic cuts, replacing Von Stroheim's original titles and preserving only the skeleton of his original conception in a final release print under three hours in length. For all that, the version of *Greed* remaining today is still a film of tremendous visceral power, its final desperate scenes of murder and betrayal under the pitiless Death Valley skies a searing indictment of human avarice. Because the studio burned all the cut material to extract the silver nitrate in the film stock, the full amplitude of Von Stroheim's work cannot be assessed.

GREENAWAY, Peter

Born 5 April 1942, London, UK.

Trained as a painter, Greenaway first exhibited at the Lord's Gallery in 1964. Employed at the Central Office of Information (1965–76) he worked as an editor and began making his own short films whose bald titles like *Train* (1966), *Tree* (1966) and *Windows* (1975) conceal their idiosyncratic approach to such subjects as numerology and taxonomy.

He later gained a reputation for originality and invention on the international festival circuit with such works as *A Walk Through H* (1978) and *The Falls* (1980), which contains brief biographies of 92 people who have fallen victim to a Violent Unknown Event involving birds. The characters all share a first name beginning with Fall and the film stands

Helen Mirren and Michael Gambon in *The Cook, The Thief, His Wife and Her Lover* (1989) directed by Peter Greenaway

as a vast cinematic encyclopaedia to the many facets of his artistic concerns from ornithology to phenomenology and could be seen to represent 92 ways that the world will end or, indeed, 92 ways of making films.

He won further critical acclaim and a wider audience with *The Draughtsman's Contract* (1982), a sumptuous brainteaser of 17th century intellectual and sexual intrigue.

He has subsequently pursued a prolific career utilising ravishing visual compositions from Sacha Vierny (1919–), a painterly sense of colour and the throbbingly insistent music of Michael Nyman (1944–) to explore such preoccupations as sex, death, decay and gamesmanship.

A maker of multi-layered narratives that could inspire book-length works of explanation and interpretation, he covered the sombre topics of bereavement and the place of death in the evolutionary chain with *A Zed & Two Noughts* (1985), in which twin brothers grieve for the wives they have lost in a car crash caused by an escaped swan, and *The Belly of an Architect* (1987), in which American architect Brian Dennehy's (1940–) physical and spiritual decline is contrasted with his wife's pregnancy.

Criticised for the inaccessibility and intimidating intellectual rigour of his work, he has contested such charges with the grand passions and accessibility of his more recent work.

The black-edged fairytale *Drowning By Numbers* (1988) is an entertaining story of visual delight and verbal panache on the conspiracy of women, as three generations of women, all called Cissis Colpitts, systematically dispose of their male companions by drowning them in a tin bath, the sea and a swimming pool respectively. As the aquatic murders unfold, a series of numbers from one to one hundred run throughout the course of the film; hence the title.

More controversial was *The Cook, The Thief, His Wife and Her Lover* (1989), a supremely stylish Jacobean saga of betrayal and revenge played out on a broad canvas of emotions with performances, music and settings to match the ferocious intensity and swagger of the director's vision.

His latest film, *Prospero's Books* (1991) stars John Gielgud (1904–) in a meditation on themes from Shakespeare's *The Tempest*.

GREENE, Hugh Carleton (Sir)

Born 15 November 1910, Berkhamsted, Hertfordshire, UK.
Died 19 February 1987, London, UK.

After studying at Merton College, Oxford, Greene, the younger brother of novelist Graham Greene (1904–91), moved to Germany to work as a foreign correspondent, first for the *Daily Herald* and later for *The Daily Telegraph* (1934–39).

In 1940, he joined the BBC to work on propaganda broadcasts to Germany which ultimately led to his 1946 appointment as controller of broadcasting in the British zone of Germany, in which capacity he helped rebuild the country's peacetime radio service.

He was Controller of the BBC's Overseas Service from 1952 to 1956 and was also the BBC's first director of news and current affairs (1958–60) before being chosen as Director-General.

In this position from 1960 to 1969, he injected fresh vigour into the BBC encouraging it to compete with the Independent Television network and creating a liberal climate in which programme-makers flourished. Among the many notable series to emerge under his leadership were the satirical late-night revue ▷ *That Was The Week That Was* (1962–63), ▷ *Dr Who*, various innovative situation-comedies and the controversial ▷ *Till Death Do Us Part*.

GRIERSON, John

Born 26 April 1898, Deanston, Kilmadock, Stirlingshire, UK.
Died 19 February 1972, Bath, UK.

A serving member of the Royal Naval Volunteer Reserve (1917–19), Grierson later graduated from Glasgow University with a degree in philosophy. Winning a Rockefeller Scholarship to study American politics and newspapers, he travelled to the University of Chicago and developed an interest in the science of mass communication.

His observations found an application back home when he returned to join the Empire Marketing Board Film Unit and set about devising the most efficacious means of communicating and promoting the various facets of the British Empire.

The individual most responsible for the development of ▷ documentary, he had applied the term to the work of ▷ Robert Flaherty and now set about emulating his attempts to capture aspects of the human character revealed in arduous work and the pursuit of a livelihood. He directed *Drifters* (1929) about herring fishing in the North Sea, and began building the EMB Film Unit into a sympathetic home for directors who would use their talents to convey to citizens the workings of the government and give them a sense of participating in the life of the nation. His efforts laid the foundations of the British documentary movement and also brought a strain of poetry to the realisation of a social purpose in such films as *Song of Ceylon*

(1934) and *Night Mail* (1936), which he produced and co-wrote.

When the EMB was dissolved and transferred to the General Post Office, he became head of the GPO Film Unit (1933–37) and later served as the Film Commissioner of Canada (1939–45) establishing the National Film Board of Canada.

From the late 1930s, the work he produced had a more cutting political edge as the documentary movement began to seek independent funding for sponsored films depicting the evils of the day, such as poor educational standards in *Children at School* (1937) and pollution in *The Smoke Menace* (1939). He was also influential in securing non-theatrical distribution of documentaries so that film could more directly feature in the fabric of everyday lives via a presence in church halls, schools and factories.

Director of Mass Communications for UNESCO (1946–48) and film controller at the Central Office of Information in London (1948–50), he was an executive producer on such features as *The Brave Don't Cry* (1952) and *Laxdale Hall* (1952), when the National Film Finance Company established the short-lived experimental body Group Three.

He served as a member of the Films of Scotland Committee from 1954, participating in the scripting of the Oscar-winning documentary *Seaward the Ships* (1959) and became familiar to television viewers as the doughty presenter of the documentary series *This Wonderful World* (1957–65).

His book *Grierson on Documentary* (1946) is a standard text.

GRIFFITH, D(avid) W(ark)

Born 23 January 1875, La Grange, Kentucky, USA.
Died 23 July 1948, Los Angeles, California, USA.

A touring actor with a variety of regional stock companies and a journalist with the *Louisville Courier* (1897–99), Griffith later appeared in vaudeville and on the legitimate New York stage before being hired to write motion picture scenarios for the Edison Company and starring in the film *Rescued from the Eagle's Nest* (1907).

Hired by the Biograph Studio in 1908 as a scriptwriter and director, he made around five hundred one and two-reel films over the next five years, a collection of westerns, romances and adventure yarns ranging from *The Adventures of Dolly* (1908) to the early work of ▷ Mary Pickford in *The Violin Maker of Cremona* (1909), *A Romance of the Western Hills* (1910) and many others, and *The Unseen Enemy* (1912), a thriller that introduced Dorothy (1898–1968) and ▷ Lillian Gish to the screen.

An innovator in the use of film technique, he is generally regarded as the first American film artist to make expressive use of dramatic devices like close-ups,

D. W. Griffith

flashbacks and ▷ cross-cutting that were to become standard elements in narrative construction.

Influenced by the lavish scale and ornateness of the Italian epic productions then being seen in America, he made the longer form *Judith of Bethulia* (1914) and then the revolutionary three-hour ▷ *The Birth Of A Nation* (1915). A story of North and South during the Civil War that impressed with the panoramic sweep of its storytelling, tumultuous battle scenes and flowing narrative pace, it was marred by a racist perspective that rendered the Ku Klux Klan in a heroic light.

He followed this with ▷ *Intolerance* (1916), another major advance in the art of American filmmaking that inter-cuts four stories from different historical periods to build a catalogue of injustice and iniquity through the ages.

He subsequently worked with ▷ cinematographer Billy Bitzer (1874–1944) and actress Lillian Gish on simple, tenderly sentimental and luminously beautiful romantic stories of hardship, tragedy and self-sacrifice like *Hearts of the World* (1918), *Broken Blossoms* (1919) and *Way Down East* (1920).

In 1919, together with ▷ Charles Chaplin, ▷ Douglas Fairbanks and ▷ Mary Pickford, he formed United Artists, but his Victorian sensibility and tendency to preach sound moral attitudes were already becoming outmoded as audiences embraced the jazz age and the work of ▷ Cecil B. De Mille and ▷ Erich Von Stroheim. Such films as the epic *America* (1924) and *Drums of Love* (1928) were not successful and he made only two sound films *Abraham Lincoln* (1930) and his last, *The Struggle* (1931).

He received a special Oscar in 1936 for his 'lasting contribution to the progress of the motion picture arts' but he was unable to secure backing for any projects during the last fifteen years of his life and his final involvement with the medium he had helped to maturity was as a consultant on *One Million Years B.C.* in 1939.

GRIP A stagehand whose tasks can be many and varied but generally comprise shifting scenery, moving props, setting up heavy equipment or laying the tracks used in ▷ dolly shots. The physically demanding nature of the job and the need for a 'firm grip' explains the job title. The individual in charge of these stagehands or 'grips' is called the key grip.

GUINNESS, Alec (Sir)
Born 2 April 1914, Marylebone, London, UK.

Receiving a scholarship to study acting at the Fay Compton Studio of Dramatic Art, Guinness made his film debut as an extra in *Evensong* (1933) and his stage debut in *Libel* (1934).

A leading light in the Old Vic company of John Gielgud (1904–), he remained exclusively a theatre actor until his breezy appearance as Herbert Pocket in *Great Expectations* (1946), the first of six films with director ▷ David Lean.

He immediately proved his versatility as a grotesque Fagin in *Oliver Twist* (1948) and won popularity for his eight turns as members of the D'Asocyne family in the black comedy ▷ *Kind Hearts and Coronets* (1949).

Associated with the comedies of Ealing Studios, his acute sense of characterisation and a diffident manner that proved highly appealing were well to the fore as the timid bank clerk in ▷ *The Lavender Hill Mob* (1951), the bumbling, thoughtless inventor known as *The Man in the White Suit* (1951), and as the leader of the inept crooks in *The Ladykillers* (1955).

A chameleon, content to submerge his own personality within the individual traits and emotional requirements of any given character, his dramatic abilities were also seen as a tortured, obdurate cardinal in *The Prisoner* (1955), in the intensity he brought to the obsessive, single-mindedness of misguided officer Nicholson in *The Bridge on the River Kwai* (1957) (Oscar) and as the bombastic, fox-terrier braggart of a military bully in *Tunes of Glory* (1960), an uncharacteristically expansive performance that ranks alongside his finest work.

The latter stages of his career have been distinguished by a range of supporting assignments and character roles from Prince Feisal in ▷ *Lawrence of Arabia* (1962) to intergalactic mentor Obi-Wan Kenobi in ▷ *Star Wars* (1977) (Oscar nomination).

However, attempts to explore the parameters of his ability have resulted in some poorly received roles like Koichi Asano in *A Majority of One* (1961), Professor Godbole in *A Passage to India* (1984) and the title character in *Hitler: The Last Ten Days* (1972) which bordered on the unintentionally comical.

He refined his minimalist, self-effacing style of performance even further on television as enigmatic spy master George Smiley in *Tinker, Tailor, Soldier, Spy* (1979) and *Smiley's People* (1982) and his most recent film work includes *Little Dorrit* (1987), *A Handful of Dust* (1989) and the forthcoming *Kafka* (1991).

Stage work over the years includes *The Cocktail Party* (1950), *Ross* (1960), *Habeas Corpus* (1973), *The Old Country* (1977–78) and *A Walk in the Woods* (1989).

He was knighted in 1959 and he received a special Oscar in April 1980 for his 'host of memorable and distinguished performances'. He had been Oscar nominated as the writer of the screenplay for *The Horse's Mouth* (1958) and more evidence of his unassuming skill with a pen was given in his gentle autobiography *Blessings in Disguise* (1985).

GÜNEY, Yilmaz

Born 1 April 1937, near Adana, Turkey.
Died 9 September 1984, Paris, France.

A student of law in Ankara and economics in Istanbul, Güney worked for a film distribution company before moving into the film industry as a screenwriter and assistant director on such films as *Alageyik* (The Hind) (1958), *Clum Perdesi* (The Screen of Death) (1960) and *Dolandiriclar* (The King of Thieves) (1961).

Arrested and imprisoned for writing the novel *Equations With 3 Strangers* (1961) which was deemed pro-communist, he established himself as a heroic matinee idol Cirkin Kral ('The Ugly King') in popular romantic adventures.

He made his directorial debut with *At Avrat Silah* (The Horse, The Woman and The Gun) (1966) and later formed his own production company.

Gradually, his films came to embody the political concerns of a man enraged by the injustice and oppression he witnessed around him and he dealt in bleakly realistic terms with high-ranking corruption, social evils and class conflicts.

He was arrested in 1972 on charges of harbouring anarchist students and again in 1974 when he was sentenced to 24 years hard labour for the alleged murder of a judge. Allowed to write and plan films from the confines of prison, his work finally reached wider audiences with *Sürü* (The Herd) (1978), an uncompromising account of the arduous life endured by a nomadic herdsman and his family.

He subsequently supervised *Düsman* (The Enemy) (1979), a plea for justice on behalf of the poor and downtrodden couched in the story of a labourer who finds his only hope of economic survival in employment as a poisoner of stray dogs.

His greatest film *Yöl* (The Way) (1982) is a vividly powerful condemnation of all forms of state oppression in Turkey which are dramatically detailed in the experiences of five prisoners on a week's parole.

In 1981, he escaped to Switzerland and, with French finance, made *Duvar* (The Wall) (1982), a searing account of the Fascism that provoked the true events of a 1976 Ankara prison riot. He died from cancer shortly after the film's release.

GUNSMOKE

USA 1955–75

This classic Western serial set in the Dodge City of the 1880s began life on radio, with William Conrad (1920–) (later to play portly detective *Cannon* (1971–75)) in the starring role. The TV series was syndicated by CBS, in half-hour and hour-long slots, with the longer shows going out as *Marshall Dillon*, titled after the town's fearless but slightly grim lawman Matt Dillon, played by James Arness (1923–).

Dillon apart, there were four notables in Dodge City. The tough-talking Kitty Russell (Amanda Blake (1929–89)) ran the Longbranch Saloon, and maintained an interest in Dillon while remaining independent (though her character was softened in later episodes). Constantly brewing 'a mean cup of coffee' was Dillon's limping deputy, Chester Goode (Dennis Weaver (1924–), later *McCloud* (1970–76)), while hillbilly Festus Haggen (Ken Curtis (1916–)) added engaging goofiness. On hand when the bullets hit home was Galen 'Doc' Adams (Millburn Stone (1904–80)).

Stylistically *Gunsmoke* tipped its hat to the sparse documentary style of ▷ *Dragnet*, and made intelligent use of the Western genre – replacing action with human dilemmas and the odd bout of existential angst. Adult problems gave the show family appeal – by 1958 it was the USA's most popular show, and it remained in the Top 30 up to its cancellation in 1975.

As part of the backward looking nostalgia of network television in the 1980s, Arness and Blake returned to recreate their original roles in the feature-length television films *Gunsmoke: Return to Dodge* (1987) and *Gunsmoke: The Last Apache* (1990).

HACKMAN, Gene

Real Name Eugene Alden Hackman
Born 30 January 1930, San Bernardino, California, USA.

The son of a veteran newsman, Hackman joined the Marines when he was sixteen and travelled the world before settling in New York and attempting to pursue his father's profession at the School of Radio Technique. He subsequently worked as a television floor manager before committing himself to the notion of an acting career and studying at the Pasadena Playhouse where his classmates included ▷ Dustin Hoffman.

He secured his first major television role in *The Little Tin God* segment of *The US Steel Hour* in April 1959 and built a burgeoning career as a theatre performer and television guest star before making his film debut as a cop in *Mad Dog Coll* (1961).

He doggedly awaited a lucky break, amassing such theatre credits as *Any Wednesday* (1963) and small film roles in the likes of *Lilith* (1964) and *Hawaii* (1966).

1967 proved to be a key year in his career; fired from the role of Mr Robinson in ▷ *The Graduate* (1967) he was cast in ▷ *Bonnie and Clyde* (1967) as Clyde's brother Buck and secured audience empathy for a man torn between conformity and individualism. The performance earned him a Best Supporting Actor Oscar nomination.

Soon a heavily-employed character actor of dependability and authority, his subsequent appearances include *The Gypsy Moths* (1969), *Downhill Racer* (1969) and *I Never Sang For My Father* (1970) (Oscar nomination).

Many others were considered for the role of 'Popeye' Doyle in *The French Connection* (1971) before it was assigned to him, but he made it his own, creating a compulsive crimebuster, delineated in detail from his crummy pork pie hat to the plodding cop's feet. The role made him a star at 40 and the recipient of a Best Actor Oscar.

An unglamorous figure, his greatest strength is his ordinariness and ability to convey human vulnerability or fallibility with utter conviction. Others may be more charismatic but he has become the cinematic personification of the common man. His notable star performances include the unconventional man of God in *The Poseidon Adventure* (1972), the lusty, hot-tempered drifter in *Scarecrow* (1973), a hilarious cameo as the blind hermit in *Young Frankenstein* (1974) and his exceptional portrayal of the fastidious, conscience-stricken surveillance expert in *The Conversation* (1974).

Disillusioned by public apathy to such challenging work, he began to sell out, judging scripts by the size of the fee rather than the quality of the writing. The dispiriting results, including *The Domino Killings* (1976) and *March or Die* (1977), provoked a self-disgusted retirement from the screen.

However, a pleasure of cinemagoing in the 1980s was his re-emergence as one of the screen's most prolific and reliable stars. Outstanding among his many performances were the beleagured tycoon in *Eureka* (1982), the veteran reporter in *Under Fire* (1983), the football coach in *Hoosiers* (UK: Best Shot) (1986), the tenderhearted romantic in *Another Woman* (1988) and his slyly circumspect, justice-seeking FBI agent in *Mississippi Burning* (1988) (Oscar nomination).

His most recent films include the thriller *Narrow Margin* (1990) and *Postcards from the Edge* (1990).

HANCOCK, Tony

Full Name Anthony John Hancock
Born 12 May 1924, Small Heath, Birmingham, UK.
Died 25 June 1968, Sydney, Australia.

The son of a hotelier and part-time professional entertainer, Hancock briefly pursued employment as a civil servant before enlisting in the RAF (1942). Overcoming extreme stage-fright, he tried his hand as a stand-up comic with ENSA (Entertainments National Service Association) and touring gang shows. After demobilisation, he made his professional stage debut in *Wings* (1946).

Pantomimes, cabaret and radio appearances in *Educating Archie* (1951) contributed to his growing popularity and he made his film debut in *Orders Is Orders* (1954). The radio series *Hancock's Half Hour* began in 1954 and allowed him to refine his lugubrious comic persona of Anthony Aloysius Hancock – a pompous, belligerent misfit whose capricious social aspirations, pretensions and blinkered patriotism are frequently thwarted or belittled.

The series transferred to television in 1956 where his down-at-heel appearance, bulldog-like features and expostulatory delivery further enriched a memorable character. Over the next five years, the series gained record-breaking viewing figures and rare public affection.

Dispensing with regular co-stars like Sid James (1913–76) and the crucial skills of writing team Ray Galton (1930–) and Alan Simpson (1929–), he made ill-advised attempts at solo projects and serious 'artistic' endeavours in poorly-received films like *The Rebel* (1960) and *The Punch and Judy Man* (1963).

A chronic alcoholic beset by self-doubt and unable to reconcile his ambition with his talent, he spent his final years in a self-destructive round of aborted projects and unsatisfactory appearances, contributing supporting performances to the films *Those Magnificent Men In Their Flying Machines* (1965) and *The Wrong Box* (1966). He committed suicide while in Australia, attempting a comeback on television there.

HANKS, Tom
Born 9 July 1956, Concord, California, USA.

The son of a divorced chef, Hanks moved home and school with tiresome regularity as a child and claims his consequent shyness and lack of deep roots as the perfect preparation for his career as an actor.

A student at California State University in Sacramento, he worked as a stage manager and actor in campus productions and later spent three seasons performing the classics with the Great Lakes Shakespeare Festival in Ohio.

He made an inauspicious film debut in the slasher thriller *He Know's You're Alone* (1981) but was more warmly received in the television comedy show *Bosom Buddies* (1980–82) and the television film *Mazes and Monsters* (1982).

A regular guest star in popular prime-time series, his appearance on *Happy Days* brought him to the attention of actor-turned-director Ron Howard (1954–), who subsequently cast him as a businessman who falls in love with a mermaid in *Splash* (1984).

The film's unexpected popularity boosted his career, and he was soon established as a prolific comedy star of such generally uninspired fare as *Bachelor Party* (1984), *The Man With One Red Shoe* (1985) and *Volunteers* (1985).

Possessed of a relaxed manner, his boy-next-door appeal, mischievous grin and rubber ball features made him one of America's most personable young performers and his sure comic touch earned him comparisons with ▷ James Stewart and ▷ Cary Grant.

He attempted to expand his range as a glib advertising executive stopped in his tracks by his parents' divorcing in *Nothing in Common* (1986) and as an American pilot in the maudlin romance *Every Time We Say Goodbye* (1986), but gave one of his finest performances in the body-swap comedy *Big* (1988) conveying, with great charm, the energy, anxiety and guilelessness of a young boy trapped in a confusing adult world.

The role earned him a Best Actor Oscar nomination and he underlined his versatility as the compulsive, unlikable but understandable stand-up comic in *Punchline* (1988).

Now a firm public favourite, his drawing power has brought financial success to such flawed comic endeavours as *The 'burbs* (1989), *Turner and Hooch* (1989) and *Joe Versus the Volcano* (1990).

His most recent film is the controversial adaptation of *Bonfire of the Vanities* (1990) and he is scheduled to remake the 1950 ▷ film noir classic *Night and the City* (1990).

HANNA-BARBERA
William Denby Hanna. *Born* 14 July 1910, Melrose, New Mexico, USA.
Joseph Barbera. *Born* 1905, New York City, USA.

Winners of 7 Oscars and 7 Emmy Awards, Hanna-Barbera are best know as the purveyors of mass-produced children's television animation and as the creators of feuding cat and mouse adversaries Tom and Jerry.

Hanna had been an engineer before working as a story editor and assistant at M-G-M and making his solo debut as a director of animation on *Blue Monday* (1938). Barbera had been a banker with the Irving Trust in New York and a storyboard writer for the Van Beuren

Studio before joining the same animated shorts department.

Their collaboration began on *Puss Gets the Boot* (1940), which originated the characters of Tom and Jerry and was nominated for an Oscar. Small gems of violent, anarchic mayhem, expertly timed slapstick and wordless visual humour, their Oscar-winning Tom and Jerry shorts comprise *Yankee Doodle Mouse* (1943), *Mouse Trouble* (1944), *Quiet Please!* (1945), *The Cat Concerto* (1946), *The Little Orphan* (1948), *Two Mouseketeers* (1951) and *Johann Mouse* (1952).

They also furthered the matching of live-action and animation when Jerry the Mouse was seen to dance with ▷ Gene Kelly in *Anchors Aweigh* (1945) and contributed similar effects to such features as *Neptune's Daughter* (1949) and *Invitation to the Dance* (1956).

As the major studios began to run down their animation departments, they moved to television with the series *Ruff and Ready* (1957–60) and have enjoyed a consistent run of popular shows like *Huckleberry Hound* (1958–62), *Yogi Bear* (1960–62), *The Flintstones* (1960–65), *The Jetsons* (1962–63 & 1985–), *Scooby Doo* (1969–86), *The Smurfs* (1981–) and countless others. However, the sheer bulk of their output and use of cost-cutting semi-animation techniques has inevitably diluted the quality of their work, particularly when compared to the richness and invention of their past endeavours at M-G-M.

They have also been responsible for the feature-length animated films *Hey There, It's Yogi Bear* (1964), *The Man Called Flintstone* (1966) and *Charlotte's Web* (1972) and such live-action dramas as *The Gathering* (TV) (1977) and *The Stone Fox* (1987).

HARLOW, Jean

Real Name Harlean Carpenter
Born 3 March 1911, Kansas City, Missouri, USA.
Died 7 June 1937, Los Angeles, USA.

After attending the Hollywood School for Girls, Harlow eloped with a local business tycoon at the age of 16 and moved to Los Angeles.

She made her film debut in *Moran of the Marines* (1928) and appeared with ▷ Laurel and Hardy in *Double Whoopee* (1929) and as an extra in ▷ *City Lights* (1931) before being signed to a contract by ▷ Howard Hughes and featuring in *Hell's Angels* (1930). Roles in *Platinum Blonde* (1931), *Red-Headed Woman* (1932) and *Red Dust* (1932) established her screen image as a fast-talking, wise-cracking predatory blonde who gave as good as she got and brazenly flaunted her sexuality.

Under contract to M-G-M from 1932, she proved a memorable sparring partner for the studio's top male stars and developed into a deft comedienne, able to find humour in her glamorous image, in films like *Dinner at Eight* (1933), *Bombshell* (1933) and *Libelled Lady* (1936).

Her death at the age of 26 from cerebral oedema followed a life blighted by ill-health and personal problems, including three failed marriages.

Her final role in *Saratoga* (1937) was completed with the aid of an insultingly obvious double.

In the 1960s, she was impersonated in two unsuccessful screen biographies; *Harlow* (1965) starring Carroll Baker (1931–) and *Harlow* (1965) starring Carol Lynley (1942–).

HART, William S(urrey)

Born 6 December 1862/5, Newburgh, New York, USA.
Died 23 June 1946, Horseshoe Ranch, Newhall, California, USA.

An employee of the New York Post Office, Hart studied acting and toured the country with numerous troupes before Broadway successes in *Ben Hur* (1899), in which he originated the role of Messala, *The Squaw Man* (1905) and *The Virginian* (1907–08).

He made his film debut in *The Fugitive* (1913) and, although initially cast as villains, he went on to enjoy great popularity in a series of westerns as a mature, solemn-faced defender of truth, justice and the honour of good women. He often devised the original story and directed his moralistic adventures, whose best-remembered titles included *Wild Bill Hickok* (1923), *Singer Jim McKee* (1924) and *Tumbleweeds* (1925), a classic of the western genre set during the Cherokee landrush of 1889 and noted for its sense of spectacle.

When his image grew old-fashioned,

he retired from the screen and had his last fleeting involvements with the industry as a consultant on *Billy the Kid* (1930), a guest star in *Show People* (1928) and the short *The Hollywood Gad-About* (1934) and in an eight-minute prologue made to accompany the 1939 re-issue of *Tumbleweeds* for sound-era audiences.

He published an autobiography entitled *My Life East and West* (1929) and also wrote several volumes of colourful sagebrush yarns including *Told Under A White Oak Tree* (1922), *Hoofbeats* (1933) and *The Law on Horseback* (1935).

HAWKS, Howard Winchester

Born 30 May 1896 Goshen, Indiana, USA.
Died 26 December 1977, Palm Springs, California, USA.

A graduate in mechanical engineering, Hawks worked briefly as a prop man on films like *A Little Princess* (1917) before serving in the US Army Air Corps. He later drifted back into silent films, writing the scripts for *Quicksands* (1923) and *Tiger Love* (1924) and making his directorial debut with *The Road to Glory* (1926).

In a career that stretched over four decades, he proved himself a master storyteller who unobtrusively explored his own preoccupations in a variety of genres. Noted for his use of freewheeling, overlapping dialogue and ability to convey mood or emotion through gesture or lighting, he enjoyed early successes with the aerial drama *The Dawn Patrol* (1930) and the ferocious gangster chronicle *Scarface: The Shame Of A Nation* (1932).

Adept at ▷ screwball comedies and evenly matched battle-of-the-sexes sparring matches, his many riotous outings in this field include *Twentieth Century* (1934), *Bringing Up Baby* (1938) and *His Girl Friday* (1940).

He made frequent use of tough action subjects to explore an exclusive male camaraderie and how it reacts to the possible intrusion of a female presence. Notable examples of this include *Only Angels Have Wings* (1939) and the comedy *Ball of Fire* (1941).

He introduced Lauren Bacall (1924–) to the screen in *To Have and Have Not* (1944) and featured her partnership with husband ▷ Humphrey Bogart in the memorably bedazzling thriller *The Big Sleep* (1946). In his post-war career, he worked frequently with ▷ John Wayne, using his heroic image to question the obsessive masculine values of the old West in films like *Red River* (1948), *Rio Bravo* (1959) and *El Dorado* (1967).

A director with a keen eye for fresh talent, his many popular entertainments also include *I Was A Male War Bride* (1949), *Gentlemen Prefer Blondes* (1953) and his final film *Rio Lobo* (1970).

He received an honorary Oscar in 1975 as 'a master American filmmaker whose creative efforts hold a distinguished place in world cinema'.

HAY, Will

Born 6 December 1888, Aberdeen, UK.
Died 18 April 1949, London, UK.

An apprentice engineer who entertained at charity shows before turning professional, Hay worked extensively in music halls and on the radio before making his film debut in the short *Know Your Apples* (1933).

He appeared in *Those Were The Days* (1934) and *Dandy Dick* (1935), both adapted from plays by Arthur Pinero, before bringing his own creation to the screen in *Boys Will Be Boys* (1935). A blustering, blundering braggart and cheat whose hopeless incompetence plunges him into inevitable chaos, he frequently played disreputable figures of authority with no rightful claim to the position.

As a seedy teacher in *Good Morning Boys* (1937), stationmaster in *Oh, Mr Porter!* (1937), prison governor in *Convict 99* (1938), or fire chief in *Where's That Fire?* (1939), he created British comedy classics that traded on his character's shabby, sniffing gentility, delusions of adequacy and shifty, petty larceny. Often partnered with Graham Moffatt (1919–65) and Moore Marriott (1885–1949) he was one of the country's top box-office attractions from 1937 to 1942.

He played a character role in *The Big Blockade* (1941) and co-directed his final wartime comedies *The Black Sheep of Whitehall* (1941), *The Goose Steps Out* (1942) and *My Learned Friend* (1943).

Ill-health caused him to retire from the screen although he remained a popular radio panellist until his death.

A respected amateur astronomer, he published *Through My Telescope* in 1935.

HAYAKAWA, Sessue

Real Name Kintaro Hayakawa.
Born 10 June 1890, Nantaura, Chiba, Japan.
Died 23 November 1973, Tokyo, Japan.

A graduate of the University of Chicago, Hayakawa immediately joined the Japanese Theatre in Los Angeles where he was noticed in a production of *Typhoon* (1913) and offered an opportunity to appear in motion pictures.

He starred in a cinema version of *Typhoon* (1914) and subsequently signed a contract with the Jesse Lasky Company. A handsome, graceful, authoritative figure, he gained stardom as the larcenous protagonist of *The Cheat* (1915) and enjoyed a decade of success as exotic villains and sensuous screen lovers in scores of films, including those produced by his own company Haworth Pictures Corporation. The many titles include *The City of Dim Faces* (1918), *The Beggar Prince* (1920) and *Sen Yan's Devotion* (1924).

A high-living, flamboyant figure off-screen, he established a unique position as the most popular and best known Japanese actor in the history of American cinema. Returning to the theatre in the 1920s, he also founded a Zen study hall in New York.

Working abroad, he played *Hamlet* (1935) in Tokyo and settled in Paris during the Second World War making infrequent appearances in films like *Patrouille Blanche* (1941) and *Le Cabaret du Grand Large* (1946). He returned to Hollywood productions with *Tokyo Joe* (1949) and received an Oscar nomination as Best Supporting Actor for his portrayal of the unyielding Japanese commander in *Bridge on the River Kwai* (1957). He made his last film appearance in *The Big Wave* (1960).

A renaissance man who wrote plays, directed for the theatre, exhibited his watercolours and was ordained as a Zen Buddhist priest, he also published a novel *The Bandit Prince* (1926) and an autobiography *Zen Showed Me The Way* (1960).

THE HAYS CODE Worried that allegations of Hollywood hedonism and the sensational coverage of the 1921 Arbuckle scandal (in which a young actress had died) would lead to the imposition of independent state and federal censorship, the American film industry formed the Motion Picture Producers and Distributors of America (MPPDA) organisation in 1922 to pre-empt such an eventuality by establishing their own self-regulatory code of conduct.

Former Postmaster General Will H. Hays (1879–1954) served as the first president from 1922 to 1945 and worked to bring the industry favourable national coverage and to officially ensure that the motion pictures portrayed a sensible set of moral values. Renewed pressure in 1930 led to the declaration of an advisory Motion Picture Production Code, or Hays Code, which became mandatory in 1943. Setting rigorous standards of what was considered permissible on screen, the Code forbade any form of explicit depiction or discussion of sexual matters, profanity, extreme violence or immorality and could arbitrate as to the proper length of a kiss or the appropriate behaviour of a married couple – who were generally relegated to separate beds and permitted intimate physical contact only if both feet were kept on the ground.

The code was administered by Catholic newspaperman Joseph Breen (1890–1965) and its influence was total, from advising at the scriptwriting stage to giving official sanction to the finished film. No film could be exhibited in cinemas belonging to MPPDA members without the Code's Seal of Approval and defiance would incur a penalty of $25,000.

Allowing ▷ Clark Gable to utter 'damn' in ▷ *Gone With the Wind* proved a major concession. A 1952 Supreme Court ruling on the film *L'Amore* (1948) freed films from censorship on religious grounds and brought the medium under the protection of the First Amendment. When ▷ Otto Preminger challenged the Code by releasing the films *The Moon is Blue* (1953), with its daring mention of the words virgin and pregnant, and *The Man With the Golden Arm* (1955), about drug addiction, bereft of the Seal of Approval, the Code's days were numbered.

Renamed the Motion Picture Association of America (MPAA) in 1945, the former MPPDA revised the Code in 1956 and 1966 but abandoned it altogether in 1968, replacing it with a ratings system fore-warning potential viewers of the nature of the film but permitting filmmakers total liberty in what they chose to portray.

HAYWARD, Susan
Real Name Edythe Marrenner
Born 30 June 1917, Brooklyn, New York, USA.
Died 14 March 1975, Los Angeles, USA.

Once described as a 'compulsive ham' in amateur dramatics, Hayward enrolled at the Feagan School of Dramatic Arts in 1936 and subsequently pursued a modelling career. A series of photos in the *Saturday Evening Post* brought an offer to test for the role of Scarlett O'Hara in ▷ *Gone With the Wind* and, in 1937, she left New York for Hollywood.

Although not chosen for Scarlett, she remained in Hollywood, signing a contract with Warner Brothers and appearing as an extra or bit player in such films as *Hollywood Hotel* (1937) and *The Sisters* (1938). A new contract with Paramount brought a better quality of supporting role in such rousing adventures as *Beau Geste* (1939) and *Reap the Wild Wind* (1942), whilst her dramatic potential was revealed as the spoilt beauty in *Adam Had Four Sons* (1941) and as the spiteful bitch in *The Hairy Ape* (1944).

A further contract with independent producer Walter Wanger (1894–1968) brought the kind of material that exploited her abrasive personality and hankering for larger-than-life emotional torments and she received her first Oscar nomination as an alcoholic in *Smash-Up: the Story of a Woman* (1947).

A major star at Twentieth Century-Fox, her best roles were as fiery as her own trademark flame-red hair and she brought a mixture of pugnacity, careful technique and raw emotion to such powerhouse characterisations as handicapped singer Jane Froman in *With A Song in My Heart* (1952), alcoholic singer Lillian Roth in *I'll Cry Tomorrow* (1955) and accused murderess Barbara Graham in *I Want to Live!* (1958). The latter's sincere approach to its subject, including the harrowing execution scenes, assisted Hayward's realistic portrayal of the first woman in California to be sent to the gas chamber; she received her fifth Oscar nomination and the award itself.

A happy second marriage outside the film industry curtailed the drive of her career and in the 1960s she appeared mainly in lavish 'soap-opera' films that required little of her beyond professionalism and an ability to suffer in splendour. The more popular titles included *Where Love Has Gone* (1964) and *Valley of the Dolls* (1967).

In 1968 she appeared on stage in a Las Vegas production of *Mame*. Her final film was the western *The Revengers* (1972). She spent her last years in a long battle with multiple, malignant brain tumours making a final, typically defiant, public appearance at the 1974 Oscar ceremony.

HAYWORTH, Rita
Real Name Margarita Carmen Cansino
Born 17 October 1918, New York City, USA.
Died 14 May 1987, New York City, USA.

A distant cousin of Ginger Rogers (1911–), Hayworth followed in her father's footsteps (literally) as a dancer and made an early film debut in the short *La Fiesta* (1926).

Part of The Dancing Cansinos nightclub act, she began her serious Hollywood career when she signed a contract with Twentieth Century-Fox and made minor appearances in *Charlie Chan in Egypt* (1935) and *Dante's Inferno* (1935). Groomed for stardom, she was utilised as little more than decoration in second features and B-pictures before securing a good secondary role as the head-turning wife in *Only Angels Have Wings* (1939).

The quality of her roles improved as she revealed an attractive way with comedy in *Strawberry Blonde* (1941). A ravishing red-head, she was well-liked in *Blood and Sand* (1941) and made an ideal partner for ▷ Fred Astaire in the musicals

You'll Never Get Rich (1941) and *You Were Never Lovelier* (1942).

She enjoyed her greatest successes in the joyous musical *Cover Girl* (1944) and as the adventuress in *Gilda* (1946) where her dazzling allure and sinuous performance of 'Put The Blame on Mame, Boys' forever typed her as a love goddess of incendiary appeal.

Married to ▷ Orson Welles from 1943 to 1948, he perversely manipulated her beauty as the brutally cropped blonde femme fatale in the bravura thriller *The Lady from Shanghai* (1948) and after a further marriage to Prince Aly Khan (1949–53) her screen appearances became less frequent.

Rita Hayworth in *Gilda* (1946)

She made claim to consideration as a character actress of merit with her portrayals of mature beauties in *Separate Tables* (1958) and *They Came to Cordura* (1959) but most of her subsequent screen work was negligible. She made her final appearances in *The Wrath of God* (1972) and *Circle* (1976).

During the last decade of her life she suffered from Alzheimer's Disease and a biography published after her death revealed a tragic tale of childhood abuse, chronic insecurity and a star saddled with a screen image that bore no relationship to the private woman.

HEAVEN'S GATE

Michael Cimino (USA 1980)
Kris Kristofferson, Christopher Walken, Isabelle Huppert, John Hurt, Jeff Bridges.
November 1980: 219 min original 70mm version opens and closes in New York and Toronto after one day. April 1981: recut 70 & 35mm prints open across the USA at 148 mins. September 1983: after favourable response to screenings at London's National Film Theatre, British distributor reissues original 70mm version. col.

Fresh from the Oscar-winning success of *The Deer Hunter* (1978), his grandiose, poetic and violent examination of the Vietnam experience, director ▷ Michael Cimino got the go-ahead from United Artists for a new movie based on the little-known 1892 Johnson County War. From historical events surrounding the massacre by Wyoming's wealthy Stock Growers' Association of the immigrant community then farming the land, Cimino was to weave together *Heaven's Gate*, a huge yet financially disastrous epic that remains one of the most controversial offerings to come out of the latterday Hollywood. With a thirst for authenticity which extended to the transportation of a period locomotive halfway across the continent, Cimino's shooting schedule soon fell by the wayside and his budget expanded from an initial (rather unlikely) $7.5 million to a highly-publicised final tally of around $36 million. In the hope that the final product would yet be a huge blockbuster UA continued to support his efforts, but were surely unprepared for the degree of critical vitriol the revisionist four-hour spectacular would attract on its initial engagement. After a swift withdrawal and the reissue of a somewhat mangled shorter cut, American public response continued to be disdainful of Cimino's apparent profligacy, but a highly favourable reappraisal of the film by European commentators apportioned at least part of the disastrous US reception to *Heaven's Gate*'s unpalatable ideological stance. While one French newspaper called it 'the first socialist western', American audiences (and critics) seemed unwilling to accept the film's passionate account of Yankee entrepreneurism's human toll. The financial failure of the film precipitated the end of UA as a viable production company, and led to the sale of the studio. Steven Bach's book *Final Cut* chronicles the travails of making the film.

HEIMAT (Homeland)

West Germany 1984
Director Edgar Reitz

Director Reitz's response to the Hollywood account of Nazi Germany given in the regrettable TV mini-series *Holocaust* (1978), was this 16-hour story of three related families in the rural village of Schabbach, in the Hunsruck area of Germany. Though Heimat literally means homeland, the experience of Hitler and the post-war division of Germany gave the word a deeper, untranslatable nuance.

The drama runs from 1919 to 1982, and pivots round Maria (Marita Breuer) who was born in 1900. The emphasis is on the smallness of everyday life rather than grander political themes, and the narrative is developed through the cumulative use of detail. Yet *Heimat* builds into a compulsive political and emotional history of the German people. The rise of the Nazis is not condemned – though only the least likeable of the townsfolk have anything to do with the Party.

Filmed over an 18-month period, its success in Germany briefly turned Hunsruck into a reluctant tourist resort. Though Reitz conceived it as a piece for cinema, *Heimat* was serialised for TV. In 1986 it was shown in 11 episodes over a fortnight on BBC2, and attracted five times the normal audience for a subtitled drama and a ▷ BAFTA special award. Worthy of note is Gernot Roll's photography, which veers between colour and black and white, between serene beauty and stark realism.

In 1990, Reitz returned to Hunsruck for the filming of *Heimat II* for screening in 1991.

HENRY V

Laurence Olivier (UK 1944)
Laurence Olivier, Robert Newton, Leslie Banks, Esmond Knight.
153 mins; some prints run 137 mins. col.
Academy Award: *Special Oscar to* Laurence Olivier *for his Outstanding Achievement as Actor, Producer and Director in bringing* Henry V *to the screen.*

With the all-star pageantry of the Max Reinhardt (1873–1943) *A Midsummer Night's Dream* (1935) perhaps the most notable effort to have come out of Hollywood, by the mid-Forties the screen had already seen its fair share of Shakespearean adaptations. As the first film however, to offer an idiomatic rendering of the Bard's enduring verse and to use the resources of the cinema to capture the scope of Shakespeare's historical conception, the release of ▷ Laurence Olivier's *Henry V* marked an instant milestone. The success of his 1939 characterisation of Heathcliff in director ▷ William Wyler's version of *Wuthering Heights* had served to win over Olivier to the merits and opportunities offered by the film medium, but his desire to bring to the screen the intensity of his highly-acclaimed Shakespearean roles on stage at London's Old Vic was to be temporarily thwarted by the outbreak of war and his subsequent military duties with the Fleet Air Arm. By 1944 however, with the Allied forces' victory looking increasingly secure, Olivier began to feel that a hawkish film interpretation of *Henry V* would make a perfect contribution to the ongoing propaganda effort, and the final result used a cut text to place the emphasis on the famous English victory at Agincourt. Although he originally wanted Wyler to direct, Olivier the filmmaker makes a confident debut as he sweeps the camera across the fields in the film's stirring battle scenes and intriguingly situates the action within the framework of an actual Elizabethan performance in the original Globe theatre. Young actor/director Kenneth Branagh's (1960–) muddily impressive recent *Henry V* (1989) tones down the wartime jingoism of its earlier counterpart to offer a more sobering account of the price of conflict, but in the famous 'Saint Crispin's Day' oration the leonine fire of the Olivier performance still retains a charge that carries all before it.

HENSON, Jim

Full Name James Murray Henson
Born 24 September 1936, Greenville, Mississippi, USA.
Died 16 May 1990, New York City, USA.

A member of a high school puppet club,

Henson later secured a job on local television in Washington DC and was host of his own show whilst still a theatre arts student at the University of Maryland. The series *Sam and Friends* (1955–61) won an Emmy in 1958 as best local entertainment show.

His creations, known as Muppets because they represent a cross between marionettes and puppets, began to make appearances in various television shows and commercials before becoming a regular attraction on *The Jimmy Dean Show* (1963–66).

Intrigued by television technology and the potential of the medium, he continued to refine his endearing characters and, in 1969, launched ▷ *Sesame Street*, a series that playfully educated pre-school tots in the alphabet and their numbers.

Many of his long-established puppets, like Kermit the Frog and Miss Piggy, gained phenomenal popularity in the light entertainment series *The Muppet Show* (1976–81) which reached an estimated 235 million viewers in more than 100 countries and attracted a who's who of celebrity guest-stars. The characters were subsequently utilised in a series of films including *The Muppet Movie* (1979) and *The Muppets Take Manhattan* (1984) and a Grammy-winning album called 'The Muppets' (1979).

The recipient of two Emmy awards for 'outstanding individual achievement in children's programming', he continued to make innovative television programmes combining live action and increasingly elaborate puppetry including *Fraggle Rock* (1983–90) and *The Storyteller* (1897–90).

Nominated for an Oscar as the writer, producer and director of the experimental short film *Timepiece* (1965), he also diversified into feature-film production, bringing inventive special-effects and creatures to *The Dark Crystal* (1984) and *Labyrinth* (1986) and contributing his technical expertise to such diverse ventures as *Dreamchild* (1985) and *The Witches* (1989). He had been negotiating the sale of his company to ▷ Walt Disney at the time of his sudden death.

HEPBURN, Audrey

Real Name Edda van Heemstra Hepburn-Ruston

Born 4 May 1929, Brussels, Belgium.

A ballet student at the Arnhem Conservatory of Music in Amsterdam and at the Marie Rambert school in London, Hepburn made her film debut in *Nederland in 7 Lessen* (1948). She made her London stage debut in the chorus of *High Button Shoes* (1948) and studied acting with Felix Aylmer (1889–1979) whilst securing minor roles in such British films as *The Lavender Hill Mob* (1951).

Spotted by the French writer Colette whilst filming *Monte Carlo Baby* (1951), she was given the lead in the Broadway production of *Gigi* (1951). Hollywood cast her as an incognito princess wooed by unsuspecting journalist Gregory Peck (1916–) in *Roman Holiday* (1953) and her beguiling performance won an Oscar.

Audrey Hepburn in *My Fair Lady* (1964)

Her screen image was very much that of a fairytale princess and she proved herself an enchanting, pencil-slim actress of coltish grace, Peter Pan-youthfulness and inimitable diction. One of the major stars of the 1950s and '60s, her greatest successes include *Sabrina* (1954) (Oscar nomination), *Funny Face* (1957), *The Nun's Story* (1959) (Oscar nomination) and *Breakfast at Tiffany's* (1961) (Oscar nomination).

Perhaps because her singing voice was dubbed and it was felt that ▷ Julie Andrews, the stage Eliza Dolittle, had merited the screen role, *My Fair Lady* (1964) was less of a triumph, despite her stunning appearances in the Cecil Beaton finery.

The steel beneath her tomboyish refinement was displayed in two contrasting performances: as one half of the bickering married couple in the modish *Two For the Road* (1967), and as the blind girl terrorised by hoodlums in *Wait Until Dark* (1967) (Oscar nomination).

She retired to Rome in 1968 citing the paramountcy of family responsibilities over the dictates of a full-time career. She returned to the screen in *Robin and Marian* (1976) but her few subsequent appearances have been in material of little distinction – although she was captivating as ever portraying the pixieish guide and mentor from the afterlife in *Always* (1989).

On television, she appeared in a 1957 production of *Mayerling* and returned to the medium for an indifferent romantic thriller *Love Among Thieves* (1987). More recently, she has travelled extensively as a goodwill ambassador for UNICEF.

HEPBURN, Katharine

Born 9 November 1909, Hartford, Connecticut, USA.

Educated at Bryn Mawr College in Pennsylvania, Hepburn made her professional stage debut in a Baltimore production of *Czarina* (1928) and continued to pursue a career in the theatre until an appearance in *The Warrior's Husband* (1932) led to offers of film work and her screen debut in *A Bill of Divorcement* (1932).

A vivid screen presence, with an inimitably cadenced speech pattern and indomitable spirit, she was often cast as rebellious, strong-willed young women and prototype feminists, roles reflecting something of her off-screen disdain for convention.

She received the first of twelve record-breaking Best Actress Oscar nominations, and the first of four wins, for her role as a struggling actress in *Morning Glory* (1933). Other early performances of note include Jo in *Little Women* (1933), a pioneering aviatrix in *Christopher Strong* (1933) and smalltown girl *Alice Adams* (1935) (Oscar nomination).

Her spry comic technique was well displayed in a series of films with ▷ Cary Grant that include *Sylvia Scarlett* (1936), *Holiday* (1938), and the classic ▷ screwball farce *Bringing Up Baby* (1938).

Inexplicably labelled 'box-office poison' she refused the demeaning offer of *Mother Cary's Chickens* (1938) and returned to Broadway in triumph in ▷ *The Philadelphia Story* in 1938. Retaining control of the screen rights and casting approval, she enjoyed one of her greatest successes when it transferred to the screen in 1940 with Cary Grant and ▷ James Stewart.

Woman of the Year (1942) began her long personal and professional association with ▷ Spencer Tracy and, over the next twenty-five years, they would star together in a series of sophisticated battle-of-the-sexes comedies that includes *Adam's Rib* (1949), *Pat and Mike* (1952) and *The Desk Set* (1957).

The tendency to cast her as eccentric spinsters and idiosyncratic matriarchs began in the 1950s when she starred as the missionary in *The African Queen* (1951) (Oscar nomination), holidaying schoolteacher in *Summertime* (UK: *Summer Madness*) (1955) (Oscar nomination), spinster in *The Rainmaker* (1956) (Oscar nomination) and domineering mother in *Suddenly Last Summer* (1959) (Oscar nomination).

After *Long Day's Journey Into Night* (1962) (Oscar nomination), she was absent from the screen, tending to the needs of the ailing Tracy, but she returned for a poignant farewell appearance with him in the sentimental drama *Guess Who's Coming To Dinner* (1967) (Oscar).

Tracy's death and a further Oscar for her performance as Eleanor of Aquitaine in *The Lion in Winter* (1968) brought a flurry of professional activity that culminated in her personal triumph with the Broadway musical *Coco* (1969).

Despite suffering from Parkinson's disease for many years, she has continued to act, winning a further Oscar for the warmth and sincerity of her performance as the loyal, loving wife in *On Golden Pond* (1981) and finding gainful television employment in such specially created vehicles as *Love Among the Ruins* (1975), *Mrs Delafield Wants To Marry* (1986) and *Laura Lansing Slept Here* (1988).

HIGH NOON

Fred Zinnemann (USA 1952)
Gary Cooper, Thomas Mitchell, Lloyd Bridges, Grace Kelly.
84 mins. b/w.

Academy Awards: *Best Actor* Gary Cooper; *Best Editing* Elmo Williams and

Harry Gerstad; *Best Score* Dimitri Tiomkin; *Best Song* 'High Noon (Do Not Forsake Me, Oh My Darlin')' (music by Dimitri Tiomkin and lyrics by Ned Washington, sung by Tex Ritter).

As a tall, handsome hero of silent era shoot-'em-ups ▷ Gary Cooper was just one of the many performers and filmmakers whose efforts established the western as the location for frequently simplistic six-gun tussles between the good guys in white hats and the bad guys in the black hats. As the years wore on however, not all the genre's offerings were to prove quite so simplistic, with the Fifties series of films by directors ▷ Anthony Mann and Budd Boetticher (1916–), for example, managing to invest familiar frontier tales with troubling psychological conflicts. Having started life as a low-budget independent offering, the eventual public and critical success of Fred Zinnemann's *High Noon* stood as another step in the western's growing maturity, and much of the film's impact came from the manner in which genre icon Cooper was depicted as a courageous man riven by very real fears. Set in the everyday western town of Hadleyville, the middle-aged star is extremely convincing as Sheriff Will Kane, whose marriage celebrations are interrupted by the news that a fearsome villain he had personally put behind bars has recently been released from prison and is due back in town on the twelve o'clock train to seek his revenge. Screenwriter Carl Foreman (1914–84) was blacklisted by the House Un-American Activities Commission shortly after the film's completion, leading some to read Cooper as the hunted man who remains true to his conscience and the cowardly townsfolk who find every excuse not to join him in his task as the American people turning a blind eye to the witch-hunt. Still, Zinnemann's inspired use of real time and a repeating song motif to build up unbearable tension leaves the audience to nervously root for the single-minded protagonist rather than ponder the political ramifications of his actions.

HILL, Benny

Real Name Alfred Hawthorne
Born 21 January 1925, Southampton, UK.

An irrepressible school clown, Hill's early career included spells as a milkman, coal weigher and musician with 'Ivy Lillywhite and Her Boys'.

Eventually he secured employment as a prop boy and stage manager, performing as a straight man before wartime service with the Royal Engineers as a driver-mechanic.

He appeared in *Stars in Battledress* and adopted the name Benny Hill as a mark of deference to his idol ▷ Jack Benny. Demobbed, he followed the traditional comic's route of working men's clubs, revues and end-of-the-pier shows.

An early convert to the potential of television, he appeared in *Hi There* (1949), *The Service Show* (1952) and *Show Case* (1953) and was named TV Personality of the Year in 1954. A beaming, round-faced individual with a mischievous leer, he combines the mind of a seasoned sinner with the look of a guileless cherub. His routines consist of finely-observed impersonations, ribaldry, tongue-twisting ditties and smutty innuendo.

He made his film debut in the Ealing comedy *Who Done It?* (1956) but the majority of his few big screen appearances have been in comic supporting roles in the likes of *Those Magnificent Men in Their Flying Machines* (1965) and *Chitty, Chitty Bang Bang* (1968).

His enduring British popularity rests on the BBC's *Benny Hill Show* (1957–66) and the subsequent series and hour-long specials he has written, devised and performed at Thames Television since moving there in 1969. Brimming with quick-fire sketches, scantily clad girls, merciless lampooning of public figures and such established characters as Fred Skuttle and Herr Otto Stumpf, they are the television equivalent of saucy seaside postcards.

Edited into fast-moving half-hour packages of mainly visual slapstick, his shows were successfully exported to America in the late 1970s where he soon developed a cult following, denounced by *Village Voice* for 'pornographic grubbiness' but often scheduled every night of the week as the epitome of bawdy British humour.

He has refused all offers to film in America and, aside from one benefit appearance, has not appeared on stage since 1959.

Seen in a 1964 television production of *A Midsummer Night's Dream* as Bottom he also enjoyed a hit record with 'Ernie, the Fastest Milkman in the West' which was the British Christmas No. 1 in 1971.

HILL STREET BLUES

USA 1980–86

Executive producer Steven Bochco

Now regarded as the show which re-wrote the rule-book for popular TV drama, *Hill Street* met with audience indifference on its launch and escaped the axe after its first season only by virtue of the nine Emmys it was awarded in 1981. A typical product of the independent production company MTM (also responsible for ▷ *Lou Grant* and *St Elsewhere*) it was, said producer, co-

Hill Street Blues: Daniel J. Travanti and Michael Conrad

creator (with Michael Kozoll) and sometime-writer Bochco (1943–), not so much a cops and robbers show, as a show about the cops themselves. Set in a tough urban precinct in an unnamed East coast American town, it showed the police battling for some kind of liberal justice in the face of near-anarchy.

Each episode began with the roll call by the precinct's Sergeant Phillip Freemason Esterhaus (Michael Conrad, (1925–83)), but the leader of the pack was patriarchal Captain Frank Furillo (Daniel J. Travanti, (1940–)) – said by some to have been modelled on Steve Carella, from Ed McBain's 87th Precinct novels. The programme was a mix of genres, blending humour and soapy character interest with a documentary approach to crime. Much of the latter was just style – the result of subdued lighting, a muddy soundtrack and fast-cut hand-held cameras (there were plans to shoot the whole show this way, but the hand-held shots were restricted to what director Robert Butler called 'certain heightened sequences'). The substance came from a narrative which supported 13 regular characters, and up to six plot-lines per episode.

Bochco fell into dispute with MTM over budgets, and left before the fifth (penultimate) season, where the innovations were diluted. He then moved directly into the mainstream with the more glamorous *LA Law* and, less spectacularly, the comedy-drama *Hooperman*.

HITCHCOCK, Alfred Joseph (Sir)

Born 13 August 1899, Leytonstone, London, UK.

Died 29 April 1980, Los Angeles, USA.

After studying drawing and design at London University, Hitchcock was a technical clerk for a telegraph company before entering the film industry with Famous Players-Lasky as the designer of title cards for such silent films as *The Great Way* (1920) and *Beside the Bonnie Brier Bush* (1922).

He made his directorial debut on the uncompleted *Mrs Peabody* (1922) (also known as *No. 13*) and worked as a script-writer and assistant director before trying again with *The Pleasure Garden* (1926). He first made his name as a master of suspense with the thriller *The Lodger* (1926) which makes use of German ▷ Expressionist techniques and Soviet ▷ montage to create a vividly atmospheric story of a man suspected of being Jack the Ripper. The film also features the director in the first of a series of cameo roles that would become a trademark of his career.

His first talking film ▷ *Blackmail* (1929) was also Britain's first talkie and made innovative use of sound in its compelling story of a woman blackmailed for a murder committed in self-defence.

He followed this with work in a variety of genres, including the musical *Waltzes from Vienna* (1933), before establishing his dominance in the field of thrillers with *The Man Who Knew Too Much* (1934), *The Thirty-Nine Steps* (1935) and *The Lady Vanishes* (1938), all noted for their excellent sense of pace and disregard for the logic of plot in favour of characterisation, romanticism and meticulously pre-planned set-pieces of suspense.

After an overripe costume melodrama, *Jamaica Inn* (1939), he moved to America, creating such classics as *Rebecca* (1940) (Best Director Oscar nomination), the propagandist *Foreign Correspondent* (1940) and *Shadow of A Doubt* (1943), in which a kindly uncle is the last man one would expect to be guilty of such unpleasantness as mass murder.

Alfred Hitchcock

Always willing to experiment with form and technique, he confined himself to the restricted locale of *Lifeboat* (1944) (Oscar nomination), collaborated with Salvador Dali for surreal dream sequences in *Spellbound* (1945) (Oscar nomination), made *Rope* (1948) using continuous ten-minute takes, and utilised ▷ 3-D for the original version of *Dial M for Murder* (1954). However, some of his finest work from the period was that involving the least trickery and merely melding a perfect blend of romance, jeopardy, black humour, plots known as 'the MacGuffin' and strong star performances. These include *Notorious* (1946) and *Rear Window* (1954) (Oscar nomination).

Drawn to stories of everyman heroes and dignified blondes who reveal a passionate nature beneath a cool exterior, his work was often accused of misogyny, a charge that could be levelled at the extraordinarily perverse romance at the heart of ▷ *Vertigo* (1958), or the relationship between boss and kleptomaniac in *Marnie* (1964).

After the polished excitement of *North By Northwest* (1959), he gave full rein to his macabre sense of humour with the gothic chiller ▷ *Psycho* (1960) (Oscar nomination) and the unsettling study of ornithological psychopaths *The Birds* (1963).

The later years of his long career were marked by some rather dull tales of espionage like *Torn Curtain* (1966) and *Topaze* (1969) but there was more of his flair in *Frenzy* (1972) and his last amiable mystery *Family Plot* (1976).

The recipient of the 1979 ▷ American Film Institute Life Achievement Award, he was knighted in 1980.

HOFFMAN, Dustin

Born 8 August 1937, Los Angeles, USA.

The son of a furniture-designer, Hoffman briefly studied to be a doctor before enrolling at the Pasedena Playhouse (1956–58), where he shared the distinction with classmate ▷ Gene Hackman of being voted the person least likely to succeed. Undeterred, he moved to New York and pursued a career on stage and television, interspersed with a variety of bill-paying odd jobs.

He made his Broadway debut in *A Cook for Mr. General* (1961) and came to prominence with his performance as a hunchbacked German homosexual in *Harry Noon and Night* (1964) and as a Russian misanthrope in *Journey of the Fifth Horse* (1966).

Following a modest film debut in *The*

Tiger Makes Out (1967), he earned his first Best Actor Oscar nomination as the amusingly confused and inexpert title character in ▷ *The Graduate* (1967).

A notoriously exacting perfectionist, he became the cinematic rarity of a character actor who also commanded loyalty as a star in his own right, a feat proved by his performances as a romantic leading man in *John and Mary* (1969) and detailed impersonation of pathetic cripple Ratso Rizzo in *Midnight Cowboy* (1969) (Oscar nomination), and as adventurous centenarian in *Little Big Man* (1970).

A diminutive, dark-haired, adenoidal chameleon his versatility has been displayed in his choice of demanding and magnetic screen characters like the Devil's Island inmate in *Papillon* (1973), scabrous, self-destructive comedian Lenny Bruce in *Lenny* (1974) (Oscar nomination), dogged investigative journalist Carl Bernstein in *All the President's Men* (1976) and vulnerable father coping with sudden single parenthood in *Kramer Vs. Kramer* (1979) (Oscar).

Dustin Hoffman in *Tootsie* (1982)

His performance in *Tootsie* (1982) (Oscar nomination) as an unemployed actor who desperately clutches at female impersonation to gain a role in a soap opera was a comic tour de force that also allowed him to make fun of his own intensity.

Less committed to the screen in the 1980s, he returned to Broadway in *Death Of A Salesman* (1984), winning an Emmy for the 1985 television reprise of his performance as Willy Loman. He also tackled his first Shakespearean role as Shylock in a London production of *The Merchant of Venice* (1989), which he subsequently repeated on Broadway.

Returning to the screen with the ill-fated comic flop *Ishtar* (1987), he emphasised his skill at richly rounded and emotionally involving characterisations as the autistic savant in *Rain Man* (1988) (Oscar).

Back with a vengeance, he did a droll turn as Mumbles in *Dick Tracy* (1990), starred as gangster Dutch Schultz in *Billy Bathgate* (1991) and prepared to play the title role in *Hook* (1991), ▷ Steven Spielberg's version of Peter Pan.

HOMELAND *see* HEIMAT

THE HONEYMOONERS

USA 1955–56
Jackie Gleason

Originally developed for a sketch in Jackie Gleason's (1916–87) hour-long variety show, the character of Brooklyn bus-driver Ralph Kramden existed as a sitcom in its own right for just one season. Ralph (Gleason), who drove his bus on the Madison Avenue line, was a fatter, less-attractive ancestor of ▷ Phil Silvers's Sergeant Bilko. Engaging in an endless quest for a fast buck, he would steel himself for the impact of his inevitable disappointment with the catch-phrase 'One of these days, one of these days . . . Pow! Right in the kisser!' Puncturing Ralph's optimism was wife Alice (Audrey Meadows), whose utterances were largely expressions of marital disgust. Neighbour Ed Norton (Art Carney 1918–)) was a sewage worker ('where time and tide wait for no man') and acted as a straight man to Gleason's fool.

Though *The Honeymooners* originally only ran to 39 episodes, sketches from *The Jackie Gleason Show* (1966–70) were cut into 13 hour-long episodes. Four specials were shot between 1976 and 1978.

The revival was unwise, as Broadway-style musical numbers were required to fill out the episodes to an hour. Sharp writing, strong characterisation and inspired comedy acting were the key to the show's initial success. All were absent from the revivals, but live on in frequent and welcome repeats of the first series.

HOPE, Bob

Real Name Leslie Townes Hope
Born 29 May 1903, Eltham, London, UK.

The son of a stonemason, Hope moved to Cleveland in 1907 and later worked in a variety of menial jobs before trying out in vaudeville as a purveyor of 'songs, patter and eccentric dancing'. He made his Broadway debut in *The Sidewalks of New York* (1927) and was soon featured prominently in such popular shows as *Roberta* (1933) and *Red, Hot and Blue* (1936).

Acquiring a growing following as a sassy, wise-cracking radio comedian, he had made his film debut in the short *Going Spanish* (1934); the feature-length *The Big Broadcast of 1938* gave him his lifelong theme tune 'Thanks for the Memory'.

The vast popularity of *The Cat and The Canary* (1939), with its blend of haunted house chills and hilarious verbal repartee firmly established the potency of his screen persona as the cowardly braggart, smug and vainly conceited until faced with the possibility of heroic action or danger.

Famed for his ski-slope nose, lop-sided grin, battery of gainfully-employed scriptwriters, and impeccable timing, he was one of the most popular performers of the 1940s with the radio programme *The Bob Hope Pepsodent Show* (1939–48) and a string of hit films at Paramount including *The Ghost Breakers* (1940), *My Favourite Blonde* (1942), *The Princess and The Pirate* (1944), *Paleface* (1948) and the enduring *Road to* ... series that revelled in the innovative ad-libbing and in-jokes of his bantering relationship with friend and co-star ▷ Bing Crosby and unrequited lust for sarong-clad Dorothy Lamoür (1914–). The Road ran from *Singapore* (1940) to *Hong Kong* (1961).

He was effective in slightly more dramatic material like the biographies *The Seven Little Foys* (1955) and *Beau James* (1957) and the comedy *The Facts of Life* (1960) but his large-screen career drizzled to a dismal conclusion with a string of poor, formulaic star vehicles like *I'll Take Sweden* (1965), *Boy Did I Get A Wrong Number* (1966) and *Cancel My Reservation* (1972), his last starring role for the cinema.

A star of television, a court jester to various American Presidents, an indefatigable entertainer of American troops, a noted golfer and humanitarian, he has become a showbusiness institution and one of its wealthiest success stories. His receipt of countless international distinctions has also won him a place in the *Guinness Book of Records.*

Still active in television specials and personal appearances, he was a guest artist in *The Muppet Movie* (1979) and starred in the tired television film *A Masterpiece of Murder* (1986).

His several humorous volumes include *I Never Left Home* (1944), *I Owe Russia $1200* (1963) and *Confessions of A Hooker* (1985).

HOPPER, Dennis

Born 17 May 1936, Dodge City, Kansas, US.

A distant relative of gossip columnist Hedda Hopper, Hopper attended high school in San Diego and later studied acting with Dorothy McGuire (1918–) at the Old Globe Theatre. He is credited with an appearance in *Johnny Guitar* (1954) although ▷ *Rebel Without A Cause* (1955) is more often cited as his film debut.

In subsequent films like *Giant* (1956), *Gunfight at the O.K. Corral* (1957) and *From Hell to Texas* (UK: Manhunt) (1958) he was frequently cast as hot-headed young malcontents or whingeing weaklings. In the 1960s, he became associated with films like *The Trip* (1966) that cast a sympathetic eye over the drug counterculture and he made his directorial debut with ▷ *Easy Rider* (1969). An influential road movie of two bikers' disheartening quest for the real America, it seemed to embody the doubts and disillusionment of a generation and its vast popularity challenged the conservatism of prevailing Hollywood filmmaking techniques.

He then directed *The Last Movie*

(1971), a pretentious and often incoherent rumination on filmmaking itself set in a small Peruvian village. He continued to act in a wide variety of international productions including *Mad Dog* (1976) and *The American Friend* (1977) and experienced a resurgence in his career after an appearance in ▷ *Apocalypse Now* (1979).

He returned to direction with *Out of the Blue* (1980) an edgy exploration of misunderstanding between generations. He then became one of the American cinema's most busily employed character actors, bringing an intensity, energy and fragments of autobiography to bear on a gallery of obsessives, psychotics, burnt-out idealists and anachronistic hippies. His many recent films include *Rumble Fish* (1983), *River's Edge* (1986), *Texas Chainsaw Massacre 2* (1986), *Hoosiers* (UK: Best Shot) (1986) which earned him an Oscar nomination as Best Supporting Actor, and *Blue Velvet* (1986), in which his ferociously foul-mouthed villain seemed evil incarnate.

Concentrating on direction, he enjoyed box-office success with *Colors* (1988) a gritty, controversial account of the warfare between Los Angeles street gangs. After outside interference in the editing of *Backtrack* (UK: Catchfire) (1989) he disowned the film, but swiftly moved on to the critically acclaimed *The Hot Spot* (1990) a moody, atmospheric thriller set in a sleepy Texas town.

HOWARD, Leslie

Real Name Leslie Howard Stainer
Born 3 April 1893, London, UK.
Died 1 June 1943 en route from Lisbon to London.

Of Hungarian origin, Howard had been a bank employee before making his film debut in *The Heroine of Mons* (1914).

He turned to full-time dramatics after being invalided home from the Western Front during World War 1, making his stage debut in *Peg o' My Heart* (1917) and making his last cinema appearances for a decade in the British comedies *Five Pounds Reward* (1920) and *Bookworms* (1920).

He concentrated on the theatre, appearing in such Broadway productions as *Just Suppose* (1922), *The Green Hat* (1925) and *Her Cardboard Lover* (1928) and he made his Hollywood debut in *Outward Bound* (1930).

Much in demand, his sensitive bearing and gentlemanly manner saw him grow to personify an archetypically tweedy Englishman: scholarly, idealistic, witty and courageous.

He received a Best Actor Oscar nomination as the dashing time-travelling romantic in *Berkeley Square* (1933) and his many lauded performances from the era include the crippled doctor in *Of Human Bondage* (1934), the foppish aristocrat Sir Percy Blakeney in *The Scarlet Pimpernel* (1934) and the intellectual in *The Petrified Forest* (1936) whose poetic, stoic manner contrasted effectively with the snarling hoodlum of ▷ Humphrey Bogart.

He had harboured ambitions to move into film production and behind the cameras, and his own company made a definitive version of *Pygmalion* (1938) (Oscar nomination) which he co-directed with Anthony Asquith (1902–68). His Higgins was authoritative and suitably insufferable but always endearing.

After reluctantly playing the effete Ashley Wilkes in ▷ *Gone With the Wind* (1939), he returned to Britain to assist in the war effort by producing and directing such patriotic fare as *Pimpernel Smith* (1941) and *The First of the Few* (1942), in which he also starred, and *The Gentle Sex* (1943) and *The Lamp Still Burns* (1943), which he co-directed and produced respectively.

When he was returning from a lecture trip to Lisbon in 1943, his plane was shot down by the Nazis, who believed Winston Churchill to have been on board.

His son Ronald Howard (1918–) also became an actor, appearing in such films as *Queen of Spades* (1948), *No Trees In The Street* (1959) and *The Hunting Party* (1971). In 1982, he published *In Search of My Father: A Portrait of Leslie Howard.*

HOWARD, Trevor Wallace

Born 29 September 1916, Cliftonville, Kent, UK.
Died 7 January 1988, London, UK.

A student at RADA, Howard made his stage debut in *Revolt in a Reformatory* (1934) and acted exclusively in the theatre until World War II, including a long spell in the West End production of *French Without Tears* (1938–39).

Invalided out of the Royal Artillery, he enjoyed further West End success in *The Recruiting Officer* (1943) and *A Soldier for Christmas* (1944), before making his film debut in *The Way Ahead* (1944).

The delicate pitch of his performance as the lovestruck doctor engaged in furtive hope of adultery with Celia Johnson (1908–82) in ▷ *Brief Encounter* (1945) brought him to the forefront of British performers, a position confirmed by the thrillers *I See A Dark Stranger* (1946) and *Green for Danger* (1946).

A stocky, pipe-smoking, cricket-loving Englishman, he became a dependable leading man of reliable professionalism, but gave his most interesting performances when called upon to portray less heroic or straightforward characters. Notable among his many credits are the sarcastic liaison officer in ▷ *The Third Man* (1949) and the fallible, weak-willed police commissioner in *The Heart of the Matter* (1953).

He received his only Best Actor Oscar nomination for his powerful study of the morose, unyielding patriarch in *Sons and Lovers* (1960) and appeared in a number of large-budget ventures; he failed to erase memories of ▷ Charles Laughton as Bligh in *Mutiny on the Bounty* (1962), but made a welcome contribution to the World War II hokum *Von Ryan's Express* (1965), and to *The Charge of the Light Brigade* (1968), in which he played Lord Cardigan.

A character actor of international stature, most of his appearances in his later years were in secondary roles and cameos, building a gallery of crusty, ruddy-featured military men, comic eccentrics and blustering authority figures. The few roles of any substance include his village priest in *Ryan's Daughter* (1970), Wagner in *Ludwig* (1972) and the Red Indian chieftain in the offbeat western *Windwalker* (1980).

His final film appearances were in *White Mischief* (1988) and *The Unholy* (1988).

His rare stage performances over the years include *The Taming of the Shrew* (1947), *The Cherry Orchard* (1954) and *The Father* (1964). He also reminded audiences of his under-used talents in television work of a higher calibre, winning an Emmy for *The Invincible Mr. Disraeli* (1963) and appearing in *Catholics* (1973), *Staying On* (1980), in which he was reunited with Johnson, and the popular American film *Christmas Eve* (1986).

HUGHES, Howard

Born 24 December 1905, Houston, Texas, USA.
Died 5 April 1976, en route from Acapulco to Houston, Texas, USA.

At the age of eighteen, Hughes inherited his father's oil-drilling equipment company and, in 1926, began to involve himself and his profits in Hollywood films.

He produced a number of interesting films, including *The Front Page* (1931) and *Scarface: The Shame of A Nation* (1932) and also directed *Hell's Angels* (1930), a slow-moving World War I drama distinguished by its impressive aerial sequences.

Already known as an eccentric, he suddenly left Hollywood in 1932 and, after working for a short while as a pilot under an assumed name, turned his entire attention to designing, building and flying aircraft. Between 1935 and 1938 he broke most of the world's air speed records, was awarded a congressional medal from Washington, and then abruptly returned to filmmaking, producing and directing his most controversial film *The Outlaw* (1943), a version of events in the life of Billy the Kid famed for its prolonged history of conflict with the censor and for the excessive attention directed to the mammary glands of Jane Russell (1921–).

He continued his involvement in aviation by designing and building The Hercules, an oversized wooden seaplane that was completed in 1947, flew only once, but yielded valuable technical knowledge to the aviation industry.

His dalliance with the film industry continued when he acquired a controlling share of the stock of R-K-O in 1948 and proceeded to mismanage and neglect their output until eventually selling the company in 1955.

The later films he produced include *Jet Pilot*, made in 1950 but released in 1957, and *The Conqueror* (1956).

After severe injuries sustained in an air-crash in 1946 his eccentricity increased and he eventually became a recluse, from 1966 living in complete seclusion while still controlling his vast business interests from sealed-off hotel suites, and giving rise to endless rumour and speculation. In 1971 an 'authorised' biography was announced but the authors were imprisoned for fraud and the mystery surrounding him continued until his death.

He has been portrayed on screen by Tommy Lee Jones (1946–) in the television mini-series *The Amazing Howard Hughes* (1977), and by Jason Robards Junior (1920–) in the picaresque tale *Melvin and Howard* (1980), whilst ▷ Warren Beatty has frequently expressed an intention to star in a biography.

From 1956 to 1971, he was married to actress Jean Peters (1924–), whose brief film career had included leading roles in *Anne of the Indies* (1951), *Pickup on South Street* (1953) and *Three Coins In A Fountain* (1954).

HUMPHRIES, (John) Barry

Born 17 February 1934, Camberwell, Melbourne, Australia.

A student at Melbourne University, Humphries made his theatrical debut at the Union Theatre Melbourne (1953–4) and also appeared with the Phillip Street Revue (1956).

In Britain from 1959, he made his London stage debut in *The Demon Barber* (1959) and subsequently appeared in the long-running musical *Oliver!* (1960, 1963 and 1968).

He created the Barry MacKenzie comic strip in *Private Eye* (1964–73) and his many one-man stage shows include *A Nice Night's Entertainment* (1962), *A Load of Olde Stuffe* (1971), *An Evening's Intercourse With the Widely Liked Barry Humphries* (1981–82) and the record-breaking *Back With A Vengeance* (1987–89).

A frequent television performer, he is a sharp-witted observer of human nature whose finest characterisations have offered crude commentary on Australian mores and stereotypes and include the grossly repellent cultural attaché Sir Les Patterson and acid-tongued superstar housewife Dame Edna Everage. The latter made a splendidly unorthodox host of the popular celebrity chat show *The Dame Edna Experience* (1987–89, and subsequent one-off specials) indulging in brazenly inventive vulgarity, sexual innuendo, sly insult, quick-witted ad-libbing and incongruous musical routines with the likes of Lauren Bacall (1924–), Tony Curtis (1925–) and a star-studded cast.

Rarely seen without the mask of an assumed character, his infrequent film appearances include *Bedazzled! (1967)*, *Barry MacKenzie Holds His Own* (1975), a dramatic role in *The Getting of Wisdom* (1977) and the disappointing *Les Patterson Saves The World* (1987).

His many humorous books include *Treasury of Australian Kitsch* (1980) and *The Traveller's Tool* (1985).

HUPPERT, Isabelle Anne

Born 16 March 1955, Paris, France.

The youngest of four sisters, Huppert had chosen Russian studies at the University of Paris before transferring to the Conservatoire National D'Art Dramatique to study acting. After several inconsequential television roles, she made her film debut with a small part in *Faustine et le Bel Été* (Faustine) (1971).

Over the next few years, she appeared in numerous films including *César Et Rosalie* (César and Rosalie) (1972), *Les Valseuses* (1973) and *Rosebud* (1975). However, she earned major international acclaim for her performance in *La Dentellière* (The Lacemaker) (1977) as the chronically shy, unworldly young girl whose fragile existence is destroyed by a failed love affair.

She followed this by winning the Best Actress prize at the Cannes Film Festival for *Violette Nozière* (1978), the true story of a flighty Parisian teenager whose double life of wild promiscuity and domestic demureness led her to make an attempt on the lives of her mother and stepfather.

Adept at capturing the mix of good and evil in one person, the well-scrubbed beauty of her freckled features has served her well in the depiction of ambiguous characters with a surface innocence.

One of France's leading actresses, she made her official English-language debut as an immigrant bordello keeper in the ill-fated ▷ *Heaven's Gate* (1980), but found more challenging roles as the bourgeois half of an incongruous couple in *Loulou* (1980) and as the aloof country girl turned city prostitute in ▷ Godard's *Sauve Qui Peut (La Vie)* (Slow Motion) (1980).

Working with some of the most respectable international directors, she made *Coup de Torchon* (Clean Slate) (1981) with ▷ Bertrand Tavernier, *La Truite* (1982) with ▷ Joseph Losey, *Coup de Foudre* (At First Sight) (1983) with Diane Kurys (1947–), *La Femme De Mon Pote* (My Best Friend's Girl) (1983) with ▷ Bertrand Blier and *The Possessed* (1987) with ▷ Andrzej Wajda (1987).

She has made tentative ventures into English-language features as the young woman facing blindness in *Cactus* (1986) and as the compromised witness to murder in *The Bedroom Window* (1987), but some of her best recent roles have been with director ▷ Claude Chabrol and comprise *Une Affaire des Femmes* (1988) and the title role in *Madame Bovary* (1991).

HURT, William M.

Born 20 March 1950, Washington DC, USA.

The son of a State Department Official, Hurt took an abiding interest in amateur dramatics whilst a theology major at Tufts University in Massachusetts and, upon graduation, moved to New York to study acting at the prestigious Juilliard School.

Subsequently travelling the country, he appeared in the Oregon Shakespeare Festival's production of *Long Day's Journey Into Night* (1975) and began to build a substantial reputation in a succession of off-Broadway productions.

Tall, blonde and muscular he has described himself as 'a character man in a leading man's body' and has sought to subvert his blandly handsome masculinity in a series of unpredictable and testing acting choices.

He worked on television in the mini-series *The Best of Families* (1977) and the film *Verna: USO Girl* (1978) before making his cinema debut in *Altered States* (1980), bringing much needed conviction to a psychophysiologist whose mind-expanding experiments cause his regression to a state of ape-like primitivism.

He followed this with the routine thriller *Eyewitness* (UK: The Janitor) (1981) and the steamy latterday ▷ film noir *Body Heat* (1981) in which his square-jawed handsomeness perfectly suited the gullible, priapic lawyer obsessed with devious femme fatale ▷ Kathleen Turner.

He was the dour, dogged Russian homicide detective in *Gorky Park* (1983), then impressed with the emotional detail and substance of his performance as the troubled Vietnam veteran in the popular ensemble piece *The Big Chill* (1983).

Eschewing the play-safe choices of many stars, he won a Best Actor Oscar for *Kiss of The Spiderwoman* (1985), an obvious and mannered performance as the flamboyant homosexual whose incarceration with a political prisoner stirs his own conscience.

He won further Oscar nominations as the compassionate and dogged teacher of the deaf in *Children Of A Lesser God* (1986) and the soulless television anchorman in *Broadcast News* (1987).

His recent film roles include the emotionally repressed travel writer in *The Accidental Tourist* (1988), a GI in the overblown romance *A Time of Destiny* (1988), a less successful comic turn as a dopey hippie in *I Love You To Death* (1990) and as the husband in ▷ Woody Allen's *Alice* (1990). He will shortly be seen in ▷ Wim Wenders *Till The End of the World* (1991) and *The Doctor* (1991).

Frequently returning to the stage between film assignments, he won particular praise for his performance as the misanthropic, drug-addicted Eddie in *Hurlyburly* (1984–5), which earned him a Tony nomination as Best Supporting Actor.

HUSTON, John Marcellus

Born 5 August 1906, Nevada, Missouri, USA.
Died 26 August 1987, Newport, Rhode Island, USA.

The son of actor Walter Huston (1884–1950), Huston's characteristically colourful early background involved spells as an amateur lightweight boxer in California, painter, actor, honorary member of the Mexican cavalry and reporter for the New York *Daily Graphic*.

In Hollywood from 1929, he played small roles in films like *The Shakedown*

(1929) and *The Storm* (1930) before learning his trade as a screenwriter or collaborator on such films as *Murders in The Rue Morgue* (1932), *Jezebel* (1938), *High Sierra* (1941) and *Sergeant York* (1941) (Oscar nomination).

Eventually, he was allowed to direct one of his own scripts and the result was the classic detective yarn ▷ *The Maltese Falcon* (1941), expert evidence of his skills as a grand storyteller and master of mood and performance. It also highlighted a consistent theme in a remarkable body of work in its tale of a fruitless quest and the greed of men beset by delusions of grandeur.

He made a number of highly regarded wartime documentaries including *The Battle of San Pietro* (1944) and *Let There Be Light* (1945) and swiftly re-established his dominance with dramatic material as the director of the cracking gangster melodrama *Key Largo* (1948) and the incisive study of greed and immorality *The Treasure of the Sierra Madre* (1948) which earned him Oscars for Best Direction and Best Screenplay and brought his father one for Best Supporting Actor with his performance as a grizzled veteran prospector.

At the time it must have seemed as if the director could do no wrong as he illustrated his versatility, virtuosity and exuberance of spirit with a succession of hit films: the crime classic *The Asphalt Jungle* (1950) (Oscar nomination), the poetic anti-war production *The Red Badge of Courage* (1951), ▷ *The African Queen* (1951) (Oscar nomination), which earned an Oscar for his longtime colleague ▷ Humphrey Bogart, and the rich-hued beauty of *Moulin Rouge* (1952) (Oscar nomination).

However, his subsequent output proved more variable with successes like the well-acted *The Misfits* (1960), *Freud* (1962) and *Night of the Iguana* (1964) balanced by such rum nonsense as *The Barbarian and The Geisha* (1958) and *Casino Royale* (1967) for which he can only accept partial blame as one of five co-directors.

The variability continued throughout the latter years of his career, but a splendid burst of autumnal creativity included the gritty boxing story *Fat City* (1972), the rousing, boisterous Kipling adventure *The Man Who Would Be King* (1975), the Southern gothic grotesquerie of *Wise Blood* (1979), the elegant black comedy of Mafia murder *Prizzi's Honour* (1985) (Oscar nomination) and the near perfection of his cinematic translation of James Joyce's *The Dead* (1987), his final film.

He had acted on stage in *The Lonely Man* (1935) and occasionally in his own films but, after a supporting role in *The Cardinal* (1963), he became an increasingly familiar figure as a cinema actor leading his roguish, avuncular air and bear-like growl to an eclectic selection of projects from *Myra Breckinridge* (1970) to *Battle for the Planet of the Apes* (1973) and *The Wind and the Lion* (1975). His best role was as the menacingly rapacious patriarch at the centre of the intrigue in *Chinatown* (1974).

A Hemingway-like figure in real life who was master of the hounds in his Irish home at Galway, married five times and lived life to the full, he recounted his story in the autobiography *An Open Book* (1980). He has been followed into the film industry by several of his children, most notably Anjelica (1949–) who won a Best Supporting Actress Oscar for *Prizzi's Honour* (1985) and has been seen in *The Dead* (1987), *Enemies A Love Story* (1989) (Oscar nomination), *The Witches* (1989), *Crimes and Misdemeanours* (1989), *The Grifters* (1990) (Oscar nomination) and *The Addams Family* (1991).

He died during the filming of *Mr. North* (1988), which was directed by his son Danny (1962–).

ICHIKAWA, Kon

Real Name Uji Yamada.
Born 20 November 1915, Ise, Mie Prefecture, Japan.

An unpredictable and versatile talent, Ichikawa began work in the cinema in 1933 as a member of the animation department at the J.O. Studios. He later became an assistant director and directed *Musume Dojoji* (A Girl At Dojo Temple) (1946), a puppet version of a Kabuki play.

His first feature film was *Toho Senichi-Ya* (1,001 Nights with Toho) (1947) and his early directorial career consists of a number of mordant comedies and melodramas offering a satirical view of

Japanese mores and the artistic concerns of someone living in the post-nuclear age. Significant among these are *Pu-San* (Mr. Pu) (1953), and *Okuman Choja* (A Billionaire) (1954) in which a family destroy themselves with a meal of radioactive tuna.

With a large percentage of his films unreleased in the West it is unwise to offer generalisations on his vast body of work, but he has been drawn to the illumination of bleak and painful themes which are often leavened with a black, incisive wit and conveyed in expertly composed monochrome or colour images that show an inventive use of ▷ Cinemascope.

He explored 'the pain of our age' in *Biruma No Tageto* (The Burmese Harp) (1955) in which a soldier musician becomes obsessed with a mission to bury as many of the war dead as possible, *Enjo* (Conflagration) (1958), an adaptation of a Mishima novel in which a young man is driven to destroy a holy temple in order to prevent its further desecration by the tourist hordes, and *Nobi* (Fires on the Plain) (1959), the gruesome story of a retreating soldier's encounters with death, deprivation and cannibalism.

In the 1960s, his work included *Yukinojo Henge* (An Actor's Revenge) (1963) a technically accomplished study of a kabuki theatre female impersonator and his revenge on the three men who caused the death of his parents, and *Tokyo Orimpikku* (Tokyo Olympiad) (1965) a portrait of the human heroism of those athletes competing in the 1964 Olympic Games achieved with the efforts of 194 cameramen and 232 different lenses and marking a new, more intimate approach to the making of sports films.

Since then, he has concentrated more on the documentary form but returned to features with such unapologetically commercial ventures as the romance *Ai Futatabi* (To Love Again) (1971) and the suspense thriller *Inugami-ke No Ichizoku* (The Inugami Family) (1976). His most recent film is *Amanogawa Densetsu Satsujin Jiken* (1991).

IM LAUF DER ZEIT *see* KINGS OF THE ROAD

IMAMURA, Shohei

Born 15 September 1926, Tokyo, Japan.

Once described as the 'cultural anthropologist of Japan', Imamura's work is unusual in its anarchic tone, visceral impact, black humour and susceptibility to western influences.

A graduate in Occidental History from Waseda University, he was an actor and writer of student plays before working as an assistant-director to ▷ Ozu on films like *Bakushu* (Early Summer) (1951) and *Tokyo Monogatari* (Tokyo Story) (1953).

He made his directorial debut with the cynical comedy *Nusumareta Yokujo* (Stolen Desire) (1958) and continued to pursue his own career whilst writing scripts for other directors. Notable early titles in his filmography include *Nianchan* (My Second Brother) (1959), a gentle study of poverty-stricken orphans, *Buta–to Gunkan* (Pigs and Battleships) (1961), about the gangsters who feed off an American naval base at Yokosuka, and *Nippon Konchuki* (The Insect Woman) (1963), a sweeping account of forty arduous years in the life of one woman that provoked inevitable parallels with Japan's own struggles with progress, repression of individuality and exploitation of women.

His ability to blur the distinctions between fiction and documentary and his fascination with the more offbeat aspects of post-War Japanese society were evident in *Jinruigaku Nyumon* (The Pornographers: Introduction to Anthropology) (1966), a bittersweet plunge into the world of men who make blue movies.

After *Nippon Sengoshi: Madamu Omboro No Seikatsu* (History of Post-War Japan as Told By A Bar Hostess) (1970), he retired from the cinema to teach and to work on a number of television documentaries, returning in 1978 to mastermind a series of powerful works that include *Fukushu Suru–wa Ware–ni Ari* (Vengeance Is Mine) (1979) and Cannes Palme D'Or winner *Narayama Bushi-Ko* (The Ballad of Narayama) (1983) a shocking portrait of the primitive social order in a remote mountain village of the last century.

He has continued to seek an illumination of the present through a dramatic interpretation of the past in *Zegen* (1985), the story of a notorious South Seas procurer who used patriotism as a defence of his barbarous treatment of women, and the monochrome *Kuroi Ame* (Black Rain) (1988), an attempt to address the emotional legacy of Hiroshima and Nagasaki.

INBETWEENER In terms of cartoons and animation work, the 'inbetweener' is the person responsible for creating the drawings that fill in the action between the main drawings made by a more senior animator.

INDUSTRIAL LIGHT AND MAGIC (ILM) The special-effects company formed in 1975 by George Lucas (1945–) to lend technical expertise to the making of ▷ *Star Wars*, ILM has become one of the most respected special-effects organisations in the film industry and was influential in the development of motion control, an electronically programmed camera system that can exactly repeat movements and shots and is efficacious in the miniature and matte processes. The many films to benefit from ILM effects include the *Star Wars* sequels *The Empire Strikes Back* (1980) and *Return of the Jedi* (1983), the *Indiana Jones* films, *Labyrinth* (1986), *Innerspace* (1987), *Arachnophobia* (1990) and others. Lucas has also been responsible for developing the power and clarity of the THX sound reproduction system now heard in selected cinemas.

INGRAM, Rex

Real Name Reginald Ingram Montgomery Hitchcock
Born 15 January 1893, Dublin, Ireland.
Died 21 July 1950, Hollywood, California.

Educated at Saint Columba's College, Dublin and a student of sculpture at the Yale School of Fine Arts, Ingram emigrated to America in 1911 and developed an interest in the cinema that led to his employment with the Edison Company.

He worked as an actor in such films as *Beau Brummel* (1913) and *Witness to the Will* (1914) and wrote the scripts for a variety of films, including *Hard Cash* (1913) and *Should A Mother Tell?* (1915), before making his directorial debut with *The Great Problem* (1916).

He worked consistently over the next few years although little of his output from this period still survives. However, the phenomenal popularity of *The Four Horsemen of the Apocalypse* (1921), which grossed in excess of $4½ million, brought him to the front ranks of silent directors and consolidated the idol status of ▷ Rudolph Valentino. He was responsible for a number of highly popular swashbucklers that combined sweeping adventure, pictorial beauty, strong narratives and star performances. His many successes include *The Prisoner of Zenda* (1922), *Scaramouche* (1923) and *The Arab* (1924).

He subsequently moved to the Victorine Studios in Nice and made *Mare Nostrum* (1925) which is considered to be his best film. He left M-G-M in 1926 and made only a few further features; *The Garden of Allah* (1927), *The Three Passions* (1929) and *Baroud* (Love in Morocco) (1931).

He then settled in Egypt before returning to Hollywood in 1936 and becoming a Muslim. A writer, sculptor and traveller he was married to actress Alice Terry (1889–1987) who appeared in many of his films.

INTOLERANCE

D. W. Griffith (USA 1916)
Lillian Gish, Mae Marsh, Robert Harron, Constance Talmadge, Miriam Cooper.
Original running time around 220 mins; existing prints generally run around 130 mins. New score by Carl Davis premiered at 1988 London Film Festival.

In the autumn of 1914 while ▷ *Birth of a Nation* was at the editing stage, ▷ D. W. Griffith began work on his next film, a contemporary saga of social deprivation set against the backdrop of labour agitation, which he had more or less completed by the time his epochal Civil War story opened to widespread adulation. Spurred on by the favourable response to *Birth of a Nation*'s epic scope, he now proposed to expand the material he'd just shot into a new film that would be his most ambitious project to date: 'The purpose of the production' he explained, 'is to trace a universal theme through various episodes of the race's history.' With an emphasis on editing as a ruling aesthetic principle that was to prove a central influence on the Soviet school of the Twenties, *Intolerance* pictured the mother of ages rocking the cradle of history as the linking footage between selected moments in time, each of which in some way manifested a struggle against oppression. The climax of the film brought together the various strands; the modern couple's fight for justice thus elides into 16th century France and the St Bartholomew's day massacre of the Huguenots, and glimpses of the Passion intertwine with the betrayal of Babylonian king Belshazzar to the invading Persian forces. Despite notable setpieces (the Babylonian sets are amongst the largest ever constructed), *Intolerance* continues to divide critical opinion: for some, its formal complexity and thematic richness make the film an American masterpiece, while others share the feelings of confused audiences of the time that Griffith's obtrusive cross-cutting does not adequately convey the film's meaning. Unfortunately no prints of Griffith's original version have survived.

INVASION OF THE BODY SNATCHERS

Don Siegel (USA 1956)
Kevin McCarthy, Dana Wynter, Larry Gates, King Donovan.
80 mins. b/w.

Fantasy specialist George Pal's (1908–80) *Destination Moon* (1950) may have opened the cinema screen to the technological possibility of space travel but its spirit of optimism was short-lived: the same year saw both Robert Wise's (1914–) *The Day The Earth Stood Still*, in which a higher intelligence threatened to destroy the planet unless man started to behave himself, and ▷ Howard Hawks' production of *The Thing From Another World*, whose Antarctica-set mayhem climaxed with a portentous warning for us to 'Watch the Skies!'. Prophetic advice it was too, as the next few years saw dear old terra firma menaced by all manner of malevolent visitors, these paranoid cinematic fantasies typifying the Cold War insecurities in the face of looming nuclear destruction. A consistent anxiety in many of the films however was the notion that aliens might be able to replace the human form with their own subservient replicas, and Don Siegel's (1912–91) *Invasion of the Body Snatchers*, easily the most compelling film of the sub-genre, follows the disturbing plight of plucky protagonist Kevin McCarthy (1914–) as he discovers that his friends, neighbours and even his lover are no longer their former selves but unfeeling identical copies sprung from mysterious extraterrestrial pods. Released in the wake of the McCarthy witch-hunts, the film's vision of highly organised zombies presents a highly ambiguous vision of American society: on the one hand, the network of emotionless pod people certainly plays up to contemporary fears of 'the Red peril', yet, when shorn of the framing scenes added by producer Walter Wanger (1894–1968) – Siegel's original cut explores the displacement of individual endeavour by social unanimity of thought and action, exactly the kind of environment that allowed the injustices of the anti-communist crusade to go ahead in the first place. The film was remade in 1975 by director Philip Kaufman (1936–) with Don Siegel and Kevin McCarthy in cameo roles.

IRIS SHOT A camera shot in which the boundary of the image portrayed is in the form of a circle. Seen most often in films from the silent era, its uses were many. To begin a scene the aperture could gradually increase the size of the circle until the full image was seen or, alternatively, a scene could be terminated with the circle decreasing in size until the screen was black. These procedures are known as iris-in and iris-out. As a dramatic device it was used to suggest events being seen through a narrow space (e.g. telescope or key-hole), to focus attention on a particular character or action, or for irony in comedy by tightly focusing on an individual and then opening out to reveal the entire scene and context of his actions. Little used in contemorary cinema, it has served to evoke the spirit of the silent era in nostalgic dramas and contemporary comedies.

IRONS, Jeremy John
Born 19 September 1948, Cowes, Isle of Wight, UK.

The son of a chartered accountant, Irons' childhood ambition was to become a veterinarian but such extra-curricular school activities as playing drums in a rock group and appearing as Mr Puff in *The Critic* clearly presaged his artistic bent.

A social worker in London and part-time busker, he eventually decided to formally pursue a career in drama and secured employment as an assistant stage manager at the Marlowe Theatre in Canterbury. Later a student with the Bristol Old Vic and a member of their company for three years, he made his London debut as John the Baptist in the long-running rock musical *Godspell* (1971–72).

He developed his craft as a stage actor appearing with the Royal Shakespeare Company and winning widespread plaudits for his performances in *Wild Oats* (1976–77) and *Rear Column* (1978).

A jack of all trades who had appeared in everything from children's television to sherry commercials and the drama series *The Pallisers* (1975), he rose to stardom with his performances as the archetypal stiff upper-lip Englishman Charles Ryder in the acclaimed television series *Brideshead Revisited* (1981) and in the dual roles of obsessive Victorian romantic and contemporary thespian in the multi-dimensional romance *The French Lieutenant's Woman* (1981).

He had made his film debut with a small role as a choreographer in *Nijinsky* (1980) and now established his versatility and eschewal of the crasser fringes of the film world with a succession of performances in either low-budget or foreign-language productions. His most notable work includes the forlorn, well-meaning Polish exile in *Moonlighting* (1982) and the romantically deceitful literary agent in *Betrayal* (1982).

A tall, lanky figure with elegantly chiselled features and a plaintive air, he has been adept at portraying emotional repression in a succession of often gloomy or humourless characterisations.

The toast of Broadway for his Tony-winning role as the buoyant dramatist crushed by his wife's romantic infidelity in *The Real Thing* (1984), his subsequent, selectively chosen film work, includes a moving performance as a Jesuit priest who selects faith and peaceful protest as his weapons to oppose rampant colonialism and genocide in *The Mission* (1986), and a subtly impressive delineation of twin gynaecologists in the unsettling psychological thriller *Dead Ringers* (1988).

Committed to the theatre, a 1986–87 season with the RSC brought performances in the *The Rover*, *The Winter's Tale* and *Richard II*. His most recent film credits include the family drama *Danny, Champion of the World* (1989) and *A Chorus of Disapproval* (1989) in which he revealed a lighter side as a reasonable farceur.

He won a Best Actor Oscar for his mannered but effective portrayal of the chilly, aristocrat Claus Von Bulow on trial for murder in *Reversal of Fortune* (1990), and, after declining the role of Peron in the forthcoming Evita, will next be seen in *Kafka* (1991).

I-SPY
USA 1965–68
Bill Cosby, Robert Culp.

This NBC espionage adventure story was responsible for making a household name of ▷ Bill Cosby, who was the first black actor to perform in a nationally networked American drama. A successful stand-up comedian, he won the part impressing producer Sheldon Leonard with a 1965 appearance on the *Tonight Show With Johnny Carson*. Cosby's role, not specifically designed for a black, was that of Alexander Scott, who played coach and undercover partner to tennis star Kelly Robertson (Robert Culp (1930–)).

I-Spy added to the espionage genre by presenting rounded characters who had the ability to laugh at themselves and their situation. The scripts led to international locations, and a series of lively, often humorous adventures. Though the introduction of black characters was new, it was not treated as an issue, and there were supporting roles for the likes of Eartha Kitt (c.1928–), Cicely Tyson (1933–) and Leslie Uggams (1943–).

Cosby's Scott won him Emmys in 1966, 1967 and 1968.

Donna Reed and James Stewart in *It's A Wonderful Life* (1946)

IT'S A WONDERFUL LIFE
Frank Capra (USA 1946)
James Stewart, Donna Reed, Lionel Barrymore, Thomas Mitchell, Henry Travers.
129 mins. b/w.

The current gulf between critical disdain and public affection for the work of ▷ Frank Capra is nowhere more powerfully exposed than in the example of *It's A Wonderful Life*. Apparently the director's favourite of his own films, its unstoppable appeal for generations of moviegoers has made this wholesome parable a yuletide staple of television and repertory house schedules. In a typically winning performance, ▷ James Stewart's good-hearted George Bailey gives up his own career ambitions as an architect to run a loan company expressly devoted to helping out the good citizens of his home town, yet when accounting problems drive him to ruin he is saved from suicide only by the heavenly intervention of guardian angel Henry Travers (1874–1965), who shows him how harsh and materialist the lives of everyone around him would have been were it not for his vital contribution, thus persuading him that, yes, it really is a wonderful life. Although the film's central section as Stewart piles self-sacrifice upon self-sacrifice only to plough deeper into financial frustration suggests that Capra's faith in the American dream might be tempered by the merest soupçon of ambiguity, the overwhelming climactic explosion of sentimentality – or 'Capracorn' as several wags have termed it – represents his reactionary post-New Deal optimism at its most staggeringly simplistic. Yet our willingness to reject such a naive fable is hardly stronger than Capra's ability to make us believe in it. Assured in his handling of Hollywood's finest character actors and tremendously gifted in his ability to mould audience response, the director's enormous emotional conviction to his material simply steamrollers over all our scepticism.

IVENS, Joris
Real Name Georg Henri Anton Ivens
Born 18 November 1898, Nijmegen, Holland.
Died 28 June 1989, Paris France.

Nicknamed the 'Flying Dutchman', Ivens spent a lifetime as a documentary filmmaker and posterity's witness to some of the most turbulent revolutionary events of the twentieth century.

Born into a family of photographers, his first film *Wigwam* (Flaming Arrow) (1911) used chocolate powder to disguise family members as Red Indians.

Educated at the Rotterdam School of Economics and a student of photo-chemistry in Germany, he returned to Holland in 1926 to run the family business. However, his attention was consumed by the Amsterdam film club, Film Liga, and he directed two documentaries *De Brug* (The Bridge) (1928) and *Regen* (Rain) (1929), strong pictorial portraits of a railway bridge and downpour respectively.

Critically admired by the French ▷ avant-garde, he received an invitation to Russia and began to display the social conscience that would inform his life's work in films like *Komsomol* (1932), *Misère Au Borinage* (Borinage) (1933), covering the pitiful living conditions of Dutch miners, and *New Earth* (1934), which focused on land reclamation.

He reported on the shelling of Madrid during the Spanish Civil War in *The Spanish Earth* (1937), narrated by Ernest Hemingway (1899–1961), championed the oppressed Chinese people in *The 400 Million* (1938) and made *The Power and the Land* (1940), about rural electrification, for the American Department of Agriculture.

Appointed as Film Commissioner to the Dutch East Indies, he typically resigned in protest against Dutch colonialism which he denounced in *Indonesia Calling* (1946).

Based in Paris from 1957, he taught filmmaking in Poland, Cuba and Chile and worked with his wife and fellow documentarist Marceline Loridan on projects throughout the world, filming *The 17th Parallel* (1968), for instance, in the 30-foot catacombs in which Vietcong peasants took refuge from the American bombing of Vietnam.

Latterly fascinated by the struggles and peoples of China, he made *Comment Yukong Deplaça Les Montagnes* (How Yukong Moved the Mountains) (1976), a twelve-film cycle on the daily life there. He had been a lifelong asthmatic but showed remarkable vigour in completing *Une Histoire Du Vent* (A Story of the Wind) (1988) a poetic autobiographical fable filmed under the most arduous of conditions in the Gobi desert and other remote locations. A filmmaker's quest to capture one of the world's greatest winds on film, it stands as an eloquent testimony to a life spent in communicating impalpable human values.

Active until the time of his death, he had participated in the protest marches against the Beijing massacre of students only days before his demise. His autobiography, *The Camera and I* was published in 1974.

IVORY, James

Born 7 June 1928, Berkeley, California, USA.

Educated in architecture and fine arts at the University of Oregon, Ivory completed his studies with a filmmaking course at the University of Southern California in Los Angeles where he directed a number of short films including *4 in the Morning* (1953) and *Venice: Theme and Variations* (1957).

The success of *The Sword and The Flute* (1960) brought him a commission from the Asia Society of New York to make a documentary on Delhi and led to a first encounter with his future partner Ismail Merchant (1936–). Together, they formed Merchant-Ivory Productions and, in frequent collaboration with screenwriter and novelist Ruth Prawer Jhabvala (1927–), it has become one of the most enduring and productive of independent associations.

Their first international success came with *Shakespeare Wallah* (1965), which follows the peregrinations of an English acting troupe in India. Further examinations of Indian culture and the clashes between eastern and western sensibilities were explored in such films as *Bombay Talkie* (1970), *Autobiography of a Princess* (1975) and *Heat and Dust* (1982).

A return to America revealed the diversity of Ivory's interests and resulted in such films as the allegorical *Savages* (1972), *The Wild Party* (1974), a robust recreation of Hollywood in the late 1920s, and *Roseland* (1977), a poignant trilogy of tales about the romantic dreamers at the famous New York ballroom.

The Europeans (1979), a fastidious version of the Henry James (1843–1916) novel, again focused on an abrasive meeting of cultures (sophisticated Europeans and God-fearing, plain living New Englanders) and began the partnership's on-going attachment to refined literary adaptations. Notable works include *The Bostonians* (1984), *A Room With A View* (1985) (Oscar nomination), *Maurice* (1987) and *Mr. and Mrs. Bridge* (1990).

Their hallmarks of a literate, precise script, ironic humour, scrupulous attention to period detail and design, and impeccable performances have led their critics to accuse them of shallowness and suffocating in good taste but beneath the veneer of beautiful compositions and languorous pacing lurks a good deal of passion over issues of class conflict and personal liberty.

JACKSON, Glenda

Born 9 May 1936, Liverpool, UK.

After a stint as a shop assistant in Boots and enthusiastic amateur actress Jackson became a student at RADA, making her theatrical debut in a production of *Separate Tables* (1957) at Worthing and her London debut the same year in *All Kinds of Men* (1957).

She made her film debut in *This Sporting Life* (1963), but for many years, remained ostensibly a stage performer in the likes of *Alfie* (1963), and with the Royal Shakespeare Company in *Hamlet* (1965) and *Marat/Sade* (1967).

A plain-looking woman with a declamatory style of acting, she won a Best Actress Oscar as Gudrun in *Women in Love* (1969) and was soon established as a fearless portrayer of liberated sexuality in such films as *Sunday, Bloody Sunday* (1971) and *The Music Lovers* (1971).

Seen on television in the BBC series *Elizabeth R* (1971), her good-humoured appearance in a ▷ Morcambe and Wise skit landed her a role in the comedy *A Touch of Class* (1973) and a second Best Actress Oscar.

For someone so highly honoured her screen career has only rarely combined a good role in a worthwhile production; however she was Oscar-nominated again as *Hedda* (1975) and gave a moving and sensitive performance as poet Stevie Smith in *Stevie* (1978).

At her best in roles that attempt to convey the everyday realities of a human situation or offer some sense of social purpose, her most effective screen performances include the title role in the television biography *The Patricia Neal Story* (1981), a concerned television documentary-maker in *Giro City* (1982) and a shop manageress fighting sexual harrassment in the work place in *Business As Usual* (1987).

A very active and much lauded trans-Atlantic stage career includes *Rose* (1980), *Strange Interlude* (1983), *The House of Bernarda Alba* (1986), *Macbeth* (1988) and *Mother Courage* (1990).

A prospective Parliamentary candidate for the Labour Party, her political aspirations may signal the end of her acting career, but a recent hectic schedule has included such films as *The Rainbow* (1989) and *Doombeach* (1990) and the television series *A Murder of Quality* (1991).

JANCSÓ, Miklós

Born 27 September 1921, Vac, Hungary.

A graduate in law from Kolozsvar University in Romania, Jancsó subsequently studied at the Budapest Academy of Dramatic and Film Art before working on numerous short films and documentaries during the 1950s.

He co-directed *Kezunbe Vettuk A Béké Ugyét* (We Took Over the Cause of Peace) (1950) and later made a number of documentaries on China including *Dél-Kína Tájain* (The Landscapes of Southern China) (1957) and *Pekingi Palotái* (Palaces of Peking) (1957).

He made his feature film debut with *A Harangok Römába Mentek* (The Bells Have Gone to Rome) (1958), won the Hungarian Critics Prize for *Cantata* (1963) and began to engage international audiences with *Igy Jötem* (My Way Home) (1964), a stark tale of a wartime friendship between a young Russian soldier and a Hungarian boy that established his skill with fluid camera movement and recurring preoccupations with the interaction of man and landscape and the relentless futility of mutually destructive armed aggression.

His distinctive style developed further with *Szegénylegények* (The Round-Up) (1965), in which Austro-Hungarian troops torture and brutalise a group of peasants in order to secure the identity of a parisian leader, and *Csillagosok, Katonák* (The Red and the White) (1967) a sweeping historical drama of Hungarian involvement in the struggle between Red and White Army forces in the Russia of 1918. Capturing a feel of wide open spaces and beautiful tableaux, he makes expressive use of precisely choreographed camera movement and views events from a perspective that avoids the depiction of heroes and individuals to focus on man as a distant pawn in a senseless game.

Other notable work from this period includes his first colour feature *Fényes Szelek* (The Confrontation) (1969), in which Hungarian students of 1947 attempt to convert a Catholic school to their cause, and *Még Kér A Nép* (Red Psalm) (1971) a sensuous, balletic portrayal of the struggles between farm workers and a wealthy count.

Latterly, he has worked extensively in Yugoslavia and Italy on films of much less potency, although the mild eroticism of *Vizi Privati, Pubbliche Virtù* (Private Vices and Public Virtues) (1976) guaranteed it a wider audience.

Less prolific of late, his more recent Hungarian work includes *A Zsarnok Szíve Avagy Bocaccio Magyarországon* (The Tyrant's Heart) (1981).

JARMAN, (Michael) Derek

Born 31 January 1942, Northwood, Middlesex, UK.

A student of English and history at Kings College, London (1960–63), Jarman subsequently studied at Slade School of Fine Art (1963–67) and was exhibited as part of the Young Contemporaries at the Tate Gallery in 1967.

A set and costume designer for productions at the Royal Opera House, he entered the film industry as a set designer for the ▷ Ken Russell films *The Devils* (1971) and *Savage Messiah* (1972) and began to make his own Super 8mm home movies and shorts including *Miss Gaby* (1971) and *Andrew Logan Kisses The Glitterati* (1972).

He made his debut as a feature film director with *Sebastiane* (1976), a lyrical, homoerotic interpretation of the saint's martyrdom shot in Latin, and followed this with *Jubilee* (1977), a provocative and anarchic punk-inspired futuristic vision of England.

Subsequent work has shown him to be an imaginatively inventive director using all forms of filmmaking as a means of intimate self-expression to record and convey his distress at the social and moral decline of modern Britain and explore his artistic and political concerns as an openly homosexual man in a homophobic society.

His elliptical, free-form style can often render his work inaccessible and forbidding but his best films include a distinctive version of *The Tempest* (1979), the more conventional, painterly biography of *Caravaggio* (1985), the lyrically visualised *War Requiem* (1988) and *The Garden* (1990), a complex, very personal reinterpretation of the Passion story that explores centuries of religious persecution and bigotry towards homosexuals with the title referring both to Gethsemene and the director's own garden that he is constructing at Dungeness.

Other work includes *The Angelic Conversation* (1985), a romantic, plotless reading of 14 Shakespeare sonnets to fuzzily romantic images of young men, and the highly controversial *The Last of England* (1987), a bitterly scathing attack on the bankrupt values of the British establishment.

He has also made pop videos for The Pet Shop Boys, The Smiths and Bob Geldof and is currently preparing a version of *Edward II*.

He has written an autobiography entitled *Dancing Ledge* (1984).

JARMUSCH, Jim

Born 22 January 1953, Akron, Ohio, USA.

A major in English and American literature at New York's Columbia University, Jarmusch grew up on a diet of cinema that ranged from James Bond to Japanese horror films; it was a trip to

John Lurie, Tom Waits and Roberto Benigni in *Down By Law* (1986) directed by Jim Jarmusch

Paris and exposure to world cinema at the Cinématheque Française that fostered his love of ▷ Ozu, ▷ Mizoguchi and ▷ Dreyer.

Accepted for the graduate department of film studies at New York University, he became a teaching assistant to director ▷ Nicholas Ray and worked as a production assistant on ▷ Wim Wenders's documentary *Lightning Over Water* (1979).

A keyboard player and singer with the new wave band Del-Byzanteens, he made his directorial debut with the student film *Permanent Vacation* (1980). The next few years were spent in securing the finance to expand a short film into *Stranger Than Paradise* (1984) which won the Cannes Film Festival Camera d'Or for the best first feature. Shot in bleached black-and-white and hailed as 'lightweight Beckett', the minimalist storyline relates incidents in the lives of two aimlessly inert New York hustlers and shows how their comfortable routine is disrupted by the arrival of a teenage cousin from Hungary.

The more exuberant *Down By Law* (1986) intertwines the lives of three idiosyncratic convicts as they escape into the Louisiana bayou. Described by its director as a 'neo-beat-noir-comedy' it is an endearingly offbeat fable full of deadpan humour and laconic performances, shot in luminous monochrome.

Continuing to create work that seemed the cinematic equivalent of jazz music, he then directed *Mystery Train* (1989), a quirky trio of stories featuring characters whose lives are linked by a mysterious gunshot, a love of Elvis Presley (1935–77) and their presence in a sleazy Memphis hotel. A prizewinner at the Cannes Festival, it was the first of his films to receive less than universal praise.

Committed to creating a new storytelling tradition that avoids the manipulation of mainstream fare and the perceived emotional repression of much European cinema, Jarmusch has also appeared as an actor in the film *Straight to Hell* (1987).

THE JAZZ SINGER

Alan Crosland (USA 1927)

Al Jolson, Warner Oland, Eugenie Besserer, Myrna Loy, Mae McAvoy.

89 mins. b/w. Synchronised musical numbers.

Academy Awards: Special Oscar to Warner Brothers for Producing *The Jazz Singer*.

Although silent films were usually accompanied by at least a piano and often a house band playing specially composed music, the first synchronised sound system to come into general use was the recorded disc apparatus developed by the Vitaphone company for Warner Brothers in the mid-Twenties. The ▷ John Barrymore swashbuckler *Don Juan*, directed by Alan Crosland (1894–1936), which premiered in August 1926, was the first to benefit from the new technology with a musical soundtrack (but without as yet synchronised speech). A number of test shorts designed to show off the movies' new toy were soon on show, with Fox Movietone launching their sound-on-film system in January 1927, but the initial feature with synchronised music and dialogue did not appear until October 1927 when ▷ Al Jolson's per-

Al Jolson in *The Jazz Singer* (1927)

formance in *The Jazz Singer* (again directed by Crosland) created a box office sensation and almost overnight ensured the new technology would be here to stay. Jolson was the original inspiration for the basic story of a Jewish cantor's son torn between family tradition and the call of the footlights. George Jessel (1891–1981), successful in the role on Broadway, was initially cast in the film before being replaced by Jolson, to whom audiences had responded more favourably in test shorts. Thus it was that Jolson's stage catchphrase 'Wait a minute, wait a minute, you ain't heard nothin' yet' was to be the first natural dialogue ever uttered on screen, and Jolson's vibrant rendition of 'Mammy' which was to transcend the creakingly melodramatic material and remain irrevocably imprinted on the public imagination. The black-faced star's personal impact was certainly to prove more substantial than that of the Vitaphone equipment which enabled him to burst into song in the first place – a year later and it was gone, and Western Electric's sound-on-film system established as the new standard.

▷ Michael Curtiz's forgettable 1953 remake starred Danny Thomas (1914–), while the even less distinguished 1980 Richard Fleischer (1916–) version had pop artist Neil Diamond (1941–) in the central role.

JENNINGS, Humphrey

Born 23 May 1907, Walberswick, UK.
Died September 1950, Greece.

An exceptional scholar, Jennings was a writer, painter and photographer who was involved with the ▷ Surrealist movement. He joined the General Post Office (GPO) Film Unit in 1934 and worked as a scenic designer and editor, making his directorial debut with the film *Post Haste* (1934), which covers three hundred years of Post Office history in prints and drawings. He subsequently made *Locomotive* (1934) on the early development of the steam locomotive, and *Birth of A Robot* (1935), a puppet fantasy advertising Shell Oil accompanied by the music of Gustav Holst.

His lyrical, emotionally charged talent really flourished during the wartime years when he directed a number of documentaries unrivalled in their sense of composition and editing.

London Can Take It (1940), *Listen to Britain* (1942), the feature-length *Fires Were Started* (1943), conveying the work of an auxiliary fire station in Dockland, and *Diary for Timothy* (1944–5) are unique documentary records of the British at war that affectionately celebrate the quiet heroism, fortitude and spirit of those on the Home Front as the nation underwent its darkest hour. Evocative, eloquent mosaics of images and sounds, they earned Jennings his reputation as the 'only true poet of British cinema'.

He made a number of documentaries after the War, including *A Defeated People* (1946) about the British occupied zone of Germany, and *Family Portrait* (1950) and was scouting locations for a film in *The Changing Face of Europe* series when he was fatally injured in a fall from a cliff.

JOLSON, Al

Real Name Asa Yoelson
Born 20 May 1886, St Petersburg, Russia.
Died 23 October 1950, San Francisco, USA.

One of the Broadway legends of the 20th century, Jolson's niche in cinema history was secured with his starring role in the first part-talkie ▷ *The Jazz Singer* (1927), which illustrates the exuberant theatricality of his persona whilst offering a few songs and the oft-repeated line of dialogue 'You ain't heard nothing yet!'

An inveterate performer, he ran away from home to join a circus and later made his stage debut in *Children of the Ghetto* (1899). His career advanced via vaudeville and minstrel shows until he became established as a stellar Broadway attraction and one of the first million seller recording artistes.

In 1923, he began work with ▷ D. W. Griffith on an abortive film project entitled *Mammy's Boy* which ended acrimoniously in an exchange of law suits between the two men. He made his official debut in *April Showers* (1926), a short film with sound in which he sang three songs.

The Jazz Singer tells a lachrymose story of a Jewish cantor's son caught between devotion to tradition and his love of sacrilegious jazz music. George

Jessel (1898–1981) was scheduled to repeat his Broadway success on film until protracted negotiations over his salary provoked his withdrawal. Others, including Eddie Cantor (1892–1964) and Harry Richman (1895–1972), are reputed to have rejected the role before it was offered to Jolson.

The novelty of synchronised sound, his vibrant vocals and larger-than-life personality appealed enormously to the public and variations on the formula of sentimentality and songs proved equally popular in the short term; *The Singing Fool* (1928), for instance, established box-office records that stood for a decade.

He returned to Broadway in *Wonder Bar* (1931) which was filmed in 1934, but cinema audiences seemed to quickly tire of star vehicles like *Mammy* (1930) and *Big Boy* (1931) and subsequent film appearances grew increasingly infrequent and were often restricted in duration.

He remained active on radio and on stage and gained fresh popularity entertaining the troops during the Second World War.

He provided the vocals for actor Larry Parks (1919–75) impersonation of him in two highly successful (and highly romanticised) film biographies *The Jolson Story* (1946) and *Jolson Sings Again* (1949), which ensured renewed interest in his career at the time of his death.

JONES, Chuck (Charles M.)

Born 21 September 1912, Spokane, Washington, USA.

A student at the Chouinard Art Institute in Los Angeles, Jones has spent over half a century as one of the most inventive of American animators bringing high standards, a strong sense of characterisation and comic timing to his work on a variety of famous creations at a diversity of Hollywood studios.

He joined Warner Brothers in 1935 and remained there until 1962, apart from a brief sojourn with ▷ Walt Disney in the mid-1950s. Coming under the influence of ▷ Tex Avery, he was encouraged to explore the comic possibilities of the short form cartoon and was a key figure in the creation and development of such characters as amorous skunk Pepe Le Pew, irascible Daffy Duck, Bugs Bunny and the frustrated Wile E. Coyote in the Roadrunner series, who never quite manages to capture his prey, despite an arsenal of weaponry, gadgets and a never-say-die attitude.

Showing an irreverent approach towards the cartoon form itself, he won Oscars for *For Scent-Imental Reasons* (1949) and *So Much for So Little* (1950) and his classic Bugs Bunny and Daffy Duck cartoons include *Rabbit Seasoning* (1952), *Duck Amuck* (1953), *Duck Dodgers in the 24½ Century* (1953) and *What's Opera Doc?* (1957).

At M-G-M from 1963 to 1967, he worked on such contemporary Tom and Jerry cartoons as *Ah Sweet Mouse-Story of Life* (1965), *Bad Day At Cat Rock* (1965) and *Jerry, Jerry Quite Contrary* (1966).

He won a further Oscar for *The Dot and The Line* (1965) and has since worked on a number of television and feature-length specials including *The Phantom Tollbooth* (1969), *How The Grinch Stole Christmas* (1970) and *Riki-Tiki-Tavy* (1975).

Some of his best work has appeared in recent compilation features like *The Bugs Bunny Road Runner Movie* (1979), *The 3rd Bugs Bunny Movie: 1001 Rabbit Tales* (1982) and *Porky Pig in Hollywood* (1986).

An acknowledged influence on the visual style of live-action director Joe Dante (1946–), he made a guest appearance in his film *Innerspace* (1987).

JORDAN, Neil

Born 25 February 1950, County Sligo, Eire.

After reading history and literature at University College Dublin, Jordan worked at various jobs in London before returning home and helping to form the Irish Writers Co-operative in 1974.

His first collection of stories, *Night in Tunisia* (1976), earned the Guardian Fiction Prize and was followed by the acclaimed novels *The Past* (1980) and *Dreams of the Beast* (1983).

Interested in exploring cinema as a visual means of story-telling, he was given the opportunity to work as a script consultant on ▷ John Boorman's *Excalibur* (1981) and made his directorial debut with *Angel* (1982), a Northern Ireland thriller of dream-like intensity in which a witness to murder discovers his own capacity for violence.

His subsequent work has boldly emphasised the fairy-tale and fantasy elements in *The Company of Wolves* (1984), a mature psychological and sexual interpretation of Little Red Ridinghood, and *Mona Lisa* (1986), the touching gang-land saga of an ex-con's redemption which featured an excellent, Oscar-nominated performance from Bob Hoskins (1942–).

Recently, he turned his hand to comedy with singularly unsuccessful results in the witless haunted house farrago *High Spirits* (1988) and the handsomely mounted but lumbering remake of *We're No Angels* (1989).

The Miracle (1991) marks a return to his home country and more dramatic concerns.

LE JOUR SE LEVE (Daybreak)

Marcel Carné (France 1939)
Jean Gabin, Jacqueline Laurent, Arletty, Jules Berry.
85 mins. b/w.

A key moment in that strand of Thirties French cinema commonly termed 'poetic realism', *Le Jour Se Lève* marks a peak in the fruitful collaboration between director ▷ Marcel Carné and frequent screenwriter Jacques Prévert (1900–77). Often considered in tandem with its 1938 companion piece *Le Quai des Brumes* (Port of Shadows), the two Carné/Prévert classics evince a world-weariness that struck a deep chord with the French public and was very much a part of the national psyche as events tumbled towards the Fall of 1940. The indications of the usual terminology their 'poetic realism' is, in fact, anything but realist; true, *Le Jour Se Lève* takes place amidst the grime of shadowy tenement blocks, but Carné's visuals are not so much an on-screen outpouring of social conscience as the outward expression of the protagonist's resignation to inescapable destiny. The film's circular structure underlines the compelling mood of fatalism, beginning as a shot rings out and the murderer barricades himself in an attic, his only future to spend the night pondering his deed before a second shot signals the breaking dawn. ▷ Jean Gabin's performance as the doomed killer epitomises his rough magnetism, allowing him to reprise the proletarian anti-hero type so evocatively paraded in Julien Duvvier's (1896–1967) *Pepé Le Moko* (1936), while Prévert's dialogue – at times criticised for sentimentalising the working class or lapsing into pretentious bathos – translates the banalities of the street into emotive shards of tough-guy lyricism. The overpowering melancholy of the final result is a tribute to the chemistry of the various creative participants, a blend whose delicately poised aura of sadness is completely lacking in *The Long Night* (1947), the disastrous Anatole Litvak (1902–74) Hollywood remake with ▷ Henry Fonda.

JULES ET JIM

François Truffaut (France 1961)
Jeanne Moreau, Oskar Werner, Henri Serre, Vanna Ubino.
105 mins. b/w.

▷ François Truffaut's *Jules et Jim* remains for many viewers the most affectionately remembered film of the flowering of the French ▷ Nouvelle Vague. Based on a little-known autobiographical novel published by Henri-Pierre Roche (at the age of 75) in 1955 which immediately attracted the attention of the then film critic Truffaut, this vital, mature and cinematically inventive chronicle of an ever-shifting ménage à trois across the decades from La Belle Epoche to the depression-hit Thirties is the most exquisite demonstration of the way in which its late director, out of all the post-*Cahiers du Cinema* generation, has carried the torch for the particularly Gallic lyricism which fired the earlier work of ▷ Jean Renoir and ▷ Jean Vigo. Surprisingly Truffaut's film retains its emotional resonance while riskily flaunting a plethora of cinematic devices – ▷ freeze frames, ▷ jump cuts, ▷ iris shots, ▷ stock footage et al – but the various effects provide an appropriate stylistic framework for the exuberance of the half-comic, half-tragic narrative. Although the complex relationship between ▷ Jeanne Moreau's mercurial Catherine and her respective suitors, Oskar Werner's Jules and Henri Serre's Jim, swerves unpredictably from one configuration to another, Moreau's entrancing central performance lends the swirling patterns of frustration and fulfilment both coherence and conviction. As all things to both men, her bewitching capriciousness still registers much more vividly than the best efforts of Margot Kidder (1948–) in Paul Mazursky's (1930–) spottily arresting American homage/remake *Willie & Phil* (1980), a film which demonstrates the associative pitfalls in ambitiously quoting your source material.

JUMP CUT An editing technique that involves the intentional deletion of part of the continuous action within a scene to make some form of dramatic point, whether it is to shock the audience with the unexpected, disturb the rhythm of a scene, or swiftly 'jump' from the cause to the effect without the interim narrative niceties. When the effect is unintentional the reason is usually poor editing, poor filming or the inferior quality of a print where wear and tear has caused the deletion of essential material.

DAS KABINETT DES DR CALIGARI *see* THE CABINET OF DR CALIGARI

KARLOFF, Boris

Real Name William Henry Pratt
Born 23 November 1887, Dulwich, South London, UK.
Died 2 February 1969, Midhurst, Sussex, UK.

The youngest child of a distinguished civil servant, Karloff was educated at Merchant Taylors' School and London University with the original intention of pursuing a diplomatic career.

Instead, he emigrated to Canada in 1909 and worked on a farm before touring with numerous theatrical troupes and making his way into the burgeoning California film industry as an extra and bit part player in such silent films as *The Dumb Girl of Portici* (1916), *His Majesty the American* (1919) and *The Last of the Mohicans* (1920).

Over the next decade he appeared in dozens of generally undistinguished film roles before returning to the stage as the convict turned killer in *The Criminal Code* (1930), a role he repeated in the film version a year later. Character roles in a number of films immediately preceded his appearance as the monster in *Frankenstein* (1931).

Tall and gaunt with a lumbering gait and soulful eyes, he brought a dignity and pathos to this scientific miscreant consumed by inarticulate rage and confusion over his cruel fate.

He repeated the role in *Bride of Frankenstein* (1935) and *Son of Frankenstein* (1939) and was to remain happily typecast within the fantasy genre for the rest of his career. A kindly, cricket-loving Englishman in real life, he used his lugubrious, velvety tones and an outwardly menacing air to enliven a gallery of monsters, ghouls, resurrected mummies and misguided scientists. His best films include *The Old Dark House* (1932), *The Mummy* (1932), *The Raven* (1935), *The Bodysnatchers* (1945) and *Bedlam* (1946).

His performances outwith this field were often in villainous roles but his more noteworthy other work includes a melodramatic turn as a religious maniac in *The Lost Patrol* (1934), a series of Mr Wong mysteries, and the killer in *The Secret Life of Walter Mitty* (1947).

Later stage work includes the black farce *Arsenic and Old Lace* (1941–4), *The Linden Tree* (1948) and Captain Hook in *Peter Pan* (1950–1). He was also gainfully employed on television as *Colonel March of Scotland Yard* (1956–8) and the avuncular host of the anthology series *Thriller* (1960–2).

Appearing in some decidedly inferior nonsense during the last decade of his career, he usually managed to wearily transcend the crudity of the material and found roles with humour and some distinction in *The Raven* (1963) and *Targets* (1967) in which his dignified performance as an ageing, anachronistic star of horror films was a singularly effective swansong.

KAURISMAKI, Aki

Born 1957, Finland.

An obsessive film buff as a child, Kaurismaki was an amateur artist, dishwasher, builder, postman and machinist before joining with his brother Mika (1955–) to become a key figure in a new generation of Finnish filmmakers.

He was the co-writer and assistant director of such films as *The Liar* (1980), *The Worthless* (1982) and *The Clan-Tale of The Frogs* (1984) whilst collaborating with Mika on the 'rockumentary' *Saimma-ilmio* (The Saimaa Gesture)

(1981) and directing his first feature *Rikos Ja Rangaistus* (Crime and Punishment) (1983).

A self-styled iconoclast, dismissive of his own work and the pretensions of the film industry, his films are austere works of idiosyncratic minimalist subversion full of morose characters living desperate lives on the outskirts of society. His narratives are pared to the bone, mordantly humorous and often just plain incomprehensible. His second film *Calamari Union* (1984) was shot without a script, features eighteen main characters (all but one of whom are called Frank) and has no real story; even the director terms it illogical.

He then began his 'Proletarian Trilogy' with *Varjoja Paratiisissa* (Shadows in Paradise) (1986) which features, among other musical eccentricities, tangos sung in Finnish as the background to a story of the romance between a garbage collector and a supermarket assistant. The trilogy was completed with *Ariel* (1988) and *Tulitkkutehtaan* (The Match Factor Girl) (1989), a story 'designed to make ▷ Robert Bresson seem like a director of epic action pictures', telling of a factory worker who exacts revenge for an unhappy home life, the oppression of her family and employers, and a doomed love affair.

His other films include *Hamlet Liikemaailmassa* (Hamlet Goes Business) (1987), the laconic oddball musical road movie *Leningrad Cowboys Go America* (1989) and *I Hired A Contract Killer* (1990), a quirky salute to the spirit of the Ealing comedy.

KAZAN, Elia

Full Name Elia Kazanjoglou
Born 7 September 1909, Constantinople (now Istanbul), Turkey.

Kazan's father established a carpet business in New York and his family joined him there in 1913. A student at Williams College, Massachusetts, and Yale drama school, Kazan worked in the theatre as a property manager and later as an actor in such productions as *Men in White* (1933), *Waiting for Lefty* (1935) and *Golden Boy* (1937).

He made his film debut as an actor in *Pie in the Sky* (1934) and had small roles in *City for Conquest* (1940) and *Blues in the Night* (1941) whilst also directing the short films *The People of the Cumberlands* (1937) and *It's Up To You* (1941).

Briefly a member of the Communist Party, he worked with a number of left-wing theatre companies including the League of Workers Theatres, but acquired his reputation as the stage director of such Broadway productions as *A Streetcar Named Desire* (1947) and *Death of A Salesman* (1949) and was a co-founder of the Actor's Studio in 1948.

His first major directorial credit on film was *A Tree Grows in Brooklyn* (1945) and he won a Best Director Oscar for *Gentleman's Agreement* (1947) which now seems a somewhat contrived study of rampant anti-Semitism.

A director with a social conscience, he tackled the issue of racism in *Pinky* (1949) and combined the elements of an exciting thriller with concern for public health matters in *Panic in the Streets* (1950).

However, his best work came in the 1950s when he provided purposeful guidance for many of the actors at the forefront of the Method school of performing. He transferred ▷ *A Streetcar Named Desire* (1951) (Oscar nomination) to the screen with the animal magnetism of ▷ Marlon Brando's Kowalski intact and he worked with Brando again on *Viva Zapata!* (1952) and on his portrayal of the martyred longshoreman in ▷ *On the Waterfront* (1954) (Oscar). He also showcased such talents as ▷ James Dean in *East of Eden* (1955) (Oscar nomination) and Lee Remick (1935–) in *A Face in the Crowd* (1957) and *Wild River* (1960), which also starred ▷ Montgomery Clift.

Always maintaining a parallel career in the theatre, he gradually withdrew from the cinema, directing films of his semi-autobiographical novels *America, America* (1963) (Oscar nomination), about an immigrant's dreams of America, and *The Arrangement* (1969) a flashy, modish view of an advertising executive's mid-life crisis.

He abandoned the cinema completely after the elegant, star-studded but somewhat stately version of *The Last Tycoon* (1976).

His novels also include *The Assassins* (1972), *The Understudy* (1974), *Acts of Love* (1978) and *The Anatolian* (1982). His autobiography, *My Life*, published in 1988 gave an exhaustive account of his career and the soul-searching over his 1952 appearance before the HUAC (House Un-American Activities Committee) in which he named those who had previously been active in the Communist movement with him.

KEATON, Buster

Born 4 October 1895, Piqua, Kansas, USA.
Died 1 February 1966, Woodland Hills, California, USA.

Born into a showbusiness family, Keaton was incorporated into his parents' vaudeville act at the age of three and The Three Keatons were a stellar attraction throughout America.

Stage experience honed his remarkable agility, making him a skilled and fearless physical performer who brought an athletic grace to the most dangerous rough and tumble part of the act.

Buster Keaton

Invited to watch the making of a Fatty Arbuckle (1887–1933) comedy, *The Butcher Boy* (1917), he was asked to join the cast and became instantly enamoured of filmmaking, subsequently appearing in several one and two-reel comedies with Arbuckle.

He made his feature debut with *The Saphead* (1920) and developed his solo career in such comedies as *One Week* (1920), *Hard Luck* (1921) and *Cops* (1922).

Named 'The Great Stone Face' for his solemn, resolutely unsmiling features, he was a lithe, game figure in a flat hat who stoically persevered against the slings and arrows of an uncaring world, offering meticulously executed physical daring, pace and a melancholy romanticism in place of ▷ Chaplin's maudlin sentimentality. Posterity has judged him the greatest of the silent clowns on the grounds of his superior film technique, wit and enduring ability to provoke appreciative laughter.

At his most creative and uninhibited from 1923, his golden period of matchless comic masterpieces includes such titles as *The Three Ages* (1923), *Sherlock Junior* (1924), *The Navigator* (1924), ▷ *The General* (1926), *College* (1927) and *Steamboat Bill Junior* (1928).

He then relinquished his independence to work for M-G-M and the initial pleasing results of *The Cameraman* (1928) and *Spite Marriage* (1929) were soon followed by mismanagement of his career, an unhappy sound debut in *Free and Easy* (1930), his own slide into alcoholism and disheartening attempts to make him part of a team with the ebullient Jimmy Durante (1893–1980) in such feeble efforts as *The Passionate Plumber* (1932) and *What No Beer* (1933).

Fired from M-G-M, penniless and tarred with an unwarranted industry reputation for being difficult, he made some dispiriting low-budget comedies in Europe and worked in America on shorts for Educational Films and Columbia. Glimpsed in small supporting assignments in the likes of *Hollywood Cavalcade* (1939), *In The Good Old Summertime* (1949) and ▷ *Sunset Boulevard* (1950), he also worked at M-G-M as a behind-the-scenes comedy supervisor and deviser of gags for the likes of ▷ The Marx Brothers and Red Skelton (1910–).

Appreciation of his work and talents grew again in the 1950s with retrospectives and screenings of his best films and he began to work more often, stealing scenes from ▷ Chaplin in ▷ *Limelight* (1952) and finding himself the subject of an inaccurate but lucrative biography, *The Buster Keaton Story* (1957), in which he was portrayed by Donald O'Connor (1925–).

He received a special Oscar in April 1960 for 'his unique talents which brought immortal comedies to the screen'.

Honoured throughout the world, he received a twenty-minute standing ovation at the 1965 Venice Film Festival and his screen career flourished anew with appearances in *The Railrodder* (1965), *Film* (1965), *Keaton Rides Again* (1965), *A Funny Thing Happened On The Way To The Forum* (1966) and *Sergeant Deadhead* (1966).

KELLY, Gene

Full Name Eugene Curran Kelly
Born 23 August 1912, Pittsburgh, Pennsylvania, USA.

A graduate in economics from the University of Pittsburgh, Kelly ran a dance studio and performed a double act with his brother Fred before venturing to New York and finding work in the chorus of *Leave it To Me* (1938). Several stage roles followed before he was cast in *Pal Joey* (1939) and signed to a Hollywood contract that led to his film debut in *Me and My Girl* (1942).

An athletic, muscular dancer and choreographer with a brash ebullience and breezy disposition, he was cast in dramas like *The Cross of Lorraine* (1944) and *Christmas Holiday* (1944) whilst beginning to leave a terpsichorean legacy in such musicals as *Cover Girl* (1944) and *Anchors Aweigh* (1945), which combined live-action and animation when he appeared to dance with Jerry the Mouse and earned him a Best Actor Oscar nomination.

Before his service in the U.S. Navy (1944–7) he appeared opposite ▷ Fred Astaire in *Ziegfeld Follies* (1944, released 1946) and later described Astaire as the aristocrat of dance in comparison to his proletarian approach.

Returning to musicals with *The Pirate* (1948) and *Take Me Out To The Ball Game* (1948), he formed a partnership with director Stanley Donen (1924–) that brought many innovations in style and technique to the film musical, using outdoor locations and integrating dance into the narrative as a dramatic device to reveal character or advance the storyline rather than as a production number. Their greatest successes are ▷ *On the Town* (1949) and the peerless ▷ *Singin' In the Rain* (1952). Kelly's own supreme artistic achievement was the balletic climax of ▷ *An American in Paris* (1951) which earned him a special Oscar in appreciation of his 'brilliant achievements in the art of choreography on film'.

He continued to essay an occasional dramatic role, making a zestful D'Artagnan in *The Three Musketeers* (1948) and playing a gangster in *The Black Hand* (1950), and subsequent musicals include *Brigadoon* (1954), *It's Always Fair Weather* (1955), *Les Girls* (1957) and *Invitation To The Dance* (1956), another fusion of dance and animation which he directed.

When the old-style M-G-M musical passed from favour, he worked as a dramatic actor in such films as *Marjorie Morningstar* (1958) and *Inherit the Wind* (1960) and turned director on a number of lightweight comedies, later achieving plaudits at the helm of *Hello, Dolly!* (1969).

He has directed on Broadway, devised a ballet for the Paris opera, and worked frequently on television in the series *Going My Way* (1962–3) and such recent all-star mini-series as *North and South* (1985) and *Sins* (1987).

He was the 1985 recipient of the ▷ American Film Institute Life Achievement Award.

KERR, Deborah

Full Name Deborah Jane Kerr-Trimmer
Born 30 September 1921, Helensburgh, UK.

Trained as a dancer, Kerr made her stage debut in the corps-de-ballet of a Sadler's Wells production of *Prometheus* (1938). However, choosing to make acting her career, she appeared in repertory in Oxford and secured a small role as a cigarette girl in the wartime thriller *Contraband* (1940).

Her appearance was consigned to the cutting room floor but she was seen as a Salvation Army girl in the film of *Major Barbara* (1940) and soon graduated to leading roles in *Love on the Dole* (1941) and *Hatter's Castle* (1942).

An association with director ▷ Michael Powell brought her excellent opportunities and she was radiant as three generations of women in *The Life and Death of Colonel Blimp* (1943) and authoritative as the nun in *Black Narcissus* (1947).

Signed to a Hollywood contract with

Deborah Kerr with Yul Brynner in *The King and I* (1956)

M-G-M, she made her American debut in *The Hucksters* (1947) and received her first Best Actress Oscar nomination for *Edward My Son* (1949). A redheaded woman with noble features and a dignified bearing she was invariably cast in ladylike roles that made the most of her 'innate gentility'. However, she sensationally strayed from her well-bred image as the embittered, adulterous army wife in *From Here to Eternity* (1953) (Oscar nomination) and her daring seashore love scenes with co-star ▷ Burt Lancaster made cinema history.

Now offered more challenging assignments, she brought vivacity and warmth to the role of the governess in the musical *The King and I* (1956) (Oscar nomination), proved an effective partner for ▷ Robert Mitchum in *Heaven Knows Mr Allison* (1957) (Oscar nomination) and ▷ Cary Grant in *An Affair to Remember* (1958) and subdued her natural radiance as the mousy spinster in *Separate Tables* (1958) (Oscar nomination).

The best of her later roles include the Australian wife in *The Sundowners* (1960) (Oscar nomination), the governess in *The Innocents* (1961) and the poet in *Night of the Iguana* (1964). She revealed an uninhibited sense of comedy as Agent Mimi in the otherwise unworthy *Casino Royale* (1967) but, discouraged by the lack of good roles for actresses in her age range, she retired from the screen after *The Arrangement* (1969).

Returning to the stage, she has been seen in such productions as *Seascape* (1975), *Long Day's Journey Into Night* (1977), *Overheard* (1981) and *The Corn Is Green* (1985). Also active in television, her appearances in the medium include the films *Witness for the Prosecution* (1982) and *Reunion at Fairborough* (1985) as well as the vastly popular mini-series *A Woman of Substance* (1984) and its sequel *Hold the Dream* (1986).

After rumours of such films as *The Lonely Passion of Judith Hearne* (1972) and *An Eagle Has Landed* (1976), she finally returned to the cinema screen in the modest *The Assam Garden* (1985) as a stuffy colonial widow who strikes up a friendship with an Indian neighbour.

KIESLOWSKI, Krzysztof

Born 27 June 1941, Warsaw, Poland.

A student at the School of Cinema and Theatre in Lodz, Kieslowski graduated in 1969 with his diploma film, the short documentary *Z Miasta Lodzi* (From the City of Lodz).

Whilst still a student, he had made his directorial debut with the television documentary *Urzad* (The Job) (1967) and he continued to work extensively for Polish television over the next decade on documentaries and the occasional feature. His cinema career began with *Blizna* (The Scar) (1976) and he attracted attention with *Amator* (Camera Buff) (1979), a satire in which a factor worker's obsessive interest in making home movies ultimately places him at odds with his bosses.

He followed this with *Przypadek* (Blind Chance) (1982) and *Bez Konca* (No End) (1984) a solemn drama set during the imposition of martial law in 1982 and showing the ghost of a dead lawyer watching over the struggles of his loved ones and legal colleagues.

However, the director's growing international reputation rests on his creation of *Dekalog* (Decalogue) (1988–89), a series of films for Polish television, each one dramatically illustrating one of the ten commandments. Two of the films were shot as features and released to cinemas. *Krótki Film O Zabijaniu* (A Short Film About Killing) (1988), winner of several prizes including the Best Foreign Film Oscar, is a detailed, chillingly austere story of an alienated youth's senseless, unmotivated killing of a taxi driver and the barbarous punishment that follows. Shot in drained colours, its unflinching eye made for compelling drama and a potent criticism of the relationship between law, compassion and justice.

His economy of style also served *Krótki Film O Miłości* (A Short Film About Love) (1988), a sensitive excursion into the tenderly ambiguous relationship that develops between a nineteen year-old and his neighbour.

As Poland has embraced western-style democracy in recent years, the economic reality of making films there has proved unpromising and his latest feature work has been in France.

KIND HEARTS AND CORONETS

Robert Hamer (UK 1949)
Dennis Price, Alec Guinness, Joan Greenwood, Valerie Hobson.
106 mins. b/w.

Robert Hamer's (1911–63) *Kind Hearts and Coronets* remains perhaps the most remarkable offering to have come out of Britain's Ealing Films, but it is also one of the company's least typical productions. The Ealing phenomenon was largely the creation of studio head ▷ Michael Balcon, who must take credit for developing its pool of bright young cinematic talents and encouraging them to contribute to each others' work. With a gallery of variously distinctive character actors (among them ▷ Alec Guinness and ▷ Alistair Sim) from which to draw, this typically British form of celluloid group enterprise gave rise to an identifiable house style in the perennially popular run of Ealing comedies. Ever delighting in a sense of English quaintness, here in films like Charles Crichton's (1910–) *The Lavender Hill Mob* (1951) and Henry Cornelius's (1913–58) *Passport to Pimlico* (1948) is a world peopled by eccentrics and driven by whimsical passions, yet in the incisive work by director ▷ Alexander Mackendrick (see ▷ *Whisky Galore!*) and in Hamer's elegant masterpiece *Kind Hearts and Coronets* we see the exceptions to such uniform cheeriness. Hamer expressed his intention as: 'Firstly, that of making a film not noticeably similar to any previously made in the English language. Secondly, that of using this English language, which I love, in a more varied and interesting way. Thirdly, that of making a picture which paid no regard whatever to established, although not practised, moral convention.' Remarkably, his story of dispossessed haberdasher Dennis Price (1915–73) murdering his way through the aristocratic d'Ascoyne family towards the dukedom denied him by his mother's lowly marriage was a success on all three counts. Its Wildean verbal wit and poised visual humour achieved a rare piquancy and Guinness's performance as all eight upper class victims offered an exercise in comic observation to savour.

KING KONG

Merian C. Cooper & Ernest B. Schoedsack (USA 1933)
Robert Armstrong, Fay Wray, Bruce Cabot, Sam Hardy.
100 mins. b/w.

Billed on its initial release as 'The Eighth Wonder of the World', *King Kong* continues to hold an impressive sway over the moviegoing imagination. For all its technical brilliance, the film's ongoing fascination lies in the richness of the myth it has indelibly etched on the movie screen. The giant ape Kong, who is taken from his remote island by Robert Armstrong's (1890–1973) showman Carl Denham, exhibited as a circus attraction in New York before rampaging through the city, and finally slain by airborne firepower on top of the Empire State building, is both sympathetic tragic figure and troubling psychological fantasy. Certainly he is savage nature brutally tamed by man's murderous technology, but as Denham himself finally remarks 'It wasn't the airplanes, it was Beauty killed the Beast' – the famous image of Kong's huge fingers unpeeling heroine Fay Wray's (1907–) clothing points to the statuesque primate as the outpouring of all our socially repressed sexual and violent urges, a release which of course must be stemmed when the ape is finally destroyed by the forces of the big city. Still, we would hardly consider any of these notions were not the on-screen execution of this tall tale so utterly credible. Recent advances in ▷ back-projection and ▷ matte work allowed effects maestro Willis O'Brien (1886–1962) to integrate the actors and studio settings with the 16-inch ▷ stop motion model of the fierce gorilla himself – though full scale versions of Kong's face and hands were required for close-up sequences that still retain their power to disturb.

Ernest B. Schoedsack (1893–1979) himself directed a dismal quickie sequel *Son of Kong* (1933) and a virtual reprise in *Mighty Joe Young* (1949), which at least has its moments. Highly unofficial Japanese versions are *King Kong vs. Godzilla* (1963) and *King Kong Escapes* (1968), while we have maverick producer Dino De Laurentiis (1919–) to thank for the disappointing 1976 big-budget remake *King Kong* and a further rather desperate follow-up in *King Kong Lives* (1986).

King Kong (1933)

KINGS OF THE ROAD (Im Lauf der Zeit)

Wim Wenders (W. Germany 1976)
Rudigler Vogler, Hans Zischler, Lisa Kreutzer, Rudolf Schundler.
176 mins. b/w.
Cannes Film Festival: *International Critics Prize.*

With the legacy of Nazism, the division of their homeland and the subsequent 'economic miracle', West German filmmakers operate under a more complex historical burden than most. It is the richness and diversity of their responses to it – from ▷ Rainer Werner Fassbinder's baroque melodramas or Werner Herzog's (1942–) epic loftiness, to the more ascetic approach of Alexander Kluge (1932–) and Hans Jurgen Syberberg's (1935–) controversial operatic revisionism – which make the loose grouping generally termed the New German Cinema one of the most intriguing filmic developments to come out of the late Sixties and early Seventies. A remark in ▷ Wim Wender's 1976 film *Im Lauf der Zeit* (Kings of the Road) on the other hand, where one character reflects that 'The Americans have colonised our unconscious', points to the distinctive approach of a director whose own fascination for Hollywood movies and rock music led him to wonder just where the imported cultural forms and values of contemporary German society ended and the repressed reality of the country's own post-war experience actually began. Couched in terms of an American road movie but shot in the desolate countryside along the East/West German border, *Im Lauf der Zeit* follows the fortunes of an itinerant cinema equipment repair man and the recently divorced pediatrician who joins him on his travels. Wenders' camera sedulously records the significances of everyday events to mount a languorous elegy for both a lost German film industry and a lost sense of place.

KLIMOV, Elem Germanovich

Born 9 July 1933, Volgograd, USSR.

A student of aeronautical engineering at Moscow's Higher Institute of Aviation, Klimov subsequently worked as an engineer and a foreign correspondent before studying at Moscow's VGIK film school.

He made his directorial debut with the television film *Attention, Platitude* (1959) and completed a number of short films before making his graduation feature *Welcome or No Entry for Unauthorised Persons* (1964), the story of a young boy's mischief-making presence in a Young Pioneer Summer Camp. Judged to be anti-Khruschev, it was banned, but subsequently released when given an official seal of approval by Khruschev himself.

The Adventures of a Dentist (1965) encountered similar travails when it was viewed as an allegorical attack on the constraints the Soviet system placed on individual creativity. The story of a dentist who can extract teeth painlessly, it shows his initial welcome as a miracle-worker and subsequent downfall at the hands of his disgruntled colleagues.

His satirical tone continued in *Sport, Sport, Sport* (1970), a surreal collage that confirmed his style of blending naturalism and fantasy.

He completed *And Nonetheless I Believe* (1974) after the death of its director Mikhail Romm (1901–71) and then changed course dramatically with the epic *Agonia* (Agony) (1975), a feverish account of the final days of the Romanov dynasty that featured the Soviet cinema's first depiction of Rasputin and remained on the shelf for a decade because of its allegedly sympathetic portrait of the Tsar.

When his wife Larisha Shepitko (1938–79) died in a car crash, he completed her feature *Proschanie* (Farewell) (1981), a lament for the passing of peasant traditions that unfolds in the story of the resettlement of a Siberian village that progress dictates must make way for a hydro-electric project.

He returned to the epic format with *Idi I Smorti* (Come and See) (1985), which he began in 1976 and resumed in 1983. An overwhelming account of a teenage partisan's harrowing encounters with Nazi atrocities during the Second World War, it is a visceral, visually-striking assault on the senses creating an unforgettable catalogue of carnage through a surreal juxtaposition of images and emotions.

A frequent victim of state censorship, arbitrariness and creativity stifling bureaucracy, he was one of the artists to benefit from the liberalisation of the glasnost era; *Agony* finally received a Russian release in 1984, *Come and See* took the first prize at the Moscow Festival in 1985 and, in 1986, he was elected President of the Soviet Filmmakers Union.

KORDA, Alexander (Sir)

Real Name Sandor László Kellner
Born 16 September 1893, Puszta, Turpósztó, Hungary
Died 23 January 1956, London, UK.

A journalist and writer, Korda began in the Hungarian film industry as a secretary and then a writer of title cards for imported silent films. He also founded the film journal *Pesti Mozi* (Budapest Cinema) (1909) and later made his directorial debut with *A Becsapott Újságíró* (The Duped Journalist) (1914).

He was a prolific force behind the camera over the next few years and formed his own production company in 1917. Two years later, he fled the country for Vienna and subsequently worked throughout Europe before being tempted to Hollywood where he made a number of films including *The Private Life of Helen of Troy* (1927), *Love and The Devil* (1929) and *Lillies of the Field* (1930).

Returning to Europe, he directed *Marius* (1931), the first of the celebrated Marseilles trilogy by Marcel Pagnol (1894–1974) and then moved to Britain where he established London Films and directed *The Private Life of Henry VIII* (1932), a rumbustious and lustily irreverent portrait of the much-married monarch that found international favour and won a Best Actor Oscar for the star ▷ Charles Laughton.

Establishing himself as a movie mogul, his charm and legendary Hungarian chuztpah made him virtually unique in his ability to persuade the British government and financial institutions to underwrite his ambitious plans and lavish lifestyle. He directed infrequently, but the generally superior results include *Rembrandt* (1936), the morale-boosting *That Hamilton Woman* (1941), the romantic *Perfect Strangers* (1945) and, his last, *An Ideal Husband* (1947).

As a producer and studio owner, his name on a film usually signalled high production values and handsome entertainment and, despite several brushes with bankruptcy, he remained a major force in the British film industry for many years, raising general standards and increasing its international standing with such productions as *Things to Come* (1936), *Elephant Boy* (1937), *The Thief of Bagdad* (1940), ▷ *The Third Man* (1949) and *Richard III* (1956).

Married to the actress Merle Oberon (1911–79) from 1939 to 1945, his nephew Michael has chronicled a flamboyant life and times in *Charmed Lives* (1980) and the fictionalised *Queenie* (1985) which was filmed in 1987 as a television miniseries with Kirk Douglas (1916–) as the Korda character.

KUBRICK, Stanley

Born 26 July 1928, New York, USA.

An apprentice and, later, staff photographer with *Look* magazine, Kubrick made his debut as the director of the documentary shorts *Day of the Fight* (1950) and *Flying Padre* (1951) and made the low-budget independent features *The Seafarers* (1953) and *Fear and Desire* (1953) before gaining critical recognition for the taut suspense and offbeat qualities of the character-based crime yarns *Killer's Kiss* (1955) and *The Killings* (1956).

He subsequently directed ▷ *Paths of Glory* (1957), a starkly realistic, fluidly structured, angry indictment of military politics, injustice and the horrors of World War I trench warfare. The star Kirk Douglas (1916–) then hired him to direct the epic *Spartacus* (1960) but despite the many intelligent virtues of the film, Kubrick disclaimed authorship of the final version.

Malcolm McDowell in *A Clockwork Orange* (1971) directed by Stanley Kubrick

Fired from the direction of *One-Eyed Jacks* (1961), he has subsequently pursued an independent course in which his hard-won artistic freedom has been put at the disposal of stories that explore man's loss of humanity.

His major successes from the 1960s comprise the blackly humorous version of *Lolita* (1962), his inventive and funny satire on the nuclear age *Dr. Strangelove: Or How I Learned to Stop Worrying and Love the Bomb* (1964) and ▷ *2001: A Space Odyssey* (1968), an imposing science-fiction speculation on the history of mankind.

He followed this with a bleak, thought-provoking version of *A Clockwork Orange* (1971) in which man's increasing savagery can only be halted by a dehumanising process of thought control. A controversial film in its day, it has been withdrawn from circulation in Britain since 1973 at the director's request.

Increasingly prone to long gaps between pictures as he has carefully chosen his subject matter, embarked on painstaking planning and a production period shrouded in secrecy, he has made only three films since then: the visually ravishing 18th century epic *Barry Lyndon* (1975), the grandiose horror yarn *The Shining* (1980) and his elaborate Vietnam saga *Full Metal Jacket* (1987).

Although a new Kubrick film is now an event, the lengthy waits have not always been worthwhile, as his over-fussy style tends to result in earnest pictures of flawless technique but somewhat aloof, manipulative narrative content.

KUROSAWA, Akira

Born 23 March 1910, Tokyo, Japan.

A student at the Doshusha School of Western Painting, Kurosawa worked as a painter and illustrator before answering a newspaper advertisement seeking assistant directors. He learned all aspects of filmmaking, including editing and scriptwriting, before making his directorial debut with *Sanshiro Sugata* (Judo Saga) (1943), the story of the establishment of judo as a dominant martial art in the 19th century.

Its popularity led to a sequel, and he was a prolific and successful director in Japan before beginning to attract wider interest with *Yoidore Tenshi* (Drunken Angel) (1948), a gangster melodrama that began his long collaboration with actor ▷ Toshiro Mifune.

▷ *Rashomon* (1950) won the Best Film Prize at the Venice Film Festival and was the first Japanese film to gain wide exposure in the West. The story of a rape and murder recounted from the differing perspectives of witnesses and those involved, it used a strong visual sensibility and storytelling flair to comment on the subjective quality of truth.

A visual stylist and humanitarian drawn to stories that involve some form of master-pupil relationship, he has found constant inspiration in a diversity of Western literature. His successes in the 1950s include adaptations of Dosto-

Kagemusha (1980) directed by Akira Kurosawa

evsky's *Hakuchi* (The Idiot) (1951) and Gorky's *Donzoko* (The Lower Depths) (1957), *Ikiru* (Living) (1952) a touching, low-key tale of a dying man's sense of social responsibility, and the sweeping epic ▷ *Shichinin No Samurai* (The Seven Samurai) (1954) remade by Hollywood as *The Magnificent Seven* (1960).

The latter was cinematically innovative in its use of long lenses and multiple cameras to capture the chaos, mud and violence of the battle scenes and he has eagerly embraced the use of wide-screen techniques, multi-track sound and ▷ Panavision to enhance the realism and power of his later work.

Particularly attracted to the blood-drenched sense of drama in Shakespeare's finest works, he brought the techniques of Noh theatre to his version of Macbeth, *Kumonosu-Jo* (Throne of Blood) (1957) and his other tales of the honour and savagery in the samurai code like *Yojimbo* (1961) and *Sanjuro* (1962) were direct precursors of the graphically violent, laconic approach that Italian filmmakers would take to the ▷ spaghetti western.

Best known for his samurai epics, his versatility also extended to gripping versions of Ed McBain thrillers like *Warui Yatsu Yoku Nemuru* (The Bad Sleep Well) (1960) and *Tengoku To Jigoku* (High and Low) (1963), and the sentimental social drama *Akahige* (Red Beard) (1965), his last film with Mifune.

After working in America on a number of abortive projects, including *Tora! Tora! Tora!* (1970), he returned to Japan for *Dodes Ka-Den* (1970), conveying the lives and fantasies of the oddball residents of a shanty town. Its financial failure led to a suicide attempt, but he has survived and endured to produce such excellent work as the Siberian adventure *Derzu Uzala* (1975), the 16th century epic *Kagemusha* (1980) and his version of King Lear, *Ran* (1985).

His most recent films have been *Dreams* (1990) which gave visual voice to a variable collection of dreams generally depicting his fears for the future of mankind, and the forthcoming *Rhapsody in August* (1991) which features ▷ Richard Gere.

LADRI DI BICICLETTE *see* BICYCLE THIEVES

LANCASTER, Burt

Full Name Stephen Burton Lancaster
Born 2 November 1913, East Harlem, New York, USA.

A student at New York University on an athletic scholarship, Lancaster left to form an acrobatic double-act with the diminutive Nick Cravat (1911–) and worked extensively in vaudeville and circuses before a hand injury forced his retirement. Subsequently employed as a salesman, floorwalker and singing waiter, he joined the United States Army in 1942 and performed in a number of Army shows before turning professional and making his Broadway debut in *A Sound of Hunting* (1945).

The play was short-lived but led to offers from Hollywood and thus he made his screen debut as the doomed, ex-prize-fighter in *The Killers* (1946). An imposing, muscular figure with the cinema's most dazzling set of teeth, he was cast in brawny action man roles in the likes of *Brute Force* (1947) and *Criss Cross* (1949) but also signalled his intention to explore other facets of his abilities as the weak husband in *Sorry, Wrong Number* (1948) and embittered son in *All My Sons* (1948).

One of the first actors to form his own independent production company, he also proved a lithe swashbuckler, combining agility and braggadocio in films like *The Flame and the Arrow* (1950) and *The Crimson Pirate* (1952).

Throughout subsequent decades his star career has been an astute mixture of commercial fare and challenging character parts whilst Hecht-Hill-Lancaster productions were responsible for the Oscar-winning *Marty* (1955) and *The Bachelor Party* (1957) in which he did not appear. His string of successes include the sergeant in *From Here To Eternity* (1953) (Best Actor Oscar nomination), the rousing western *Vera Cruz* (1954), the circus melodrama *Trapeze* (1956) and ▷ *Sweet Smell of Success* (1957), in which he offers a chilling portrait of megalomania as a feared gossip columnist.

He won a Best Actor Oscar as the swaggering, bible-thumping evangelist in *Elmer Gantry* (1960) and displayed versatility and range as the murderer turned ornithological authority Robert Stroud in *The Birdman of Alcatraz* (1962) (Oscar nomination), the Italian nobleman in ▷ *Il Gattopardo* (The Leopard) (1963) and the military conspirator in *Seven Days in May* (1963).

Over four decades he has conducted his career with an intelligence and sense of responsibility rare in the film industry, lending his name to works of social and political intent and showing an openness to work in all film cultures. *The Professionals* (1966) and *Airport* (1969) were among the biggest financial successes of his career but later roles of note include the distressed suburbanite in *The Swimmer* (1968) and the grizzled Indian scout in *Ulzana's Raid* (1972). He also served as the director of *The Kentuckian* (1955) and co-director of *The Midnight Man* (1974).

More recently, he brought a mature grace and finesse to the smalltime hood relishing a late autumn of action and romance in *Atlantic City* (1980) (Oscar nomination), the Texas oilman with his head in the stars in *Local Hero* (1983), the ageing patriarch in *Rocket Gibraltar* (1988) and the smalltown doctor in *Field of Dreams* (1989).

His television work over the years includes the mini-series *Moses* (1975), *Marco Polo* (1982) and *On Wings of Eagles* (1986) and his rare stage appearances comprise *Knickerbocker Holiday* (1971), *The Boys in Autumn* (1981) and *Handy Dandy, A Comedy But ...* (1984).

LANG, Fritz

Born 5 December 1890, Vienna, Austria.
Died 2 August 1976, Beverly Hills, California, USA.

The son of an architect, Lang studied engineering and travelled extensively as a young man before serving with distinction in the Austrian army during World War I. Discharged to an army hospital in 1916, he began to write screenplays and later moved to Berlin to work as a reader and story editor.

He made his directorial debut with *Halbblut* (The Halfbreed) (1919) and was singled out as a director of promise after *Der Müde Tod* (Destiny) (1921), an expansive production on the historical inevitability of death. He began his series of Dr Mabuse films with *Dr. Mabuse, Der Spieler* (Doctor Mabuse the Gambler) (1922) and displayed his versatility in the confident handling of large themes with the epic two-part *Die Niebelungen* (1923) and the futuristic ▷ *Metropolis* (1926).

Fascinated by violence and cruelty, his films vividly explored the nightmares of urban living, making expressive use of light and shade and psychological insights into the criminal mind to create dark, fantasy-like dramas where inescapable jeopardy and inhumanity lurked around every shadowy corner or towering set. His later German work includes the espionage story *Spione* (Spies) (1927), the stark, sober ▷ *M* (1931) with Peter Lorre (1904–64) giving an outstanding performance as the child killer, and *The Testament of Dr. Mabuse* (1932) a thinly veiled critique of Nazism.

Offered the post of head of the German film industry when the Nazis came to power he immediately fled to Paris, making *Liliom* (1934), before journeying further to settle in California.

He made his American debut with *Fury* (1936), a searing indictment of mob rule, and followed this with *You Only Live Once* (1937) a strikingly poetic account of lovers on the run. He experimented with colour in *The Return of Frank James* (1940) and *Western Union* (1941) and made a series of astringent wartime thrillers, among them *Manhunt* (1941), *Hangmen Also Die* (1943) and *Ministry of Fear* (1944).

Subsequently active in a number of genres, he made such stylised, nightmarishly fatalistic ▷ film noir as *The Woman in the Window* (1944) and *Scarlet Street* (1945), the flamboyant western *Rancho Notorious* (1952), the 18th century smuggling yarn *Moonfleet* (1955) and a number of bleak, pessimistic thrillers including *The Big Heat* (1953) and *Human Desire* (1954).

He abandoned filmmaking in America after 1956, completing only two further projects in India and a final return to Germany for *Die Tausend Augen Des Dr. Mabuse* (The Thousand Eyes of Dr. Mabuse) (1960).

He appeared as himself in *Le Mépris* (Contempt) (1963), but failing eyesight prevented the contemplation of any further projects.

LANGE, Jessica

Born 20 April 1949, Minnesota, USA.

Once described by ▷ Jack Nicholson as a 'delicate fawn crossed with a Buick', Lange was studying Fine Art at the University of Michigan when she met Spanish photographer Paco Grande (1945–) who became her mentor and her first husband.

They travelled throughout America and Europe and she later settled in Paris, studying mime and dance at the Opera Comique. Returning to New York, she worked as a waitress and model before being cast by Dino De Laurentiis (1919–) as the screaming blonde in his lavishly-budgeted remake of *King Kong* (1976).

The critical hostility and public indifference to the film damaged her career aspirations but she resumed work as the breathtakingly beautiful Angel of Death in *All That Jazz* (1979) and the suburban housewife in the modest comedy *How To Beat The High Cost of Living* (1980).

She silenced all her detractors with a performance of emotional complexity and raw sensuality as the passion-inflamed murderess in the new version of *The Postman Always Rings Twice* (1981) and immediately laid claim to greatness as an actress portraying with harrowing conviction the travails and anguish of rebellious, self-destructive 1930s star Frances Farmer (1914–70) in *Frances* (1982) (Best Actress Oscar nomination), and contrasting this with a charismatic romantic-comedy turn as the soap-opera star in the popular *Tootsie* (1982) which earned her a Best Supporting Actress Oscar.

A finely boned, natural blonde she has been drawn to roles of tortured and anguished women and has lent her star name to work that reflects some of her political and environmental concerns. She gave one of her finest performances as a woman struggling to maintain her family and farm in *Country* (1984) (Oscar nomination), which she also co-produced, and was highly effective as country-western singer Patsy Cline in *Sweet Dreams* (1985) (Oscar nomination).

Since starring as Maggie in a television version of *Cat on A Hot Tin Roof* (1985), she has been seen in such diverse roles as a Southern sister in the overheated *Crimes of the Heart* (1986), as the daughter seeking reconciliation with her father in the rural drama *Far North* (1988), homecoming queen in the sprawling *Everybody's All-American* (1988) and as the suburban widow picking up the pieces of her life in *Men Don't Leave* (1990).

She received a further Oscar nomination as the righteously indignant lawyer facing the devastating realisation of her father's Nazi past in *The Music Box* (1989) and will shortly be seen in the remake of *Cape Fear* (1991) and in *Blue Sky* (1991).

The longtime companion of playwright and actor Sam Shepard (1943–) she also has a child from a previous liaison with dancer Mikhail Baryshnikov (1948–).

LANSBURY, Angela

Full Name Angela Brigid Lansbury
Born 16 October 1925, London, UK.

The granddaughter of labour politician George Lansbury, Lansbury was evacuated to America in 1940 and worked briefly on a nightclub act before being cast as the cockney maid in *Gaslight* (1944), a performance that earned her a Best Supporting Actress Oscar nomination.

Signed to a contract with M-G-M, she was seen in *National Velvet* (1944) and was Oscar-nominated again as the saloon singer in *The Picture of Dorian Gray* (1945). A singer and actress of great versatility and presence, her home studio failed to develop her talents and wasted her in supporting roles and secondary assignments although she was excellent as the honky tonk singer in *The Harvey Girls* (1946) and icily accomplished as the sharp-tongued career woman in *State of the Union* (1948).

Frequently cast as women older than her years, she eventually secured a niche as a character actress of intelligence and perspicacity, winningly seen as the mistress in *The Long Hot Summer* (1958), the comforting best friend in *The Dark at the Top of the Stairs* (1960) and exceptional as the devious black widow of a matriarch in the dazzling political thriller *The Manchurian Candidate* (1962) (Oscar nomination).

She had made her Broadway debut in the farce *Hotel Paradiso* (1957) and was to become one of the first ladies of the musical stage with much lauded performances in *Anyone Can Whistle* (1964), *Mame* (1966), *Dear World* (1969), *Gypsy* (1974) and *Sweeney Todd* (1979).

Theatrical stardom brought her the lead in the jolly Disney extravaganza *Bedknobs and Broomsticks* (1971) and the wan Miss Marple mystery *The Mirror Crack'd* (1980) and she proved an outrageously eccentric scene-stealer in the all-star *Death on The Nile* (1978).

Her television credits include such lavish mini-series as *Little Gloria ... Happy At Last* (1982), *Lace* (1984) and *Rage of Angels: The Story Continues* (1986), the dramas *A Gift of Love: A Christmas Story* (1983) and *The Shell Seekers* (1989) and the long-running primetime mystery series *Murder She Wrote* (1984–) in which her sane, sensible sleuth Jessica Fletcher has won her enduring popularity long after the careers of M-G-M contemporaries have faded and passed.

LAST TANGO IN PARIS (Ultimo Tango a Parigi)

Bernardo Bertolucci (Italy/France 1972)
Marlon Brando, Maria Schneider, Jean-Pierre Leaud, Massimo Girotti.
129 mins. col.

A milestone in the frank treatment of sexuality on screen, ▷ Bernardo Bertolucci's *Last Tango in Paris* was an act of liberation for the entire medium. Its highly favourable box office reception something of a *succès de scandale*, the film nevertheless proved that audiences (and censors) would accept the most shockingly intimate discussion and depiction of physical relations outside the context of mere pornography. A towering, self-revelatory central performance by ▷ Marlon Brando undoubtedly helped; cast as a middle-aged widower whose near-wordless, increasingly degraded affair with much younger Parisienne Maria Schneider (1935–) forces him to confront the emptiness of his marriage and his wife's recent suicide, his at times brutish, at times tragic presence raised film acting to a new peak of soulful credibility. Unfortunately, however, Schneider's efforts are no match for him and although Bertolucci's script ostensibly puts the focus on her choice between carefree physicality in sexual possession

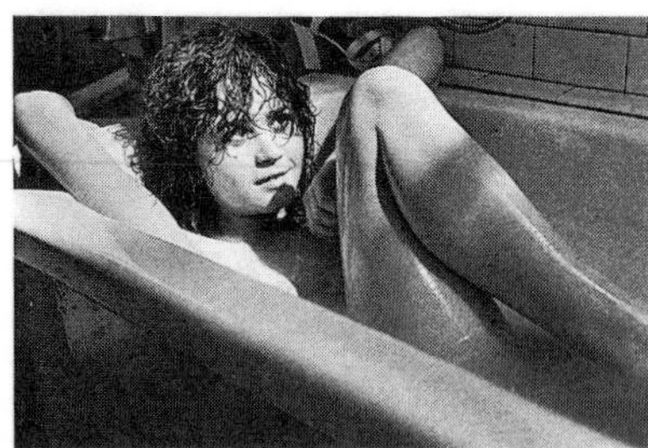

Maria Schneider in *Last Tango in Paris* (1972)

by Brando or conventional courtship and marriage as recorded by incorrigible filmmaker boyfriend Jean-Pierre Leaud (1944–), we learn less about the anxieties of a young woman trying to preserve her sense of self than we do of the older man's self-deluding search for communion in the darkest recesses of sensuality. Still, even with its flaws, *Last Tango in Paris* burns with an emotional truthfulness, its unforgettable scenes like Brando's tearful collapse in front of his spouse's corpse or the troubling moments of mutual depravity in which the couple indulge representing for each and every viewer the kind of contemporary epiphany only too rare in filmic art.

LAST YEAR AT MARIENBAD (L'Année Dernière à Marienbad)

Alain Resnais (France/Italy 1961)
Delphine Seyrig, Giorgio Albetazzi, Sacha Pitoeff, Francoise Bertin.
100 mins. b/w.
Venice Film Festival: *Golden Lion.*

Dismissed in some quarters as impenetrable pretension, many commentators regard this one-off collaboration between Resnais and Nouveau Roman doyen Alain Robbe-Grillet (1922–), combining as it does the director's sweeping formal élan and his interest in the cinematic representation of time and memory with the writer's cerebral dissection of the nature of fictive identity, one of the fundamental landmarks of modernist cinema. Set in an imposing baroque palace surrounded by formal gardens, the identifiable action appears to involve a man known only as 'X' and his attempts to convince a woman, 'A', in the glacial form of Delphine Seyrig (1932–90), to fulfil the promise to elope with him that she may have made the year before in Marienbad. Note the adjective 'identifiable' however, for Resnais's abstract approach to plot and characterisation involves frequent unsignposted shifts in temporal location, the disjunction of narration from image, and alienating, anti-naturalist performances. It is puzzling for the viewer expecting a straightforward romance, but by exploiting the unique present tense quality of the film medium to the full, Resnais's exploration of reality as a subjective or artistic construct cleared the way for subsequent filmmakers – notably Britain's ▷ Nic Roeg in films like *Don't Look Now* (1973) and *Bad Timing* (1980) – to move between truth and fantasy or past, present and future, as if such freedom had always been part of the established grammar.

LAUGH IN

USA 1968–73
Dick Martin, Dan Rowan

Few shows produced more catchphrases than the eclectic fast-paced platform of comics Dick Martin (1923–) and Dan Rowan (1922–). There was Martin's 'You bet your sweet bippy,' the German accented 'Verrry interesting ... but stupid,' of Arte Johnson (1934–), or the all-purpose 'Hi, sports fans'. Most popular of all was 'Sock it to me!', which earned actress Judy Carne (1939–) a face-full of custard or a dousing in water every time she said it. A mark of the show's popularity was its ability to attract unusual guests. During his 1968 campaign for the Presidency, Richard Nixon appeared, and six times implored the nation to sock it to him, while never managing to sound relaxed.

Coming from 'Beautiful downtown Burbank', *Laugh In* was the first variety show to fully exploit television. While others updated music hall traditions, it framed a barrage of non-sequiturs, visual puns and one-liners with slick videotape editing and blacked-out screens. In keeping with the irreverence of the times there was a satirical element but the targets were rarely politically specific.

Rowan said the show was inspired by one-frame cartoons from *Punch* and *Playboy*. Though largely formless, *Laugh In* was built on a number of segments – The Party, where a mass of ensemble Watusi dancers was interrupted by gags, the topical Laugh-In Look At The News, the Mod, Mod World section which punctured the pretension of modern trends, and the Joke Wall, where comics were catapulted out of trapdoors to deliver their lines. Female comics were central, among them Goldie Hawn (1945–) – she played a dumb blonde – and Lily Tomlin (1939–), whose character of Ernestine the unhelpful telephone operator, became a nationally-shared joke. *Laugh In* was America's most popular show and elements of it have peppered TV variety ever since. But the attempts of George Schlatter, the show's auteur, to cash in with *Turn On*, foundered when ABC axed it after just one week. The new show was hosted by a computer and was faster still. Schlatter had stretched the form too far; some viewers were said to have been physically disturbed by his innovation.

LAUGHTON, Charles
Born 1 July 1899, Scarborough, UK.
Died 15 December 1962, Hollywood, USA.

The son of a hotelier, Laughton was a gold medal winner at RADA before embarking on a distinguished stage career in the West End. He first dabbled with the cinema by appearing in two short films *Bluebottles* (1928) and *Daydreams* (1928) alongside his future wife Elsa Lanchester (1902–86).

Stage success in America soon led him to Hollywood where he impressed with a series of vivid characterisations, none more so than his primitive Mussolini-like Emperor Nero in *The Sign of the Cross* (1932).

His exuberant portrayal of the lusty monarch in the satirical *The Private Life of Henry VIII* (1932) won him an Oscar and initiated a remarkable run of performances unrivalled in their consistent excellence, range and ability to delineate character through the subtlest of gestures and broadest of strokes. His many notable roles include the tyrannical paterfamilias in *The Barretts of Wimpole Street* (1934), the spritely English manservant out West in the comedy *Ruggles of Red Gap* (1935), the naval martinet Captain Bligh in *Mutiny on the Bounty* (1935) (Oscar nomination), Javert in *Les Miserables* (1935), a beautifully modulated portrayal of *Rembrandt* (1936) and a hauntingly pitiful Quasimodo in *The Hunchback of Notre Dame* (1939).

Charles Laughton in *Mutiny on the Bounty* (1935)

Riddled by self-doubt and insecurities about his lack of good looks and homosexuality, he showed a tendency towards lipsmacking excess when faced with inferior material but he did find some character roles worthy of his abilities, particularly in *Hobson's Choice* (1954), as the irascible barrister in *Witness for the Prosecution* (1957) (Oscar nomination) and as the wily Southern senator in *Advise and Consent* (1962).

His later stage work included a collaboration with Bertholt Brecht (1898–1956) on *Galileo* (1947), a 1958 season with the RSC that included *King Lear* and Bottom in *A Midsummer Night's Dream* and extensive tours of readings from the Bible and noted literature.

His one film as director, the baroque drama ▷ *Night of the Hunter* (1955) showed him to be a master of ▷ expressionist composition, mood and a sensitive judge of performance. Its commercial failure discouraged him from tackling further projects including a planned version of *The Naked and The Dead* by Norman Mailer.

LAUREL & HARDY

Stan Laurel
Real Name Arthur Stanley Jefferson
Born 16 June 1890, Ulverston, Lancashire, UK.
Died 23 February 1965, Santa Monica, California, USA.

Oliver Hardy
Real Name Norvell Hardy Junior
Born 18 January 1892, Harlem, Georgia, USA.
Died 7 August 1957, Hollywood, California, USA.

Enchanted by toy theatres, puppetry and magic lanterns, Laurel made an early decision to pursue a career in the entertainment world and made his debut at 16 billed as 'Stan Jefferson–He of the Funny Ways'. He toured extensively in America with the Fred Karno troupe, understudying ▷ Chaplin at one point, and made his film debut in *Nuts in May* (1917). He subsequently played the character of Hickory Hiram in a series of shorts and appeared in *The Lucky Dog* (1919) which coincidentally also featured Oliver Hardy.

Hardy was a boy soprano in minstrel shows, who began acting in films as the heavy in a series of comedies produced in Jacksonville, Florida between 1914 and 1917. In the 1920s he co-directed a number of comedies starring Larry Semon (1889–1928) and joined the Comedy All-Stars team assembled by ▷ Hal Roach.

Laurel had starred in a number of silent comedies that lampooned the dramatic hits of the day with titles including *Mud and Sand* (1922), *The Soilers* (1923) and *Monsieur Don't Care* (1925). He too then joined with Roach and the two men appeared in a number of films together before officially being launched as a team in *Putting Pants on Philip* (1927).

Averaging one short film a month, they gradually refined their familiar screen personae as the simple-minded, gentlemanly, bowler-hatted buffoons. Ollie – fat, pretentious and blustering – fiddled with his tie and appealed to the camera for help, Stan – thin, bullied and confused – scratching his head, looked blank and dissolved into tears. Their simple, honest slapstick and disaster-packed predicaments made them a universally popular comic duo whose best short films include *The Battle of the Century* (1927), *Two Tars* (1928), *The Perfect Day* (1929), *Laughing Gravy* (1931) and *The Music Box* (1932) which won an Oscar in the Best Short Film category.

They began to concentrate on feature films from *Pardon Us* (1931) and enjoyed a string of successes, including *Bonnie Scotland* (1935), *Way Out West* (1937) and *The Flying Deuces* (1939) that also allowed them to incorporate some sublime musical numbers into their routines.

However, they later signed agreements to work for M-G-M and Twentieth Century-Fox and found themselves creatively constrained and unflatteringly showcased in work of an inferior quality that included *Great Guns* (1941), *The Big Noise* (1944) and *The Bullfighters* (1945). They subsequently toured Britain and Europe with a music-hall act and found

Laurel and Hardy in *Chumps at Oxford* (1940)

public affection undimmed by the passing of the years and their recent lack of top notch material.

Hardy appeared on his own in *The Fighting Kentuckian* (1949) and *Mr. Music* (1950) and they made one last film together, the dismal *Atoll K* (also known as *Robinson Crusoeland*) (1950). Individual and collective ill-health kept them from the studios and their last joint appearance was in a December 1954 edition of *This Is Your Life*.

Laurel, generally acknowledged as the more creative part of the team lived to see their work hailed afresh, and in April 1961 Danny Kaye (1913–87) accepted an honorary Oscar on his behalf for 'creative pioneering in the field of cinema comedy'.

LAWRENCE OF ARABIA

David Lean (UK 1962)
Peter O'Toole, Jack Hawkins, Alec Guinness, Anthony Quinn, Omar Shariff.
Premièred at 222 mins; original release prints 202 mins; 1970 US reissue at 187 mins; 1989 restored version released, fine cut by Lean himself to 216 mins. col.
Academy Awards: *Best Picture; Best Director; Best Cinematography* Frederick A. Young; *Best Score* Maurice Jarre; *Best Editing* Anne V. Coates; *Best Art Direction* John Box, John Stoll, Dario Simoni; *Best Sound* John Cox.

Cited as a formative influence by major American filmmakers ▷ Martin Scorsese and ▷ Steven Spielberg (each of whom hold his control of narrative in high esteem), ▷ David Lean's critical reputation suffered in the wake of *Doctor Zhivago* (1965) and *Ryan's Daughter* (1970) – both representing an increasing propensity for elephantine construction and vacuous romanticism – to the point where he was looked on as little more than a highly skilled technician. The restoration and reissue of his 1962 epic *Lawrence of Arabia*, a remarkable project requiring the original cast to redub footage trimmed after the first screenings, demonstrated however that in this case at least the expansive pictorialism was indeed tempered by a resonant psychological insight into the complex motivations of its protagonist, leader of the Arab revolt against German-allied Turkey during World War I, the enigmatic English soldier and scholar T. E. Lawrence. Held together by a performance of riveting intensity by then screen newcomer ▷ Peter O'Toole and frequent collaborator Robert Bolt's (1924–) detailed, unpatronising screenplay, Lean's film puts the desert landscape to rich use as both battlefield for predictably well-drilled military encounters and site of Lawrence's ongoing crisis of identity; as he seeks to cast himself anew in this alien environment the previously repressed darker urges in the Lawrence psyche are only to drive him to greater excesses of strategic brutality. It may offer exactly the kind of spectacle we have come to expect from big-budget movies of its kind, but Lean's multi-Oscar winner literally turns the genre inside out, its underlying impulse a challenging interior odyssey.

Peter O'Toole in *Lawrence of Arabia* (1962)

LEAN, David (Sir)

Born 25 March 1908, Croydon, UK.
Died 16 April 1991, London, UK.

The youngest son of a chartered accountant, Lean was expected to follow in his father's footsteps but displayed such a passion for the cinema that he was permitted to seek employment within the film industry. He began as a tea boy at Gaumont studios and, with application and enthusiasm, progressed through the ranks to become a camera operator, assistant director and editor.

He was a highly respected editor for *Gaumont Sound News* (1930) and *British Movietone News* (1931–2) before moving on to fictional features like *Escape Me*

Never (1935), *As You Like It* (1936) and *Pygmalion* (1938).

He worked as an assistant director on *Major Barbara* (1940), and co-directed *In Which We Serve* (1942) with Nöel Coward (1899–1973). He made his solo directorial debut with *This Happy Breed* (1944) and was a skilled interpreter of Coward's work, bringing a cinematic finesse to his sophisticated comedy *Blithe Spirit* (1945) and the restrained romantic heartache of ▷ *Brief Encounter* (1945).

He then made two of the finest screen adaptations of Charles Dickens work: *Great Expectations* (1946) (Best Director Oscar nomination) and *Oliver Twist* (1948), which perfectly illustrate his skill with mood, setting, vivid characterisation, composition and fine judgement in editing that can create the most striking juxtaposition of images and seamless, fluid narratives.

Concentrating on more contemporary British drama, his less successful ventures include *The Passionate Friends* (1948) and *Madeleine* (1949) but he enjoyed successes with the patriotic *The Sound Barrier* (US: *Breaking the Sound Barrier*) (1952) and the comedy *Hobson's Choice* (1954).

After the bittersweet romance of *Summer Madness* (US: *Summertime*) (1955) (Oscar nomination) he was increasingly drawn towards work on an epic and grandiose scale, winning Best Director Oscars for *Bridge On The River Kwai* (1957) and ▷ *Lawrence of Arabia* (1962), the most intelligent and keenly crafted of all the large-scale productions of that era.

Working frequently with scriptwriter Robert Bolt (1924–) and composer Maurice Jarre (1924–), his technique was as compelling as ever in subsequent productions but both the immensely popular *Dr. Zhivago* (1965) (Oscar nomination) and *Ryan's Daughter* (1970) were less satisfactory, inflating their slender storylines to a scale that seemed inappropriate.

Stung by the first consistently negative reviews of his illustrious career, he became disheartened and this dejection, combined with the failure to realise a project on the mutiny aboard the H.M.S. Bounty, kept him away from the cinema for fourteen years.

He returned with *A Passage to India* (1984) (Oscar nomination), a work as elegantly crafted as ever but suffering under the weight of a still persistent need to inflate modest subject matter into a big-screen epic.

In 1990, he was the first non-American recipient of the ▷ American Film Institute Life Achievement award.

LEAR, Norman Milton

Born 27 July 1922, New Haven, Connecticut, USA.

Following his wartime service in the U.S. Army Air Force, Lear worked for a New York publicity firm, sold furniture door-to-door and ran a short-lived company specialising in novelty ashtrays before embarking on a writing career.

Selling jokes and comedy sketches to established performers, his reputation soon earned him permanent employment on a number of television series including *The Martha Raye Show* (1955–6) and *The George Gobel Show* (1955), which he wrote and directed.

In 1959, he formed Tandem Productions with director Bud Yorkin (1926–) and together they were responsible for a number of noted television specials, including *An Evening With Fred Astaire* (1959) and *Henry Fonda and the Family* (1963), as well as such largely comic feature films as *Come Blow Your Horn* (1963), *Never Too Late* (1965) and *Divorce, American Style* (1967), for which Lear's screenplay received an Oscar nomination.

Lear also wrote the film *The Night They Raided Minskys* (1968) and made his directorial debut with the satirical comedy *Cold Turkey* (1971).

In the 1970s, he turned his attention to network television, winning praise and a modicum of right-wing condemnation for his endeavours to create situation-comedies that more accurately reflected the ethnic mixture in contemporary American culture and were also unafraid to tackle controversial social issues like abortion, rape and racism. His many top-rated successes include ▷ *All in The Family*, *Maude* (1972–7), *Sanford and Son* (1972–6), *The Jeffersons* (1975–83) and *Mary Hartman, Mary Hartman* (1976).

A vocal advocate of civil liberties and freedom of speech, he has been active in the organisation 'People for the American Way'. More recently, he was an executive producer of the television movie *Heartsounds* (1984) and the motion picture swashbuckler spoof *The Princess Bride* (1987).

LEE, Bruce

Real Name Lee Yuen Kam
Born 27 November 1940, San Francisco, California, US.
Died 20 July 1973, Hong Kong.

As a child actor, Lee appeared under the name of Lee Siu Lung in a variety of Hong Kong films including *The Birth of Mankind* (1946), *An Orphan's Tragedy* (1955) and *A Goose Alone in the World* (1961). However, it was his adult prowess as a master of martial arts that was to bring him international popularity.

In America, he appeared in such television series as ▷ *Batman* and *Ironside* before securing the role of Kato in *The Green Hornet* (1966–67). For the cinema, he was featured in *Marlowe* (1969) and worked as a karate advisor on the Matt Helm adventure *The Wrecking Crew* (1969).

Returning to Hong Kong, he developed his own style of martial arts known as Jeet Kune Do and became the hero of such kung fu spectaculars as *The Big Boss* (1971), *Fist of Fury* (1972) and *Enter the Dragon* (1973).

A diminutive, slightly-built figure with the screen persona of a shy innocent, he could explode into expertly choreographed bursts of bone-crunching violence that underlined his physical strength, agility and skill as a fight arranger. As the vogue for kung fu films reached its apex in the 1970s, he became a cult figure.

He turned to direction with *The Way of the Dragon* (1973) and was at work on *Game of Death* at the time of his mysterious demise from kidney failure. The project was completed with a double and released in 1978.

LEE, Spike

Full Name Shelton Jackson Lee
Born 20 March 1957, Atlanta, Georgia, USA.

The eldest son of jazz musician Bill Lee, Lee majored in mass communications at Morehouse College (1975–79) where he developed an interest in Super 8mm filmmaking and made such early efforts as *Last Hustle in Brooklyn* (1977).

Studying for his master's degree at New York University's Institute of Film and Television he provoked controversy with *The Answer* (1980), a pointed attack on the racism in ▷ *The Birth Of A Nation* (1915), but earned a student Oscar for his graduation film *Joe's Bed-Stuy Barbershop: We Cut Heads* (1982), an hour-long portrait of ghetto life.

Spike Lee

Struggling to support himself, he sank his energies and personal finances into the low-budget independent feature *She's Gotta Have It* (1986), a visually and dramatically accomplished, naturally witty story of a young woman's involvement with three men of very differing personalities.

Awarded the Camera D'Or at the Cannes Film Festival, the film established him internationally as a commercially viable young talent of great promise.

Declaring that his intention is to express the vast richness of black culture, he has found continued box-office favour with an approach of stylish visuals, pointed soundtracks, strong performances and sometimes controversial narratives. His subsequent films comprise *School Daze* (1988), a richly textured college musical focusing on the splits within the black community, *Do The Right Thing* (1989), a blistering assault on racism, and *Mo' Better Blues*

(1990), the story of a self-obsessed jazz musician's personal growth. *Jungle Fever* will be released in 1991.

An engaging actor in his own productions, he has also appeared in television commercials for Nike baseball shoes which he directed and has assisted in the 1988 Presidential Campaign of Jesse Jackson, directed music videos for the likes of Public Enemy, Tracey Chapman and Miles Davis and co-authored books on the making of his films like *Spike Lee's She's Gotta Have It: Inside Guerrilla Filmmaking* (1987), *Uplift The Race: The Construction of School Daze* (1988) and *Do The Right Thing: A Spike Lee Joint* (1989).

LEIGH, Mike

Born 20 February 1943, Salford, Lancashire, UK.

The son of a doctor, Leigh studied as an actor at RADA (1960–62) before attending the Camberwell and Central Art Colleges and the London Film School.

For the theatre, he directed the original six-hour production of *Little Malcolm and His Struggle Against the Eunuchs* (1965) and built a reputation for an individual and influential style of working that rested on guided group improvisations. Notable theatrical productions include *Babies Grow Old* (1974), *The Silent Majority* (1974) and *Goose Pimples* (1981).

His methods transferred successfully to television where he was able to create acutely observant improvised texts that used humour and compassion to explore the minutiae of working-class mores: everyday lives, aspirations, searches for a decent system of values and feelings of loneliness or despair. The results were both entertaining and moving and widely applauded for their verisimilitude, although also criticised as patronising and condescending. Notable BBC *Plays for Today* include *Hard Labour* (1973), *Nuts in May* (1976), *Abigail's Party* (1977) and *Who's Who* (1978).

He made his cinema debut with *Bleak Moments* (1971) but, aside from the short film *Short and Curlies* (1987), he remained gainfully employed by the theatre and television until *High Hopes* (1988) which offered mercilessly witty attacks on the newly monied classes and a more attractive portrait of a fond working-class couple who manage to reject the Victorian 'family values' espoused by the Thatcher era to assert their own caring priorities of tending to an elderly relative and considering whether to bring a child into the world.

He followed this with *Life is Sweet* (1990), a free-flowing emotional mosaic of the tensions and interplay in a typical family featuring his wife and frequent interpreter Alison Steadman (1946–).

LEIGH, Vivien

Real Name Vivian Mary Hartley
Born 5 November 1913, Darjeeling, India.
Died 7 July 1967, London, UK.

A student at RADA, Leigh made her professional debut in the film *Things Are Looking Up* (1934). Her stage debut followed in *The Green Sash* (1935) and she was an overnight sensation the same year in the comedy *The Mask of Virtue*.

She signed a contract with ▷ Alexander Korda, and made her first appearance with future husband ▷ Laurence Olivier in the costume drama *Fire Over England* (1936). A charming, vixenish actress of great beauty, she lent a bright presence to some flimsy comedies and folderols like *Storm in A Teacup* (1937) and *21 Days* (1938) before securing one of the most coveted roles in the history of the cinema – Scarlett O'Hara in ▷ *Gone With The Wind* (1939).

Stunningly photographed, she was able to bring the right mixture of wilfulness, guile and vulnerability to one of the screen's most tragic romantic heroines and her performance won a Best Actress Oscar.

After the remake of *Waterloo Bridge* (1940) and the morale-boosting *That Hamilton Woman* (1941) with Olivier, she was absent from the screen until *Caesar and Cleopatra* (1945) and a poor version of *Anna Karenina* (1948). However she was excellent recreating her stage role as Blanche Du Bois in *A Streetcar Named Desire* (1951) touching grand, electrifying emotions as the fluttering Southern belle drawn moth-like to the hulking Kowalski of ▷ Marlon Brando.

Ill-health and frequent bouts of manic depression curtailed her later career and she appeared in only three further films: *The Deep Blue Sea* (1955), and as fading beauties who have known better times in

The Roman Spring of Mrs Stone (1961) and *Ship of Fools* (1965).

Highlights from her lengthy and prolific stage career include *A Streetcar Named Desire* (1949), *The Sleeping Prince* (1954), *Tovarich* (1960) and *Ivanov* (1966).

She was divorced from Olivier in 1960 and died after a recurrence of tuberculosis.

LEMMON, Jack

Full Name John Uhler Lemmon
Born 8 February 1925, Boston, Massachusetts, USA.

A graduate of Harvard, Lemmon served in the navy during World War II and before entering films garnered a wealth of showbusiness experience as everything from a singing waiter to a Broadway actor in *Room Service* (1953) and a veteran of an estimated 400 television plays and series episodes.

He made his film debut in *It Should Happen To You* (1954), won a Best Supporting Actor Oscar for his role as wily Ensign Pulver in *Mr. Roberts* (1955) and was soon established as one of the screen's most dynamic comedy performers. The classic ▷ *Some Like It Hot* (1959) (Best Actor Oscar nomination) began a seven-film collaboration with director ▷ Billy Wilder that further illustrated his deft, bittersweet touch as a contemporary underdog caught in everyday moral dilemmas.

His sure touch with the physical gestures and timing of comedy and his warm, empathetic personality were seen to great effect as the careerist junior executive in ▷ *The Apartment* (1960) (Oscar nomination), the alcoholic in *Days of Wine and Roses* (1962) (Oscar nomination) and the injured television cameraman cajoled into seeking vast legal redress by Walter Matthau (1920–   ) in *The Fortune Cookie* (UK: *Meet Whiplash Willie*) (1966).

Matthau would become one of his closest friends and colleagues and they were teamed with memorable results as *The Odd Couple* (1967); he starred for Lemmon in the actor's only directorial credit *Kotch* (1971), a sentimental but winning tale of a septuagenarian.

Whilst still seen in such comedies as *The Front Page* (1974) and *The Prisoner of Second Avenue* (1975), he has increasingly turned towards dramatic roles, winning a Best Actor Oscar as the clothes manufacturer undergoing a mid-life crisis in *Save The Tiger* (1973) and impressing with the deep-felt sincerity and humanity he brought to the portrayals of the conscience-striken engineer in *The China Syndrome* (1979) (Oscar nomination), the cancer-ridden press agent in *Tribute* (1980) (Oscar nomination), the patriotic American father gradually awakened to his country's complicity in Chilean politics in *Missing* (1981) (Oscar nomination), and the suddenly spry seventy-eight-year-old in *Dad* (1989).

His television roles over the years include *The Day Lincoln Was Shot* (1955), *The Entertainer* (1976) and the mini-series *The Murder of Mary Phagan* (1988) and he has retained his commitment to the theatre with a body of work ranging from *Face of A Hero* (1961) to *Juno and The Paycock* (1974), *Tribute* (1978), *Long Day's Journey Into Night* (1988) and *Veteran's Day* (1990).

He was the 1988 recipient of the ▷ American Film Institute Life Achievement Award.

THE LEOPARD (Il Gattopardo)

Luchino Visconti (Italy-France 1963)
Burt Lancaster, Alain Delon, Claudia Cardinale, Paolo Stoppa.
Original running time 205 mins; US and UK release prints cut to 161 mins; 1983 reissue restored to 186 mins. col.

Distinctively combining nostalgia for the elegance of the past with a Marxian determinist's approach to the inevitability of social change, *Il Gattopardo* (The Leopard) remains perhaps the quintessential film by Milan-born aristocrat and avowed communist sympathiser ▷ Luchino Visconti. The publication in 1958 of the original novel by fellow nobleman Prince Giuseppe Tomasi di Lampedusa provided Visconti with the perfect material, enabling him to return to the key historical moment of the Risorgimento he had already visited in *Senso* (1953). Against the background of Garibaldi's expedition to Sicily to depose the Bourbon King Francis II and unite the island with Italy, ▷ Burt Lancaster's Prince Fabrizio is forced to reflect on the inexorable decline of the old ruling classes by the marriage of his impecunious nephew Tancredi (Alain Delon, (1935–)) to the bourgeois mayor's daughter Angelica, played by Claudia Cardinale (1939–). Lancaster's superlative performance exudes dignified disdain for the emergent new order, and at times it appears that Visconti expects us to be fully in sympathy with him. Although the intermingling of the classes during the marriage ball marks the fullest disruption of the status quo to date by the emergent new order, the detailed finery of this spectacular sequence – which took 36 days to shoot and takes up the last third of the film – seems rather more marked by elegiac resignation than any note of celebratory optimism. As Visconti's camera lingers over the authentic ballgowns and exquisite furnishings we are more than a little inclined to doubt the conviction or rigour of his leftist analysis, but the contradictory impulses in the director's work are precisely what makes it, at its most fully achieved, so fascinating.

LEONE, Sergio

Born 3 January 1929, Rome, Italy.
Died 30 April 1989, Rome, Italy.

A law student in Rome, Leone first worked as an assistant and small part actor on ▷ *Ladri Di Biciclette* (Bicycle Thieves) (1948). His father Vincenzo was one of Italy's pioneering film directors and Leone followed in his footsteps working as an assistant to directors like ▷ Raoul Walsh, ▷ William Wyler and Mervyn LeRoy (1900–87) during the 1950s heyday of American-made Biblical epics.

He moved on to scriptwriting and assumed the directorial reins of *Gli Ultimi Giorni di Pompeii* (The Last Days of Pompeii) (1959) when the original director fell ill. He made his official directorial debut with the similar sword and sandal adventure *Il Colosso di Rodi* (The Colossus of Rhodes) (1961).

After serving as second-unit director on *Sodom and Gomorrah* (1962), he made his mark with ▷ *Per Un Pugno Di Dollari* (A Fistful of Dollars) (1964). Inspired by the ▷ Kurosawa film Yojimbo (1961), it revitalised the western by taking a fresh, meticulously-researched and highly stylised approach to the savagery behind the mythology of this most popular of American genres. Making an international star of ▷ Clint Eastwood as the enigmatic 'Man With No Name', it combined stark landscapes, a minimum of dialogue, extreme close-ups, graphic violence and the wailing, bestial music of ▷ Ennio Morricone for a harshly poetic, primitive vision of the brutality and venality of the (often morally indistinguishable) heroes and villains who made the west.

The worldwide success of the film initiated the ▷ spaghetti western genre, and was followed by the sequels *Per Qualche Dollaro in Piu* (For A Few Dollars More) and *Il Buono, Il Brutto, Il Cattivo* (The Good, The Bad, and The Ugly) (1966) before his increasingly sophisticated perspective on the West matured into *C'Era Una Volta Il West* (Once Upon A Time in the West) (1968), an operatic reflection on the corruption and power struggles involved in the arrival of civilisation and the railways that offers hope only in the mother figure who attempts to instil a sense of community values.

After the more lightweight *Gui La Testa* (Duck You Sucker!) (1972), he became a producer of such films as *Un Genio, Due Compari, Un Pollo* (The Genius) (1975), *Il Gatto* (The Cat) (1978) and *Fun Is Beautiful* (1979).

His final film as director, *Once Upon A Time in America* (1984) is a sumptuously crafted gangster chronicle, both panoramic and intimate, that expertly weaves several generations of American experience into the lives of four neighbourhood children and concludes with one character's search for grace and self-knowledge.

At the time of his death, he was at work on *The 900 Days*, a $70 million Italian-Russian co-production about the wartime siege of Leningrad that was projected to star ▷ Robert de Niro.

LEVINSON, Barry

Born 6 April 1942, Baltimore, Maryland, USA.

A student in broadcast journalism at the American University in Washington D.C., Levinson had worked part-time as an assistant-director at a local television station before dropping out and making his way to California. He enrolled in some acting classes and developed a comedy act that brought him to the attention of local television shows; subsequently he was hired as a comic writer for *The Tim Conway Show* and *The Carol Burnett Show*, winning three Emmys for his efforts.

▷ Mel Brooks engaged him as one of the contributors to the scripts of his spoof comedies *Silent Movie* (1976) and *High Anxiety* (1977) and also encourage him to turn the amusing anecdotes he related from his adolescence into a full-length script. The result was *Diner* (1982) which also marked his directorial debut. Set in the Baltimore of 1959, it provided a warmly sympathetic view of the group of male friends who hang-out at the local diner and discuss the latest emotional and professional traumas of their young adult lives. Imbued with many subtle pleasures from the sharply-characterised main figures, to the refreshingly believable dialogue and situations, it became a ▷ sleeper hit of the year and earned him an Oscar nomination for Best Screenplay.

After collaborating on the scripts for *Best Friends* (1982) and *Unfaithfully Yours* (1984), he became a full-time writer-director displaying many pleasingly unfashionable virtues in his concentration on strong storytelling, worldly-wise behaviour, trenchant dialogue and warmly engaging human characters.

Unafraid to work in all genres, he made the 'adult fairy tale' *The Natural* (1984) starring ▷ Robert Redford as an enigmatic baseball player of legendary prowess, and the children's adventure yarn *The Young Sherlock Holmes* (1985).

He then returned to his beloved Baltimore for another witty excursion into the foibles of modern man in *Tin Men* (1987), a tit-for-tat romantic comedy set in 1963 and starring Richard Dreyfuss (1947–) and Danny De Vito (1945–).

He subsequently enjoyed his greatest commercial successes with the comedy *Good Morning Vietnam* (1987), which showed an astute handling of ▷ Robin Williams's explosive personality in the role of an irreverent armed forces disc-jockey, and the sentimental buddy movie *Rain Man* (1988) which earned him a Best Director Oscar.

Repeating the earlier pattern of his career, he created *Avalon* (1990) (Oscar nomination), the kaleidoscopic autobiographical drama of a close-knit immigrant family's upwardly mobile experiences over fifty years of Baltimore history. His latest film is the gangster melodrama *Bugsy* (1991).

LEWTON, Val

Real Name Vladimir Ivan Leventon
Born 7 May 1904, Yalta, Russia.
Died 14 March 1951, Hollywood, California, USA.

A journalist and prolific author of novels and pseudonymous pornography, Lewton worked for five years in the publicity department at M-G-M and, in 1933, became a story editor for ▷ David O. Selznick.

However, his greatest renown was earned as a producer at R-K-O where he masterminded a series of highly imaginative, low-budget horror films that provided opportunities for a variety of directors, including Mark Robson (1913–78), Jacques Tourneur (1904–77) and Robert Wise (1914–).

The cycle began with *Cat People* (1942) which created its sense of fear and unease by implicit, suggestive means; the 'monster' is never graphically depicted but there is an unsettling mood created by the use of lighting and shadow, the juxtaposition of gruesome events with a credible, even mundane, contemporary setting and a taut screenplay.

Irrational fears, bizarre phenomena and the occult were explored in films that often took their inspiration from highly-regarded literary sources. Although set in the West Indies and concerned with voodoo *I Walked With A Zombie* (1943) shamelessly steals from Jane Eyre, the innocent stumbling on a witch's coven in *The Seventh Victim* (1944) borrows its theme from a sonnet by John Donne, whilst *The Body Snatchers* (1945) was an adaptation of Robert Louis Stevenson's macabre tale of grave-robbing in 19th century Edinburgh.

Lewton stamped his creative authority on a body of diverse work within the genre that also includes *The Leopard Man* (1943) and *Curse of the Cat People* (1944). He also used the pseudonym of Carlos Keith to contribute to the screenplays of *The Body Snatchers* and *Bedlam* (1946).

Later films outwith the horror field, including the romance *Please Believe Me*

(1950) and the western *Apache Drum* (1951), proved less noteworthy.

LEWIS, Jerry
Real Name Joseph Levitch
Born 16 March 1926, Newark, New Jersey, USA.

Born into a showbusiness family, Lewis worked on a nightclub act before joining with singer and straight man Dean Martin (1917–) to form a comic double-act that gained popularity in cabaret and on television.

Signed to a contract by producer Hal Wallis (1898–1986), they made their film debut in *My Friend Irma* (1949).

Never critically well-admired, their unsophisticated madcap antics found audience favour and their high-grossing hits include *That's My Boy* (1951), *Scared Stiff* (1953), *Artists and Models* (1955) and, their last together, *Hollywood or Bust* (1956).

Pursuing a separate career after an acrimonious split, Lewis developed into a popular comedian in the silent tradition whose primarily visual humour relied on a set of facial contortions, an element of physical dexterity, crude slapstick and a comic image that bordered on arrested development and sought to milk the maximum pathos from his unloveable, sometimes grotesque, childlike loners and misfits. Working frequently with the sympathetic direction of ▷ Frank Tashlin, his films include *Cinderfella* (1960), *Who's Minding the Store?* (1963), *The Patsy* (1964) and *The Disorderly Orderly* (1965).

He had directed a number of spoof comedy shorts before making his feature-length directorial debut with *The Bellboy* (1960) and was soon known as the total filmmaker responsible for the writing, directing, music and performance in each film. If the results failed to please, as they increasingly did, then there was clearly only one person to blame. His greatest accomplishment was *The Nutty Professor* (1963) a witty and disciplined variation on *Dr. Jekyll and Mr Hyde*, and he was seen in more conventional material opposite Tony Curtis (1925–) in the farce *Boeing, Boeing* (1965).

A series of financial flops curtailed his screen activities, and after the abandonment of a European production, *The Day The Clown Cried* (1972), he concentrated on personal appearances and his tireless telethon charity campaigning on behalf of muscular dystrophy. He was seen on Broadway in the brief run of *Hellzapoppin'* (1976) and returned to the cinema as the director and star of *Hardly Working* (1979).

In an excellent piece of casting, he gave a fine dramatic performance as the shallow television talk show host in *The King of Comedy* (1982) and was also effective in a straight role in the television drama *Fight for Life* (1987) as the dentist father of a girl stricken with epilepsy.

Lauded in Europe as a comic genius and ▷ auteur, his most recent film appearances include the comedies *Retenez-moi Ou Je Fais Un Malheur* (1983), *Par Ou T'Es Rentré On T' A Pas Vu Sortir?* (1984) and a small role in *Cookie* (1989).

LIBRARY SHOT Also known as a stock shot, this is a scene or shot from a film that is literally available from stock, material that exists already and is inserted in another production as a cost-saving exercise. It may be a battle scene employing many extras or a particularly expensive location. *Raid on Rommel* (1971), for instance, uses desert footage that was originally filmed for *Tobruk* (1967), whilst *Storm Over the Nile* (1955) incorporates material from the previous version of *The Four Feathers* made in 1939. Several television series in the 1970s, like *The Incredible Hulk* and *The Six Million Dollar Man* regularly cannibalised the vast reserves of material from feature films owned by their production company Universal.

THE LIFE AND DEATH OF COLONEL BLIMP
Michael Powell and Emeric Pressburger (UK 1943)
Roger Livesey, Anton Walbrook, Deborah Kerr, Roland Culver.
Original running time 163 mins; cut to circa 140 mins for UK release; further cut to 120 mins for US release; 1985 restored print reissued with all previous cuts reinstated to 163 mins. col.

Although it was the fifth film in the ongoing collaboration between English director ▷ Michael Powell and Hungarian-born screenwriter Emeric Pressburger (1902–88), *The Life and Death of Colonel Blimp* was doubly significant as the first film to open with the distinctive target logo announcing their unique writer/producer/directorship *The Archers*, and their initial venture into the magical world of Forties Technicolor. The duo's previous monochrome propaganda efforts *49th Parallel* (1941) and *One of Our Aircraft Is Missing* (1942) were indeed little preparation for the uniquely eccentric, defiantly anti-realist investigation of Tory Englishness that was hereby to follow, establishing The Archers as the great outsiders in the British film canon. Loosely based on satiric cartoonist David Low's ageing military reactionary, Powell and Pressburger's approach to their Colonel Blimp, in the shape of Roger Livesey's (1906–76) General Wynne-Candy, is altogether more richly ambiguous, for while the film's central thesis affirms that the English notion of honour at all costs (as represented by old duffers like Blimp) must give way to the harsh realpolitik required to counter the Nazi threat, it's undercut by a good deal of nostalgic affection for the traditionalist's sense of fair play. With Deborah Kerr (1921–) quite radiant as Wynne-Candy's thrice-lost love, an air of romantic pessimism runs through the complex flashback structure, but it's in the General's life-long friendship with the characteristic 'good' German, Theo Kretschmar-Schuldorff (Anton Walbrook, (1900–67)), that we find the perfect image of the Powell/Pressburger partnership and – of course – the main catalyst of wartime prime minister Winston Churchill's famous disdain which found expression in his attempts to prevent the film's production.

THE LIKELY LADS

UK 1965–9, 1973
Written by Dick Clement and Ian La Frenais

Like all the best British sitcoms, *The Likely Lads* found its humour in the foibles of the class system, which came iced with occasional pathos. Not unlike a comic version of a kitchen-sink-drama, the show grew from a chance meeting between BBC producer Clement (1937–) and unemployed salesman La Frenais (1938–), and was initially conceived as a sketch. It concerned the high ambitions and limited horizons of Terry Collier and Bob Ferris, two young men in Northumberland, an area where the Sixties were, apparently, less than Swinging. Terry (James Bolam (1938–)) was assured, but directionless, while Bob (Rodney Bewes (1937–)) was keen to better himself by working his way up the career ladder. Despite lengthy discussions over numerous pints of beer, both men viewed the approach of the other with ill-conceived dismay. The show was a deflation of working class aspirations and a subtle critique of British society in a time when such notions were less than fashionable.

Though the initial run was a success, the idea really matured with the 1973 follow-up series *Whatever Happened To The Likely Lads?* Set four years later, Terry's cynicism had hardened through a spell in the army and a disappointing marriage, while Bob, now engaged to the fearsome Thelma (Brigit Forsyth), was hanging uncertainly to a new, middle-class identity. But the show was far lighter and sharper than a sociological analysis might suggest – one classic episode chronicled Terry and Bob's efforts to survive without finding out the result of a football match to be televised that evening. Like most of their schemes it ended in tears. A disappointing *Likely Lads* feature film, directed by Michael Tuchner (1934–), was released in 1976.

LINE PRODUCER More concerned with the day to day running of a film during its actual production period than broader creative or administrative matters, the line producer will generally be a highly experienced figure who takes responsibility for supervising most aspects of shooting a film, ensuring that filming is progressing on schedule and within the budgetary limits set. He or she may work under an executive producer who is further removed from ground level activities.

LLOYD, Harold Clayton

Born 20 April 1893, Burchard, Nebraska, USA.
Died 8 March 1971, Beverly Hills, California, USA.

Stagestruck from an early age, Lloyd studied acting in San Diego and worked extensively with regional stock companies before entering the film industry as an ▷ extra.

Attempts were made to build a comic character for him and he appeared in a series of comedies featuring Willie Work and subsequently Lonesome Luke, a figure heavily influenced by ▷ Chaplin's success. Numerous titles include *Lonesome Luke* (1915), *Luke the Candy Cut-Up* (1916), *Lonesome Luke on Tin Can Alley* (1917) and dozens of others.

From *Over the Fence* (1917) he began to develop an individualistic screen persona that would crystallise into the shy, sincere, bespectacled boy-next-door, anxious to make good and perennially involved with hair-raising stunts from the top of tall buildings.

He made his feature-length debut with *Grandma's Boy* (1922) and created a series of thrilling comedies that made him the equal in popularity and inventiveness of ▷ Keaton and ▷ Chaplin. His most famous film remains ▷ *Safety Last* (1923) in which he dangles perilously from the hands of a high-rise clock face; however such silent films as *Why Worry?* (1923), *Girl Shy* (1924), *The Freshman* (1925) and *Speedy* (1928) are consistently funny and revive well today.

He made the transition to sound with *Welcome Danger* (1929) and *Feet First* (1930) but his attempts to work within more conventional, narrative-driven comedy frameworks were less successful and his popularity began to slide with each increasingly infrequent release. His meagre output from the 1930s includes *Movie Crazy* (1932), *The Cat's Paw* (1934) and *Professor Beware* (1938). Tempted by an offer from ▷ Preston Sturges to make a sequel to *The Freshman*, he returned to the screen for *The Sin of Harold Diddlebock* (1947), but even when retitled *Mad Wednesday* it found only sparse audiences and was to be his last film.

He produced *A Guy, A Girl and A Gob* (1941) and *My Favourite Spy* (1942) and later released two popular compilations of the choicest moments from his heyday as one of the daredevil kings of comedy – *Harold Lloyd's World of Comedy* (1962) and *Harold Lloyd's Funny Side of Life* (1966).

His autobiography *An American Comedy* was published in 1928 and he was awarded an honorary Oscar in March 1953 affectionately dedicated to a 'master comedian and good citizen'.

LOACH, Kenneth

Born 17 June 1936, Nuneaton, UK.

After studying law at Oxford, Loach spent some time as an actor before training as a television director and joining the BBC.

He directed various early episodes of ▷ *Z Cars* before making his name in the Wednesday Play series where his emphasis on realism and naturalistic performances dramatically illuminated social ills like homelessness in ▷ *Cathy Come Home* (1966).

His first feature film *Poor Cow* (1967) attempted to illuminate the mundane nature of ordinary lives as a young married woman falls into an adulterous affair as an escape from the drudgery of her domestic routine. He followed this with *Kes*, (1969), a sensitive exploration of another wasted life as a teenage boy channels his energy and enthusiasm into the training of a kestrel who grows to embody a futile flight from the banality of his existence. *Family Life* (1971), perhaps his best film, proved a stark, grippingly acted study of the pressures within the family unit.

His deceptively simply-crafted dramas offer a committed political perspective rare in British cinema and a heightened naturalism that is almost poetic. He has consistently explored social issues and questioned socialist history in such television work as the four-part *Days of Hope* (1975) and *The Price of Coal* (1977) and the moving film *Looks and Smiles* (1981).

During the Thatcher era he worked on a number of controversial documentaries, including the series *Questions of Leadership* (1983) which was prohibited from transmission on political grounds.

More recently he has returned to the cinema with the dour *Fatherland* (1986), exploring the differences and similarities between Eastern and Western European political systems, and the taut thriller *Hidden Agenda* (1990), exploring the role of the British forces in Northern Ireland and accusations of a right-wing conspiracy within the British establishment.

His most recent film is *Riff Raff* (1991) looking at the lives of casually employed manual labourers.

LOCATION SCOUT The individual who scours the globe in search of suitable places, either indoors or outdoors, that most closely conform to the director's vision of the film and provide adequate facilities that allow for the shooting of that film outwith the confines of a studio. In 1953, producer Arthur Freed (1894–1973) came to Scotland scouting locations for his forthcoming production of *Brigadoon* (1954). His travels took him to some of the country's most picturesque spots from Culross to Dunkeld and Braemar but, unimpressed, he returned to California and decided to build his Highland village at M-G-M studios declaring, 'I went to Scotland but could find nothing that looked like Scotland.' Those seeking locations for the 1990 production of *Hamlet* were more successful, selecting Dunottar Castle near Stonehaven as one of their sites. One of the most imaginative uses of locations in recent British cinema has been ▷ Stanley Kubrick's *Full Metal Jacket* (1987), in which a disused North Thames gas works was transformed to represent areas of Vietnam.

LOMBARD, Carole

Real Name Jane Alice Peters
Born 6 October 1908, Fort Wayne, Indiana, USA.
Died 16 January 1942, near Las Vegas, Nevada, USA.

Whilst resident in California, the young Lombard was spotted by director Allan Dwan (1885–1981) and cast as a tomboy in the film *A Perfect Crime* (1921).

After completing her studies, she resumed her career with roles in *Marriage in Transit* (1925) and *Hearts and Spurs* (1925) and her blonde hair and beauty made her a decorative addition to many comedy shorts for ▷ Mack Sennett. Signed to a long-term contract with Paramount in 1930, her roles gradually improved and she revealed a delicious comic flair in *Twentieth Century* (1934).

A glamorous, sophisticated woman, unafraid to play for laughs, her witty, wacky effervescence made her the perfect heroine of ▷ screwball comedies like *My Man Godfrey* (1936) (Best Actress Oscar nomination), *True Confession* (1937) and *Nothing Sacred* (1937). Quieter, dramatic abilities were revealed as the dedicated nurse in *Vigil in the Night* (1940) and as the mail-order bride in *They Knew What They Wanted* (1940), but her sharply honed comic skills were to the fore in her final films *Mr and Mrs Smith* (1941) and the classic black comedy *To Be Or Not To Be* (1942).

Married to ▷ Clark Gable in 1939, she was one of Hollywood's most popular stars at the time of her death in an air crash.

LOREN, Sophia

Real Name Sofia Villani Scicolone
Born 20 September 1934, Rome, Italy.

Attempting to escape the dire poverty of her upbringing, Loren was encouraged to become a teenage beauty queen and model to gain work as an extra in *Cuore Su Mare* (Hearts Upon The Sea) (1950) and the American production *Quo Vadis* (1951).

Sophia Loren with Marcello Mastroianni in *Yesterday, Today, Tomorrow* (1963)

Under contract to Carlo Ponti (1913–), she blossomed into a breathtakingly beautiful star with a talent for earthy drama and vivacious comedy. Building her popularity within Europe with films like *Miseria E Nobilitia* (Poverty and Nobility) (1954) and *L'Oro Di Napoli* (The Gold of Naples) (1955), she then moved into the international English-language market with a string of middling adventure stories like *The Pride and The Passion* (1957) and *Legend of the Lost* (1957), but did show a warmth and gaiety in the comedy *Houseboat* (1958).

Revealing hitherto unexpected depths she won a Best Actress Oscar for *La Ciociara* (Two Women) (1961), as the young mother coping with numerous hardships in war ravaged Italy. Her career then alternated between epic international ventures like *El Cid* (1961), *The Fall of the Roman Empire* (1964) and *Operation Crossbow* (1965) and a succession of Italian comedies in which she was often happily teamed with ▷ Marcello Mastroianni. Titles include *Ieri, Oggi, E Domani* (Yesterday, Today, Tomorrow) (1963) and *Matrimonio All'Italiana* (Marriage-Italian Style) (1964) (Oscar nomination).

Radiant in the fairytale *C'Era Una Volta* (Cinderella Italian Style) (1967) her debatable box-office power waned with such poorly received work as *Man of La Mancha* (1972), *Verdict* (1974) and a misguided remake of *Brief Encounter* (1974) in which she co-starred with ▷ Richard Burton. However, she gave one of her best performances opposite ▷ Mastroianni, touchingly conveying the weary melancholy of a dowdy 1930s housewife in *Una Giornata Particolare* (A Special Day) (1977).

In 1979, she published *Sophia Loren: Living and Loving* (with A. E. Hotchner) which was filmed for television as *Sophia Loren: Her Own Story* (1980) with the actress playing both herself and her mother. Her career continues with television films like *Aurora* (1984), *Courage* (1986), *The Fortunate Pilgrim* (1988), the remake of *Two Women* (1989) and *Saturday, Sunday, Monday* (1990). In March 1991 she received an honorary Oscar.

LOSEY, Joseph

Born 14 January 1909, La Crosse, Wisconsin, USA.
Died 22 June 1984, London, UK.

Graduating from Harvard University with an M.A. in English Literature, Losey wrote arts reviews and directed for the New York stage before studying film with ▷ Eisenstein in Moscow.

Working with the Federal Theater Project, he entered filmmaking as the producer of industrial documentaries and made his directorial debut with the marionette short *Pete Roleum and His Cousins* (1939). His entry in the M-G-M Crime Does Not Pay Series, *A Gun in His Hand* (1945) won an Oscar nomination as Best Short subject.

Active in War Relief shows and a writer for radio, he directed *Galileo* (1947) for the stage in Los Angeles and New York before making his feature-length film debut on *The Boy With Green Hair* (1948), a charming fantasy with a deftly incorporated social message. His other Hollywood credits included *M* (1951) and *The Prowler* (1951) before he fell foul of McCarthy-era hysteria and found himself blacklisted.

He moved to Europe, directing a number of films under the pseudonyms of Victor Hanbury and Joseph Walton, and took time to find sympathetic material when resuming work under his own name. He worked well with actor Stanley Baker (1928–76) on the gritty thrillers *Blind Date* (1959) and *The Criminal* (1960) and the more characteristic *Eva* (1962), an exploration of a writer's troubled private and creative worlds.

Intrigued by the nature of good and evil and the role of a social outsider as a catalyst in a community or in an established relationship, he began a fruitful collaboration with Harold Pinter (1930–) on *The Servant* (1963), a sinister account of the power struggles between a spoiled, indolent master and his apparently charming manservant, played by ▷ Dirk Bogarde. The notion of one person living a vicarious fantasy life through the unhealthy dominance of a stronger personality would recur in films like *Secret Ceremony* (1969) and *Mr. Klein* (1976).

Other collaborations with Bogarde include *The Sleeping Tiger* (1954), the anti-war courtroom drama *King and Country* (1964), the cod spy spoof *Modesty Blaise* (1966) and *Accident* (1967), a thoughtful character study of class and a complex web of amorous entanglements.

Often accused of succumbing to heavy-handed pretentiousness and obscurantism, his later work varied enormously although *The Go-Between* (1971) won the Palme d'Or at the Cannes Film Festival and his handsome version of *Don Giovanni* (1979) was much liked. His last films were *The Trout* (1982) and *Steaming* (1984), completed shortly before his death.

LOU GRANT

USA 1977–82
Ed Asner

The character of Lou Grant began life as

the newsroom boss in *The Mary Tyler Moore Show*. At the end of that show's run he was fired, only to return in his own series as the city editor of the *Los Angeles Tribune*. The previous spin-offs from the Tyler Moore show, *Rhoda* and *Phyllis*, had retained the essence – comedy with well-loved characters – of their parent. *Lou Grant* was different, being an hour-long drama with a social conscience. It was also the show which helped make the reputation of MTM, the production company, as a purveyor of popular, quality television (subsequent efforts included ▷ *Hill Street Blues* and *St Elsewhere*).

Robert Walden and Ed Asner in *Lou Grant*

Developed by Gene Reynolds who also worked on the pilot of ▷ *M*A*S*H*), James Brooks and Allan Burns, *Lou Grant* was a blend of genres – soap, sitcom and workplace drama – with several parallel plotlines in each episode. Straight cop shows were in vogue, but having been granted a free hand with the character of Lou, the writers cashed in on the popularity of ▷ Asner and the post-Watergate gratitude afforded to the journalistic profession. There was a strong pool of characters, with much play on the rivalry between post-feminist career woman Billie Newman (Linda Kelsey) and hardnosed Jack Rossi (Robert Walden). Plotting an individual course between the two was the hip-piesque photographer Animal (Daryl Anderson). The plots were bold, dealing seriously with social and moral issues.

Controversially, *Lou Grant* was cancelled by CBS at the end of its fifth season. Ratings were beginning to flag, though the show's prestige might have been expected to see it through. Asner had attracted right wing criticism as President of the Screen Actors' Guild, and this may have been a factor.

LOUMA A valuable assistant in capturing shots that are inconceivable with manual operation of the camera, a Louma is a portable crane for a lightweight camera that replays video pictures of the exact image that will be captured on film and thus allows the camera to be operated by remote control with the director totally informed as to his choice of shots. The additional flexibility allows access to cramped and restricted spaces and an added fluidity in the camera movement when used in tandem with a mobile crane.

LOWE, Arthur

Born 22 September 1914, Hayfield, Derbyshire, UK.
Died 15 April 1982, Birmingham General Hospital, UK.

Leaving school at sixteen, Lowe worked as a salesman and served with the cavalry in Palestine during the Second World War, acquiring an interest in performing with the entertainments division of the armed forces.

After a brief course at RADA, he made his first stage appearance in *Bedtime Story* (1945) at Hulme and his London debut in *Larger Than Life* (1950). His film debut in *London Belongs To Me* (US: *Dulcimer Street*) (1948) was the first of many supporting roles in the likes of ▷ *Kind Hearts and Coronets* (1949), *The Green Man* (1956) and *This Sporting Life* (1963).

His subsequent theatre work included *Call Me Madam* (1952), *Pal Joey* (1954) and *The Pajama Game* (1955), but it was television that brought him his greatest popularity, first as the irascible Leonard Swindley in ▷ *Coronation Street*, a role he played for six years, and then as the bumbling Captain Mainwaring in ▷ *Dad's Army* (1968–77).

A rotund, owlish, balding figure, he was particularly adept at portraying pompous officials and comic duffers and starred in many television series.

His later stage work included *Inadmissible Evidence* (1964), *The Tempest* (1974) and *Bingo* (1974), while he took more substantial film roles in the big screen version of *Dad's Army* (1971), as the manservant Tucker in *The Ruling Class* (1972), in a triple role in the rancorous, surreal satire *O, Lucky Man* (1973) and as Charters in *The Lady Vanishes* (1979).

His last screen roles were in *Britannia Hospital* (1982) and the television epic *Wagner* (1982), and he collapsed on stage during a performance of the comedy *Home at Seven*, dying shortly thereafter.

LOY, Myrna

Real Name Myrna Adele Williams
Born 2 August 1905, Radersburg, Montana, USA

In 1937, when ▷ Clark Gable was voted the King of Hollywood, Myrna Loy was chosen as Queen. A top M-G-M star, considered the embodiment of the screen's perfect society wife, she elegantly swapped martinis and an equally dry wit with the studio's most prized male idols. The intelligence and warmth of her chic urbanity remains one of the most enduring and underrated delights of Hollywood's so-called Golden Age.

Of Welsh ancestry, she moved with her family to Los Angeles in 1919 and subsequently performed in the chorus of prologues for movie houses like the Egyptian Theater. Spotted by ▷ Rudolph Valentino, she was tested as his potential leading lady for *Cobra*. She did not secure the part but did make her debut in *Pretty Ladies* (1925) and worked as an extra on *Ben Hur* (1926). Signed to a five-year contract by Warner Brothers in 1925, she appeared in scores of silent features and was quickly typecast as Oriental vamps, spies and seductresses.

Gaining a reputation for professionalism and hard work, she freelanced briefly in the early 1930s before signing a contract with M-G-M in 1932. Her role as the sadistic daughter in *The Mask of Fu Manchu* (1932) was to be the last Oriental villainess. In its place she was allowed to develop a flair as a polished comedienne and dramatic performer of incisiveness and humanity. Her appearance opposite William Powell (1892–1984) as the husband-and-wife detectives Nick and Nora Charles in *The Thin Man* (1934) began a long and happy partnership in which their impeccable timing, witty bantering, teasing relationship and evident affection contributed to a more sophisticated dramatic view of the married couple. Their thirteen films together include several Thin Man sequels, *The Libelled Lady* (1936), *Double Wedding* (1937) and *Love Crazy* (1941).

Equally successful in such major productions as *Test Pilot* (1938), she extended her range as the feckless socialite in *The Rains Came* (1939). During the Second World War, she worked extensively for the Red Cross. In 1946, she attended the first meetings of the United Nations and later served as an American representative with UNESCO.

She placed less emphasis on her career, but affectingly portrayed the loyal, level-headed homefront wife in ▷ *The Best Years of Our Lives* (1946) and brought her smart presence to such popular comedies as *Mr Blandings Builds His Dream House* (1948) and *Cheaper By the Dozen* (1950).

Later a discriminating character actress, she appeared as the alcoholic mother of ▷ Paul Newman in *From the Terrace* (1960) and accepted modest roles in such films as *April Fools* (1969), *The End* (1978) and *Just Tell Me What You Want* (1980). Active on television and the theatre, she made her belated Broadway debut in *The Women* (1974) and made her last acting appearance to date in the television film *Summer Solstice* (1981) opposite ▷ Henry Fonda.

Her autobiography *Being and Becoming* (1987) (written with James Kotsilibas-Davis) displayed the qualities of wit, candour and charm that had been evident throughout her career.

In March 1991, she finally received an honorary Oscar in recognition of her long career.

LUBITSCH, Ernst

Born 28 January 1892, Berlin, Germany.
Died 29 November 1947, Hollywood, California, USA.

A teenage actor in Max Reinhardt's theatre company, Lubitsch starred as the ethnic 'Dummkopf' character, Meyer, in a popular slapstick series before beginning his directorial career with *Fräulein Seifenschaum* (1914).

A specialist in comedy, he also worked on numerous costume dramas, frequently featuring actress Pola Negri (1897–1987). Among the titles are *Carmen* (1918), *Madame Dubarry* (1919) and *Medea* (1920).

He was invited to Hollywood by ▷ Mary Pickford, whom he directed in *Rosita* (1923), and stayed to become an acknowledged master of light, sophisticated comedies of sexual mores, graced with the 'Lubitsch touch', a mixture of wit, urbanity and visual elegance. Offering a refreshingly healthy perspective on romance and sex, he avoided heavyhanded moralising and created well-rounded female characters who were ahead of their time, in screen terms at least.

His many subtle confections from the silent era include *The Marriage Circle* (1924), *Forbidden Paradise* (1924) and *The Patriot* (1928), for which he received a Best Director Oscar nomination.

At the onset of the sound era, sparkling, witty dialogue enhanced his basic approach and he brought a fluidity to the musical genre in productions like *Love Parade* (1929), *One Hour With You* (1932) and *The Merry Widow* (1934).

Trouble in Paradise (1932) represents the scintillating apotheosis of his style and flair, but a bittersweet melancholy deepened the qualities of such later comedies as *Ninotchka* (1939), *The Shop Around the Corner* (1940), the black *To Be Or Not To Be* (1942) and *Heaven Can Wait* (1943) (Oscar nomination).

He suffered a massive heart attack in 1943 and completed only one further film filming of *That Lady in Ermine* (1948) which was completed by ▷ Otto Preminger.

He received a special Oscar scroll in March 1947 for his 'distinguished contributions to the art of the motion picture'.

LUMET, Sidney
Born 25 June 1924, Philadelphia, Pennsylvania, USA.

The son of an actor, Lumet made his performing debut opposite his father in a 1928 performance at the Yiddish Art Theatre in New York. Also in the children's radio series *The Adventures of Helen and Mary*, he made his Broadway debut in *Dead End* (1935) and subsequently featured in *The Eternal Road* (1937) and *My Heart Is in The Highlands* (1938). He also appeared in two films *One Third Of A Nation* (1939) and *Journey to Jerusalem* (1940).

A member of the Army Signal Corps from 1941 to 1946, he replaced ▷ Marlon Brando as the lead in the stage play *A Flag is Born* (1946) and formed an off-Broadway acting group the following year. An actor, director and a lecturer at the High School for the Performing Arts, he joined CBS Television in 1950 and between 1951 and 1953 directed an estimated 150 episodes of the weekly series *Danger* and 26 *You Are There* shows. From 1953, he directed scores of television dramas and his talents were forged in the high-pressure world of tight schedules, low budgets and live transmissions.

He made his cinema debut as the director of *Twelve Angry Men* (1957) (Oscar nomination), a taut courtroom drama that signalled his fascination with the workings of the judicial system and fondness for central characters of individual conscience who will stand up for a point of principle, regardless of the personal consequences.

A prolific and professional director, noted for his speed, efficiency and particular empathy with actors, he has made almost forty films in just over three decades. A meticulous investigator of American institutions, mores and the moral complexities of crime and punishment, his many well-crafted gritty urban thrillers include *Serpico* (1973), *Dog Day Afternoon* (1975) (Oscar nomination), *Prince of the City* (1981), *The Verdict* (1982) (Oscar nomination) and *Q & A* (1990), for which he also wrote the script.

His fondness for the theatrical community was evident in *Stage Struck* (1958) and he has also proved a faithful, cinematic interpreter of such diverse stage works as *Long Day's Journey Into Night* (1962), *The Seagull* (1968), *Equus* (1977) and *Deathtrap* (1982).

His liberal concerns have surfaced in such films as *Daniel* (1983) and *Running on Empty* (1988), whilst his versatility has extended to less characteristic work like the elegant Agatha Christie mystery

Murder on the Orient Express (1974) and the all-black musical *The Wiz* (1978).

The acting in his films is never less than excellent and stars to have given some of their finest performances under his direction include Rod Steiger (1925–) in *The Pawnbroker* (1965), ▷ Faye Dunaway in *Network* (1976) (Oscar nomination), ▷ Paul Newman in *The Verdict* (1982) and such frequent collaborators as ▷ James Mason, and ▷ Sean Connery, who has appeared in five of his films, most notably *The Hill* (1965) and *The Offence* (1972).

THE LUMIÈRE BROTHERS

Auguste Lumière

Born 19 October 1862, Besançon, France.
Died 10 April 1954, Lyons, France.

Louis Lumière

Born 5 October 1864, Besançon, France.
Died 6 June 1948, Bandol, France.

Employed at their father Antoine's photographic supplies factory at Lyons, the Lumière brothers made significant advances in the development of dry photographic plates and later expanded the factory activities into the manufacture of photographic paper and roll film.

Louis was a trained scientist, whilst Auguste managed the business. A demonstration of ▷ Edison's Kinetoscope in Paris in 1894, and their father's retirement, prompted them to concentrate on the production and projection of moving film. In February 1895, they patented the Cinematographe, initially a camera and projector in one, which improved on previous work in the field with its use of sprocket holes and the addition of a metal claw system which moved the film along.

The first public performance of the Cinematographe took place on December 28th, 1895 at the Salon Indien in the Grand Cafe on the Boulevard des Capucines in Paris. Spectators saw a programme of ten films like *La Sortie des Usines* (1895) and *Le Déjeuner de Bébé* (1895) which captured fragments of everyday life. *L'Arroseur Arrosé* (1895) in which a gardener is soaked by a garden hose was a precursor of the earliest slapstick comedies.

Louis displayed little faith in the future of his invention and quickly ceased making films, later devoting himself to improving raw stock and the development of colour photography and stereoscopic cinema. Auguste founded the Clinique Auguste Lumière and devoted himself to research work in the fields of biology and physiology.

LUPINO, Ida

Born 4 February 1914 (or 1918), London, UK.

The daughter of popular comedian Stanley Lupino (1893–1942), Lupino was born into a strongly theatrical family, trained at RADA and was still a teenager when she made her leading role debut in *Her First Affaire* (1932).

Working prodigiously in the British cinema of the period, she then signed a contract with Paramount and moved to Hollywood in 1933. Her early years there were spent in a succession of thankless roles as decorative ingenues; she had to fight for the part of the scheming tramp in *The Light That Failed* (1939) and surprised many with the complexity of her characterisation.

The authority of her adulterous murderess in *They Drive By Night* (1940) confirmed her new stature and she signed a long-term contract with Warner Brothers. Although featured in comedies and musicals, she left an indelible impression delineating the darker side of the female psyche in a series of roles where inner torment, repression and malevolence were expressed with an intensity and energy that bordered on hysteria. Her most memorable roles include the gangster's moll in *High Sierra* (1941), the murderous housekeeper in *Ladies in Retirement* (1941), the compulsively ambitious sister in *The Hard Way* (1943) and the torch singer in *Road House* (1948).

She left Warner Brothers in 1947 and subsequently formed her own company, producing and co-writing *Not Wanted* (1949) which she also directed when the original director fell ill. She continued to act but increasingly devoted her energies to direction, tackling the issue of rape in *Outrage* (1950) and bigamy in *The Bigamist* (1953).

In 1952, with ▷ Charles Boyer, ▷ David Niven and Dick Powell (1904–

63), she formed Four Star Productions and worked extensively for television, directing the series *On Trial* (1956) and appearing in *Mr. Adams and Eve* (1957–58) with her third husband Howard Duff (1917–90).

She directed one more feature, *The Trouble With Angels* (1966), and ended her acting career with roles in such television movies as *Women in Chains* (1972) and *The Letters* (1973) and supporting assignments in such cinema releases as *Junior Bonner* (1972), *Food of the Gods* (1976) and *My Boys Are Good Boys* (1978).

LYNCH, David K.
Born 20 January 1946, Missoula, Montana, USA.

Described by ▷ Mel Brooks as 'Jimmy Stewart from Mars', Lynch's boyish, courteous demeanour conceals one of America's most darkly original filmmakers in much the same way that the idyllic, picket-fenced surfaces of his fictional small towns are prone to reveal a subculture of sexual and criminal intrigue.

Graduating from Washington D.C.'s Corcoran School of Art in 1964, he travelled briefly in Europe before studying painting at the School of the Museum of Fine Arts in Boston and at the Pennsylvania Academy of Fine Arts in Philadelphia (1965–69).

He began to work in film with *The Alphabet* (1967) and subsequently enrolled as a fellow in the Centre for Advanced Film Studies at the ▷ American Film Institute in Los Angeles where he made the short film *The Grandmother* (1970), in which an emotionally deprived and abused little boy grows a grandmother from seed.

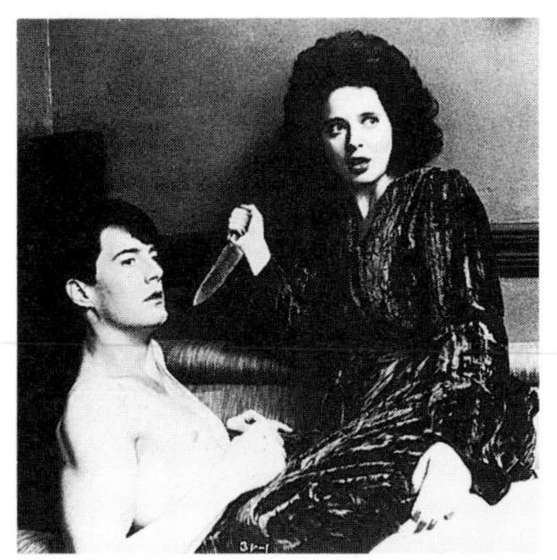
Kyle MacLachlan and Isabella Rossellini in *Blue Velvet* (1986) directed by David Lynch

He then spent a number of years at work on the startling feature *Eraserhead* (1976), a murky, monochrome urban nightmare that paints a visual and aural picture of decay as a backdrop to the bizarre story of a sensitive daydreamer burdened with a mewling mutant baby and variegated offspring that he disposes of by hurling against a wall.

Allied to the more conventional narrative of carnival freak John Merrick's life story in *The Elephant Man* (1980), he produced a grimy, threatening Dickensian view of Victorian England and a moving testimony to the dignity of the human spirit that earned him a Best Director Oscar nomination.

His bloated, big-budget version of *Dune* (1984), whilst ripe with visual delights and grotesque characters, sank in long-winded incoherence and was a major box-office flop. He was Oscar nominated again for *Blue Velvet* (1986), a remarkable surrealistic thriller in which a young man's discovery of a severed human ear leads him to uncover a fetid underworld of corruption, violence and a perversely appealing sadomasochistic relationship between an obscenity-spewing psychotic hoodlum and a troubled singer.

After a lengthy absence, during which he masterminded the cult television series ▷ *Twin Peaks* (1989–), he returned with *Wild at Heart* (1990) a love story and perverse homage to ▷ *The Wizard of Oz* that recounts the high-octane relationship between a two-fisted Elvis devotee in a personality-defining snakeskin jacket and his ever-lovin' sex-crazed good time gal. A mixture of heightened realism, visual alchemy and outrageous characterisations it is a virtuoso art of celluloid creation that became one of the more sensationalistic talking points of the film year.

M
Fritz Lang (Germany 1931)
Peter Lorre, Gustaf Grundgens, Ellen Widman, Inge Landgut.
118 mins. b/w.

While the later crime thrillers of his

American tenure, *The Woman in the Window* (1944) and *The Big Heat* (1953) among them, were to focus on dark cityscapes populated perhaps almost entirely by psychopathic felons and disaffected innocents, the finest expression of ▷ Fritz Lang's uniquely pessimistic vision could well be *M*, the last film he made in Germany before fleeing the Nazi regime in 1933. Blending ▷ Expressionist and Realist styles, this chilling account of a murderous paedophile is set in troubled Weimar Germany just after World War I. The central figure of Peter Lorre's (1904–64) Hans Beckert, the petit bourgeois who turns uncontrollable serial killer, emphasises the tension between weakening order and burgeoning chaos that marked the political events of the era, a period whose decadence, depression and confusion was eventually to give rise to the dominance of Hitler, *M* marked the first time Lang had worked with sound, but his mastery of the new medium is evident in the way he pushes the narrative along with both visual and aural information – it takes a blind flower seller, for instance, to recognise the murderer's whistling a snatch of Grieg, while the chalking of the letter M on his hand indicates the guilty party to his trackers – and in the innovative use of overlapping dialogue to draw parallels between both the police and the local underworld who are each attempting to apprehend the fiend in their midst. Moving towards a memorable climax Lang has the trapped Beckert, claiming that he's simply unable to help himself, pleading for clemency to the assembled gallery of ne'er-do-wells before the police finally arrive on the scene – typically, the lines between innocence and guilt, law and disorder have become more blurred than ever.

MACKENDRICK, Alexander (or Sandy)

Born 1912, Boston, Massachusetts, USA.

Born to Scottish parents and educated at the Glasgow School of Art, Mackendrick worked as an animator and commercial artist before joining the script department at Pinewood where he wrote *Midnight Menace* (1937).

During the Second World War, he made propaganda films for the Ministry of Information and worked in Rome on documentaries and newsreels. He joined Ealing Studios in 1946 and made his directorial debut with ▷ *Whisky Galore* (1948). His work at Ealing, including *The Man in The White Suit* (1951), *The Maggie* (1954) and *The Ladykillers* (1955), illustrates his deft comic touch, interest in multi-character stories and ability to weave social criticism and a cynical view of the world into ostensibly populist slices of entertainment.

A gifted director of children, he achieved exemplary performances from his young actors in *Mandy* (1952), *Sammy Going South* (1962) and the subtle pirate adventure *High Wind in Jamaica* (1965).

In 1956, he signed a contract with ▷ Burt Lancaster's production company and made ▷ *Sweet Smell of Success* (1957). A corrosive portrait of greed, venality and corruption in the urban jungle, it features a memorable jazz soundtrack, atmospheric visuals of New York at night, an unrelentingly bitter view of human nature, matchless dialogue and excellent performances from Lancaster as an all-powerful gossip columnist and Tony Curtis (1925–) as a self-serving, smarmy, small-time press agent. It remains the director's finest achievement.

He was dismissed from Lancaster's production of *The Devil's Children* (1959) but worked with Curtis again on the breezy satirical comedy *Don't Make Waves* (1967). Thereafter, he retired from active filmmaking, choosing to lecture at the California Institute of Arts where he was Dean of the Film Department from 1969 to 1978.

MACLAINE, Shirley

Real Name Shirley MacLean Beaty
Born 24 April 1934, Richmond, Virginia, USA.

MacLaine took dancing lessons from the age of three to strengthen her weak ankles and entered showbusiness whilst still a teenager when she applied for a job in a New York revival of *Oklahoma* (1950) and was accepted for the chorus. Over the next few years, she worked on television, as a model, and in the chorus of *Me and Juliet* (1952). An understudy on

The Pajama Game (1954), she moved centre stage when the original star broke her leg. Acclaimed for her performance, she signed a long-term contract with producer Hal B. Wallis (1898–1986).

She made her film debut in Alfred Hitchcock's *The Trouble With Harry* (1955) and made her first appearance for Wallis as a romantic foil for ▷ Jerry Lewis in *Artists and Models* (1955). Adept at light comedy, her star rose when she was cast in the epic adventure *Around the World in 80 Days* (1956).

In an era populated by pneumatic blondes, tomboyish teenagers or genteel ladies, she became an unconventional leading lady; her impish good-humour and waif-like manner all the more appealing because of their freshness and frankness. Evidence of her dramatic potential came with a touching portrayal of a cheerful, resilient floozie in *Some Came Running* (1958), which earned her a Best Actress Oscar nomination.

Her growing status was confirmed with the popularity of the bright comedy *Ask any Girl* (1959) and further Oscar nominations for the bittersweet romance ▷ *The Apartment* (1960) and *Irma La Douce* (1963).

Throughout the 1960s she was often asked to carry some lavish and inappropriate large-scale productions that overstretched her talents and proved commercial flops, but she returned to favour with a dynamic, showstopping performance as a lovelorn prostitute in the innovative musical *Sweet Charity* (1968).

A disappointing reaction to her work in the contemporary drama *Desperate Characters* (1971) and the failure of her television series *Shirley's World* (1971–72) caused her to retreat from the screen to concentrate on political activity, writing and her record-breaking one woman song'n'dance show.

She returned again in excellent form as a housewife who has sacrificed her promising ballet career for the rewards of domesticity in *The Turning Point* (1977) (Oscar nomination) and finally won an Oscar for her amusing and broadly drawn portrait of a mother finding middle-aged love and heartbreak in the superior melodrama *Terms of Endearment* (1983).

In 1987, she portrayed herself in a television mini-series based on one of her autobiographical books *Out On A Limb*. Subsequent film appearances, including *Madame Sousatzka* (1988) and *Steel Magnolias* (1989) have indulged an unattractive propensity for eccentric, larger-than-life character acting, although she returned to a more fully-rounded role as the alcoholic mother in *Postcards from the Edge* (1990).

A strong and sincere believer in reincarnation and out of body experiences, her many books on the subjects include *You Can Get There From Here* (1975), *Dancing in the Light* (1985) and *Going Within* (1989). Her brother is the filmmaker ▷ Warren Beatty.

MCLAREN, Norman

Born 11 April 1914, Stirling, UK.
Died 1987.

A student of interior design at the Glasgow School of Art, McLaren became an avid cineaste through his involvement with the school film society; it was his exposure to Russian cinema and German abstract animation that led to his own first endeavours like *Seven Till Five* (1934), *Camera Makes Whoopee* (1934) and *Hell Unlimited* (1934), a mixture of live-action and drawn and object animation.

He also made live-action advertising shorts for a local butcher and painted directly on the surface of film stock for *Hand-Painted Abstraction* (1935). His work brought him to the attention of ▷ John Grierson who employed him at the GPO Film Unit on a series of documentaries and *Love on the Wing* (1939), which was drawn directly with pen and ink frame-by-frame on to raw film stock.

In 1939, he moved to New York and continued to experiment with his hand-drawn technique, discovering that music could be made by marking the soundtrack area of a film with pen or ink.

In 1941, he emigrated to Canada and worked with Grierson at the National Film Board on animated war efforts like *V for Victory* (1941), *Dollar Dance* (1943) and *Keep Your Mouth Shut* (1944).

Running an animation department at the Board, his virtuosity led to *C'Est L'Avion* (1944), which uses a travelling zoom to create the impression of travelling down a river, *Le Haut Sur Les Montagnes* (1945), which creates a chiaroscuro effect on a pastel-drawn landscape, and further hand-drawn films like *Hoppity Hop* (1946), *Fiddle-de-Dee* (1947) and *Begone Dull Care* (1949), an almost totally frameless film that matches a riot of colour with the piano jazz of Oscar Peterson.

After visiting China, he worked in ▷ 3-D with *Around Is Around* (1952) an anti-War fable made in his 'pixillation' technique of animating human actors.

He continued to work in a diversity of forms including paper cut-out films like *Alouette* (1944), *Rhymetic* (1956) and *Le Merle* (The Blackbird) (1958), the 16mm totally abstract *Serenal* (1959) and *Lines Vertical* (1960) in which vertical lines were animated by engraving directly with a stylus and ruler onto black film. The latter technique was further explored in *Lines Horizontal* (1961) and *Mosaic* (1965).

Towards the end of his career, he made further advances in matching sound and visuals with *Synchromy* (1971), completed a series of animation instruction films *Le Mouvement Image Par Image* (1976–78) and created a trio of live-action dance studies that culminated in *Narcissus* (1981–83). He retired in 1983.

MCQUEEN, Steve

Full Name Terence Steven McQueen
Born 24 March 1930, Slater, Missouri, USA.
Died 7 November 1980, Juarez, Mexico.

A rough and tumble background, including a spell in a reform school and a stint in the marines, eventually brought Steve McQueen to New York where, in between jobs as a bartender, docker and bookie's runner, he studied at the Neighbourhood Playhouse and Uta Hagen School.

He appeared in summer stock and made his Broadway debut in *The Gap* (1954), later taking over from Ben Gazzara (1930–) in *A Hatful of Rain* (1956).

Following numerous television appearances, he made his film debut as an extra in *Somebody Up There Likes Me* (1956) and soon graduated to leading roles in lesser fare like *Never Love A Stranger* (1958) and *The Blob* (1958).

A highly-rated western television series *Wanted–Dead Or Alive* (1958) brought him to national prominence and led to a co-starring role in *The Magnificent Seven* (1960).

A relaxed and natural-seeming screen leading man with wintry blue eyes and a coolness all of his own, he created an indelible image as a taciturn loner, chivalrous, aloof and a master of his own destiny. His most successful roles include the indomitable p.o.w. in *The Great Escape* (1963), the gambler in *The Cincinnati Kid* (1965) and the jaded cop in the vastly popular *Bullitt* (1968). He

Steve McQueen in *The Great Escape* (1963)

also displayed some sensitivity as an actor in less characteristic roles like the psychopathic wartime pilot in *The War Lover* (1962), the feckless jazz musician in *Love With the Proper Stranger* (1963) and as the cynical American marine in the epic *The Sand Pebbles* (1966), for which he received his only Best Actor Oscar nomination.

He was at the peak of his popularity as the devious business executive in *The Thomas Crown Affair* (1968), the violent fugitive in *The Getaway* (1972), the Devil's Island prisoner in *Papillon* (1973) and the resolute fireman in *The Towering Inferno* (1974).

He more or less abandoned his acting career during a marriage to actress Ali

McGraw (1939–) from 1973 to 1978, emerging only for a creditable attempt at Ibsen in *An Enemy of the People* (1977). He then shed much of the excess weight that he had accrued and returned to the kind of action man roles that his fans clearly preferred, bringing a weary dignity and charm to the Cavalry scout in the western *Tom Horn* (1980) and a real-life bounty hunter in the routine thriller *The Hunter* (1980), his final screen appearance.

His death came after a long struggle against cancer.

MAGNANI, Anna

Born 7 March 1908, Alexandria, Egypt.
Died 26 September 1973, Rome, Italy.

Raised in Rome, Magnani studied for the stage at the Corso Eleanora Duse at Santa Cecilia and began her career as a nightclub singer and theatrical performer. She is credited with an appearance in the film *Scampolo* (1927), but her major debut is more widely acknowledged as being in *La Cieca de Sorrento* (The Blind Woman of Sorrento) (1934).

Later, established as one of Italy's leading stage actresses, her film roles remained largely comedic in tone, and small or insignificant in scope, which left audiences unprepared for the force of her performance in ▷ Robert Rossellini's ▷ *Rome–Citta Aperta* (Rome–Open City) (1945). A key film in the ▷ neo-realist movement, set during the last days of the German occupation of Italy, it won her worldwide acclaim for the apparent naturalness and authenticity she displayed as a pregnant woman shot down by the Nazis for harbouring a member of the underground.

A matronly, earthy figure, her style of acting consisted of a passion and overwhelming vitality that were then widely welcomed as the antithesis of Hollywood glamour.

Consolidating her European stardom, she subsequently won the Best Actress Prize at the Venice Film Festival for *L'Onorevole Angelina* (Angelina) (1947) as an indomitable Rome housewife battling for better conditions for her tenement neighbours. She was reunited with Rossellini for *Amore* (1948) and other notable performances include the boisterous comedy *Bellissima* (1951), directed by ▷ Luchino Visconti, and ▷ Renoir's *La Corrozza D'Oro* (The Golden Coach) (1952) a celebration of her commedia dell'arte heritage.

She made her American debut in *The Rose Tattoo* (1955), especially written for her by Tennessee Williams (1911–83), and won a Best Actress Oscar for her fiery tempestuousness as a mournful widow, wooed and won by ▷ Burt Lancaster. She earned a further Oscar nomination for *Wild Is The Wind* (1957) and also starred in Williams's *The Fugitive Kind* (1960).

Her sorties from Italy were rare and, inexplicably, in the 1960s good roles on her home territory grew increasingly scarce. Her later stage work included *La Lupa* (1965) and *Medea* (1966). She made her last Hollywood appearance, as a typically larger-than-life villager, in *The Secret of Santa Vittorio* (1969) and Fellini's *Roma* (1972) was her final film. Her death from cancer provoked national mourning in Italy.

THE MAGNIFICENT AMBERSONS

Orson Welles (USA 1942)
Joseph Cotten, Dolores Costello, Anne Baxter, Tim Holt, Agnes Moorhead.
88 mins. b/w.

To follow the artistic triumph but disappointing box office reception of ▷ *Citizen Kane*, ▷ Orson Welles originally had in mind a screen version of Charles Dickens' *The Pickwick Papers* to star ▷ W. C. Fields. When the star's schedule rendered the project unfeasible Welles turned his attentions instead to the adaptation of Booth Tarkington's 1919 Pulitzer Prize-winning novel *The Magnificent Ambersons*, the story of an aristocratic family's decline and fall during the period of increasing mechanisation around the end of the nineteenth century. Eschewing the overt sentimentality of his source, Welles begins this chronicle of ill-fated romance between Amberson matriarch Isobel (Dolores Costello, (1905–79)) and debonair horseless carriage manufacturer Eugene Morgan (Joseph Cotten, (1905–)) as a jaunty picture of period foibles before darkening the tone to one of characteristically elegiac regret, a lament not so much for an era but for a whole set of values – although Welles's suavely omniscient narration retains some element of dispassionate distance on the dynasty's self-destruction. Having edited his completed version to 148 and then 131 minutes, Welles ventured off to Brazil to start work on the abortive *It's All True* project but, unfortunately, his absence coincided with a change of regime in the studio boardroom. Before long the new management – whose slogan 'Showmanship instead of genius: a new deal at RKO' was a clear declaration of intent – responded to an unfavourable preview of the film by cutting it to 88 minutes, inserting a mawkish final scene not even directed by Welles himself, and opening the result on the lower half of a double bill with the Lupe Velez (1908–44) vehicle *Mexican Spitfire Sees a Ghost.* His reflective masterpiece partially destroyed, the so-called 'genius''s turbulent relationship with the Hollywood system had only just begun.

MAKAVEJEV, Dusan

Born 13 October 1932, Belgrade, Yugoslavia.

Educated in psychology at Belgrade University and a student at the city's Academy for Theatre, Radio, Film and Television, Makavejev began his directorial career with the short *Jatagan Mala* (1953). Over the next decade, in between his military service and playwriting commitments, he learnt the craft and grammar of his chosen profession as the director of numerous experimental short films and documentaries, including *Slikovnica Pcelara* (Beekeeper's Scrapbook) (1958), *Sto Je Radnicki Savjet?* (What Is A Worker's Council?) (1959) and *Ljepotica 62* (Miss Yugoslavia 62) (1962).

He made his feature debut with *Covek Nije Tica* (Man Is Not A Bird) (1965), an account of a hairdresser's casual affairs in a desolate industrial town. This was followed by the more provocative *Ljubavni Slucaj* (The Switchboard Operator) (1967) in which a young woman pays for love with her life, and then the positively anarchic *Nevinost Bez Zastite* (Innocence Unprotected) (1968), a salute to the 1942 Serbian feature of that title combining stock footage, interviews and newsreels.

Displaying a particularly mordant, idiosyncratic black humour and a penchant for physical explicitness, he has been interested in dissecting the emotions of people both liberated and destroyed by their loss of sexual inhibitions. His works reveals him as an unorthodox mosaic-maker, reliant on odd juxtapositions of apparently unrelated images.

W.R.: Mysteries of the Organism (1971), his best known film, was an abrasive mixture of documentary and drama, exploring the radical theories of sexologist Wilhelm Reich, that also condemns Stalinism and the hollowness of American sexual politics. The frequently banned *Sweet Movie* (1974) brutally contrasts the dehumanising of a Miss World with an allegorical tale of a ship under the command of a whore who murders her many lovers. More accessible was *Montenegro* (Or Pigs and Pearls) (1981), his most successful attempt to make a readily comprehensible conventional narrative as a bored middle-class housewife rediscovers sexual passion at a boisterous bar frequented by immigrant workers.

Changing times have inevitably blunted the iconoclasm of his style and concerns; recent work like *The Coca-Cola Kid* (1985) and *Manifesto* (1988) have displayed only distant echoes of his dark comic inventiveness and ability to affront.

MALLE, Louis

Born 30 October 1932, Thumeries, France.

Announcing his intention to become a filmmaker at the tender age of 13, Malle received a resounding slap across the face from his outraged mother. His resolve was only stiffened and, after studying at the Institut des Hautes Etudes Cinematographiques, he embarked upon a career that has few rivals in terms of its richness and diversity.

An assistant and cameraman to Jacques-Yves Cousteau for two years, he co-directed *Le Monde du Silence* (The Silent World) (1956) with him. He worked as an assistant on ▷ Robert Bresson's *Un*

Condamné a Mort c'est Echappé (A Man Escaped) (1956) and was a cameraman on ▷ Jacques Tati's *Mon Oncle* (My Uncle) (1958) before making his first striking contributions to the ▷ nouvelle vague with the jazzy, atmospheric thriller *Ascenseur Pour L'Echafaud* (Lift to the Scaffold) (1957), and the then scandalous erotic satire *Les Amants* (The Lovers) (1958), both of which established ▷ Jeanne Moreau as a stellar attraction.

Throughout his career, he has shown a sensitivity to the problems of adolescence, an understanding of world affairs and a desire to express the sensuality he discerns in most topics. Perceptive in his choice of actors, he used ▷ Brigitte Bardot effectively as the harried film star in the autobiographical *Vie Privée* (A Very Private Affair) (1961) and teamed her with Moreau in the popular western spoof *Viva Maria!* (1965).

Disenchanted with the filmmaking process by the late 1960s, he left France and his restless curiosity took him to Calcutta, Delhi and Bombay and resulted in a television series and feature-length documentary *Phantom India* (1969).

He returned to drama with *Le Souffle Au Coeur* (Dearest Love) (1971) a sensitive story of adolescent anguish and incest that earned him an Oscar for the screenplay. Claiming to find an attraction in disturbing topics, he showed a similarly discreet and sympathetic handling of such potentially exploitative subjects as the guilt-ridden legacy of wartime collaboration in *Lacombe Lucien* (1974) and child prostitution in *Pretty Baby* (1978).

His American career was distinguished by *Atlantic City* (1980), featuring a glorious performance from ▷ Burt Lancaster as the has-been who never was; he swayed unpredictably from documentaries to such features as *Crackers* (1983) and the conversation piece *My Dinner With Andre* (1981).

Returning to France, he was widely honoured for *Au Revoir Les Enfants* (1987), the moving wartime memory of his tragic friendship with a Jewish schoolboy betrayed to the Nazis, and he continued to mine autobiographical elements of his past with the mild *Milou en Mai* (1990) vaguely set against the revolutionary fervour of 1968 in which Malle himself had enthusiastically participated.

He has been married to the actress Candice Bergen (1946–) since 1980.

THE MALTESE FALCON

John Huston (USA 1941)
Humphrey Bogart, Mary Astor, Sidney Greenstreet, Peter Lorre, Elisha Cook Jr.
100 mins. b/w.

Few today recall the first filming of Dashiell Hammett's 1929 crime thriller *The Maltese Falcon* two years later under the same title by Roy Del Ruth (1895–1961) or its 1936 incarnation as William Dieterle's (1893–1972) *Satan Met A Lady*, yet ▷ John Huston's scintillating debut with his 1941 version of *The Maltese Falcon* has long been established as a classic of Hollywood cinema. Rewarded with a shot behind the camera after many years service as a reliable screenwriter, Huston took his chance to turn in a model of narrative concision that also bore out his maxim that 'the trick is in the casting'. Here ▷ Humphrey Bogart grasped the leading role that George Raft (1895–1980) turned down, and as Sam Spade created the archetypal trench-coated private dick, a sardonic yet romantic anti-hero that established his screen persona for years to come, the experienced Mary Astor (1906–87) exuded alluring maliciousness, and the pairing of Hungarian-born Peter Lorre (1906–64) and 62 year-old English stage actor Sydney Greenstreet (1879–1954) created a delicious partnership in crime that was to run throughout the Forties. While the dark wit, constant deceit and rampant (perhaps justifiable?) paranoia combined with the downbeat urban setting and brooding presentation to trace the first steps into the ▷ film noir genre that was to become increasingly baroque as the decade wore on, the subject matter – a motley crew of diverse types greedily pursue a fabulous prize in the shape of a fabled jewelled falcon, only to discover that the treasure is to prove illusory – offered the first exploration of territory to which Huston, in the likes of *The Treasures of the Sierra Madre* (1947) and *The Asphalt Jungle* (1950), would fruitfully return.

MAMOULIAN, Rouben
Born 8 October 1898, Tiflis, Georgia, Russia.
Died 4 December 1987, Woodland Hills, California, USA.

A student at the University of Moscow and in Paris, Mamoulian began his stage career by directing in London in 1920. Later, he moved to New York, achieving the degree of Broadway renown that led to an offer of work from Paramount.

His first films as director, *Applause* (1929) and *City Streets* (1931) are notable for his early liberation of the camera and innovative use of sound for dramatic effect. He would remain a technical pioneer and an artist dedicated to finding a rhythm and poetry in the mechanics of filmmaking. He was also a sensitive actor's director as evidenced by the outstanding performances of ▷ Fredric March in *Dr Jekyll and Mr Hyde* (1931) and Greta Garbo in ▷ *Queen Christina* (1933).

In 1935 he directed ▷ *Becky Sharp*, the first feature to utilise the three-strip ▷ Technicolor process, and he proved his versatility by masterminding a number of popular entertainments including the musical *High Wide and Handsome* (1937), *Golden Boy* (1939) which introduced William Holden (1918–81) to the screen, and the exhilarating swashbuckler *The Mark of Zorro* (1940).

He maintained a parallel stage career, directing the original productions of *Porgy and Bess* (1935), *Oklahoma!* (1943) and *Carousel* (1945).

After resigning from *Laura* in 1944 he completed only two further films, the colourful musical *Summer Holiday* (1948) and the elegant *Silk Stockings* (1957), a musical remake of Ninotchka (1939).

Fired from the filming of *Porgy and Bess* in 1958, he resigned from the production of *Cleopatra* in 1960 and never worked for the cinema again. He remained active in the theatre, directing his own adaptation of *Hamlet* for the University of Kentucky in 1966.

THE MAN FROM UNCLE
USA 1964–8

This cultish, part-spoof spy drama was conceived over a lunch in London. In attendance were three men looking to capitalise on the success of the James Bond films – producer Norman Felton, Bond creator Ian Fleming and an unnamed NBC executive. Fleming's contribution was the surname Solo, borrowed from a character in *Diamonds Are Forever*, to which Felton added Napoleon. Incredibly, with little more to go on, NBC bought the idea. The concept was fleshed out by Sam Rolfe, who added the show's sense of self-mockery (he had previously worked on *Have Gun Will Travel*). Rolfe wrote the pilot, and the series, at this stage still called *Solo* – was confirmed.

The easy-going secret agent Solo (Robert Vaughn (1932–)) was joined by the distant, polo-neck wearing Illya Kuryakin (David McCallum (1933–)), signalling that the show was to combine, with its espionage theme, elements borrowed from buddy movies. Unusually, the battle lines between good and evil were drawn along post-Cold War lines. Behind the shopfront of a Manhattan tailor lurked the futuristic headquarters of UNCLE (United Network Command for Law Enforcement), the suave, sophisticated grouping which was in constant battle with the criminal organisation THRUSH (initially to be called WASP, the name was changed to avoid causing offence to White Anglo-Saxon Protestant viewers). A further refinement of the Bond idea was the introduction of a 'normal' person – a trucker or a housewife, for example – into each adventure, making the underworld seem more accessible. The show's hardware was especially inventive – UNCLE communicated through radio-receiver pens, while THRUSH waged their struggle with vapourising machines, ageing chemicals and will gases.

Such gimmickry heightened the show's appeal, half a million Americans joined the UNCLE fan club at its peak, proudly wearing badges saying 'Flush THRUSH' and 'All the way with Illya K' (an adapted Kennedy campaign slogan, which underlined the utopian era in which it was launched). Among the people involved were writer ▷ Robert Towne (who later scripted *Chinatown*) and director Richard Donner (1939–). A sequel series *The Girl From UNCLE*, with Stefanie Powers (1942–) in the title role, was broadcast in 1967.

The success of the series in Britain resulted in a number of spin-off films released to cinemas including *One Spy Too Many* (1966), *The Karate Killers* (1967) and *The Helicopter Spies* (1968).

In 1983 Vaughn and McCallum were reunited for the television film *The Return of the Man from UNCLE.*

MANKIEWICZ, Joseph Leo

Born 11 February 1909, Wilkes-Barre, Pennsylvania, USA.

A reporter for *Variety* in Berlin in 1928, Mankiewicz used his skill with words to gain employment at the U.F.A. Studios as a translator of silent movie title cards into English. Back in America the following year, his brother Herman (1897–1953) found him a job as a junior writer at Paramount where he contributed dialogue or title cards to such films as *Fast Company* (1929), *Only Saps Work* (1930) and *The Gang Buster* (1931).

As a screenwriter, he collaborated on many successful scripts including *If I Had A Million* (1932), *Alice in Wonderland* (1933) and *Manhattan Melodrama* (1934). Under contract to M-G-M from 1933, he became a producer with a distinguished track record that includes *Fury* (1936), *Three Comrades* (1938), ▷ *The Philadelphia Story* (1940) and *Woman of the Year* (1942).

He made a belated directorial debut with the Gothic chiller *Dragonwyck* (1946) and achieved a unique feat as a double Oscar winner for the script and direction of both *A Letter To Three Wives* (1949) and ▷ *All About Eve* (1950). Grandly-acted, compelling narratives, they are distinguished by diamond-sharp dialogue and verbal sparring that is dipped in acid, witty and urbane.

He then tackled a variety of ambitious and literate projects including the creditable *Julius Cesar* (1953), *The Barefoot Contessa* (1954), the musical *Guys and Dolls* (1954) and *Suddenly Last Summer* (1959).

The prolonged and frustrating production history of *Cleopatra* (1963) was hardly justified by the tedium of the end results and he withdrew from filmmaking for a number of years. He returned to form with the cheerfully amoral western *There Was A Crooked Man* (1970) and a masterly adaptation of *Sleuth* (1972) that perfectly suited the grandiloquent theatricality of his style and earned him another Best Director Oscar nomination.

Despite the announcement of further projects, he has not filmed since. A noted wit and raconteur with a fund of insightful anecdotes from Hollywood's Golden Age, he also directed the television production *Carol For Another Christmas* (1964).

MANN, Anthony

Real Name Emil Anton Bundsman
Born 30 June 1906, San Diego, California, USA.
Died 29 April 1967, Berlin, Germany.

Educated in New York, Mann went directly from school into the theatre, progressing from production manager to director in the 1930s before ▷ David O. Selznick hired him as a talent scout and screen test director in Hollywood.

An assistant to ▷ Preston Sturges and others at Paramount, he made his directorial debut with the B picture mystery *Dr. Broadway* (1942) and continued to work on low-budget ventures before making his name on stylish, punchy crime thrillers like *T-Men* (1947) and *Raw Deal* (1949).

Winchester '73 (1950) began a fruitful association with actor ▷ James Stewart that helped to revitalise the western genre. Making dramatically expressive use of landscape, his westerns centred on desperately driven protagonists seeking revenge or an escape from a less than honourable past. Achieving redemption through brutal confrontations with intriguingly evil alter egos, Mann's characters blur the traditional psychological distinctions between hero and villain and display a propensity for unusually harsh violence that would influence directors like ▷ Sam Peckinpah. His best westerns include *The Naked Spur* (1953), *The Man from Laramie* (1955) and *Man of the West* (1958).

He also worked with Stewart on more conventional material, including the warmhearted musical biography *The Glenn Miller Story* (1953).

In the 1960s, he turned his attention to intelligent historical epics like *El Cid* (1961) and *The Fall of the Roman Empire*

(1964) and died whilst filming *A Dandy in Aspic* (1968), which was completed by its star Laurence Harvey (1928–73).

MARCH, Fredric
Real Name Frederick Ernest McIntyre Bickel
Born 31 August 1897, Racine, Wisconsin, USA.
Died 14 April 1975, Los Angeles, California, USA.

One of the most respected of screen actors, March began his career with small stage roles, modelling work and as an extra in such films as *Paying the Piper* (1921) and *The Great Adventure* (1921).

Stage success in Los Angeles brought him a long-term contract with Paramount and he established a reputation as a handsome, dependable and somewhat self-effacing leading man. Recreating his boisterous stage impersonation of ▷ John Barrymore in *The Royal Family of Broadway* (1930) earned him a Best Actor Oscar nomination and he won the award for his effective dual characterisation as *Dr. Jekyll and Mr Hyde* (1931).

A popular casting choice for costume epics, he gave a commanding performance as Valjean in *Les Miserables* (1935) and his unaffected naturalism and emotional honesty greatly enhanced the role of the fading movie idol in ▷ *A Star Is Born* (1937) (Oscar nomination). A diligent performer, equally at home in swashbuckling adventure or drawing-room drama, he also displayed a flair for comedy in films like *Nothing Sacred* (1937), *Bedtime Story* (1941) and *I Married A Witch* (1942).

He won a second Oscar for ▷ *The Best Years Of Our Lives* (1946), credible and moving as the wartime veteran readjusting to civilian life. His performance as Willy Loman in *Death of a Salesman* (1952) brought a further Oscar nomination and, despite a tendency to overact in films like *Inherit the Wind* (1960), he proved an incisive character actor in films like *Seven Days in May* (1964) and *Hombre* (1967). He made his final appearance in *The Iceman Cometh* (1973).

Stage work over the years, often in partnership with his second wife Florence Eldridge (1901–88), includes *The Skin Of Our Teeth* (1942), *Years Ago* (1946) and an acclaimed production of *Long Day's Journey Into Night* (1956).

THE MARCH OF TIME
Louis de Rochemont (USA 1935–51)
Narrator Westbrook van Voorhis
Each monthly issue circa 20 mins. b/w.
Academy Award: Special Oscar in 1936 for having revolutionised one of the most important branches of the industry, the newsreel.

An offshoot of the magazines *Time* and *Life* and their radio programme of the same name, from February 1935 onwards *The March of Time*'s monthly issues were the cinematic precursor of today's current affairs television. Supervised by the committed hand of producer Louis de Rochemont (1899–1978), their twenty-minute sections fell somewhere between the newsreel and the feature documentary, a consistent structure – exposition of a topical issue, examination of its historical causes, exploration of its current ramifications and prognostication of future prospects – seeking not only to report newsworthy events but to look at the conflicts that created them. Using a stentorian narrator to lend weight to the highly assertive commentary and often reconstructing situations for the benefit of the camera, the series reached a huge audience all over the US and beyond, and while its liberal tendencies tended to be only intermittently forthright when dealing with domestic American matters it brought a campaigning edge to its coverage of Thirties international affairs – swift to focus attention on the activities of Stalin and Mussolini, 1938's instalment *Inside Nazi Germany* was a detailed look at Hitler's propaganda machine that came at a time of pronounced US isolationism and was banned in Britain because Germany at the time was still theoretically a friendly power. De Rochemont himself left the unit in the mid-Forties to produce a trio of fictionalised semi-documentaries, of which Henry Hathaway's (1898–1985) Nazi conspiracy exposé *The House on 42nd Street* (1945) was the first and the most influential on Hollywood's eventual motion towards more extensive use of location shooting.

MARTIN, Steve

Born 14 August 1945, Waco, Texas, USA.

At the age of ten, Martin moved to Garden Grove in Orange County, California which is only miles removed from Disneyland and Knott's Berry Farm where he would spend much of his youth as a guide, magician and neophyte performer.

Steve Martin and Daryl Hannah in *Roxanne* (1987)

A major in philosophy at Long Beach College, he then studied theatre at UCLA and began perfecting his stand-up comedy act in a variety of nightclubs. He subsequently left college to become a comedy writer for television, winning an Emmy Award for *The Smothers Brothers Comedy Hour* (1968) and a nomination for *Van Dyke and Company* (1975).

He then perfected his live performances until his popularity was such that he could regularly sell-out venues of 10,000–20,000 seating capacity. On television he became a regular guest on *The Tonight Show* and a frequent host of *Saturday Night Live*; he won consecutive Grammy Awards for his albums *Let's Get Small* (1977) and *Wild and Crazy Guy* (1978) which sold in excess of seven million copies, and his first book, *Cruel Shoes* (1979), reached number seven in the New York Times bestseller list.

He made his cinema debut in *The Absent-Minded Waiter* (1977) which received an Oscar nomination for Best Short Film, and, after cameo roles in *Sgt. Pepper's Lonely Hearts Club Band* (1978) and *The Muppet Movie* (1979), he took the title role in the proudly asinine comedy *The Jerk* (1979).

A performer of great physical and verbal dexterity, manic inventiveness and skilled characterisation, his early films are rudely healthy mixtures of crazed fantasy, non-sequitur jokes, vulgarity and downright idiocy. Notable titles include *Dead Men Don't Wear Plaid* (1982) and *The Man With Two Brains* (1983).

He displayed different sides of his talent in the lavish dramatic musical *Pennies from Heaven* (1981) and the more conventional comedy *The Lonely Guy* (1983) and won the New York Film Critics' Best Actor Award for his performance of inspired lunacy in *All of Me* (1984).

He has continued to provide joyous moments of unadulterated nonsense in films like *Little Shop of Horrors* (1986) and *Dirty Rotten Scoundrels* (1988) but has broadened his appeal with less hysterical and more character-based work in such popular successes as *Plains, Trains and Automobiles* (1987) and *Roxanne* (1987), a witty and inventive contemporary reinterpretation, which he also wrote, of *Cyrano de Bergerac*.

In 1988 he appeared on the New York stage with ▷ Robin Williams in *Waiting for Godot*. Returning to the cinema, his role as the endearingly over-anxious father in *Parenthood* (1989) accorded him his biggest box-office hit to date and he followed this with the comedies *My Blue Heaven* (1990) and *L.A. Story* (1991), which he also wrote.

He has been married to British actress Victoria Tennant (1946–) since 1986.

THE MARX BROTHERS

Chico

Born **Leonard Marx**, 22 March 1886, New York City, USA.
Died 11 October 1961, Los Angeles, California, USA.

Harpo

Born **Adolph Marx**, 23 November 1888, New York City, USA.
Died 28 September 1964, Los Angeles, California, USA.

Groucho

Born **Julius Henry Marx**, 2 October 1890, New York City, USA.
Died 19 August 1977, Los Angeles, California, USA.

Zeppo

Born **Herbert Marx**, 25 February 1901, New York City, USA.

Died 30 November 1979, Palm Springs, California, USA.

The nephews of vaudevillian Al Shean (1868–1949), the brothers were encouraged to enter showbusiness by their mother Minnie. Singers and comedians, their early acts included The Four Nightingales and, with the addition of mum and an aunt, the Six Musical Mascots. By 1914, when they toured in *Home Again*, they were billing themselves as 'The Greatest Comedy Act in Showbusiness: Barring None'.

They left vaudeville for the legitimate theatre and long runs allowed them to finely-polish their technique so that by the time *I'll Say She Is* (1924) reached New York after two years on the road it was a well-honed comic extravaganza that transformed them into instant cult attractions. This was followed by equally popular stage shows *The Cocoanuts* (1925) and *Animal Crackers* (1928).

In 1926 they made their film debuts in the privately produced *Humorisk* which was never publicly released. In 1929, they signed a three-picture contract with Paramount, making their official debuts in a film version of *The Cocoanuts* (1929).

An anarchic quartet, whose humour was physically daring, anti-authoritarian, surreal and often risqué, they had developed well-defined individual characters. Cigar-chomping and falsely moustachioed Groucho dispensed lethal wisecracks and verbal reparteé whilst loping through a larcenous scheme or inviting boudoir and managed to woo and offend a redoubtable collection of duchesses and dowagers portrayed by that fine foil Margaret Dumont (1889–1965). Chico offered eccentric piano-playing whilst destroying the English language with his Italian-American puns, and the divinely dumb, harp-playing Harpo sported a fright wig and startled expression as he innocently pursued a destructive course of self-centred mischief and lust. By comparison Zeppo was a rather bland straight man who left the group after 1933 to become a theatrical agent. A fifth brother Gummo (Milton Marx) (1893–1977) was only a performer in the early days of the group and later served as their business manager and agent.

Their early film successes included *Monkey Business* (1931), *Horse Feathers* (1932) and *Duck Soup* (1933). After a short break, they moved to M-G-M, where the addition of musical members and a secondary love interest broadened their appeal and gave them their biggest commercial hits *A Night at the Opera* (1935) and *A Day at the Races* (1937).

Room Service (1938) was a strained attempt to confine them within an existing format and the material on offer in *Go West* (1940) and *The Big Store* (1941) was definitely inferior. After an absence, they re-appeared with flashes of their beloved inventiveness in *A Night in Casablanca* (1946) but a final team film *Love Happy* (1949) was rather derisory.

They appeared in the bizarre all-star *The Story of Mankind* (1957) with Groucho as Peter Minuit, Harpo as Sir Isaac Newton and Chico as a monk who advises Columbus. Their final joint appearance was in the television comedy *The Incredible Jewel Robbery* (1959).

Their individual careers were not insubstantial. Chico toured with his own band and played a straight role in a segment of *Playhouse 90* (1958). Harpo appeared in the film *Too Many Kisses* (1925) toured with his band, appeared in a memorable episode of *I Love Lucy* and wrote an excellent autobiography *Harpo Speaks!* (1961). Groucho enjoyed the greatest solo success, appearing in such films as *Double Dynamite* (1950), *A Girl in Every Port* (1952) and *Skidoo* (1968), co-writing the play *Time for Elizabeth* (1948), hosting the radio quiz show *You Bet Your Life* (1947–58) and its TV counterpart (1950–58) and publishing a variety of books from *Beds* (1930) to *Groucho and Me* (1959) and *Memoirs of a Mangy Lover* (1963). He also lived to see the Marx brothers lionised as universal cinema clowns and received a special Oscar in 1974 for 'the unequalled achievements of the Marx Brothers in the art of motion picture comedy'.

M*A*S*H

USA 1972–83
Alan Alda

When moviemaker ▷Robert Altman adapted Robert Hornberger's book about the exploits of a Mobile Army Surgical Hospital during the Korean War, few could have predicted that he was preparing the way for the benchmark comedy of the 1970s. The show, which began its run as the US began to realise the full horrors of the continuing Vietnam war, was a curious mix of comedy and serious drama. Using the same set as Altman's film, it chronicled the wartime efforts of a likeable collection of oddballs and eccentrics, among them Captain 'Hawkeye' Pierce (Alan Alda (1936–)), Trapper John (Wayne Rogers (1934–)), BJ Hunnicutt (Mike Farrell (1939–)), Margaret 'Hotlips' Hoolihan (Loretta Swit (1937–)) and the transvestite Klinger (Jamie Farr (1936–)).

Though funny, there was no laughter track, and a landmark show was an episode entitled 'Sometimes You Hear The Bullet' in which a friend of Hawkeye was seen to die on the operating table. CBS chiefs complained that the death was inappropriate, but eventually relented, and the tragic undertow became a regular feature. The show had high production values. Unusually for prime time, it devoted a whole day of each shooting schedule to rehearsals, and allowed multiple takes (using single film camera when multiple video cameras were the norm). The actors contributed to the story conferences, and had a hand in the development of their characters. Original writer Larry Gelbart (1925–) (who was lured away from Marty Feldman to script *M*A*S*H*) left in 1976, whereupon Alda took on the role of 'creative consultant', often writing and directing. Hawkeye became more central, and plots relied more on the banter between characters. The liberal message, which was always heavily cloaked, became less tangible as Vietnam became a memory. Having earned 14 Emmys, *M*A*S*H* bowed out with a feature-length episode which attracted 125 million viewers in the US. 'We came to tell jokes and stayed to touch the edges of art,' Alda has said, not entirely without foundation.

MASON, James Neville

Born 15 May 1909, Huddersfield, Yorkshire, UK.
Died 27 July, 1984, Lausanne, Switzerland.

A student in architecture at Marlborough College, Mason subsequently turned to acting, making a stage debut in *The Rascal* (1931) at Aldershot and appearing at the Old Vic and with the Gate Company in Dublin before his film debut in *Late Extra* (1935).

He was gainfully employed in a number of the ▷ quota quickies that then largely constituted the British film industry and attained stardom with portraits of suave, saturnine villainy in such popular costume melodramas as *The Man in Grey* (1943), *Fanny By Gaslight* (1944), *The Seventh Veil* (1945) and *The Wicked Lady* (1945).

He gave one of his finest performances as the hunted IRA gunman in *Odd Man Out* (1946) before moving to Hollywood where he struggled to establish himself on an equivalent footing. However, he won acclaim for his portrayal of Rommel in *The Desert Fox* (1951), as the spy in *Five Fingers* (1952) and as Brutus in *Julius Caesar* (1953), enjoyed box-office success with *20,000 Leagues Under the Sea* (1954) and received his only Best Actor Oscar nomination for his superb portrayal of the fading movie idol in ▷ *A Star Is Born* (1954).

A prolific, professional and always incisive performer, he brought distinction to even the poorest of screen fare and was a formidable character actor when faced with challenging material. Among his many notable credits are the drug-addicted father in *Bigger Than Life* (1956), the urbane villain in *North By Northwest* (1959), the hapless Humbert Humbert in *Lolita* (1962) and the north country businessman in *Georgy Girl* (1966) (Best Supporting Actor Oscar nomination).

He was acting until the last days of his life; highlights from he latter stages of his career include the persecuted teacher in *Child's Play* (1972), a delightful Doctor Watson to the Sherlock Holmes of Christopher Plummer (1927–) in *Murder By Decree* (1978), the silkily sinister lawyer in *The Verdict* (1982) (Oscar nomination) and the weary aristocrat in the underrated *The Shooting Party* (1984).

An occasional theatre performer, he made his Broadway debut in *Bathsheba* (1946) and also appeared in *The Faith Healer* (1978). His rare television work includes *Jesus of Nazareth* (1977). His autobiography, *Before I Forget* was published in 1981.

MASTROIANNI, Marcello
Born 28 September 1924, Fontana Liri, near Frosinone, Italy.

The son of a carpenter, Mastroianni trained as a draughtsman before his internment in a Nazi labour camp. A cashier in post-War Rome, he developed an interest in amateur dramatics as a hobby and made his film debut in *I Miserabli* (1947).

From 1948 he was employed by ▷ Luchino Visconti's theatrical troupe in productions of such contemporary American dramas as ▷ *A Streetcar Named Desire* and *Death of a Salesman* and worked industriously in the indigenous film industry even if few of the end results were deemed worthy enough to bother exporting.

Peccato Che Sia Una Canaglia (1955) and *La Bella Mugnaia* (1955) began an enduring and prolific partnership with actress ▷ Sophia Loren, whilst his performances in Visconti's *Le Notte Bianche* (White Nights) (1957) and ▷ Federico Fellini's ▷ *La Dolce Vita* (1960) made his name outwith Italy and he was soon fêted as an international star.

His career continued to embrace work of the highest quality and, although stereotypically perceived as the archetypal 'latin lover', he was capable of a wide diversity of characterisations that showed his subtle grasp of comedy, masterful mime-like use of his body and warm humanity. He received a Best Actor Oscar nomination for *Divorzio All'Italian* (Divorce-Italian Style) (1962) and was also noteworthy as the professor in *I Compagni* (The Organiser) (1963).

His few films in English, including the comedy *Diamonds for Breakfast* (1968) and the gloomy romance *A Time for Lovers* (1969), did not fare well and his Italian career appeared to flounder in the doldrums of conventional comedies and star vehicles until he starred as an aristocrat in *Allonsanfan* (1974) and received a further Oscar nomination as the homosexual radio journalist seduced by housewife Loren in *Una Giornata Particolare* (A Special Day) (1977).

He has subsequently brought a gaiety and grace to such roles as the aged dancer in *Ginger E Fred* (Ginger and Fred) (1986), the spineless, incurable romantic in *Oci Ciornie* (Dark Eyes) (1987) (Oscar nomination), the cinema owner in the nostalgic *Splendor* (1989) and the aged patriarch on a journey to visit the dispersed members of his family in *Stanno Tutti Benne* (Everybody's Fine) (1990).

One of Europe's busiest actors with over 130 roles to his credit, he will shortly be seen in *Verso Sera* (Toward's Evening) (1990) and *The Suspended Step of the Stork* (1991).

MATTE SHOT Utilised in special-effects work as a means of combining different images, a matte shot uses a mask, or 'matte', to partially restrict the image area exposed during filming or part of the aperture in the printer. Another image will subsequently be placed in the unexposed area to create a composite effect.

The matte may be a card or metal cut-out mounted in front of the lens during photography, or a strip of film with opaque and transparent areas used during printing.

A MATTER OF LIFE AND DEATH
Michael Powell and **Emeric Pressburger** (UK 1946)
David Niven, Roger Livesey, Kim Hunter, Marius Goring, Raymond Massey.
104 mins. b/w.

A sub-genre of Forties Hollywood cinema was the series of light comedies, including Alexander Hall's (1894–1968) *Here Comes Mr Jordan* (1941) and ▷ Frank Capra's ▷ *It's A Wonderful Life* (1946), in which the earthly protagonists undergo a change in their situations which they can only ascribe to the real or imagined intervention by messengers from heaven; yet more sophisticated and stylishly achieved than all the rest is the British ▷ Michael Powell and Emeric Pressburger (1902–88) picture *A Matter of Life and Death* (released in the US under the title *Stairway To Heaven*). In one of his finest roles, ▷ David Niven stars as Peter Carter, a World War II RAF pilot who falls in love with American radio operator June (Kim Hunter, (1922–)) over the airways as his plane is about to crash, but who survives a parachuteless fall to meet up with her before his fight against brain damage takes the form of an extravagantly fantasised courtroom drama. Moving between this world and the next, the lush Technicolor visions of the earthbound Home Counties and their evocatively monochrome heavenly plane are separated by a vast metaphorical escalator, and the crossing point between the two spheres becomes the locus for a court case to determine the young flyer's fate. Before an audience drawn from all nations, the forces debating the issues are Peter's recently deceased English doctor (Powell/Pressburger regular Roger Livesey, (1906–76)) and Raymond Massey's (1896–1983) defiantly anti-English 18th century American patriot, the outcome is not only to forge a propaganda plea for greater understanding between the two nations but an impassioned assertion of eternal love, a statement of faith in the individual, and a visual *tour de force* as well.

MAYER, Louis B(urt)
Born 4 July 1885, Minsk, Russia.
Died 29 October 1957, Los Angeles, California, USA.

Brought to St John, New Brunswick at the age of three, Mayer was working in the junk and scrap metal business at the age of eight and later expanded his activities to ship salvaging.

In 1907, he purchased a house in Haverhill, Massachusetts, refurbished it as a ▷ nickelodeon called 'The Gem' and opened one of the earliest custom-designed cinemas. He subsequently acquired a chain of theatres in New England and displayed shrewd knowledge of commercial and artistic achievement by buying the regional rights to such popular attractions as ▷ *Birth Of A Nation* (1915).

He later moved in to film production with the formation of Metro Films (1915) and Louis B. Mayer Productions (1917), which later joined with ▷ Sam Goldwyn to form Metro-Goldwyn-Mayer (M-G-M) in 1924.

The first vice-President in charge of production, he was instrumental in the creation of Hollywood as a dream factory and the establishment of the star system. Renowned for his cruelty, capriciousness and calculation he was a supremely canny businessman and shamelessly paternalistic manipulator of careers and personalities. A lover of wholesome, patriotic fare, he scrupulously ensured M-G-M's profitability with such enduring successes as *Ben Hur* (1926), *Grand Hotel* (1932), the apple-pie Americana of the Andy Hardy series, *Ninotchka* (1939) and countless others.

He received an honorary Oscar in March 1951 for 'distinguished service to the motion picture industry' and was forced to retire from M-G-M in June of that year, a victim of changing times and a loser in the game of studio politics that he had once played with such astute precision.

A biographical portrait of this 'Czar of all the rushes' is contained in *Hollywood Rajah* (1960) by Bosley Crowther.

MEDIUM SHOT As its name suggests, a medium shot falls between the extremes of a close-up and a long shot and generally captures a character or characters from the knees or waist up rather than showing just the face or some form of distant profile. Also known as a mid-shot or three-quarters shot, it can serve as a transitional perspective on events as the attention moves in closer or begins to pull away and is also useful for detailing relationships between characters or capturing them with a small but telling illustration of the surrounding environment.

MÉLIÈS, Georges
Born 8 December 1861, Paris, France.
Died 21 January 1938, Paris, France.

Fascinated by all aspects of illusion and conjuring, Méliès bought the Théâtre Robert Houdin in Paris in 1888 and began to present his own performances of sleight of hand and magic lantern shows.

A witness to the Cinematographe Lumière in 1895, he became devoted to the potential of the new medium called film, acquired the relevant equipment and presented the first film performance at the theatre on 4 April 1896. The same year, he made his first film *Une Partie de Cartes* and over seventy others including a three-minute vampire drama entitled *The Devil's Castle*.

In his brief and feverishly productive career, he is reckoned to have made in excess of 500 films of which barely ten percent still survive. Constructing a glass studio in the garden of his home, he experimented tirelessly with form and content, making every conceivable type of production from realistic depictions of everyday events to dramatic stories, recreations of historical incidents and mildly risqué 'stag' films.

However, his greatest legacy has been as a master showman exploring the magical wizardry of special-effects, trickery and fantastical storytelling, propelling the medium from its documentary-like origins into a world of unbounded imaginative possibilities. Credited as the instigator of technical innovations like double-exposure, ▷ dissolves and ▷ fade-outs, he created the cinematic equivalents of pantomimes and comicbooks in films like *Cinderella* (1900), *Le Voyage dans la Lune* (Voyage to the Moon) (1902), *L'Homme à la Tête de Caoutchouc* (The Man With the Rubber Head) (1902) and *Voyage à Travers L'Impossible* (The Impossible Voyage) (1904) which are crammed with imaginative set-designs, dream sequences, absurdist humour and a mischievous propensity for expanding, dismantling and even dismembering the human body.

His productions grew even more ambitious, but after such 1912 efforts as *A La Conquête du Pole* (The Conquest of the Pole) and *Cendrillon ou La Pantoufle Mysterieuse* (Cinderella or the Glass Slipper), he ceased filmmaking and returned to the stage, giving his last performance in 1920. Substantial debts forced him to sell off much of his work, but he lived to see his position as a movie pioneer acknowledged and received the Légion D'Honneur in 1933.

MELVILLE, Jean-Pierre

Real Name Jean-Pierre Grumbach
Born 20 October 1917, Paris, France.
Died 2 August 1973, Paris, France.

Evacuated to Britain after Dunkirk, Melville served with the Free French forces in North Africa and Italy before returning home, forming his own production company and making his directorial debut with the short *Vingt Quatre Heures de La Vie D'Un Clown* (1946).

After the feature-length *Le Silence de la Mer* (1947), a parable about the Resistance, he collaborated with ▷ Jean Cocteau on *Les Enfants Terribles* (The Strange Ones) (1949).

A lover of American pulp fiction and gangster thrillers, he directed *Bob La Flambeur* (Bob the Gambler) (1955), in which a retired bankrobber plots a casino heist to alleviate his temporary financial embarrassment. He would constantly return to the iconography of the ▷ film noir and his approach on this occasion of using natural locations, energetic camera moves, abrupt editing and a jazz score was to be highly influential on the directors of the ▷ nouvelle vague.

Stubbornly independent throughout his career, he pursued an obsession with the mythology of the gangster that manifested itself in *Le Doulos* (Doulos-The Finger) (1962), *Le Deuxième Souffle* (Second Breath) (1966), *Le Samourai* (The Samurai) (1967) and *Le Cercle Rouge* (The Red Circle) (1970). He enjoyed notable commercial successes with a very personal vision that explored questions of friendship, betrayal, loyalty and a code of ethics using a distinctive style of emotional restraint, atmospheric lighting and expertly tense illustrations of the mechanics of killing and robbery.

A weary police commissioner, a rare representative from the other side of the law, served as the protagonist in his final film *Un Flic* (Dirty Money) (1972). Melville also played small roles in such films as *Les Dames Du Bois De Boulogne* (1946), *Orphée* (1949) and ▷ *A Bout de Souffle* (Breathless) (1960).

MENZEL, Jiri

Born 23 February 1938, Prague, Czechoslovakia.

Educated at the Film Academy in Prague, Menzel worked as an assistant-director on *O Necem Jinem* (Something Different) (1962) before making his directorial debut with *Zločin V Divči Ŝkole* (Crime At A Girl's School) (1965).

He quickly gained recognition as a leading figure in the brief flourishing of the Czechoslovakian New Wave and found welcoming international audiences for his warm comic observations and wry indulgences of the very human foibles observed in the perennial pursuit of love, companionship and sex. He won a Best Foreign Film Oscar for *Ostrě Sledované Vlaky* (Closely Observed Trains) (1966) an assured comedy-drama about a trainee railway guard's romantic yearnings. His fine touch was also evident in the bittersweet mood of *Rozmarné Léto*

(Capricious Summer) (1968), a languorous, pastel-coloured tale of the longings provoked by a beautiful circus performer's summer sojourn in a sleepy provincial town.

Skřivánci Na Niti (Larks On A String) (1969), a satirical tale of bourgeois re-education in the 1950s, went into production just as Czechoslovakia was invaded by the Warsaw Pact and was subsequently suppressed by the authorities until its triumphant re-emergence at the Berlin Film Festival of 1990 when it shared the Golden Bear for Best Film.

Forced to disassociate himself from his early films, he did not direct again until the more conventional, state-approved *Kdo Hledá Zlaté Dno* (Who Seeks the Gold Bottom) (1975) but has continued to charm audiences with his warmly nostalgic portraits of bygone days, notably the sepia-hued *Báječní Muži S Klikou* (Those Wonderful Movie Cranks) (1978) in which an itinerant showman establishes Prague's first cinema, and *Postřižiny* (Short Cut) (1980) in which the coming of sound radio has a profound effect on the inhabitants of a small provincial town.

More recently he directed the simple rustic comedy *Vesnicko Ma Strediskova* (My Sweet Little Village) (1985) and the satirical World War One comedy of manners *Kanec Starých Ĉasů* (The End of Old Times) (1989).

He has also directed for the stage and made acting appearances in such films as *Dita Saxová* (1967) and *Hra O Jalko* (The Apple Game) (1976).

MERCER, David

Born 28 June 1928, Wakefield, Yorkshire, UK.
Died 8 August 1980, London, UK.

After studying painting at King's College, Newcastle, Mercer moved to Europe and began his writing career. His first television play, *Where The Difference Begins* (1961) signalled his interest in fusing the personal and the political in work that challenged the conventions of television drama.

Further plays, like *A Climate of Fear* (1962), *A Suitable Case for Treatment* (1962) and *In Two Minds* (1967), explored his fascination with mental health, meddlesome psychiatric practices, and his struggle to reconcile a belief in socialism with the repression revealed during his stays in Eastern Europe.

Winner of the 1965 Evening Standard Award for Most Promising Playwright, his stage work includes *Ride A Cock Horse* (1965), *Flint* (1970) and *Cousin Vladimir* (1978).

He worked only rarely for the cinema, adapting *Morgan–A Suitable Case for Treatment* (1966) for director Karel Reisz (1923–) and creating two works of stark originality and invention: *Family Life* (1970), which explored the claustrophobia of urban living and the domestic causes of mental illness, and *Providence* (1977), a remarkably stylised portrait of an ailing author's juxtaposition of reality and fiction as he attempts to complete his final novel and cope with the demands of his family.

He continued to address issues of personal alienation and the class system in later television plays like *Huggy Bear* (1976), *The Ragazza* (1978) and *Rod of Iron* (1980).

MÉSZÁROS, Marta

Born 19 September 1931, Budapest, Hungary.

Forced to leave Hungary for Russia in 1936, Meszaros endured a traumatic childhood in which her father's internment and disappearance was swiftly followed by her mother's death from typhoid. The force of such autobiographical events has played a major part in her work.

She obtained a scholarship to study at Moscow's VGIK Film School and made her directorial debut with the short subject *Ujra Mosolygonak* (Smiling Again) (1954). Over the next fourteen years in Romania and Hungary, she made over thirty documentaries and diverse works designed to popularise science.

She made her feature-length debut with *Eltávozott Nap* (The Girl) (1968), a plaintive portrait of an orphan's disillusioning search for the parents who had abandoned her.

She quickly followed this with *A 'Holludvar'* (Binding Sentiments) (1968), an account of the conflict between an ageing mother and her son's fiancée, and the semi-musical romance *Széplányok, Ne Sirjatok* (Don't Cry Pretty Girls) (1970).

Throughout her films, she has

attempted to deal with the daily realities of Hungarian life and how they affect women using carefully composed narratives to deal with their oppression and struggle for self-fulfilment and satisfying emotional relationships. Her many significant features include *Örökbefogadás* (Adoption) (1975) which explores the friendship between a widowed factory worker and an adolescent girl, *Kilenc Hónap* (Nine Months) (1976) in which an iron foundry worker retains her independence despite the jealousies and demands of her lover, and *Ök Killan* (Two Women) (1977) another tale of women finding more emotional support in their friendships with other women than in their inadequate male partners.

Naplo Gyermekeimnek (Diary for My Children) (1982) began an insightful, autobiographically-inspired trilogy on the history of post-War Hungary in which political upheavals and social changes are mirrored in the life of an orphan who returns from Russia to 1947 Budapest. This was followed by *Naplo Szerelmeimnek* (Diary for My Loves) (1987) in which the now-teenage girl attempts to trace her father, evade the interference of her despised foster mother and become a filmmaker, and *Naplo Apamnak Anya'Mnak* (Diary for My Father and My Mother) (1990), which deals with the uprising of 1956 and apportions blame for the post-war European situation between Stalinist oppression and American complacency.

METROPOLIS

Fritz Lang (Germany 1926)
Brigitte Helm, Alfred Abel, Gustav Frohlich, Rudolf Klein-Rogge.
Original running time over 150 mins; most remaining prints run 120 mins; 1984 tinted reissue with rock soundtrack produced by Giorgio Moroder re-edits footage to 83 mins. b/w. silent.

The German economy may have been on shaky foundations during the Twenties, but that failed to stop director ▷ Fritz Lang from mounting three of the largest film productions ever to have been attempted at that time, following the megalomaniac thriller *Dr Mabuse, der Spieler* (Dr Mabuse the Gambler, 1922) and the Wagnerian legend *Die Niebelungen* (1923) with a utopian fantasy *Metropolis* that was intended as both huge spectacle and message movie, would take almost eighteen months to shoot, and would bankrupt the massive German Ufa studios before its much trumpeted release in January 1927. While Lang's political sympathies placed him throughout his career on the left, his then wife and co-scriptwriter Thea Von Harbou (1888–1954) would in future become closely aligned with the Nazi party. Their scenario for *Metropolis* was simplistic enough to satisfy both of them, positing a future society where the technological luxury of a ruling elite is maintained by an enslaved underclass who revolt and almost destroy the city before love wins the day, a reconciliation assuring equal right is reached, and the moral pointed out that 'the intermediary between the hand and the brain is the heart'. The plotting and characters are however dwarfed by the scale of the undertaking and Lang's impressive grasp of visual architectonics, all lavishly stylised sets and massive crowd scenes, still ensures the power to mesmerise the viewer. With its mad scientist, evil female robot and expansive vision of futuristic technology, the film was to prove highly influential on later science fiction, while its respectful reception in the US was at length to prove Lang's ticket to Hollywood – even if his dark contemporary urban work there would eventually leave his most grandiose masterpiece looking rather atypical amidst the body of a distinguished canon.

MEYER, Russ

Born 21 March 1922, Oakland, California, USA.

As a child, Meyer made numerous home movies and later served as an Army combat photographer on wartime service in Europe. After the war, he made industrial documentaries in San Francisco and established a reputation as a photographer of nudes for magazines like *Playboy*.

A stills photographer on the films *Guys and Dolls* (1955) and *Giant* (1956), he combined his filmmaking skills with an abiding devotion to the female anatomy to create the feature-length *The Immortal Mr Teas* (1959).

The estimated return of one million dollars on a fairly minimal investment, launched him on a prolific career of dis-

Metropolis (1926)

tinctive soft-porn 'entertainments' that recycle a fairly outrageous mélange of ingredients including a plethora of large-breasted women, tongue-in-cheek moralising and a sadistic streak of humour that often alludes to Nazism. The titles include *Erotica* (1961), *Heavenly Bodies* (1963), *Faster, Pussycat! Kill! Kill!* (1966) and *Mondo Topless* (1966).

The increasingly lucrative nature of his activities brought him to the attention of Hollywood and he moved towards more mainstream filmmaking as the director of *Beyond the Valley of the Dolls* (1970) and the relatively straight *The Seven Minutes* (1971).

However, he preferred the liberties to be taken as an independent filmmaker and resumed his own endeavours with *Sweet Suzy* (1973), *Supervixens* (1976), *Up!* (1976) and, his last to date, *Beneath the Valley of the Ultravixens* (1979).

Frequently a producer, director, writer, cinematographer and editor of his films, he has been the scourge of the feminist movement and deliberately provoked them by his assertions that his work is pro-feminist in its depiction of strong, tough-minded women who control their own lives and easily outwit the dumb, plastic men in his stories. It seems more rational to accept it as flamboyant, hysterically-pitched pornography but there is also a critical fraternity that advocates Meyer as a talented ▷ auteur who has created his own inimitably overstated genre.

Recently he made an acting appearance in *Amazon Women on the Moon* (1987) and, for some years, he has been at work on a compilation documentary of his career entitled (inevitably) *The Breast of Russ Meyer*.

MIFUNE, Toshiro

Born 1 April 1920, Tsingtao, China.

Born of Japanese parents, Mifune served in the Army from 1940 to 1945 and drifted in to the cinema after the War on the advice of a friend. He made his debut with a small role in *Shin Baka Jidai* (These Foolish Things) (1946) and soon progressed to leading roles in films like *Genrei No Hate* (Snow Trail) (1947) and *Yoidore Tenshi* (Drunken Angel) (1948), which began his long and mutually beneficial association with director ▷ Akira Kurosawa.

A man of imposing physical presence, he was frequently cast as crooks, gangsters and men of action, but made his first claim on western audiences as the bandit-rapist whose crime is recalled from a variety of perspectives in ▷ *Rashomon* (1950). He was equally seductive in *Saikaku Ichidai Onna* (The Life of O. Haru) (1952) for ▷ Kenji Mizoguchi but did much of his best known work for Kurosawa bringing a gusto and virility to portraits of very human samurai warriors, both fearsome and fearless, in such classic films as ▷ *Shichi-nin no Samurai* (Seven Samurai) (1954), *Kumonosu-Jo* (Throne of Blood) (1957), *Kakushi-Toride no San Akunin* (Hidden Fortress) (1958) and *Yojimbo* (1961) for which he received the Best Actor Prize at the Venice Film Festival.

He proved his versatility in a multiplicity of other roles including a version of *Cyrano de Bergerac* entitled *Kengo no Shogai* (Samurai Saga) (1959) and as the unctuous vengeance-seeking son in *Warui-Yatsu Hodo Yoko Nemuru* (The Bad Sleep Well) (1960) before making his directorial debut with *Goju Mannin no Isan* (The Legacy of the 5,000) (1963).

He gave a performance of great power and complexity as the autocratic doctor in *Akahige* (Red Beard) (1965) which proved to be his last film with Kurosawa, although neither has confirmed rumours of a quarrel. One of the rare Japanese actors to have become a star in the West, he then began accepting some of the American offers coming his way, making his dubbed English-language debut in *Grand Prix* (1966).

A heroic figure of integrity and authority, he maintained his Japanese career whilst making international forays into films like *Hell in the Pacific* (1968) and other lesser efforts like *1941* (1979), *Inchon* (1980) and *The Challenge* (1981).

His persona of the indomitable, swashbuckling samurai has remained phenomenally popular in Japan, and he has enjoyed continued box-office success with projects like *Shinsen Gumi* (Band of Assassins) (1969), *Ningen No Shomei* (Proof Of A Man) (1977) and *Port-Arthur* (1980). After a lengthy absence acting in television, he returned to a warm welcome for *Otoko-wa Tsuraiyo Shiretiko Bojo* (1987) and *Taketori Monogatari* (Bamboo Princess) (1987).

MIAMI VICE

USA 1984–9
Executive Producer – Michael Mann

The legend that *Miami Vice* was created in response to a memo from NBC President Brandon Tartikoff which pondered the popularity of music videos and asked about the possibilities of 'MTV cops', may be apocryphal, but it neatly sums up the essence of the show. Through a jungle of fast-editing, loud rock music and attractive settings, walks Sonny Crockett (Don Johnson (1950–)), an eccentric undercover cop, who carries with him enough character quirks to support a dozen subsidiary characters. (He's a divorced Vietnam veteran, an ex-football star who lives on a houseboat, keeps a pet alligator called Elvis and drives a Ferrari Daytona – and yes, the car model *is* important). The white Crockett is partnered by the romantic sharply-dressed Black/Hispanic Ricardo Tubbs (Philip Michael Thomas (1949–)), who brings, along with his multiple ethnic stereotyping, the memory of his murdered cop brother. With conventional methods failing in gangland Miami, the duo pose variously as gangsters and drug dealers.

Many criticised *Miami Vice* for superficiality. But technically it was a bold break with tradition, and to an extent, redefined the medium. That the innovation was not supported by any narrative to speak of was disappointing. In the early shows the line between good and bad was blurred by the emphasis on conspicuous consumption, and the dubious practices of the cops. Later episodes played down the designer violence and began to edge towards a clearer moral tone. The techniques employed are those of the advertising industry. Just what *Miami Vice* was selling remains an open question.

MR. SMITH GOES TO WASHINGTON

Frank Capra (USA 1939)
James Stewart, Jean Arthur, Claude Rains, Edward Arnold, Thomas Mitchell.
130 mins. b/w.
Academy Award: *Best Original Story* Lewis R. Foster.

'I would gamble with the long-shot players who light candles in the wind and resent with the pushed-around because of race or birth. Above all, I would fight for their causes on the screens of the world.' Thus wrote ▷ Frank Capra in his 1971 autobiography, and a prime illustration of his populist and idealist faith in the 'little man' comes in a pair of similarly-styled Thirties movies, 1936's *Mr Deeds Goes to Town* – wherein ▷ Gary Cooper's small town greeting cards writer inherits a fortune and moves to New York, where high society tries to have him proved insane just because he invests in poor small farmers – and the follow-up three years later which took the theme of the democratic hero right to the heart of the American political system in *Mr Smith Goes To Washington.* Here ▷ James Stewart is appropriately-named Boy Scoutmaster Jefferson Smith, whose very gullibility sees him rigged into a seat in the Senate by a band of corrupt wheeler-dealers eager to force through a dam construction bill that will substantially line their own pockets. The querulous hick from the sticks (with a little help from hard-boiled Jean Arthur, (1905–)) proves to be made of sterner stuff and in an all-day fillibuster speech proposing that a National Boys Camp be built on the land that might otherwise be flooded his strength of idealism leads unscrupulous Senator Claude Rains (1889–1967) to confess his sins before the entire House. Cornball naivety perhaps, but the appealing performances and Capra's deft handling of the quickfire dialogue exchanges move it along quickly enough to momentarily suspend credulity, and even if Rains's last-minute change of heart takes some swallowing the film does at least voice some worthwhile disquiet at the day-to-day machinations of American political life.

MINNELLI, Vincente
Born 28 February 1910 or 1913, Chicago, Illinois, USA.
Died 25 July 1986, Beverly Hills, California, USA.

A child dancer and tumbler with the Minnelli Brothers Dramatic Tent Show, Minnelli spent his entire working life involved with some aspect of showbusiness. A stage designer at the Chicago Theatre in the 1920s, he moved to New York in 1931 and worked as a costume designer for the *Earl Carroll Vanities* of 1932 and for Radio City Music Hall between 1933 and 1935.

He made his debut as a Broadway director with *At Home Abroad* (1935). He signed a contract with Paramount in 1936 but left their employment after eight months. He returned to Hollywood in 1940, beginning a long professional association with M-G-M. Initially responsible for staging the musical sequences in films like *Strike Up the Band* (1949) and *Babes on Broadway* (1941), he made his official directorial debut with the all-black musical *Cabin in the Sky* (1943).

Using the vast resources of the Arthur Freed (1894–1973) unit at M-G-M, he created some of the studio's most lavish and colourful musicals that brim with sumptuous set design, elegant artistry, a lighter-than-air fantasy quality and a surprisingly sombre view of the world. His many musical triumphs include *Meet Me In St. Louis* (1944), *The Pirate* (1947), ▷ *An American in Paris* (1951) (Best Director Oscar nomination), *The Band Wagon* (1953) and *Gigi* (1958) (Oscar).

Adept at such domestic comedies as *Father of the Bride* (1950) and *The Long, Long Trailer* (1954), his stylised direction, bold use of colour and design achieved striking effects on such passionate, even overwrought, melodramas as *The Bad and The Beautiful* (1952), *The Cobweb* (1955) and *Lust for Life* (1956).

He left M-G-M in 1965 and made only two further films; the uneven musical *On A Clear Day You Can See Forever* (1970) and *A Matter of Time* (1976) which he disowned after outside interference. The latter starred Liza Minnelli (1945–), his daughter by a marriage to ▷ Judy Garland. In 1974, he published his autobiography *I Remember It Well.*

MITCHUM, Robert
Full Name Robert Charles Mitchum
Born 6 August 1917, Bridgeport, Connecticut, USA.

After a peripatetic youth spent as a labourer, vagrant and professional boxer, Mitchum was working at the Lockheed factory in Los Angeles when he drifted in to the film business, making his screen debut as an extra in *Hoppy Serves A Writ* (1943).

A prolific performer with some 19 appearances to his credit during the first year of his career, he gradually acquired leading man status and received his first and only Oscar nomination as a tough, heroic captain in *The Story of G.I. Joe* (1945).

After brief service in the Army, he returned to the cinema and displayed the right poise and lethargy to perfectly embody the doomed protagonists of certain classic ▷ film noir including *The Locket* (1946), *Out of the Past* (1947) and *Where Danger Lives* (1950). Even his arrest and imprisonment in 1948 on a charge of possessing marijuana did little to affect his enduring popularity.

A burly figure of sleepy-eyed charisma, he has perfected a façade of nonchalant indifference towards his own abilities and career that conceals a considerable talent and universally respected professionalism. He has sauntered through innumerable second-rate assignments but brought real skill to bear on such challenging roles as the homicidal preacher in ▷ *Night of the Hunter* (1955), the restless sheep farmer in *The Sundowners* (1960), the veteran lawman in *El Dorado* (1967) and the country teacher in *Ryan's Daughter* (1970).

Age has only served to increase affection for his tough guy persona and later roles of note include the veteran con in *The Friends of Eddie Coyle* (1973) and private detective Philip Marlowe in *Farewell, My Lovely* (1975) a role he imbued with a world-weary gallantry.

Turning to television, he has starred in such sprawling mini-series as *The Winds of War* (1983) and its sequel *War and Remembrance* (1989) and appeared in a number of television films including the sluggish mystery *One Shoe Makes It Murder* (1982) and *The Hearst and Davies Affair* (1985) in which he played newspaper magnate William Randolph Hearst.

His most recent cinema appearances include *Mr. North* (1988), *Scrooged* (1988) and a cameo role in the remake of *Cape Fear* (1991), in which he had starred in 1962.

MIX, Tom

Real Name Thomas Erwin Mix
Born 6 January 1880, Mix Run, Pennsylvania, USA.
Died 12 October 1940, Florence, Arizona, USA.

A sergeant in the United States Army Artillery and drum major with the Oklahoma City Band, Mix was also a bartender and deputy sheriff in Dewey, Oklahoma before joining a number of Wild West shows. He first became involved with the burgeoning film industry as an off-screen wrangler and supplier of cowboys and Indians to the Selig company.

He made his acting debut in *Ranch Life in the Great Southwest* (1910) appearing in scores of westerns over the next few years, turning to direction with *Local Color* (1913) and often assuming responsibility for the screenplay as well. Many of his films no longer survive but during the silent era he was one of the screen's most popular stars and has been judged a precursor to ▷ John Wayne in his embodiment of the honest, naive western hero.

Dressed in an oversized cowboy hat and white duds that emphasised his purity, he was partnered by an acutely intelligent horse called Tony. Historians have suggested that rather than crudely portraying good guys, he was in fact a comic performer who played up the excessive nobility and gaucheness of his character.

His many films include *The Heart of Texas Ryan* (1917), *Tom Mix in Arabia* (1920), *Riders of the Purple Sage* (1925), *The Great K & A Train Robbery* (1926) and *Destry Rides Again* (1932).

He made his final appearance in the serial *The Miracle Rider* (1935) and then embarked on a costly tour with the *Tom Mix Circus* (1936–38). He was said to be attempting a comeback as a character actor when he was killed in a car accident.

MIXER Apart from referring to the equipment used within the process, a mixer or dubbing mixer is the technician responsible for the dubbing together of the individual soundtracks or dialogue, music and aural effects into a single composite or 'mixed' soundtrack. For the high quality stereophonic sound of today's best cinema presentations, the final mix will be on six separate tracks for 70mm and four tracks for 35mm projections. The floor mixer is the person in charge of sound recording during the production of a film whilst the music mixer is responsible for the recording of the required music.

MIZOGUCHI, Kenji
Born 16 May 1898, Tokyo, Japan.
Died 24 August 1956, Kyoto, Japan.

An apprentice designer, painter and newspaper lay-out artist, Mizoguchi found employment as an assistant director in the early 1920s before swiftly making his directorial debut with *Ai Ni Yomigaeru Hi* (The Resurrection of Love) (1923).

A prolific director over the next three decades, only around one third of his total output still exists and nearly all of his pre-1936 oeuvre has been lost. Influenced by seeing his sister sold into the life of a geisha, he was acutely critical of the subservient position of women in Japanese society and brought a painterly sensibility, visual grace and fluid camerawork to bear on dramas of quite radical, hard-hitting political content.

In *Naniwa Hika* (Osaka Elegy) (1936) (which began his long partnership with screenwriter Yoshitaka Yoda) a woman resorts to prostitution to support her family; in *Gion No Shimai* (Sisters of the Gion) (1936) two geisha sisters are used and abused by the men in their lives; in *Zangiku Monogatari* (The Story of the Late Chrysanthemums) (1939) a wife sacrifices herself for the greater good of her husband's acting career; in *Waga Koi Wa Moenu* (My Love Has Been Burning) (1949) a 19th century woman faces imprisonment and ostracisation because of her pioneering feminist ideals.

The full extent of his artistry became known to Western audiences with the release of *Saikaku Ichidai Onna* (The Life of O-Haru) (1952), a picaresque tale of how a 17th century samurai daughter's love for a man of the lower classes brings about her tragic descent into poverty and prostitution. Painstaking in its attention to detail, the discretion of the camerawork, excellence of the acting and sympathy of the direction create a memorable testimony to the woman's suffering and dignity rather that accentuating the obvious melodrama or potential for sentimentality in the events.

The success of the film led to one of the most productive phases of his career with a string of major artistic achievements that included the lyrical ghost story *Ugetsu Monogatari* (1953), the barbaric tale of injustice and suffering *Sansho Dayu* (Sansho The Baliff) (1954) and *Chikamatsu Monogatari* (The Crucified Lovers) (1954), a moving, beautifully cinematic story of young lovers in conflict with the basest desires of an avaricious, uncaring society.

His last film *Akasen Chitai* (Street of Shame) (1956) condemned the eternal and continuing exploitation of woman and was instrumental in creating support for the Japanese ban on prostitution that was implemented the following year. He died of leukemia during the production of *Osaka Monogatari* (1957), which was completed by Kimishiburo Yoshimura.

MONROE, Marilyn
Real Name Norma Jean Baker
Born 1 July 1926, Los Angeles, California, USA.
Died 5 August 1962, Los Angeles, California, USA.

The product of an unhappy childhood spent largely in a foster home because of her mother's mental illness, Monroe became a photographer's model in 1946 with her most famous portrait being the nude calendar shot.

She made her film debut as an extra in *Scudda Hoo! Scudda Hay!* (1948). Several small parts followed including *Love Happy* (1949) in which she was chased round a desk by ▷ Groucho Marx, *The Asphalt Jungle* (1950) as the mistress, and ▷ *All About Eve* (1950) in which she features as an aspiring actress and protégé of the acid-tongued critic George Sanders (1906–72).

A stunningly beautiful blonde, who combined a fulsome sexuality with an almost childlike innocence and husky whisper of a voice, she was groomed for stardom, showing her limitations as a dramatic actress in her portrayal of a psychotic babysitter in the feeble *Don't Bother To Knock* (1952), but proving effective as the provocative wife in *Niagara* (1953).

However, she proved a delight in broad comic roles as the dumb blonde in *Gentlemen Prefer Blondes* (1953), the shortsighted gold-digger in *How To Marry A Millionaire* (1953) and the girl

Marilyn Monroe and Clark Gable in *The Misfits* (1960)

upstairs who provides the temptation for married man Tom Ewell (1919–) in *The Seven Year Itch* (1955).

She then moved to New York, to study acting with Lee Strasberg (1901–82) and returned to the screen showing a maturity and emotional vulnerability as the roadhouse singer in *Bus Stop* (1956).

In Britain she co-starred with ▷ Laurence Olivier in the leaden period romance *The Prince and The Showgirl* (1957), but showed her flair for comedy again as the showgirl Sugar in ▷ *Some Like It Hot* (1959).

Her last completed film was *The Misfits* (1960) in which she was touching as the lonely, rootless woman drawn to the rugged masculinity and reliability of ▷ Clark Gable. Increasingly said to be unreliable and temperamental, she was fired from the filming of *Something's Got To Give* (1962) after having appeared for work only twelve times in 32 days of shooting.

Insecure and increasingly uneasy with her sex symbol image, her death from an overdose of barbiturates two months later has never been satisfactorily explained and her romantic entanglements with both President John F. Kennedy and his brother Robert have cast sinister shadows over her final hours.

A compilation of her work, *Marilyn*, appeared in 1963 and she has become a 20th century icon and symbol of Hollywood's ruthless exploitation of youth and beauty.

MONTAGE Taken from the French monter, (to assemble), montage techniques in relation to film were pioneered and developed by Soviet filmmakers during the 1920s. A process of editing, it was defined in the contrasting approaches of directors Vsevolod Pudovkin (1893–1953) and ▷ Sergei Eisenstein.

Pudovkin believed that the editing of a film should serve the smooth flow of the narrative, with each shot leading unobtrusively to the next in a kind of narrative chain of links, hence the term 'linkage' that is applied to his work. Eisenstein's theory, often referred to as 'collision', saw editing as a device to create sequences that contain rapidly changing multiple images that are often superimposed, dissolved together or juxtaposed to create a visually dramatic or emotional impact. The most famous example of his theory in practice is the Odessa Steps sequence from ▷ *Battleship Potemkin* (1925).

The term has come to have a general application to editing that is consciously constructed to achieve a more impressionistic impact by the use of images in a less conventional or linear fashion than traditional filmmaking. Memorably prevalent in the social dramas and gangster epics created during the 1930s by Warner Brothers, where future director Don Siegel (1912–91) headed a montage department from 1939, its use added immeasurable flair and character to films like *The Roaring Twenties* (1939).

MONTAND, Yves

Real Name Ivo Livi

Born 13 October 1921, Monsummano Alto, Italy.

Died 9 November 1991, Senlis, France.

Raised in Marseilles, Montand was a movie lover who worked in a pasta factory and a metal factory before beginning to develop his own cabaret act as a kind of singing cowboy. He made his professional debut at the Alhambra Theatre in Marseilles in 1939 and later toured France, making his Paris debut in 1944.

A protege of Edith Piaf (1915–63), he made his film debut with her in *Etoile Sans Lumiere* (1946) and was also seen as a replacement for ▷ Jean Gabin in the unsuccessful *Les Portes De La Nuit* (Gates of the Night) (1946).

He displayed little interest in building a film career, but gained increasing renown as a singer and cabaret artist. He married actress ▷ Simone Signoret in 1951 and was influenced by her in the direction of his career and growing commitment to a variety of left-wing causes.

His film career caught fire again with *Le Salaire de La Peur* (The Wages of Fear) (1953) and he was well received in a Paris production of *The Crucible* (1954).

He made his New York performing debut in 1959 but seemed ill-at-ease in the English-language films that came his way over the next few years, including the musical *Let's Make Love* (1960), *Sanctuary* (1961) and *My Geisha* (1962).

He himself claims to have found his stride as an actor from the beginning of his collaboration with director ▷ Costa-Gavras on the thriller *Compartiment Tueurs* (The Sleeping Car Murders) (1965). Together, they made a number of incisive and moving thrillers that managed the rare feat of combining serious political content and a high entertainment value. Montand's somewhat dour impassivity and skilled underplaying lent credence and a tortured anguish to his roles as a Greek politician in ▷ *Z* (1968), a Czech official in *L'Aveu* (The Confession) (1970) and a CIA official in *Etat de Siège* (State of Siege) (1973).

Other significant film roles from this period include the political *Bien* (1972) and the popular romantic melodrama *César et Rosalie* (César and Rosalie) (1972). A further English-language appearance in the musical *On A Clear Day You Can See Forever* (1970) did nothing to discourage the impression that he was not comfortable outwith his adopted homeland.

A solo performer of legendary magnetism, he performed to sell-out houses at the Paris Olympia (1981–82) and was fêted for his week of solo performances at New York's Metropolitan Opera House in 1982. He has also become something of a distinguished elder statesman within the film industry with his performance as the wily, venal patriarch in *Jean De Florette* (1986) and *Manon Des Sources* (1986). His more recent film performances include the autobiographical musical *Trois Places Pour Le 26* (1988) and the thriller *Netchäiev Est De Retour* (1990).

MOORE, Mary Tyler

Born 29 December 1936, Brooklyn, New York, USA.

Trained as a dancer, Moore's first professional job was as the Happy Hotpoint Pixie in a series of television commercials in 1955. Small acting roles followed and she was seen as the secretary in the series *Richard Diamond, Private Eye* (1957–59) and made her film debut in *X-15* (1961).

The series ▷ *The Dick Van Dyke Show* (1966) highlighted her talent for domestic comedy and won her Emmy awards in 1964 and 1965.

Her small screen popularity was subsequently used to launch a multi-media career on Broadway with a musical of *Breakfast at Tiffanys* (1966) and in the cinema, with poorly received fare like *What's So Bad About Feeling Good* (1968) and *Change of Habit* (1969).

She returned to television with *The Mary Tyler Moore Show* (1970–77) where her charm and pert comic touch created a warmly sympathetic portrait of a self-reliant career girl whose personal and professional worries came to represent the travails of the average single woman in modern America. Among numerous awards, the series won her Emmys in 1973, 1974 and 1976.

She subsequently proved her dramatic worth with a vengeance, winning a further Emmy as a woman coming to terms with her mastectomy in *First, You Cry* (1978), a Tony for her Broadway tour de force as the paralysed artist in *Who's Life is it Anyway?* (1980) and a Best Actress Oscar nomination for her chilling portrait of a repressed, perfectionist matriarch inflicting a form of domestic tyranny in *Ordinary People* (1980).

Despite this flurry of achievement, she has made few subsequent forays into films, with the tearjerkers *Six Weeks* (1982) and *Just Between Friends* (1986) finding little favour. However, television has provided more significant opportunities in the heartwarming May–December romantic comedy *Finnegan,*

Begin Again (1984), *Heartsounds* (1984) in which she co-starred with James Garner (1928–) as a woman coping with her husband's illness, and the mini-series *Gore Vidal's Lincoln* (1988) in which she played Mary Todd.

MTM Enterprises, formed with her former husband Grant Tinker in 1970, has been responsible for such television series as ▷ *Lou Grant* and ▷ *Hill Street Blues*.

MONTY PYTHON'S FLYING CIRCUS

UK 1969–74
John Cleese, Graham Chapman, Eric Idle, Michael Palin, Terry Gilliam, Terry Jones.

Britain's answer to the fast-moving anarchy of the ▷ *Laugh-In* was, well, more fast-moving anarchy. Often the only link between the sketches was the catchphrase, 'And now for something completely different'. But if *Laugh-In* was updated variety, *Python* was something else entirely. The group came from an Oxbridge comedy background and the targets were recognisably British, petty bureaucrats being a particular favourite (hence the Cleese sketch about the Ministry of Silly Walks). Many of the jokes were about the conventions of television, though few were about TV personalities – the island swarming with ▷ Alan Whicker impersonators was a rare exception. The show frequently came with multiple false endings; the BBC logo would appear, an announcer would trail the next programme, which would then unravel into another sketch.

Despite the derision poured on authority figures the show was not satire. Though class-riddled, it was deriding 'upper-class twits' or 'Gumbies', grunting proletarians with knotted handkerchiefs on their heads. Often the point was simply that there was no point. Each member brought something different to the show, and all succeeded when the team split (Cleese left before the fourth series in 1974). Python had a tremendous impact, but though skits like the Cheese Shop and the Dead Parrot have passed into small screen legend, it was often hit and miss. The anarchy, once recognised, became a straitjacket, and though oft-repeated it depends now on familiarity for its impact, rather than the quality of the jokes, which often lacked punchlines. In 1972 *Python* won a BAFTA award for best light entertainment programme, but in retrospect Cleese's straighter work on *Fawlty Towers* (1975 & 1979) may prove to have a longer shelf-life.

The team's first film *And Now For Something Completely Different* (1971) proved to be a compendium of some of the best sketches previously seen on television but subsequent feature films proved more imaginative and comprise *Monty Python and the Holy Grail* (1974), *Monty Python's Life of Brian* (1979), *Monty Python Live at the Hollywood Bowl* (1981) and *Monty Python's Meaning of Life* (1983).

All the Pythons have enjoyed some degree of individual success outwith the team. Cleese (1939–) in films like *Silverado* (1985) and *Clockwise* (1986), and most spectacularly as the comic mastermind behind *A Fish Called Wanda* (1988) and a highly effective company that made business and industrial training films. Chapman (1941–89), perhaps least of all, in films like *The Odd Job* (1978) and *Yellowbeard* (1983). Eric Idle (1942–) spoofed the Beatles in *The Rutles* and has been seen in such films as *National Lampoon's European Vacation* (1985) and *Nuns on the Run* (1990). Palin (1943–) has starred in such film comedies as *The Missionary* (1982) and *A Private Function* (1984) and proved a personable travelling companion on the BBC series *Around the World in 80 Days* (1989). Terry Jones (1942–) has turned to film direction with *Personal Services* (1987) and *Erik the Viking* (1989) whilst Terry Gilliam (1940–) has gone on to become one of the most boldly imaginative of contemporary film directors, responsible for such inventive fantasies as *Jabberwocky* (1977) and *The Time Bandits* (1980), the visually dazzling *The Adventures of Baron Munchausen* (1988) and his dark Orwellian masterwork *Brazil* (1985).

The remaining members retain a joint interest in the company Prominent Features which backed *A Fish Called Wanda* and Palin's *American Friends* (1991).

MOREAU, Jeanne

Born 23 January 1928, Paris, France.

A student at the Conservatoire National D'Art Dramatique, Moreau made her stage debut in *La Terrasse de Midi* (1947) and began a long association with the Comédie Française (1948–52) following her appearance in *A Month in the Country* (1948). She later worked with the Théâtre National Populaire.

She made her film debut in *Dernier Amour* (1948) and would perform in a succession of rarely memorable film roles over the next decade.

Her performances of then daring sexual frankness as the adulterous provincial wife in Louis Malle's *Les Amants* (The Lovers) (1958) coupled with an earlier appearance in ▷ Louis Malle's

Jeanne Moreau in *Eva* (1962)

Ascenseur Pour L'Echafaud (Lift to The Scaffold) (1957) brought her to screen prominence and she became one of France's most respected film performers and a prime figure in the ▷ nouvelle vague.

An intense, hypnotic actress bearing physical similarities to ▷ Bette Davis, she proved capable of immersing her own personality beneath a succession of generally world-weary, sensual characterisations that displayed a complex web of emotions. Her best roles include the female at the centre of the romantic triangle in ▷ *Jules et Jim* (1961), the icy sexual catalyst in *Journal D'Une Femme De Chambre* (Diary Of A Chambermaid) (1964) and the vengeance-seeking widow in *La Mariée Était En Noir* (The Bride Wore Black) (1967).

She has worked frequently in English, most notably for ▷ Joseph Losey in *Eva* (1962) and ▷ Orson Welles in *The Trial* (1963) and *Chimes at Midnight* (1966).

In the 1970s, she began to direct, making *Lumière* (1976), a study of female friendship, *L'Adolescente* (The Adolescent) (1979) telling of a young city girl's rural romance during the summer of 1939, and the documentary on ▷ *Lillian Gish* (1984).

A welcome recent burst of activity includes performances in *Nikita* (1990) and such prestigious forthcoming work as ▷ Wim Wenders's *Till The End of the World* (1991) and ▷ Theo Angelopoulos's *The Suspended Step of the Stork* (1991).

MORECAMBE, Eric

Real Name John Eric Bartholomew
Born 14 May 1926, Morecambe, Lancashire, UK.
Died 28 May 1984, Cheltenham General Hospital, UK.

The only child of a manual worker, Morecambe left school at the age of thirteen and subsequently won a local talent contest that earned him a place in the touring show *Youth Takes A Bow* (1940). Also part of the show was Ernest Wiseman (Wise) (1925–) and, in 1941, they formed a double-act that was later dissolved when Morecambe was drafted to work in the mines.

They met up again in 1947 and resumed the partnership in *Lord George Sanger's Variety Circus* and began to make appearances together that led to their first, poorly received television series *Running Wild* (1954).

They toured extensively, all the while smoothing the brash edges of their comic personae, perfecting their timing and developing a keen rapport until *The Morecambe and Wise Show* (1961–68) firmly entrenched them in the ranks of television's most beloved double-acts.

They also appeared in a number of feature films including *The Intelligence Men* (1965), *That Riviera Touch* (1965) and *The Magnificent Two* (1967) all of which failed to capture the affection and spontaneity in their best work.

Basing their comedy on the interplay of two familiar characters, Eric the pipe-smoking, naive sceptic and Ern of the short fat hairy legs and perennially slapped cheeks, they became national institutions with a host of catchphrases, propensity for musical numbers and an ability to attract the most celebrated of showbusiness names as guest participants in their fond mayhem.

Morecambe suffered a major heart attack in 1968, but recovered to continue the show under the auspices of the BBC and later Thames Television with their Christmas Special becoming almost as much a fixture of the holiday season as mince pies and carols.

Persistently weakened by poor health, he underwent open heart surgery in 1979 and subsequently took to writing, publishing the novel *Mr Lonely* (1981) and the children's books *The Reluctant Vampire* (1982) and *The Vampire's Revenge* (1983).

His partner has attempted to pursue a solo career since 1984 and their best work is permanently on display in television compilations and video releases.

MORRICONE, Ennio

Born 10 November 1928, Rome, Italy.

Raised in a strongly musical family, Morricone was composing at the age of six and later studied at the Santa Cecilia Conservatory in Rome where he received a diploma in composition, trombone and orchestra direction.

Pursuing a diverse career as a songwriter, arranger and composer of classical chamber and symphonic works, he wrote his first film score for *Il Federale* (The Fascist) (1961).

Working in a variety of genres, he subsequently began an association with director ▷ Sergio Leone on ▷ *Per Un Pugno Di Dollari* (A Fistful of Dollars) (1964), the first ▷ spaghetti western. Utilising a diversity of sound effects, from a whiplash to ricocheting bullets, whistles and eerie cries, he created a memorably baroque score that matched the tone of swift violence and black humour in films that revitalised the western with their highly stylised approach. Notable work for Leone includes *Il Buono, Il Brutto, Il Cattivo* (The Good, The Bad and The Ugly) (1966), *C'Era Una Volta Il West* (Once Upon A Time in the West) (1968) and *Once Upon A Time in America* (1984).

Probably the most prolific of all film composers, his credits run into the hundreds and include everything from farcical comedies to historical epics, from gangster chronicles to the bloodiest of continental horror yarns and even pseudonymous contributions under the name of Leo Nichols or Nicola Piovani. Inevitably, quality and originality have sometimes been sacrificed, but he has experimented with a wide range of percussion, vocal or choral effects and such styles as jazz, rock or folk to complement the visual imagery and emotional intentions of any given director.

A frequent composer for the most distinguished of Italian directors, including ▷ Bertolucci, ▷ Pasolini and the ▷ Taviani Brothers, his numerous scores for indigenous productions include *Prima Della Revoluzione* (Before the Revolution) (1964), *Teorema* (Theorem) (1968), *Allonsanfan* (1974) and *Cinema Nuovo Paradiso* (1989).

Moving on to international assignments, he received a British Academy Award for his evocative pastoral score for *Days of Heaven* (1978) and an Oscar nomination for his majestic choral accompaniment to the epic moral conflicts of *The Mission* (1986).

His other American scores include *The Thing* (1982), *The Untouchables* (1987) and *Old Gringo* (1989).

MOVIOLA Although ostensibly a trade name, Moviola has come to be used as a term for any editing machine. The Moviola itself is an upright portable machine that can handle a single reel of motion-picture film and a single soundtrack which can be operated individually or run synchronically. The picture can be viewed on a small screen and the equipment's flexibility has made it efficacious in the editing of a film, when a reel can be examined frame-by-frame, and also for students keen to dissect all minutiae of a film's construction.

MULTIPLANE An animation technique developed by Ub Iwerks (1900–71) for the ▷ Walt Disney Studios, multiplane involves the drawing of diverse parts of a scene on separate ▷ cels made of transparent plastic or glass. The cels are then placed a certain distance apart in different planes beneath the camera and when photographed this creates the illusion of three-dimensional depth. First used in the short cartoon *The Old*

Mill (1937) its first feature-length application was in ▷ *Snow White and the Seven Dwarfs* (1937) and it brought the Disney Studio a Special Class II Plaque in the Scientific or Technical category at the 1938 Oscar ceremony.

MURNAU, F. W.

Real Name Friedrich Wilhelm Plumpe
Born 28 December 1888, Bielefeld, Germany.
Died 11 March 1931, Santa Barbara, California, USA.

A student of philology in Berlin, and art history and literature in Heidelberg, Murnau was briefly an actor with Max Reinhardt's theatre troupe before World War I.

A combat pilot, he crashlanded in Switzerland during 1917. Returning to Germany two years later, he founded the Murnau Veidt Filmgesellschaft and made his directorial debut with *Der Knabe in Blau* (The Boy in Blue) (1919).

Experimenting with the mobility of the camera, his expressive use of light and shade heightened the menace in such macabre works as *Der Januskopf* (1920) (a version of Dr Jekyll and Mr Hyde) and ▷ *Nosferatu* (1922), a chilling and faithful rendition of the Dracula myth.

After a successful trio of films with actor Emil Jannings (1885–1950), including *Der Letze Mann* (The Last Laugh) (1924), he moved to America and made ▷ *Sunrise* (1927) a lyrical tale of a young rural couple whose love is threatened by the sophistication of the big city. This won three of the first ever Oscars including Best Actress for Janet Gaynor (1906–84).

Although much of his work no longer exists, what does remain reveals a career of great accomplishment and much promise. He had just completed the poetic South Seas documentary *Tabu* (1931) and was about to sign with Paramount when he was killed in a car crash.

MURPHY, Eddie

Full Name Edward Regan Murphy
Born 3 April 1961, Brooklyn, New York, USA.

The son of a New York City policeman, Murphy was a popular prankster and mimic at school who decided to pursue a career in showbusiness after hosting a talent show at the Roosevelt Youth Center in 1976.

Eddie Murphy, Judge Reinhold and John Ashton in *Beverly Hills Cop* (1984)

A stand-up comic in Long Island night spots, he first came to national prominence on the television show *Saturday Night Live* (1980–4), where he created a repertoire of satirical characters including the militant film critic Raheem Abdul Muhammad, senior citizen Solomon and television huckster Velvet Jones.

A charismatic, self-confident humorist whose act relies on scatological observation, sharp satire, impersonation and profanity, he commanded the screen in his dynamic, scene-stealing film debut as a fast-talking con in the comedy thriller *48 HRS* (1982) and became one of the most potent box-office draws in the film industry with his work as an engaging street-smart hustler in *Trading Places* (1983), and in the title role of a Detroit cop coping with the bizarre culture clash of crime-busting California-style in *Beverly Hills Cop* (1984).

His virtually unbroken run of successes continued with *The Golden Child* (1986), *Beverly Hills Cop 2* (1987) and *Coming to America* (1988), a more traditional comedy-fantasy that allowed him to illustrate his flair for impersonating both white and black characters.

Critical appreciation has not kept pace with public affection for his work and he was roundly denounced for his sluggish directorial debut *Harlem Nights* (1989). The disappointing box-office response to that, combined with a comparatively poor showing for *Another 48 HRS* (1990) would suggest a certain audience weariness towards his now over-familiar act and mannerisms of seemingly spontaneous, improvised verbal repartée and a trademark laugh akin to a gurgling drain.

Vastly popular as a stand-up comedian, his concert film *Eddie Murphy – Raw* (1987) became one of the most successful in the history of the medium and his bestselling albums include *Eddie Murphy Comedian* (1983) and *How Could It Be?* (1984). His latest film project is entitled *Big Baby* (1991).

NAPOLEON

Abel Gance (France 1927)
Albert Dieudonne, Wladimir Roudenko, Gina Manes, Nicolas Koline.
270 mins (original length). b/w.

Despite its epic length and scope, ▷ Abel Gance's *Napoléon* is only a sketch for a much more massive, ultimately unfulfilled project. The film was hailed on its opening in Paris not only as a triumphant restoration of the French film industry destroyed by the 1914–18 war, but also as a vindication of the artistic spirit over the overt commercialism of pre-war filmmaking in France. Gance took a distinctly romantic view of great men throughout his films, and Napoleon was certainly no exception, but the film has been admired, then and now, as much for its extraordinary technical virtuosity as for its portrayal of the Emperor. That portrayal is in any case a partial one, since the film only manages to take in his boyhood, youth, and early military career; the remainder of the huge project to record his entire life which the director had envisioned was never committed to film, thereby depriving Gance of the tragic ending which was so characteristic of his work. What he undoubtedly did succeed in doing, however, was to make full and unprecedented use of the possibilities offered by the camera, which swoops and twists through scene after scene in extraordinary fashion, and of film itself, including some visionary editing and cross-cutting techniques, and a version of the ending in which the closing scenes of Napoleon's entry into Italy were projected as a heroic triptych across three screens, rather than conventionally restricted, like the rest of the film, to one. Like most silent films, however, *Napoléon* does not exist in a definitive print. Even at the time of its original release, there were several different versions, while modern audiences have the choice of an edited American print generated by ▷ Francis Ford Coppola, or Kevin Brownlow's (1938–) epic five-hour amalgamation of all extant footage. Neither can claim genuine authenticity, but Brownlow's tireless work is undeniably an impressive tribute to Gance's artistic vision.

NEIGHBOURS

Australia 1985–

The story behind the success of this unassuming tale of everyday Australian folks is – appropriately enough – filled with bizarre twists. Its creator was Reg Watson, who based Ramsay Street, the fictional Melbourne location, on the Brisbane street where he grew up. With shows like *Young Doctors*, *Sons and Daughters* and *Prisoner: Cell Block H* to his name, Watson is now a soap giant in his homeland. More important to the development of *Neighbours*, perhaps, is the work he did during a stay in Britain, where he created that other much-maligned, but curiously popular saga, *Crossroads*.

Conceived as a part-comedy, but launched as a somewhat po-faced drama on Australian Channel 7, *Neighbours* was axed after six months. Channel 10 bought it, lightened it up and introduced characters like Charlene (Kylie Minogue (1968–)), and Scott (Jason Donovan (1969–)) to give it youth appeal. Only heavy promotion stopped a second cancellation, but it caught on, and was then spotted by a BBC buyer anxious to fill yawning daytime schedules. Soon a cult hit in Britain, *Neighbours*' fate was sealed when the teenage daughter of BBC programme director Michael Grade (1943–) complained about its timeslot. He switched its repeat showing from lunchtime to early evening, and *Neighbours* leapt to the top of the UK ratings. A plethora of young characters, the easygoing lifestyle of Melbourne and the good weather have all been cited as explanations of this success. The show is also exceptionally fast-moving – life-threatening dilemmas are resolved and forgotten about within minutes, and long-lost relatives regularly appear from nowhere. The discerning viewer can thus enjoy the ham acting and the absurdity of it all – something which can be characterised as The *Crossroads* Factor. *Neighbours* launched the pop careers of Minogue, Donovan and Craig McLachlan (who left the part of Henry Ramsay for a job on rival soap *Home And Away*). Also notable is the depressingly catchy theme by Tony Hatch.

NEO-REALISM Arising in opposition to the perceived artificiality of the middle-class dramas created under the Fascists, neo-realism was a style of filmmaking adopted by Italian directors in the immediate post-war period. It involved a documentary-like approach to fictionalised, socially aware stories of ordinary people and everyday events and attempts as realistic a presentation of these events as possible through the use of authentic locations and, where possible, non-professional actors. ▷ Visconti's *Ossessione* (1942) is seen to be a precursor of the short-lived movement but ▷ Rossellini's ▷ *Roma – Citta Aperta* (Rome – Open City) (1945), conveying the experiences of life during the last days of the German occupation, is the first of the genuine neo-realist films. Other notable works include *Paisan* (1946), *Sciuscia* (Shoeshine) (1946) and ▷ *Ladri Di Biciclette* (Bicycle Thieves) (1948).

As the key directors in the movement moved on to other types of project and the Italian government discouraged such an unpatriotic view of their country, the brief flourish of neo-realism ended, although its far-reaching influence can be seen on the documentary-like thrillers produced by Hollywood in the post-war years and the later ▷ Free Cinema movement in Britain.

NEWMAN, Paul Leonard
Born 26 January 1925, Cleveland, Ohio, USA.

A radio operator on torpedo bombers during the War, Newman later studied at Kenyon College in Ohio where he appeared in numerous undergraduate productions. Determined to learn more of the acting profession, he joined a number of repertory companies and later studied at the Yale School of Drama and the Actor's Studio in New York.

Finding work in many of the live television dramas then broadcast from New York, he made his Broadway debut in *Picnic* (1953–4) and a disastrous film debut as a Greek sculptor in the costume hokum *The Silver Chalice* (1954).

He redeemed himself as the snarling killer in the Broadway production of *The Desperate Hours* (1955) and in a variety of television roles including a musical version of *Our Town* (1955), in which he co-starred with ▷ Frank Sinatra and *The Battler* (1956), in which he aged twenty years to play an over-the-hill fighter.

Paul Newman in *Cool Hand Luke* (1967)

He returned to the cinema as the heavyweight boxing champion Rocky Graziano in *Somebody Up There Likes Me* (1956) and soon combined his blue-eyed masculinity, steely intelligence and rebel-like attitude to become one of the key stars of his generation in films like *The Left-Handed Gun* (1958) and *Cat On A Hot Tin Roof* (1958) (Best Actor Oscar nomination).

He maintained his television career and also starred on Broadway as the gigolo in *Sweet Bird of Youth* (1959–60), a role he repeated on film in 1962.

Throughout the 1960s his career blended bland escapist entertainments like *The Prize* (1963) and *Lady L* (1965) with some of his finest roles as the small-time pool shark in *The Hustler* (1961)

(Oscar nomination), amoral Texas stud in *Hud* (1963) (Oscar nomination) and the cocky, anti-authoritarian prisoner in *Cool Hand Luke* (1967) (Oscar nomination).

He enjoyed some of his biggest commercial success in partnership with ▷ Robert Redford in *Butch Cassidy and The Sundance Kid* (1969) and *The Sting* (1973) and his box-office potency continued with *The Towering Inferno* (1974). However, he was frustrated by the limitations imposed by his movie star image and chose to pursue interests in motor racing, politics and food production as well as turning to displaying a sensitivity and empathy for actors as the director of such films as *Rachel, Rachel* (1968), *The Glass Menagerie* (1987) and the television film *The Shadow Box* (1980).

He had appeared disenchanted with the acting profession for some time but recently he has returned with an enthusiasm and authority in a series of character portrayals that include the honest Miami businessman suspected of illegal activities in *Absence of Malice* (1981) (Oscar nomination), the down-at-heel lawyer who finds a last chance for redemption in a malpractice suit in *The Verdict* (1982) (Oscar nomination) and a reprise of 'Fast' Eddie Felson the pool hustler in *The Color of Money* (1986), which finally won him a Best Actor Oscar.

His recent work has striven for an increasingly minimalist style of underplaying and includes the military mind behind the Atom bomb in *Fat Man and Little Boy* (UK: The Shadowmakers) (1989) and the smalltown patriarch in *Mr and Mrs Bridge* (1990) although it proved somewhat colourless for the role of roistering politician Earl Long in *Blaze* (1990).

He has been married to the actress ▷ Joanne Woodward since 1958 and was awarded a special Oscar in March 1986 'in recognition of his many memorable and compelling screen performances and for his personal integrity and dedication to his craft'.

NICHOLS, Mike

Real Name Michael Igor Peschkowsky
Born 6 November 1931, Berlin, Germany.

The son of a Russian doctor, Nichols moved with his family to New York in 1939 to escape Nazi persecution. He later attended the University of Chicago and simultaneously commuted to New York where he studied acting with Lee Strasberg (1901–82).

He was subsequently instrumental in the development of the improvisational theatre group Compass Players in Chicago (1955–57). From 1957 to 1961, he gained great popularity on radio, records and the stage with Elaine May (1932–), dissecting the American psyche through offbeat, satirical duologues.

The partnership culminated in a year-long Broadway engagement after which he turned to direction, earning the first of seven Tony Awards for the original production of *Barefoot in the Park* (1963) starring ▷ Robert Redford. Highlights from his illustrious theatrical career include *The Odd Couple* (1965), *The Real Thing* (1984) and *Waiting for Godot* (1988), starring ▷ Steve Martin and ▷ Robin Williams.

He made his film debut with the excoriating screen version of *Who's Afraid of Virginia Woolf?* (1966) which earned acting Oscars for ▷ Elizabeth Taylor and Sandy Dennis (1937–) and Nichols his first Best Director nomination.

He won the Oscar for ▷ *The Graduate* (1967) and has consistently revealed a flair for comedy, a liking for literate scripts and a talent to elicit polished performances from his casts.

Catch 22 (1970) and *Carnal Knowledge* (1971) confirmed him as an incisive chronicler of American life, social mores and sexual politics but the financial failure of the thriller *Day of the Dolphin* (1973) and the madcap comedy *The Fortune* (1975) brought about his absorption in the theatre. Apart from the concert film *Gilda Live* (1980), he did not direct for the cinema again until *Silkwood* (1983) (Oscar nomination).

His subsequent work, largely in comedy, has not displayed the same cutting edge or political awareness but has included such smoothly crafted entertainments as *Biloxi Blues* (1988), the popular corporate Cinderella story *Working Girl* (1988) (Oscar nomination) and the showbusiness saga *Postcards from the Edge* (1990).

His latest project is *Regarding Henry* (1991) starring ▷ Harrison Ford.

NICHOLSON, Jack
Born 22 April 1937, Neptune, New Jersey, USA.

An actor in high school, Nicholson moved to Los Angeles and worked as an office boy in the M-G-M cartoon department whilst studying acting with Jeff Corey (1914–).

Working on television, he was seen by producer-director ▷ Roger Corman and made his film debut as a teenage delinquent in *Cry Baby Killer* (1958). His work over the next decade was mostly in low-budget horror and biker movies, including a memorable appearance as the dentist-loving masochist in *Little Shop of Horrors* (1960).

An association with Corman also allowed him to work as a scriptwriter, contributing to such films as *Thunder Island* (1963) and *Flight to Fury* (1966).

When Rip Torn (1931–) left the cast of ▷ *Easy Rider* (1969) Nicholson stepped in and commanded the screen as a booze-loving lawyer, tragically drawn to throwing off the shackles of respectability. Nominated for a Best Supporting Actor Oscar, he was a performer whose time had finally arrived and in *Five Easy Pieces* (1970) (Best Actor Oscar nomination) his aimless, middle-class drifter seemed to personify the rudderless state of a sceptical generation no longer sure of their place in the world.

Now established as one of the most incisive portrayers of explosive nonconformists, he revealed the breadth of his talent in a gallery of characterisations virtually unrivalled in cinema history; unleashing his drive, charisma and mischievous sense of humour on the roles of a jaded disc-jockey in the underrated *The King of Marvin Gardens* (1972), the raunchy sailor in *The Last Detail* (1973) (Oscar nomination), the perplexed 1930s private eye in *Chinatown* (1974) (Oscar nomination) and the quintessential free-spirited rebel in ▷ *One Flew Over the Cuckoo's Nest* (1975) (Oscar).

Unperturbed by the need to protect his standing or enhance his status, he has also accepted supporting roles, singing as a Harley Street Specialist in *Tommy* (1974) and providing a few powerful moments as the union leader in *The Last Tycoon* (1976).

Slowing his work-rate, he was then seen on thunderous form as the maniacal patriarch in *The Shining* (1980) and brought presence to bear on his sardonic interpretation of Eugene O'Neil in the epic *Reds* (1981) (Oscar nomination).

An Oscar winner again for his comically etched performance as the over-the-hill astronaut in *Terms of Endearment* (1983), recent years have witnessed a softening of his screen image and a tendency for him to play lovable variations on his highly entertaining but now somewhat baroque style of roguish merriment. Notable recent performances include the Neanderthal Mafia hit man in *Prizzi's Honour* (1985) (Oscar nomination), and the randy Devil in *The Witches of Eastwick* (1987). However, the best of his recent characterisations has been his subdued and very moving hobo attempting to pacify the ghosts of his troubled past in *Ironweed* (1987).

More recently, his style was positively operatic in his highly lucrative incarnation as the Joker in the blockbuster *Batman* (1989) (Oscar nomination).

His occasional films as director include the jolly, indulgent western *Goin' South* (1978) and *Two Jakes* (1990) a poorly received sequel to Chinatown.

He is said to be considering a biography of Napoleon as his next venture.

NICKELODEON The earliest type of cinema, the nickelodeon derived its name from the admission price of a nickel with odeon being the Greek word for theatre. The first one to be opened was in Pittsburgh in 1905 and by 1910 there were reckoned to be some 10,000 spread over America. The rush to satisfy the public appetite for motion pictures led to the opportunistic conversion of shops, halls and arcades and the typical nickelodeon had a seating capacity of around one hundred. However, the turnover was swift with most programmes running between twenty minutes and one hour. The arrival of feature-length films, higher profit margins and a public demand for more sophisticated surroundings all contributed to the demise of the nickelodeon and the rise of the grand cathedral-style cinema thereafter.

THE NIGHT OF THE HUNTER

Charles Laughton (US 1955)
Robert Mitchum, Shelley Winters, Lillian Gish, Billy Chapin, Sally Jane Bruce.
93 mins. b/w.

The Night of the Hunter is a curious anomaly in the history of Hollywood cinema, but an undeniably fascinating one. A religious allegory of the battle between evil (▷ Robert Mitchum's false, murderous 'Preacher' Powell) and good (the pure Christian protector Rachel, played by ▷ Lillian Gish) cast in the form of a suspense thriller, it employed a self-consciously artistic approach in direction and editing which was largely inimical to

Robert Mitchum in *Night of the Hunter* (1955)

the Hollywood way, and, as it turned out, to the film's own commercial and critical success. One result of that failure was that debutant director ▷ Charles Laughton was never again given the chance to direct; history, however, has alloted a more honoured place to his unique and highly personal vision in this flawed masterpiece than his contemporaries were prepared to do. The film betrays a certain lack of coherence in welding together its diverse concerns, but overcomes that weakness with a vividly atmospheric visual imagery and highly inventive, dream-like use of light and shadow. Its greatest strength, though, lies in the acting, particularly that of Robert Mitchum in arguably his greatest on-screen performance as the evil Harry Powell, remorselessly stalking the innocent children whose mother he has killed. Only the children, still devoid of the taint of adulthood, are able to see Powell for what he is, at least until the equally innocent Rachel intercedes on their behalf. The film retains much of its dark, fairy-tale power, but remains a maverick creation, and seems closer to European art cinema than anything we would expect to find in the Hollywood canon.

NIVEN, David

Full Name James David Graham Nevins
Born 1 March 1910, London, UK.
Died 27 July 1983, Chateau D'Oex, Switzerland.

Never a man to let the exacting call of veracity impede the telling of a good story, Niven was born in London rather than the more fanciful locale of Kirriemuir in Scotland that he and others were prone to claim. As a student at Sandhurst Royal Military College, he appeared in a production of *The Speckled Band* (1928), although he showed little inclination to pursue acting as his career.

After serving with the Highland Light Infantry, he set sail for Canada and, after several dubious forms of employment including that of a whisky salesman, he drifted towards Hollywood and found work as an extra in films like *Mutiny on the Bounty* (1935).

Signed by ▷ Sam Goldwyn, he developed into a polished light comedian and gallant hero in *The Charge of the Light Brigade* (1936), *The Dawn Patrol* (1938) and *Bachelor Mother* (1939).

He served with distinction in the British Army during World War II, securing temporary releases to make such superior, morale-boosting films as *The First of the Few* (1942) and *The Way Ahead* (1944).

He returned to Hollywood and spent the next thirty years as an urbane leading man, lending much needed élan to a lot of generally pallid comedy material but perfectly cast as the gentlemanly adventurer Phineas Fogg in *Around the World in 80 Days* (1956) and playing with some discretion and skill the rare dramatic role of the bogus major with an unsavoury private life in *Separate Tables* (1958), a performance which won him a Best Actor Oscar.

The most popular of the many light comedies and high adventures in which he appeared include *Ask Any Girl* (1959), *The Guns of Navarone* (1961), *The Pink Panther* (1964), *Bedtime Story* (1964), *Casino Royale* (1967) and *Paper Tiger* (1974) in which he was again rather touching as a man facing up to a fabricated image of heroism.

A hugely personable man, much-loved celebrity and inimitable raconteur, he wrote two volumes of vastly amusing Hollywood anecdotes, *The Moon's A Balloon* (1971) and *Bring on the Empty Horses* (1975) which were international bestsellers and he also penned the fictional volumes *Round the Rugged Rocks* (1951) and *Go Slowly, Come Back Quickly* (1981).

His last few years were spent in the public eye, courageously concealing his suffering from the wasting motor neurone disease. He had a good supporting role in *Death on the Nile* (1978) but his last appearances were undistinguished and his final role in *The Curse of the Pink Panther* (1983) was dubbed by comedian Rich Little to cover the vocal deterioration caused by his fatal illness.

NOIRET, Philippe

Born 1 October 1930, Lille, France.

Encouraged to become an actor after his participation in a school play, Noiret studied under Roger Blin (1907–84) and at the Dramatic Centre of the West, making his film debut in *Gigi* (1948).

In 1953, he joined the Theatre National Populaire and remained primarily a stage actor and cabaret performer for the remainder of that decade, apart from an isolated leading role in *La Pointe Courte* (1955) by ▷ Agnès Varda.

Embracing the cinema with a vengeance from 1960, he was seen as a female impersonator in *Zazie Dans Le Metro* (1960) and soon became established as a dependable supporting actor with opportunities to shine as the boring, sanctimonious husband of *Thérèse Desqueyroux* (1962), the phoney man of the cloth in *Les Copains* (1964) and the endearingly indolent farmer in the hit comedy *Alexandre Le Bienhereux* (Alexandre) (1967).

He made his English-language debut as the philandering politician in *Lady L* (1965) and has appeared in such American productions as *Night of the Generals* (1967), *Topaz* (1969) and *Justine* (1969).

A tall, well-built character actor with a wide range he was now becoming established as one of France's most distinguished and versatile film performers, a process encouraged by his long association with director ▷ Bertrand Tavernier who first used his talents in *L'Horloger de St Paul* (The Watchmaker of St Paul) (1973) in an affecting role as a quiet widower forced to re-examine his life when his son is accused of murder. Other roles for Tavernier include an exuberant Philippe D'Orleans in *Que La Fête Commence* (Let Joy Reign Supreme) (1975), a glory-seeking magistrate in *Le Juge et L'Assassin* (The Judge and the Assassin) (1976) and the lubricious, lackadaisical colonial policeman in *Coup de Torchon* (Clean Slate) (1981).

Other notable performances include the judge in *La Grande Bouffe* (Blow-Out) (1973), *Le Vieux Fusil* (The Old Gun) (1975) for which he received the César as Best Actor, and the pornographic film director in *Il Commune Senso Del Pudore* (1976).

Adept at all genres and a gamut of emotions from reassurance to venality, comic bumbling to crushed ardour, he remains one of the world's busiest and most respected actors and enjoyed a huge success as the cheerfully corrupt veteran law enforcer in *Les Ripoux* (Le Cop) (1984).

He achieved his century of film appearances with two of his finest performances, bringing a tenderness and grace to his portrayal of the aged projectionist and mentor in *Nuovo Cinema Paradiso* (Cinema Paradiso) (1989) and the World War I veteran officer in Tavernier's *La Vie est Rien D'Autre* (Life and Nothing But) (1989) (César).

He then appeared in the much-anticipated sequel *Les Ripoux Contre Les Ripoux* (1990), the popular *Uranus* (1990) and *Especially on Sunday* (1991).

NOSFERATU (Nosferatu, Eine Symphonie des Grauns)
F W Murnau (Germany 1922)
Max Schreck, Greta Schroder, Alexander Granach, Gustav von Wangenheim.
75 mins. b/w.

The subtitle 'A Symphony of Horror' does not seem misplaced for this first and most impressively realised of all the film versions of Bram Stoker's classic novel (Mina in the book, but Nina in the film; the names were changed partly because no rights were paid for use of the book, although some later prints confusingly reverted to Stoker's names). Nina, whose own curiously emaciated appearance seems to prefigure her martyr's fate, then becomes both the victim of the vampire's desire, and the means of the vampire's destruction, inverting the book's symptomatic Victorian fear of female sexuality. Inevitably, critics have expli-

Nosferatu (1922)

of Victorian sexual repression, *Dracula*. At every stage in his career, Murnau was part of the German Expressionist movement, whose guiding lights unquestionably included the post-Victorian theories of Sigmund Freud, and his reading of the vampire myth is heavily weighted with that changing interpretation of its sexual significance. That is nowhere more apparent than in his decision to reverse the central relationship of the novel, turning the male character (Stoker's van Helsing) from strong destroyer of Dracula/Nosferatu into an ineffectual weakling, and foregrounding the female cated the Freudian resonances at great length, but for the contemporary viewer, the main interest of the film may well lie in the extraordinary visual qualities which Murnau brings to it, and the atmospheric, imaginatively ▷ Expressionist backdrops he invokes, many using real locations rather than the more customary studio sets. *Dracula* has been remade many times and in many ways, but Werner Herzog's *Nosferatu The Vampire* (1979) specifically attempted, with very mixed results, to invoke the spirit, and often the letter, of this pioneering, and arguably still greatest, version.

NOUVELLE VAGUE A term given to the 'New Wave' of young French film directors who came to the fore in the late 1950s and 1960s, the nouvelle vague was a group of individual film enthusiasts whose critical writings in the journal *Cahiers Du Cinéma* championed an ▷ auteurist view of filmmaking and proposed a cinema that discarded many of the prevailing formulae that they perceived as responsible for the stultifying state of the French industry at that time.

Eventually they put their ideas into practice behind a camera. ▷ Claude Chabrol's *Le Beau Serge* (Bitter Reunion) (1958) is considered to be the first authentic flourishing of the 'New Wave' and over the next few years ▷ Francois Truffaut directed *Les Quatre Cent Coups* (The 400 Blows) (1959), ▷ Jean-Luc Godard made ▷ *A Bout De Souffle* (Breathless) (1960) and ▷ Alain Resnais created *Hiroshima, Mon Amour* (1959).

Revelling in the freedoms afforded by lightweight, hand-held cameras they made extensive use of natural locations. Combining this with innovative storylines, elliptical editing and unconventional use of sound, they created an alternative, more loosely structured cinema that placed less emphasis on smoothly rounded and resolved linear narratives to focus on the individualistic expression of the director's artistic sensibility. Thus the manner of presentation and the recognition of the artifice in cinematic storytelling became as important a part of the cinema experience as the emotions of the story itself.

Notable titles from the early flourishing of the nouvelle vague include ▷ *Jules et Jim* (Jules and Jim) (1961), ▷ *L'Anneé Dernière A Marienbad* (Last Year At Marienbad) (1961) and *Vivre Sa Vie* (It's My Life) (1962). As the key individuals in the movement chose radically diverse paths in their subsequent careers, the term as a generic expression for the totality of their work holds little value but the influence of the nouvelle vague on all filmmaking of the past thirty years has been incalculable.

NYKVIST, Sven Vilhem
Born 3 December 1922, Moheda, Sweden.

The son of Lutheran missionaries based in the Belgian Congo, Nykvist was raised by relatives in Sweden and first developed an interest in photography from the slides his parents would send to convey their life in Africa.

A student at the Stockholm Municipal School for Photographers, he secured employment at the Sandrew Studios as a focus-puller on such films as *The Poor Millionaire* (1941) and *In the Darkest Corner of Smaland* (1943). After a brief sojourn at the Cinecitta Studio in Rome, he returned to Sweden and made his debut as a cinematographer on *The Children from Frostmo Mountain* (1945).

Gainfully employed in Swedish productions of the period, he also made the documentaries *In the Footsteps of the Witch Doctor* (1947) and *Under the Southern Cross* (1952) which he co-directed.

Gycklarnas Afton (Sawdust and Tinsel) (1953) began his long association with ▷ Ingmar Bergman, although their partnership did not flourish creatively until a reunion on *Jungfrukallan* (The Virgin Spring) (1959). He subsequently shot all of the director's work, developing a restrained style of low-key lighting and austere imagery that served as a visual partner in the creation of an all-embracing psychological mood. Their many films together include ▷ *Persona* (1966), *Skammen* (The Shame) (1968), *Hostsonaten* (Autumn Sonata) (1978) and the director's last ▷ *Fanny Och Alexander* (Fanny and Alexander) (1982). He won an Oscar for his cinematography of Bergman's *Cries and Whispers* (1972), a harrowing story of a woman dying in agony from cancer that makes striking use of an extraordinary palette of rich red hues.

In 1965, he directed the documentary *The Vine Bridge* (1965) and has brought his consummate artistry to bear on a number of international productions including *Swann in Love* (1984), *The Sacrifice* (1986), *Another Woman* (1988), *Crimes and Misdemeanours* (1989) and *The Unbearable Lightness of Being* (1988), for which he received a further Oscar nomination.

ODD MAN OUT
Carol Reed (UK 1946)
James Mason, Robert Newton, Kathleen Ryan, Robert Beatty, Cyril Cusack.
115 mins. b/w.

The indigenous British film industry

blossomed in the immediate post-World War II years, and ▷ Carol Reed was among the handful of directors who put the new possibilities to best use. *Odd Man Out* is a stylish and occasionally slightly over-heated thriller about an anti-British insurgent (▷ James Mason) on the run from the authorities after an audacious wages robbery which goes wrong. The film explores Reed's favourite theme of a lone protagonist caught in a situation over which he has no control, and which he must struggle (often vainly) to overcome. Mason is excellent as the doomed romantic hero of the Irish rebellion (it is impossible to imagine a similar contemporary treatment of an IRA or

Robert Newton and James Mason in *Odd Man Out* (1946)

INLA man), and Reed directs both the action sequences and the more suspenseful moments with immense style, although the highly symbolic ending seems a little overdone. The film remains one of the masterpieces of British cinema, and pre-figures in many respects his subsequent and even better known ▷ *The Third Man* (1949), notably in its atmospheric monochrome photography, its underworld milieu, and its lone but dubious protagonist being hunted toward a tragic end. The director adopts a near-documentary approach to the film which clearly echoes the style of wartime documentaries, allied to a developing taste for cinematic realism which was to dominate British film-making in the ensuing decades, and arguably does so yet. *Odd Man Out* was updated and remoulded to fit black Americans in the civil rights struggle as *The Lost Man* in 1969, but with a considerably less successful outcome.

OLIVIER, Laurence Kerr (Lord)
Born 22 May 1907, Dorking, Surrey, UK.
Died 11 July 1989, Steyning, Sussex, UK.

The son of a clergyman, Olivier's school appearance as Katharine in *The Taming of the Shrew* set him on an illustrious thespian career the likes of which is unlikely to be seen again; he was widely experienced on stage before his film debut in the short film *Hocus Pocus* (1930).

New York success in *Private Lives* (1930) took him to Hollywood and an unhappy sojourn as a dashingly unexceptional leading man in films like *The Yellow Ticket* (1931) and *Westward Passage* (1932).

His dismissal from *Queen Christina* (1933) in favour of John Gilbert (1895–1936) appears to have dampened his ardour for the cinema in general and Hollywood in particular.

His stage reputation in London's West End and with the Old Vic was unassailable when he returned to Hollywood as Heathcliff in *Wuthering Heights* (1939) (Best Actor Oscar nomination) and he stayed for a productive spell that included *Rebecca* (1940) (Oscar nomination), *Pride and Prejudice* (1940) and *That Hamilton Woman* (1941) in which he played Nelson.

An actor of matinée idol looks and unquenchable virtuosity, who revelled in a chameleon like approach to characterisation involving make-up and disguise, he made several distinguished contributions to the wartime British film industry, notably his superbly crafted, patriotic ▷ *Henry V* (1944) (Oscar nomination) which also marked his directorial debut.

A fearless artist, unafraid to take risks, he proved a great populariser of Shakespeare with his productions of *Hamlet* (1948) (Oscar), *Richard III* (1956) (Oscar nomination) which features one of his finest performances as the grotesquely vulpine crookback, and the less successful *Othello* (1965) (Oscar nomination).

Looking more like his off-screen self, he gave a fully-rounded account of a man's destructive amorous obsession in *Carrie* (1952) and he impressed with the repellent verisimilitude of his seedy, second rate musical hall star in *The Entertainer* (1960) (Oscar nomination).

A tower of strength in the formation and running of Britain's National Theatre, his secondary interest in the cinema was displayed with a series of character roles and supporting assignments as high ranking officers and authority figures in such all-star ventures as *The Battle of Britain* (1969) and *A Bridge Too Far* (1977).

His acceptance of a leading role as the contemptuous and manipulative author in the two-hander *Sleuth* (1972) (Oscar nomination) was welcome but his appearance as a randy septuagenarian automobile tycoon in *The Betsy* (1978) was indicative of the lucrative rubbish he sometimes accepted over the last twenty years of his career.

Toying with a variety of ethnic character parts, he always lent a relish and sometimes flamboyant theatricality to such unlikely assignments as a bald Nazi in *The Marathon Man* (1976) (Oscar nomination), the elderly Nazi hunter grappling with the even unlikelier Mengele of Gregory Peck (1916–) in *The Boys From Brazil* (1978) (Oscar nomination) and the cantor in *The Jazz Singer* (1980). He made his last film appearance in *War Requiem* (1989).

His many television appearances include *John Gabriel Borkman* (1958), *Love Among the Ruins* (1975), *Jesus of Nazareth* (1977), *Brideshead Revisited* (1981), *A Voyage Round My Father* (1982), *King Lear* (1983) and *Lost Empires* (1986).

Created a life peer in 1970, he received an honorary Oscar in April 1979 for the totality of his career.

An appreciation of his protean contributions to British theatrical life over half a century would require a separate entry in itself. He covered his life and approach to his profession in the books *Confessions Of An Actor* (1982) and *On Acting* (1986).

OLIVEIRA, Manoel de

Real Name Manoel Candido Pinto de Oliveira.

Born 12 December 1908, Oporto, Portugal.

A passionate sportsman, it was Oliveira's renown as a high-jump champion and keen competitive motor racer that led to his appearance in Portugal's first talking feature *A Cancao De Lisboa* (The Song of Lisbon) (1933).

However, his interests lay behind the cameras and he had already directed his first short documentary *Douro, Faima Fluvial* (Hard Labour on the River Duoro) (1931) and would return to this format throughout the next decade on such social documents as *Ja Se Frabricam Automoveis Em Portugal* (They Already Manufacture Cars in Portugal) (1938) and *Famalicas* (1939).

He subsequently made his first feature *Aniki-Bobo* (1942), a simple but effective account of the street urchins growing up in his home town of Oporto that won praise for a mixture of naturalistic performances and location shooting that anticipated the basic tenets of ▷ neo-realism

Politically out of favour for many years, he devoted himself to vinegrowing and running his father's textile factory before returning to filmmaking with the documentary *O Pintor E A Cidade* (The Painter and the City) (1956). Continuing to encounter financial difficulties and other obstacles, he worked sporadically, completing the features *Acto Da Primaveros* (Act of Spring) (1963) and *O Passado E O Presente* (The Past and the Present) (1971), an international breakthrough and the first in what became known as the 'Tetralogy of the Frustrated Loves' that also comprises *Benilde Ou A Virgem-Mae* (Benilde or the Virgin Mother) (1975), *Amor de Perdicao* (Love of Perdition) (1978) and *Francisca* (1981), philosophical romantic dramas with the sensibilities of 19th century literature and a very formal cinematic technique of long shots and fixed compositions.

Established as Portugal's most distinguished director, he has worked productively over the past decade on art documentaries and such features as *The Satin Slipper* (1985), *Os Canibais* (The Cannibals) (1988) and *'Non' or The Vain Glory of Command* (1990) in which the last Portuguese colonial war in Africa serves as a starting point for an action-filled account of his country's history that is both epic in scope and intimate in quirky detail.

ON THE TOWN

Gene Kelly and Stanley Donen (US 1949)
Gene Kelly, Frank Sinatra, Betty Garrett, Ann Miller, Jules Munshin, Vera-Ellen.
98 mins. col.
Academy Award: *Best Score* Roger Edens and Lennie Hayton.

On The Town marks the beginning of the modern Hollywood musical. The film, co-directed by debutants ▷ Kelly and Donen (1924–), broke much new ground, and threw aside the rather old-fashioned mould in which the Hollywood musical had been set during its classic decade of the 1930s.

The most obvious change lay in the fact that the directors were able to persuade MGM to let them shoot on the streets of New York, rather than in the omnipresent studio sets which had dominated the genre. The amorous adventures of the three sailors on a twenty-four hour leave begin and end in the Brooklyn Navy Yard, scene of the famous 'New York, New York' routine, and take in the Empire State Building and the Statue of Liberty en route. Equally significantly, the dance routines and big musical numbers were integrated into the action in a way which had not been considered necessary in the films made by ▷ Fred Astaire in the 1930s, far less those of a set-piece specialist like ▷ Busby Berkeley. Songs are never treated as an aside, but are used to advance the plot rather than stall it, or to establish relationships between characters, like ▷ Ann Miller's (1919–) 'Prehistoric Man' routine in the Museum of Natural History, tapped out for the benefit of Jules Munshin (1915–70). The women whom Kelly and ▷ Sinatra meet are no shrinking violets, either, but seem unusually outspoken and liberated for their day, and every bit as dominant as the men. The other factor which has established the film as a continuously vital one, though, does not depend on its influence over subsequent generations of American musicals. Rather, it is the sheer exuberance and vitality of the cast, music and dancing; even without its ground-breaking role, *On The Town* would rate among the handful of greatest film musicals.

ON THE WATERFRONT

Elia Kazan (US 1954)
Marlon Brando, Eva Marie Saint, Karl Malden, Lee J. Cobb, Rod Steiger.
108 mins. b/w.
Academy Awards: *Best Picture*, *Best Director*; *Best Actor* Marlon Brando; *Best Supporting Actress* Eva Marie Saint; *Best Writing – Story and Screenplay* Budd Schulberg; *Best Cinematography – Black and White* Boris Kaufman; *Best Art Direction – Black and White* Richard Day; *Best Editing* Gene Milford.
New York Film Critics Awards: *Best Picture*, *Best Direction*, *Best Actor*.
Venice Film Festival: *Silver Prize*.

Despite some dissatisfaction over the ending, *On The Waterfront* has established a place as a classic American realist drama, and boasts one of the very great cinematic performances from ▷ Marlon Brando. The actor plays Terry Malloy, a failed boxer thrust by a combination of innate good instincts and tragic circumstance into standing up to the machinations of the hoodlums who run the New York docks, partly from principle, and partly to avenge the murder of his brother. The film was written by novelist Budd Schulberg from a series of newspaper reports on violence and corruption in the docks, and directed by ▷ Elia Kazan. Both were implicated in giving evidence to the House Un-American Activities Committee, and some critics have seen Terry Malloy's rather sudden apotheosis from reviled 'stoolie' (after testifying against the gangsters in a public court) to lone hero leading the longshoreman in a defiant stand against corruption as a self-justifying redemp-

Eva Marie Saint and Marlon Brando in *On The Waterfront* (1954)

tion for his creators as well. Interestingly, Schulberg chose a completely different, distinctly non-heroic ending for the novel he subsequently published. What is certain, though, is that the filmmakers' gritty determination not to spare the audience any of the pain and squalor of the unsanitised waterfront setting, and memorable performances from the whole cast, but especially from Brando, have secured its niche in film history. The famous 'I coulda been a contender' speech which Terry delivers to his brother in the back of a cab has been much parodied over the years, but is a genuinely moving moment in the context of the film. Brando succeeds in bringing a marvellous complexity to a basically simple character, notably in his relationship with the girl ▷ Eva Marie Saint (1924–). *On The Waterfront* is a brutally clear-cut morality tale, but an enduringly powerful one.

ONE FLEW OVER THE CUCKOO'S NEST

Milos Forman (US 1975)
Jack Nicholson, Louise Fletcher, Brad Dourif, Christopher Lloyd, Will Sampson.
133 mins. col.
Academy Awards: *Best Picture, Best Director, Best Actor* Jack Nicholson, *Best Actress* Louise Fletcher, *Best Screenplay* Lawrence Hauben and Bo Goldman

It was a remarkable enough achievement that *One Flew Over The Cuckoo's Nest* should have become the first film since *It Happened One Night* in 1934 to capture all five of the top Oscars. What is even more astonishing, though, is that the normally conservative Academy should have so lavishly honoured one of the most uncompromising and radical films to have emerged from Hollywood. It would have been difficult, certainly, to ignore the claims of ▷ Jack Nicholson in the role of Murphy, a rebellious spirit admitted to a docile psychiatric ward. Once there, he proceeds to turn the well-established, repressive norms of the institution on their head, waging war against the superbly cold Nurse Ratched (Louise Fletcher (1936–) took the Best Actress award for her portrayal), and giving the unfortunate inmates a taste of freedom, both metaphorical and, eventually, literal. Based on a novel by Ken Kesey, the film postulates a marvellous triumph for the human spirit over the dehumanising forces of repression and enforcement, and retains enough of the implicitly allegorical dimension of the book (written in the 1960s by a vigorously anti-establishment writer) to appeal to the immediate post-Vietnam disillusion of American youth with their political and social systems; it would be interesting to know how it would have been received if made a decade later. In so doing, it offered a hitherto undreamed of depiction of mental illness on screen, at once sympathetic and hugely funny, hard-hitting and absurd. It is Nicholson, though, who remains most strongly in the mind. In a film crowded with memorable performances, his stands out as an astonishingly exuberant and impeccably judged vindication of the uncowed individual who will not be quashed in to a standardised mould. If Murphy is denied his final triumph in person, his example lives on in Will Sampson's (1935–87) ecstatic transformation from shambling broom-pusher to triumphant chief, making that classic American literary escape, as Huck Finn succinctly put it, by 'lighting out for the Territory'.

OPHULS, Marcel

Born 1 November 1927, Frankfurt, Germany.

The son of director ▷ Max Ophuls, Ophuls has established his own reputation as a director of distinctive documentaries that use narrative techniques to jog the collective memory and probe notions of justice, inhumanity, guilt and moral absolutes in post-war Europe.

Fleeing with his family from France to America in 1941, he spent part of his adolescence in Hollywood before serving with the Occupation forces in Japan. Returning to France to study philosophy at the Sorbonne, he used the name Marcel Wall to work as an assistant director on such films as *Moulin Rouge* (1952), *Marianne De Ma Jeunesse* (1954) and his father's *Lola Montes* (1955).

Between 1956 and 1959, he worked as a radio and television story editor in West Germany. Back in Paris, he directed the short film *Matisse, or The Talent for Hap-*

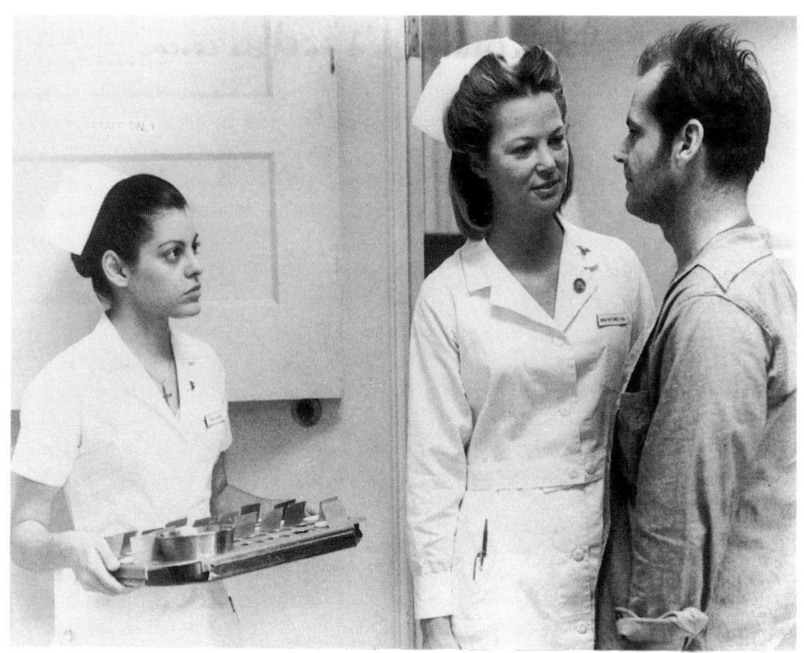

Louise Fletcher and Jack Nicholson in *One Flew Over The Cuckoo's Nest* (1975)

piness (1960), contributed one of the episodes to *L'Amour A Vingt Ans* (Love at 20) (1962) and directed the comedy *Peau de Banane* (Banana Peel) (1963) co-starring ▷ Jeanne Moreau and Jean-Paul Belmondo (1933–).

From 1966 to 1968, he worked as a television news reporter on French magazine programmes like *Zoom* and *16 Million de Jeunes*. He made his first significant historical documentary *Munich, Ou La Paix pour Cent Ans* (Munich or Peace in Our Time) (1967) before leaving Paris in 1968 to relocate in Germany.

Whilst directing numerous documentaries for German television, he completed *Le Chagrin et La Pitie* (The Sorrow and the Pity) (1970) a trenchant confrontation with the topic of wartime collaboration featuring telling interviews with the surviving inhabitants of Clermont Ferrand. Skillful in its unsparing juxtaposition of past footage, private testimony and contemporary music, it was an incisive exposé that sparked international controversy and was banned in France.

He followed this with *A Sense of Loss* (1972) on the situation in Northern Ireland and questioned who had the right to judge the inhumanity of man in *The Memory of Justice* (1976), which tackled the Nuremberg show trials and their aftermath.

Throughout recent years he has worked on documentaries for European and American television, lectured and written journalism and he returned to the full-length documentary investigation after a lengthy absence with *Hotel Terminus: The Life and Times of Klaus Barbie* (1988) which won the Oscar for Best Documentary. A compelling, thriller-like enquiry into the lives of those people who came into contact with the Nazi War criminal it also revealed Allied collusion in his subsequent flight and evasion of justice.

OPHULS, Max

Real Name Max Oppenheimer
Born 6 May 1902, Saarbrucken, Germany.
Died 26 March 1957, Hamburg, West Germany.

A stage actor and director, with around two hundred productions to his credit during the 1920s, Ophuls began his association with film as a dialogue director for Anatole Litvak (1902–74) at the Ufa Studios and made his directorial debut with *Dann Schon Lieber Lebertran* (1930).

The best known of his early films is *Liebelei* (1932) a typically graceful story of a doomed love affair between a young officer and a violinist. Leaving Germany in 1932, he spent the next decade at work throughout Europe, most notably on *La Singora di tutti* (1934) which recounts the life of a seemingly successful Italian film star whose public persona conceals heartaches from a tragic past.

The plight of women destroyed and corrupted by the power of love and their subservience in a male-dominated world was to feature in his most famous films. Resident in Hollywood from 1941, he made his American debut with the swashbuckler *The Exile* (1947) but was more at home with the supremely romantic *Letter From An Unknown Woman* (1948) which illustrates all the hallmarks of his artistry: the elegantly mobile camerawork, ornate, sensual decor and faultless composition.

After the stylish thrillers *Caught* (1949) and *The Reckless Moment* (1949), he moved to France, creating superbly crafted ruminations on the recurring theme of love's most breathless fantasies thwarted by the cruelest of circumstances.

His later achievements include *La Ronde* (1950), *Madame De* (1952) and his only colour film *Lola Montes* (1955) which makes an accomplished use of ▷ CinemaScope to convey the life of a notorious courtesan now reduced to selling herself as a circus attraction. 'Life is movement' she states and it is an appropriate philosophy for the cinema of Ophuls, whose son is the documentarist ▷ Marcel.

OPTICALS A general term for special-effects achieved by the use of previously exposed film in an optical printer. The process of duplicating the original image and placing it on a second generation of film allows for changes in lighting or the addition of new elements not present at the time of filming. Thus ▷ fades and ▷ dissolves can be achieved this way or more elaborate ▷ matte effects such as the seamless combination of live-action and animated material in a film like *Who Framed Roger Rabbit?* (1988).

ORPHEE

Jean Cocteau (France 1950)
Jean Marais, Francois Perier, Maria Casares, Marie Dea.
112 mins. b/w.
Venice Film Festival: *International Critics Award.*

The essential achievement of the cinematic strand of ▷ Jean Cocteau's diverse artistic endeavours is contained in what has become known as his *Orpheus Trilogy*, three disparate films made over a thirty year span, but linked by a remarkable continuity of vision. *Orphée* (Orpheus) stands between *Le Sang d'un Poète* (The Blood of A Poet) (1930) and *Le Testament d'Orphée* (The Testament of Orpheus) (1960), and is the most completely realised of all his films. Cocteau described the first film of this trilogy as 'a realistic documentary of unreal events,' and that encapsulation will stand equally effectively for the method and matter of *Orphée*. A wonderfully imaginative fusion of his obsession with Greek myth in general, and the story of Orpheus in the underworld in particular, with his own highly idiosyncratic ideas on the nature of the poet and the creative process, the film stands as one of the immense achievements of European cinema. The adventures of Orpheus and his wife Eurydice are depicted on two different planes, one in contemporary domesticity, the other in a fantastic underworld into which she is transported by the angel Heurtebise, who has fallen in love with her. Orpheus follows her over, having himself fallen in love with the Princess of Death, and Cocteau formally demarcates the two realms through an imaginative employment of such visual effects as reverse ▷ slow-motion, negative images, and a bath of mercury as the mirror through which they may pass over. The domestic existence of Orpheus, however, is shot quite realistically, and the division serves as an emblem of the position of the artist, caught between mundane reality and the wonders of the imagination. Always challenging and often funny, *Orphée* is Cocteau's masterpiece, and as close as he – and perhaps anyone else – came to his aim of realising film as poetry.

OSCAR The American Academy of Motion Picture Arts and Sciences was established by thirty-six leading figures in the film industry who met on 4 May 1927 and selected ▷ Douglas Fairbanks as their first President. One week later three hundred guests attended an industry banquet to discuss the Academy's aims and purposes. ▷ Louis B. Mayer suggested that the Academy should sponsor a series of awards to celebrate outstanding achievement in motion pictures. ▷ Art director Cedric Gibbons (1893–1960) sketched a statuette on a tablecloth that was later sculpted by George Stanley into the Academy Award – a goldplated figure standing on a reel of film carrying a sword. It is 13½ inches high and weighs 8½ pounds.

Barbra Streisand with her *Funny Girl* (1968) Oscar

The responsibility for nicknaming the award 'Oscar' has been claimed by many, including actress ▷ Bette Davis, but the likeliest explanation seems to be that Academy Librarian Margaret Herrick once declared it to resemble her uncle Oscar and, from 1931 onwards, the name stuck.

The first awards to honour achievement between 1 August 1927 and 1 August 1928, were announced on 18 February 1929. The presentations took place three months later at a Banquet at the Hollywood Roosevelt Hotel on 26 May. Contemporary accounts of the event detail two highlights from the evening: a portable sound projector was used to show 'talkie' footage of ▷ Adolph Zukor receiving the Best Picture Award for *Wings* (1927) from Douglas Fairbanks in New York, whilst ▷ Al Jolson was there in person to sing for the assembled crowd.

Fairbanks presented those first awards in four minutes and twenty two seconds and the whole evening had the air of a private party. The ensuing sixty years have turned the event into a gaudy, interminable spectacle dubbed, with good reason, 'the real star wars'.

Over the years the process has been streamlined; since 1935 the awards have covered a calendar year and other categories have been added including Best Foreign-Language Film, the Jean Hersholt Humanitarian Award and the Irving G. Thalberg Memorial Award in recognition of 'the most consistent high quality of production achievement by an individual producer during the year'.

The Oscar ceremony now has an estimated global audience of around one billion viewers and, apart from prestige, an award is deemed to add millions of dollars to a film's gross takings and to immeasurably advance a career, hence the lavish amounts now spent by the major studios on campaigns to support their favoured films and stars. The actual effect of winning however is debatable. Certainly a film like ▷ *Chariots of Fire* (1981) benefited from its 'Cinderella' win at the 1982 ceremony, but four Oscars could do nothing to prevent *The Right Stuff* (1983) from being a box-office flop and certain careers could not be seen to have advanced from an Oscar win. Louise Fletcher (1936–), Best Actress for ▷ *One Flew Over the Cuckoo's Nest* (1975), and F. Murray Abraham (1940–), Best Actor for *Amadeus* (1984), have both suffered from a sense of anti-climax in the afterglow of their victories.

Based in Beverly Hills, the Academy has an active membership of around 3,500 and devotes itself to research and education. Membership is by invitation only and members in specialist areas vote for nominations in their own field with the entire membership then eligible to vote from these nominations for a final winner. Since 1989 the Academy President has been actor Karl Malden (1913–).

OSHIMA, Nagisa

Born 31 March 1932, Kyoto, Japan.

An active student leader when he studied political history at Kyoto University, Oshima graduated in 1954 and spent five years as an assistant-director and scriptwriter before making his directorial debut with *Ai To Kibo No Machi* (A Town of Love and Hope) (1959).

Critical of the status quo and sympathetic to the anti-authoritarian attitudes of his contemporaries, he gained critical attention for *Nihon No Yoru To Kiri* (Night and Fog in Japan) (1961), but the commercial failure of *Amakusha Shiro Tokisada* (The Rebel) (1962) saw him retreat to television documentaries for a number of years.

He established his pre-eminence among contemporary Japanese directors with *Koshikei* (Death By Hanging) (1968), in which incompetent official attempts to hang a young Korean for the rape and murder of two Japanese girls serve as both a condemnation of capital punishment and an indictment of Japanese attitudes to racial minorities. The tone of the film was similarly inventive as it moved from grim reality towards increasing surrealism.

Equally invigorating was *Shinjuku Dorobo Nikki* (Diary of a Shinjuku Thief) (1968), a timely collage of various cinematic styles that links sexual freedom and self-expression with open revolt. Later, he directed a number of initially light dramas that darkened into critical snapshots of Japanese traditions, materialism and the drive for conformity. Notable titles include *Shonen* (Boy) (1969), *Gishiki* (The Ceremony) (1971) and *Natsu No Imoto* (Dear Summer Sister) (1972).

He courted controversy with *Ai No Corrida* (In the Realm of the Senses) (1976), a graphic exploration of the relationship between love and death that climaxes in murder and castration as a couple retreat into a private world of sexual fantasy. Less complex and more open to charges of commercial opportunism was *Ai No Borei* (Empire of Passion) (1978), in which a ghost returns to haunt a couple for their crime of passion.

In the intervening thirteen years he has made only the beautifully photographed but unremarkable prisoner of war saga *Merry Christmas Mr Lawrence* (1983) and *Max Mon Amour* (1986) an icily bizarre black farce in which a diplomat's wife transfers her affections to a chimpanzee.

O'TOOLE, Peter

Full Name Peter Seamus O'Toole
Born 2 August 1932, Connemara, Ireland.

A journalist with the *Yorkshire Evening Post*, O'Toole later served with the British submarine service (1950–52) before attending RADA and making his professional stage debut with the Bristol Old Vic in *The Matchmaker* (1955).

A West End star in *The Long and the Short and the Tall* (1959), he made his film debut in *Kidnapped* (1959) and attained screen stardom when chosen to play the mercurial *Lawrence of Arabia* (1962) (Best Actor Oscar nomination). A blue-eyed, fair-haired adonis, he captured all the wilfulness, madness and passion of this enigmatic figure and was soon much in demand as a leading man. His best performances include twin perspectives on the character of King Henry II in *Becket* (1964) (Oscar nomination) and *The Lion in Winter* (1968) (Oscar nomination), a wonderfully touching portrayal of the crusty schoolmaster in the vastly underrated musical remake of *Goodbye Mr Chips* (1969) (Oscar nomination) and the mad earl, insistent on his identity as Jesus Christ, in the savage satire *The Ruling Class* (1972).

A succession of poor scripts and ill-advised projects (including the notorious *Caligula* (1977)) meant that most of his work in the 1970s failed to live up to the illustrious heights of the previous decades; however he regained his standing as the eccentric, messianic film director in *The Stunt Man* (1980) (Oscar nomination) and with a comic performance of impeccable timing and exuberance as the sozzled ▷ Errol Flynn-like matinée idol in *My Favourite Year* (1982) (Oscar nomination).

Subsequent films, including *High Spirits* (1987) and *King Ralph* (1991), have traded on his penchant for imperious idiosyncracy and have largely been in a comic vein, but he was quietly effective as the aristocratic tutor in *The Last Emperor* (1987).

A noted roisterer in his time, his

emaciated, dissipated figure pays testimony to a full life and many struggles with ill health. He has worked on television in the mini-series *Masada* (1980) and such generally undistinguished pieces as *Svengali* (1983) and *The Pied Piper* (1989).

He has also maintained a parallel and very active stage career that includes *Hamlet* (1963), *Waiting for Godot* (1971), *Uncle Vanya* (1978), a critically-roasted *Macbeth* (1980) and a universally admired *Jeffrey Bernard is Unwell* (1989–90 and 1991).

OZU, Yasujiro

Born 12 December 1903, Tokyo, Japan.
Died 12 December 1963, Kamakura, Japan.

An inveterate cinemagoer as a youngster, Ozu worked briefly as a schoolteacher before he joined the film industry as an assistant cameraman, became an assistant director and made his fully-fledged directorial debut with *Zange No Yaiba* (The Sword of Penitence) (1927).

Proving adept at many popular genres, from light comedies to gangster thrillers, he began to specialise in 'home drama', acutely observed, compassionate snapshots of Japanese domesticity, conveying such everyday concerns as marriage, the raising of children and the care of the elderly with gentle humour and a cinematic style noted for its purity and directness. His notable pre-war work includes *Tokyo No Gassho* (Tokyo Chorus) (1931), the story of a married man's search for employment that is slightly uncharacteristic in its satirical tone and *Umarete Wa Mita Keredo* (I Was Born, But...) (1932), in which two young boys go on hunger strike in protest at their father's unseemly obsequiousness to his boss and learn some lessons of life in the adult world.

He served in China from 1937 and was later sent to Singapore to make propaganda films and briefly interned in a British P.O.W. camp. He resumed his career with *Nagaya No Shinshi Roku* (The Record of a Tenement Gentleman) (1947).

A prolific filmmaker, his work from the 1950s has been most widely seen in the West and displays his increasingly formal approach to dramatic presentations of scenes, pursuing a rigorous aesthetic that keeps camera movement to a minimum and maintains a low level perspective in the placing of a camera in relation to the object being filmed. In terms of content, his later work also reveals an underlying melancholy or sadness in his now acutely attuned observations of the life around him.

His greatest works include *Ochazuke No Aji* (The Flavour of Green Tea Over Rice) (1952), a slyly humorous insight into the strains on a long-lasting marriage that has lost some of its sparkle, *Tokyo Monogatari* (Tokyo Story) (1953) which movingly conveys the tensions within a family as an elderly couple come to see themselves as a burden to the household of their children and grandchildren and *Ohayo* (Good Morning!) (1959), a very amusing variation on *Umarete Wa Mita Keredo* in which two young boys swear a continuing vow of silence unless their parents purchase a television for the household.

His final film *Samma No Aji* (An Autumn Afternoon) (1962) is a plaintive story of a widower who finds himself alone in the world once he has successfully arranged his only daughter's marriage.

PABST, Georg Wilhelm

Born 27 August 1885, Raudnitz, Czechoslovakia.
Died 29 May 1967, Vienna, Austria.

Educated in Vienna, Pabst began an acting career in his early twenties that took him to America. Returning to Europe, he was interned in France as an enemy alien during the First World War but subsequently resumed his theatrical career in Vienna, accepting the appointment as Artistic Director of the Neuen Wiener Buhne in 1920.

His involvement with film began when he acted in *Im Banne der Kralle* (1921). An assistant and screenwriter on *Luise Millerin* (1922) he made his directorial debut with *Der Schatz* (The Treasure) (1923) and very quickly developed a reputation for his strong visual sense, concern with contemporary social issues and ability to elicit notable performances from reputedly difficult female stars. Early successes include *Die Freudlose Strasse* (The Joyless Street) (1925) with ▷ Greta Garbo and Asta Nielsen (1882–1972), *Geheimnisse einer Seele* (Secrets of

A Soul) (1926) and *Die Liebe der Jeanne Ney* (The Love of Jeanne Ney) (1927).

He then hired ▷ Louise Brooks for two memorable and sophisticated portraits of guileless female sexuality – *Die Büchse von Pandora* (Pandora's Box) (1928) and *Das Tagebüch einer Verlorenen* (Diary of A Lost Girl) (1929).

He revealed his versatility in numerous projects including the mountaineering adventure *Die Weisse Holle vom Pitz-Palu* (The White Hell of Piz-Palu) (1929), a creditable version of *Die Dreigroschenoper* (The Threepenny Opera) (1931) and two epic pacifist pleas in favour of the harmony of man: *Westfront 1918* (1931), which remorselessly details the living hell of trench warfare and *Kameradschaft* (1931), in which French and German miners unite to rescue colleagues trapped during a disaster.

He left Germany in 1933 and worked with variable success in France and Hollywood until circumstances once again found him in the wrong place at the wrong time. Trapped in Austria in 1941 and forced to make films for the Nazi régime he was later roundly criticised for collaborating.

He resumed control of his own destiny after the war, making *Der Prozess* (The Trial) (1948), a frequently powerful true story about anti-semitism from the 19th century concerning Jewish pogroms in Hungary. It won him the Best Director prize at the Venice Film Festival.

Much of his later work seemed determined to atone for his having worked for the Nazis and often sacrificed cinematic virtues for dramatic invective. His last films include *Der Letzte Akt* (The Last Ten Days) (1955), covering Hitler's final hours, *Es Geschah Am 20 Juli* (Jackboot Mutiny) (1955), a reconstruction of the 1944 attempt on Hitler's life, and his last *Durch die Walder, durch die Auen* (1956).

He retired after a mild stroke in 1956.

PACINO, Al(fred)

Born 25 April 1940, New York City, USA.

Born in East Harlem and raised in the Bronx, Pacino was a sensitive only child who used a vivid imagination and love of movies to escape from the world around him. An academic failure at the High School of Performing Arts in Manhattan, he left there in 1957 and worked in a succession of menial jobs before studying at the Herbert Berghof Studio under Charles Laughton.

Working in avant-garde theatre, he was accepted as a student at the Actors Studio in 1966 and won an Obie (an Off-Broadway award) for his performance as a bullying hoodlum in *The Indian Wants the Bronx* (1966 & 1968). Further stage success followed in *Does A Tiger Wear A Necktie?* (1969) (Tony (Antoinette Perry award for a Broadway performance), Best Dramatic Actor in a Supporting Role) and *Camino Real* (1970).

He made his film debut as a junkie in *Me, Natalie* (1969) and offered a fuller portrait of drug addiction in *The Panic in Needle Park* (1971). He received a Best Supporting Actor Oscar nomination for his subtle performance as Michael Corleone in ▷ *The Godfather* (1972), conveying the inner conflicts and turmoils of a character who progresses from a fresh-faced college boy to emotionally repressed Mafia Don.

A Method actor, dedicated to the realistic psychological portrayals of characters on an emotional knife-edge, he received Best Actor Oscar nominations as the fastidiously honest cop in *Serpico* (1973), the mature Corleone in *The Godfather, Part II* (1974), the bisexual bank-robber in *Dog Day Afternoon* (1975) and the combative lawyer in *And Justice For All* (1979).

After critical resistance to the controversial subject matter of the homophobic thriller *Cruising* (1980) and the extreme gangster chronicle *Scarface* (1983), and his miscasting in the monumental flop *Revolution* (1985), he worked more in the theatre, including a London production of *American Buffalo* (1984).

He returned, seemingly more at ease with his celebrity, to reinforce his pre-eminence as the melancholic, middle-aged cop in *Sea of Love* (1989) and offered a hilarious, scene-stealing turn as the Arturo-Ui-like villain Big Boy Caprice in *Dick Tracy* (1990) (Best Supporting Actor Oscar nomination) before resuming the Corleone saga in *The Godfather III* (1990) and appearing in *Frankie and Johnny* (1991).

For many years, he has been fine-tuning his directorial debut *The Local Stigmatic*, adapted from a playlet by Heathcote Williams (1941–), that he first performed on stage in 1969.

PAGNOL, Marcel

Born 28 February 1895, Aubagne, near Marseille, France.
Died 18 April 1974, Paris, France.

A French Infantryman during the First World War and a teacher of English, Pagnol was appointed to the Lycée Condorcet in Paris in 1922. Pursuing writing in his spare hours, he eventually found critical success and popular acclaim with a number of plays including *Les Marchands de Gloire* (1925), *Topaze* (1928) and *Marius* (1929) which began his Marseilles trilogy.

Immediately struck by the potential of talking pictures to capture theatre texts for posterity, he accepted an offer to film one of his plays, writing the script for *Marius* (1931) which was directed by ▷ Alexander Korda.

Keen to further his interest in the medium, he founded the magazine *Cahiers du Film* in 1931 and co-produced his script of *Fanny* (1932), a sequel to *Marius* directed by Marc Allegret (1900–73). In 1933, he formed his own production company and later directed *César* (1936), the concluding film of the Marseilles trilogy.

Although not perhaps the most cinematic of directors, he was an accomplished storyteller who believed in the virtues of well-rounded characterisations, colourfully evoked locales and strongly constructed narratives which he imbued with acute observations of human nature and a consistently sunny and seductive presentation of Provence and its people. His notable pre-War successes include *Jofroi* (1934), *Angele* (1935), *Regain* (1937) and *La Femme du Boulanger* (1938).

His filmmaking decreased after the War, and he served as President of the Society of French Dramatic Authors and Composers from 1944 to 1946. His most significant later films include an expansive version of *Manon des Sources* (1952) and his last, the episodic *Lettres de Mon Moulin* (Letters From My Windmill) (1954). His final work was for television but his warmhearted writing continues to inspire contemporary French directors, most recently with *Jean De Florette* (1986), *Manon des Sources* (1986), *La Gloire de Mon Pere* (1990) and *Le Chateau De Ma Mère* (1990).

PAKULA, Alan Jay

Born 7 April 1928, Bronx, New York, USA.

Educated at the Bronx High School of Science, Pakula later studied drama at Yale University before joining the cartoon department of Warner Brothers as an assistant in 1948.

From 1950 he worked as an assistant to producer-director Don Hartman (1901–58), before beginning his own career as a producer with *Fear Strikes Out* (1957). A producer on all the films of director Robert Mulligan (1925–) over the next decade, the highlights of their collaboration include *To Kill A Mockingbird* (1962) (Best Picture Oscar nomination), *Love With A Proper Stranger* (1963) and *Up the Down Staircase* (1967).

He made a belated directorial debut with *The Sterile Cuckoo* (1969) and received critical and commercial attention for the expertly handled claustrophobia and voyeurism of the thriller *Klute* (1971), which won a Best Actress Oscar for star ▷ Jane Fonda.

His subsequent work tends to split in to two different categories of taut thrillers and offbeat romances that are always underpinned by a meticulous, almost scientific, observation of their milieu, strong characterisations and a potent visual sensibility.

Fascinated by the obsessive behaviour of individuals in search of truth or justice, his best films include the political thriller *The Parallax View* (1974), the Watergate exposée *All The President's Men* (1976) (Best Director Oscar nomination) and the big business chicanery of *Rollover* (1981).

His romantic endeavours include the wry, warmhearted *Starting Over* (1979) and the deeply emotional *Sophie's Choice* (1982), in which the lead actress (this time ▷ Meryl Streep) again was guided towards a richly complex portrayal that resulted in a Best Actress Oscar win.

Thereafter, for no accountable reason, his films failed to display the sharpness and cohesion of earlier work and the implausible, little seen thriller *Dream Lover* (1986), the heavyhanded adaptation of the stage drama *Orphans* (1987) and the meandering second-time around romance *See You In the Morning* (1989) marked an uneventful period in his career. However, he was back on form with the courtroom mystery *Presumed Innocent* (1990) which placed a beautifully modulated visual style and a gallery of incisive portrayals at the disposal of a thoroughly involving tale of intrigue and passion.

PALIN, Michael *see* **MONTY PYTHON'S FLYING CIRCUS**

PAN SHOT Taken from the word panorama, a pan shot is a camera movement from a fixed vertical axis along a horizontal plane. Affording the viewer a more panoramic view of a scene it is most frequently used to follow a character or a vehicle across a landscape or give a subjective view of what is being witnessed by a character. Variations on the basic shot include a swish pan, which involves a rapid camera movement between objects that results in blurring of the image, and a search pan when the camera appears to be scanning in search of some object or individual.

PANFILOV, Gleb
Born 21 December 1933, Magnitogorsk, Urals, USSR.

A graduate in chemical engineering from the Urals Polytechnic Institute, Panfilov worked as a factory foreman before developing an interest in cinema and studying at the State Cinema Institute.

He had directed short films before making his feature-length debut with *V Ogne Broda Net* (No Ford in the Fire) (1967), the story of an army nurse praised for her solicitous ministrations to her Red Army charges but condemned by a cultural commissar for expressing her individuality as an amateur painter.

The conflict between art and life was also explored through the dilemma of the female protagonist in *Nacala* (The Debut) (1970); she is a factory worker who finds escapism and solace in amateur theatricals. His next film *Ja Prasu Slova* (I Want The Floor) (1976) follows a female mayor who lives to regret her acquiescence in official censorship.

Using strong female characters to tackle provocative themes and developing a distinctive style of long, static takes and intense emotions, he had built a promising and accomplished career. However, *Tema* (The Theme) (1979), an ironic romantic drama that addresses controversial issues of artistic freedom of expression and freedom of movement for emigration, was deemed too critical by the authorities and remained unreleased until 1987 when it won the Golden Bear at the Berlin Film Festival.

His career undoubtedly suffered as he moved to less contentious literary adaptations like *Valentina* (1981) and *Vassa* (1983), an intriguingly ambiguous account of a dominant matriarch in conflict with her politically rebellious daughter in the turbulent Russia of 1913.

After a lengthy absence, he returned with an adaptation of *Matj* (Mother) (1990) by Maxim Gorky (1868–1936), an epic tragic fresco that attempts to find contemporary relevance in Soviet history between 1894 and 1902.

PARADJANOV, Sergei Iosifovich
Real Name Sarkis Paradjanian
Born 9 January 1924, Tbilisi, Georgia, USSR.
Died 20 July 1990, Yerevan, Armenia.

A princely figure and visionary artist of immense pictorial style, Paradjanov came from a wealthy background and studied violin and vocal music at the Kiev Conservatory (1942–45) before moving to Moscow and enrolling at the VGIK state film academy.

On his graduation, he directed the short film *Moldavskaia Skazka* (Moldavian Fairy Tale) (1951) and, in 1953, began work at Kiev's Dovzhenko Film Studios making his feature-length debut with *Andriesh* (1954).

None of his first four features have been distributed in the West but international interest was aroused following screenings of *Teni Zabytykh Predkov* (Shadows of Our Forgotten Ancestors) (1964) a lyrical tale of a blood-feud between two families in a small Carpathian village that illustrated his respect for folk art and strong visual sensibility.

Sayat Nova (The Colour of Pomegranates) (1968) was a fictionalised account of events in the life of the 18th century Armenian poet Arutiun Sayadin, known as Sayat Nova (King of Song). Branded as 'hermetic and obscure' by the authorities and banned for a number of years, its static stateliness and sumptuous tableaux were in direct contrast to the feverish pace of the previous film although they share aesthetic evidence of a filmmaker spellbound by religious iconography, ancient music, dance and ornate costumes.

The hedonism and sensuality that were identified as trademarks were in stark contrast to prevailing state ideology and he became a target for the authorities. In 1973 he was arrested on charges of illegal currency dealings, homosexuality, spreading venereal disease and 'incitement to suicide'. Found guilty of the second charge, he served three years of a five-year sentence before international pressure secured an amnesty.

He celebrated his return home with a short film *Achraroumes* (Sign of the Times) (1978) and was acquitted when re-arrested in 1982 for unspecified crimes.

He resumed full-scale filmmaking with *Legenda Suramskoi Kreposti* (The Legend of the Suram Fortress) (1985), a poetic celebration of Georgia's folklore told through the lavish recreation of a legendary tale. His final film, *Ashik Kerib* (1988) was a handsomely mounted Arabian Nights adventure telling of a wandering minstrel seeking his fortune and the woman he would marry.

Whilst the subject matter of his films can prove impenetrable to western viewers, there is no denying their visual splendours nor their unique insights into the ancient regional cultures of the Soviet Union.

PARKER, Alan

Born 14 February 1944, London, UK.

Coming from a highly successful background in the advertising industry, Parker's first work for the cinema was as the scriptwriter of the youthful romance *S.W.A.L.K.* (*Melody*) (1969). He directed the short films *Footsteps* (1973) and *Our Cissy* (1973) and worked for television on commercials and the award-winning wartime tale *The Evacuees* (1974) before making his feature-length cinema debut with the musical gangster pastiche *Bugsy Malone* (1976).

Midnight Express (1978), a brutal account of a young American incarcerated in a Turkish jail, was more indicative of the films he would favour in its controversially-handled subject matter, glossy visuals and overwrought emotions. This larger-than-life approach found suitable topics in the traumas of teenage hopefuls (*Fame* (1980)) and the bitter break-up of a marriage (*Shoot the Moon* (1981)).

Frequently accused of letting style dominate the substance of his work, and a vocal opponent of social realist traditions, he has worked extensively within the American studio system on such films as *Birdy* (1985), *Angel Heart* (1987) and *Mississippi Burning* (1988), a tense Civil Rights thriller that combined a compelling narrative, strong visual sensibility and dynamic performances in a potent, popular entertainment.

The sentimental romance *Come See The Paradise* (1990) and the youth musical *The Commitments* (1991) seemed to represent a conscious move away from his previous films to a lighter more lyrical vein.

An irreverent critic of the British film establishment, he has published an anthology of his cartoons entitled *Hares in the Gate* (1982) and directed the television documentary *A Turnip Head's Guide to the British Cinema* (1984).

PARKS, Gordon

Born 30 November 1912, near Fort Scott, Kansas, USA.

One of 15 children born on a dirt farm, Parks was a busboy, piano player, bandleader, lumberjack, basketball player and itinerant before buying his first camera from a Seattle pawnshop in 1937 and learning the craft of photojournalism. In 1942, he became the first Julius Rosenwald fellow in photography and subsequently worked in the Office of War Information and the Farm Security Administration for which he travelled through Alaska documenting the 'forgotten men' of the land.

He then spent five years as a photographer with Standard Oil and joined Life magazine in 1949, gaining a reputation as one of the world's most accomplished photojournalists.

He developed an interest in motion pictures whilst on an assignment to cover the filming of *Stromboli* (1950) and later made short films and documentaries for the National Educational Television network.

The recipient of numerous international honours, in 1967 he was voted the photographer-writer who had done the most to promote understanding among the nations of the world.

A writer, composer and poet, he became the first black producer-director of a large-scale picture for a major studio when Warner Brothers hired him to direct *The Learning Tree* (1969), an adaptation of his 1963 autobiographical novel· for which he also wrote a three-movement symphony musical score.

He subsequently directed *Shaft* (1971), a tough, adult tale about a macho, sexy black private eye that initiated a cycle of slick, violent 'blaxploitation' thrillers. After the sequel, *Shaft's Big Score* (1972), he made the superior detective story *The Super Cops* (1974) and the stunning musical *Leadbelly* (1976), a biography of folksinger, musician and chaingang convict Huggie Ledbetter.

After a lengthy absence, he returned with the television film *Solomon Northup's Odyssey* (1985).

One of his sons, Gordon Parks Jnr (1948–79) directed such films as *Superfly* (1972), *Three The Hard Way* (1974) and *Aaron and Angela* (1975).

PASOLINI, Pier Paolo

Born 5 March 1922, Bologna, Italy.
Died 2 November 1975, Ostia near Rome, Italy.

Educated at the University of Bologna, Pasolini was a published poet before being conscripted in to the army during the Second World War and taken prisoner by the Germans following the Italian surrender.

A teacher and secretary of a local Communist Party, he acquired notoriety when he was charged with obscenity on the publication of *Ragazza Di Vita* (1955), a poetry collection that gave voice to his sense of rebellion and the timeless innocence of his mother's birthplace Friuli.

He worked in the cinema as a scriptwriting collaborator on such films as *La Donna Del Fiume* (1954) and *Le Notti De Cabiria* (Nights of Cabiria) (1956) before making his directorial debut with *Accattone* (1961), an unsentimental account of a young pimp's ill-fated attempt to live an honest life and his ultimate tragic return to crime.

Influenced by the work of the ▷ neorealists, and his interest in linguistics and the philosophies of Gramsci, he brought a poetic sensibility to bear on a body of work that began by romantically celebrating the struggles of the sub-proletariat classes and spiralled into a more direct engagement with the cinema's capacity for myth-making and the visual interpretation of personal and societal conflicts.

The greatest achievement of his early career is *Il Vangelo Secondo Matteo* (The Gospel According to St Matthew) (1964), an interpretation that envisaged Christ as a rebellious outcast, embodying the fight for equality and justice among the peasant sub-classes.

His subsequent work includes *Edipo Re* (Oedpius Rex) (1967), a visually entrancing adaptation of the Greek tragedy with a contemporary prologue and epilogue to stress its enduring relevance, *Teorema* (Theorem) (1968) (which again resulted in his being charged with obscenity), which made excellent use of Terence Stamp (1940–) as the enigmatic sexual catalyst who breaks down the inhibitions in a bourgeois household, and *Il Porcile* (Pigsty) (1969), an obscure if beautifully bizarre juxtaposition of two stories of a medieval cannibalist and modernday industrialist's son whose only attraction is to pigs.

He then embarked upon a series of bawdy literary adaptations, among them *Il Decamerone* (The Decameron) (1971) and *Il Fiore Delle Mille E Una Notte* (The Arabian Nights) (1974), before his final controversial work *Salo O Le Centiventi Giornate Di Sodoma* (Salo or the 120 Days of Sodom) (1975) a brutal, upsetting and clinical depiction of the worst excesses of Fascism as a group of young people are used, abused and violated in every imaginable psychological and physical manner for the delectation of a group from the ruling élite. De Sade transposed to Mussolini's Italy, it is a difficult film to watch and an uncertain one to judge, failing as work of political criticism and wallowing in that which it seeks to condemn. A profoundly bleak comment on the depravity of mankind it clearly revealed the mind of a deeply troubled individual.

A poet, artist and political activist, he was bludgeoned to death, probably as the result of a homosexual encounter, although some theories advance the notion that he was killed to silence his outspoken political opposition.

LA PASSION DE JEANNE D'ARC (The Passion of Joan of Arc)

Carl Dreyer (France 1928)
Renee Falconetti, Eugene Silvain, Maurice Schutz, Michel Simon, Antonin Artaud.
114 mins. b/w.

Although not a commercial success on its initial release, ▷ Carl Dreyer's epic account of the trial of Joan of Arc has long been considered a great masterpiece of world cinema. The film condenses the French heroine's eighteen months of interrogations in to a single day, the last of her life, but takes as its real theme the battle for her soul, and the attainment of salvation through suffering. Dreyer's relentless use of close-ups was considered revolutionary at the time, and is central to the realisation of his need to involve the audience in the suffering of the girl, as well as to reveal the machinations of her persecutors, starkly framed without any softening make-up. This reliance on the close-up has perhaps been over-emphasised in critical discussion of the film, to the detriment of its equally inventive use of other devices, notably the rapid cross-cutting as it moves into the climactic rhythm of the closing scenes. The director wrings a nakedly emotional performance from Renee Falconetti (1901–46) in her only film role as the suffering girl, and is said to have treated her mercilessly in the process. *La Passion de Jeanne D'Arc* took almost eighteen months – and much expenditure on elaborate sets – to make, and was Dreyer's last silent film. As in all his work, his suffering heroine is defeated by a cruel and heartless world, but triumphs in what he saw as the ultimate realm of the soul. His shameless manipulations of chronology and historical fact are all aimed at drawing the viewer into this timeless struggle; generations of film-goers have attested to his success.

PATHER PANCHALI

Satyajit Ray (India 1955)
Subir Bannerjee, Kanu Bannerjee, Karuna Bannerjee, Uma Das Gupta, Chunibala.
122 mins. b/w.

▷ Satyajit Ray's remarkable debut *Pather Panchali* (Little song of the road) opened up an entirely new subject matter and set of aesthetic and formal possibilities in an Indian cinema previously dominated by escapist musical adventures and romances. For one thing, the film was in the Bengali tongue rather then the predominant Hindi one, and was cast in a determinedly downbeat (but strangely lyrical) realist mode. Based on the first of a well-known trilogy of novels by the Indian writer Bhibuti Bashan Bannarjee, *Pather Panchali* tells a very ordinary story about the life of a young boy, Apu, and his impoverished family in a small rural village in Bengal. Ray, however, transforms this everyday material into something almost literally magical, but in a way totally opposed to the escapist fantasies of mainstream Indian cinema. The wonders themselves are simply part of the rich texture of the everyday world he reflects with such sureness of touch, but take on their sense of wonder through the consciousness of Apu and the other villagers. In the film's most famous scene, the boy runs with his sister through the long grass in order to catch a glimpse of a passing train, bound for the city. It is a supremely ordinary moment, but the characters' response invests it with a transforming joy which both renders it marvellous, and implicitly reveals much about the wider social and political context of their lives. This revelation of the social and political through the personal and intimately emotional is typical of the director's approach to his subsequent work, which includes the remaining two parts of the trilogy, *Aparajito* (1956) and *Apur Sansar* (1959).

PATHS OF GLORY

Stanley Kubrick (US 1957)
Kirk Douglas, Ralph Meeker, Adolphe Menjou, George Macready.
86 mins. b/w.

Paths of Glory remains among the most powerful anti-war films ever committed to celluloid, and the harbinger of a re-evaluation of the depiction of war and the military in Hollywood in the ensuing decades. Set in Verdun during the Great War of 1914–18, it is based on a novel by Humphrey Cobb. The title is deeply ironic; the action around which the film revolves, and the behaviour of the military hierarchy, reflect anything but glory. General Broulard persuades the ambitious General Mireau to launch a suicidal attack on an impregnable German position known as The Anthill; the horror of the resulting slaughter is unforgettably captured in ▷ Kubrick's broad, unflinching sweep of the camera across the wholesale destruction. Captain Rousseau refuses Mireau's order to fire on his own retreating troops, and the attack fails. In retribution, Mireau persuades Broulard to authorise the trial of a single scapegoat from each of the three companies involved, in the interests of maintaining discipline. Captain Dax, played by Kirk Douglas (1916–), is assigned to defend the soldiers against this unjust persecution. Cast in the role of a decent, compassionate counter-weight to the corrupt military machine of which he is part, Dax is the moral centre of the film. Despite his efforts, the men are condemned, but a further treacherous twist is yet to come. Kubrick pursues the logic of the script (which he co-wrote as well as directed) with a relentless and unforgiving determination, ameliorated only by the quiet dignity of Captain Dax's final commitment to his men. A classic piece of filmmaking, it pre-figured the more blackly comic anti-war sentiments of Kubrick's subsequent *Dr Strangelove (Or How I Learned To Stop Worrying And Love The Bomb)* (1964), and makes an interesting contrast with his more ambiguous reflections on the Vietnam War in *Full Metal Jacket* (1987), although the latter certainly shares the theme of the dehumanising effect of war on those drawn into its bloody compass.

PECKINPAH, Sam

Born 21 February 1925, Fresno, California, USA.

Died 28 December 1984, Inglewood, California, USA.

An enlisted man in the Marine Corps, Peckinpah was a drama major at Fresno State College before joining the Huntington Park Civic Theatre as director-producer in residence for the 1950–51 season. Subsequently a propman at KLAC-TV in Los Angeles and assistant editor at CBS, he became an assistant to director Don Siegel (1912–91) on *Riot in Cell Block 11* (1954) and four other features including ▷ *Invasion of the Bodysnatchers* (1956), in which he also essayed a small role.

Active in television from the late 1950s, he wrote episodes of the series ▷ *Gunsmoke* and helped devise and direct such series as *The Westerner* and *Rifleman*. He made his cinema debut with the creditable western *The Deadly Companions* (1961) but stamped his individual concerns on *Ride the High Country* (1962) (UK: *Guns in the Afternoon*), a tale of two aged cowboys who fall out over a shipment of gold that he managed to elevate into a moving elegy for the passing of a certain type of Westerner and a simpler code of values.

In 1964 he was fired as the director of *The Cincinnati Kid* and disowned *Major Dundee* (1965) when it was removed from his artistic control. However, ▷ *The Wild Bunch* (1969) allowed him to make a graphically violent, definitive statement on the moral code of the men who built the West and were subsequently rendered outcasts by the arrival of so-called civilising influences, commerce and politicians.

He demonstrated a gentler side to his talent with the lyrical *Ballad of Cable Hogue* (1970) and the contemporary rodeo story *Junior Bonner* (1972); however he is more immediately recalled as an influential figure in the use of explicit screen violence including slow-motion in films like *Straw Dogs* (1971) and *The Getaway* (1972). His last western, the melancholic *Pat Garrett and Billy the Kid* (1973) was, once again, the victim of a troubled production and distribution history.

His final films were a disappointing collection of tough, resolutely macho diversions like *Convoy* (1978) and *The Osterman Weekend* (1983) which lacked the richness or complexity of his westerns.

PEEPING TOM

Michael Powell (UK 1959)

Carl Boehm, Moira Shearer, Anna

Massey, Maxine Audley.
109 mins. col.

Successive generations of critical acclaim have rescued ▷ Michael Powell's disturbing and provocative shocker from the neglect into which his entire oeuvre had fallen. His account of the behaviour of a disturbed young man who takes his pleasure from murdering women with the sharpened leg of a camera tripod while filming them remains arguably the most genuinely shocking British film ever made. Unlike films in the horror genre, though, there is no escape in to the fantasy element for the captive viewer, who is forced not only to come to terms with Mark Lewis (Karl Boehm (1928–)) as a character we are invited to understand and even empathise with, but also to recognise his or her own complicity in the voyeuristic act of watching a film. Powell employs multiple films-within-films effects, including the films made by Mark, which he replays for his pleasure, and those made by his father of experiments carried out on Mark as a child, which suggest the source of his perverse behaviour. That structure reflects Powell's own obsession with the nature and form of film itself, a preoccupation which runs, albeit in a less overt manner, through most of his major works. His distinctive use of colour and lighting and the constantly shifting visual surface of the film make it the most stylistically radical of his works, and easily the most controversial. Vitriolically condemned by the critics on its release, the film is now recognised as an important contribution to contemporary cinema. If it has grown no more comfortable to watch, the passage of time and a changing social and psychological context have conspired to make it less of an oddity, and much more than the gratuitously nasty shocker it must have appeared to audiences in 1959.

PENN, Arthur
Born 27 September 1922,
Philadelphia, Pennsylvania, USA.

Enlisting in the Army in 1943, Penn later joined the Soldiers Show Company in Paris and gave acting lessons whilst studying at Black Mountain College in North Carolina. After completing his studies abroad, he worked as a floor manager and assistant director on *The Colgate Comedy Hour* (1951–52) and directed numerous television plays for such anthologies as *Philco Television Playhouse* and *Playhouse 90*.

He made his feature film directorial debut with *The Left Handed Gun* (1958) starring ▷ Paul Newman as Billy the Kid. His next film was *The Miracle Worker* (1962), a powerful emotional study of the painful progress made between teacher Anne Sullivan and the deaf, blind Helen Keller, which he had already directed for television and Broadway.

Influence by the ▷ nouvelle vague, he made *Mickey One* (1965), an elliptical sinister thriller, but enjoyed his greatest success with ▷ *Bonnie and Clyde* (1967) an exhilarating, highly cinematic evocation of the Depression-era activities of the Barrow Gang that managed to illuminate contemporary American attitudes through the events of the past.

His most interesting work has always reflected the social and political turmoils of his native land, showing disdain for the falsehoods of the American Dream and a fascination with the violence endemic to American culture. The lynch-mob mentality of *The Chase* (1966) echoed the aftermath of the Kennedy assassination, attitudes towards the American Indian in *Little Big Man* (1970) mirrored criticism of American activities in Vietnam, and the identity crisis experienced by the private eye in *Night Moves* (1975) was but a microcosm of the post-Watergate national mood of shame and confusion.

Finding himself out of step with a Hollywood intent on producing lucrative, mass-market escapism, he has worked infrequently for the cinema, completing only six features in the past twenty years. Always active in the theatre, he has directed such Broadway shows as *Sly Fox* (1976), *Golda* (1977) and *Monday After the Miracle* (1982).

His most recent films, the spy story *Target* (1985), the thriller-chiller *Dead of Winter* (1987) and the comedy *Penn and Teller* (1989), have been little more than professionally executed, conventional entertainments and evidence an alarming decline from the intellectual vigour and stylistic daring of his earlier work.

PER UN PUGNO DI DOLLARE
see **A FISTFUL OF DOLLARS**

PERSISTENCE OF VISION The medical understanding of the human capacity for seeing moving pictures from what is basically a series of projected still photos, persistence of vision is the retina's ability to retain an image long enough so that a succession of slightly differentiated images are perceived as being one continuous action. It is possible to achieve persistence of vision at a rate of sixteen frames per second, but at the cost of painful flickering. Motion pictures are traditionally projected at twenty-four frames per second with a shutter on the camera assisting the eye by concealing the intermittent movement from frame to frame. Apparently a clever trick by the optic nerve in collusion with the brain, there is no totally logical, scientific explanation of why this should be possible.

PERSONA
Ingmar Bergman (Sweden 1966)
Liv Ullmann, Bibi Andersson, Gunner Bjornstrand, Margaretha Krook, Jorgen Lindstrom.
84 mins. b/w.

In *Persona*, ▷ Ingmar Bergman created an enigmatic and multi-layered masterpiece from the most basic of cinematic materials, a one-sided dialogue between a highly talkative nurse, Alma (Bibi Andersson (1935–)), and her psychosomatically speechless patient Elizabeth (▷ Liv Ullman). Most of the film is spent in unrelenting concentration on the relationship between these two women (who may well represent dual aspects of a single woman, a reading encouraged by a frequently repeated composite shot made up from half of each woman's face). The exceptions are a brutal pre-title sexual fantasy of a young boy, a psychiatrist who talks to both women early in the film, and fills in Elizabeth's history for the viewer, and the appearance of a man towards the end who may be Elizabeth's husband, although he seems unable to tell one woman from the other. The material from the pre-title sequence also returns in a surprising mid-film incursion. The film begins with a roll of film threading into a projector; at midpoint, and in a moment of physical crisis in the relationship, the film apparently burns away, introducing this extraneous material from the adolescent fantasy. Few films have attempted to deal so directly and yet in such oblique fashion with psychoanalysis, both at the level of subject-matter and in the formal devices (there is, for example, almost no cross-cutting from one woman to the other, except in moments crucial to the evolving relationship). Ultimately, *Persona* defies any easy solution or explanation of what the viewer has encountered on screen, and that ambiguity itself is unusual in a medium historically yoked to linear narrative and action.

PERRY MASON
USA 1957–66, 1985–
Raymond Burr, Barbara Hale

Mason, the defence lawyer who never lost a case, was based on a character created by Erle Stanley Gardner (1889–1970), and had already been adapted for radio. Burr (1917–), who had a history of playing Hollywood heavies (and would later become wheelchair-bound detective *Ironside* (1967–74)) originally auditioned for the part of his principal adversary, District Attorney Hamilton (Ham to his friends) Burger. At the screen test Gardner reportedly leapt out of his chair, yelling 'That's him, that's Perry!'. Burr took the role seriously throughout his initial nine-year tenure, reading up on the law and even receiving an honorary degree from a Sacramento college.

The show followed a fairly strict formula. Burger (William Talman (1915–68)) and Lt Arthur Tagg (Ray Collins (1890–1965)) would research what seemed like an unimpeachable prosecution case, sending the doomed suspect scuttling for assistance to Mason and his faithful secretary Della Street (Barbara Hale (1922–)). Just when all seemed lost, private detective Paul Drake (William Hopper (1915–69)) would uncover vital new evidence, or a witness would crumble in the face of a Mason onslaught, invariably based on a gnawing hunch. Throughout its run the show was high in the ratings, though *The Defenders* (running from 1961 to 1965, with E.G. Marshall (1910–)) had a stronger ethical edge. In the final episode of *Perry Mason*, the lawyer defended a television actor accused of murdering his producer. Creator Erle Stanley Gardner played the judge.

In the shortlived 1973 revival, *The New Adventures of Perry Mason*, Monte Markham (1935–) took the lead. However, Burr and Hale were reunited for *Perry Mason Returns* (1985) in which Mason steps down from his position as a judge to defend Della on a charge of murder. The popularity of this one-off film has resulted in a regular series of feature-length Mason mysteries which have appeared at the rate of two to three a year. Hale's real-life son, actor William Katt (1955–) now plays Paul Drake Junior. The strongest of the titles thus far include *Perry Mason: The Case of the Notorious Nun* (1986), *Perry Mason: The Case of the Musical Murder* (1989) and *Perry Mason: The Case of the Desperate Deception* (1990).

PEYTON PLACE

USA 1964–9
Dorothy Malone, Ed Nelson, Mia Farrow, Ryan O'Neal.

The steamy bestseller by Grace Metalious (1924–64) had already seen service as a lavish Hollywood film in 1957 that earned a Best Actress Oscar nomination for Lana Turner (1920–) and was popular enough to merit an imaginatively entitled sequel *Return to Peyton Place* (1961).

This first American prime time ▷ soap opera was inspired by the success of ▷ *Coronation Street*. Gambler-turned-producer Paul Monash (1917–), having first discounted the notion of importing the cobbled streets of Weatherfield to the New World, bought the television rights to the Metalious book, which, though tame by today's standards, had been thought too hot to handle by most TV executives. The stories were linked by the roving Dr Michael Rossi (Ed Nelson (1928–)) whose work brought him in to contact with all manner of intriguing situations, in a New England setting full of wronged women and children of confused parentage.

Multi-faceted love affairs and sex were the central theme, and each episode had a cliffhanger. Screen newcomers Ryan O'Neal (1941–) and Mia Farrow (1945–) were among the beneficiaries of the serial's success. O'Neal was Rodney Harrington with Farrow playing Allison Mackenzie, the illegitimate daughter of Constance Mackenzie (Dorothy Malone (1930–)), and Elliot Carson (Tim O'Connor (1927–)) who was serving an 18-year jail term. Connie loved Rossi, but eventually married Elliot. In the meantime, Farrow's real-life romance with ▷ Frank Sinatra caused continuity problems when Ol' Blue Eyes chopped her hair off. Such erratic behaviour irked the show's producers, who responded by putting Allison in a coma, then forgetting about her.

As well as adding big budgets to US soap (which till then had been consigned to daytime schedules) *Peyton Place* was the first such show to be imported to the UK, after ITV paid £30,000 for the first 104 episodes. Two teams under different directors worked simultaneously to produce over 100 episodes a year. The production-line method led to the plot becoming convoluted and intimidating to new viewers, so ratings fell. At the show's close Dr Rossi was arrested on a murder charge. The case was left unresolved, a conclusion which irked Dutch viewers, who flew the characters to Holland to film a further ending.

Cynical reviewers nicknamed it 'Blatant Disgrace', yet the show clearly paved the way for ▷ *Dallas* and *Dynasty*. The 1972–74 revival *Return to Peyton Place* was a less ambitious effort aimed at the daytime schedules, with few of the original cast. Malone, Nelson and O'Connor however were seen in the television film *Murder in Peyton Place* (1977) which niftily explained the absence of the now exalted O'Neal and Farrow by making them the victims of strange deaths. A certain Linda Gray (1942–), later seen in *Dallas*, was also in the cast. This, in turn led to *Peyton Place: The Next Generation* (1985) which introduced Mackenzie's long-lost daughter to the heady brew.

PFEIFFER, Michelle

Born 29 April 1958, Santa Ana, California, USA.

Chosen as Miss Orange County in a 1978

beauty pageant, Pfeiffer subsequently worked in television commercials and as a model whilst taking acting lessons.

Her professional career began with roles in the television series *Delta House* (1979) and *B.A.D. Cats* (1980) and she made her film debut in *Falling in Love Again* (1980).

A typically Californian beauty with almond-shaped blue eyes and silky blonde hair, she was just another good-looking young actress in cinema and television films like *Callie and Son* (1981) and *Charlie Chan and the Curse of the Dragon Queen* (1981), although there was some fire to her sullen high-school rebel in the flop musical *Grease 2* (1982).

However, nothing in her rather vapid set of credits had prepared audiences for the conviction of her playing as the coke-sniffing wife of a drugs baron in the baroque and bloody crime melodrama *Scarface* (1983) opposite ▷ Al Pacino.

Her minxish sense of comedy was evident in *Into the Night* (1985) and *Sweet Liberty* (1985) and she enjoyed her first major commercial success as part of the ensemble in the devilish romp *The Witches of Eastwick* (1987).

Further success followed as the brassy Italian-American mobster's widow in *Married to the Mob* (1988), as the elegant love interest in *Tequila Sunrise* (1988) and in *Dangerous Liaisons* (1988). The purity and delicacy of her playing in the latter, conveying the transformation from chaste demureness to hopelessly vulnerable paramour brought her a Best Supporting Actress Oscar nomination.

Her newfound star status was thoroughly endorsed by a smouldering performance as the sultry chanteuse in *The Fabulous Baker Boys* (1989) (Best Actress Oscar nomination) in which her languorous rendition of 'Makin Whoopee' set pulses racing.

Her subsequent work includes *The Russia House* (1990), the interracial romance *Love Field* (1991) and the forthcoming *Frankie and Johnny* (1991) which is scheduled to reunite her with ▷ Pacino.

Her television work includes *One Too Many* (1985) and *Natica Jackson* (1987) and her rare stage appearances comprise *Playground in the Fall* (1981) and Olivia in *Twelfth Night* (1989).

Michelle Pfeiffer and Jeff Bridges in *The Fabulous Baker Boys* (1989)

THE PHILADELPHIA STORY

George Cukor (US 1940)

Cary Grant, Katharine Hepburn, James Stewart, Ruth Hussey.

112 mins. b/w.

Academy Awards: *Best Actor* James Stewart; *Best Screenplay* Donald Ogden Stewart.

The Philadelphia Story boasts one of the

most effective opening sequences in the history of Hollywood movies, and does so without a single line of dialogue being spoken. ▷ Cary Grant is precipitately ejected from the front door of a splendid mansion; ▷ Katharine Hepburn appears in the door, breaks a golf club over her knee, then tosses the remainder after him and slams the door; Grant rings the bell, and when she answers, pushes her in the face. The entire film, widely regarded as the quintessential example of one of Hollywood's richest historic genres, the ▷ screwball comedy, builds from that single scene. ▷ George Cukor plays mis-

James Stewart and Katharine Hepburn in *The Philadelphia Story* (1940)

chievously with the audience's expectation that these two Hollywood heroes will eventually get together again, but screws obstacles in their path, including another suitor to Miss Hepburn, and the presence of two reporters from a scandal sheet, ▷ James Stewart and Ruth Hussey (1914–). The script is witty, the richly endowed supporting cast are superb, and Cukor misses no opportunity to send up the foibles of the rich. Hepburn's haughty demeanour is perfect for the role of the spoiled Tracey Lord, but she succeeds equally well in pulling off the ultimate humanising softening of her character. Assisted by champagne and an uncharacteristically fast-talking Stewart's down-to-earth philosophy, Tracey re-discovers love and the simple pleasures of life, thereby enabling the requisite happy ending which will send everyone home in a warm glow. The film smashed box office records on release, and has never waned in popularity. A musical version *High Society*, was made in 1956, but, despite an equally stellar cast, lacks the sheer charm of the original.

PICKFORD, Mary

Real Name Gladys Mary Smith.
Born 8 April 1893, Toronto, Ontario.
Died 29 May 1979, Santa Monica, California, USA.

Appearing in theatrical melodramas from the age of five, Pickford became involved with the film industry as an extra in such ▷ D.W Griffith productions as *Pippa Passes* (1909) and *Her First Biscuits* (1909) and she had her first substantial screen role in *The Violin Maker of Cremona* (1909).

Billed as 'Little Mary' and 'The Girl With the Curls' she appeared in such films as *The Little Teacher* (1910), *The New York Hat* (1912) and *Tess of the Storm Country* (1913).

The screen embodiment of the Victorian Miss, her golden curls, beauty and ingenuous charm made her the first major star of the cinema industry and earned her the title of 'The World's Sweetheart' and the hearts of loyal international audiences. A Broadway appearance in *A Good Little Devil* (1913) is reputed to have resulted in the kind of mob adulation thought to have been invented by the fans of pop singers and a 1926 visit to Russia brought tumultuous crowds to the streets.

An innocent, high-spirited lass of sweetness and light on screen, she was a shrewd businesswoman in real-life, earning $10,000 a week by 1916 and later founding United Artists with ▷ Griffith, ▷ Charlie Chaplin and her second husband ▷ Douglas Fairbanks Senior. Her greatest silent successes include *Rebecca of Sunnybrook Farm* (1917), *Pollyana* (1920) and *Little Lord Fauntleroy* (1921).

Limited by her juvenile image and

unable to stem the flow of the years, she attempted more grown-up roles as teenagers in *Dorothy Vernon* (1924) and *My Best Girl* (1927) but they were not the 'Little Mary' that the public wished to see.

She won a Best Actress Oscar for her performance as a jazz-age flapper in *Coquette* (1929), her sound debut, and also starred with Fairbanks in *The Taming of the Shrew* (1929), but she retired from the screen after *Kiki* (1931) and *Secrets* (1933).

Over the next forty years, there were occasional rumours of a planned comeback but they amounted to nothing. She published a number of books including *Why Not Try God?* (1934) and such novels as *Little Liar* (1934) and *The Demi-Widow* (1935) and eventually became a recluse in the company of her third husband, actor Charles 'Buddy' Rogers (1904–).

She received an honorary Oscar in March 1976 in recognition of her 'unique contributions to the film industry and the development of film as an artistic medium'.

PICKPOCKET

Robert Bresson (France 1959)
Martin La Salle, Marika Green, Kassagi, Pierre Leymarie.
80 mins. b/w.

Pickpocket is a typically austere and rigorous piece of film-making from a director who eschews conventional film narrative and perspectives as a matter of course. Although it belongs to the period of the birth of the ▷ nouvelle vague in France, it takes a quite different approach to its subject matter and style than any of the films associated with that movement. Michel (Martin La Salle) is a young man who perversely defines his moral superiority through the act of stealing, not so much for gain – the door to his own unlocked room remains open throughout the film as a statement of his disregard for property – as for the act itself. Stealing has become obsessive, but has also become his way of interacting with the world and with other people, a symbol as much as an act. ▷ Bresson constantly signals this symbolic aspect of the thefts in his concentration on shots of near-disembodied hands, and in the way he momentarily but palpably suspends the stolen objects in the instant of their transferal from victim to thief, allowing the camera to linger meaningfully on them (then again, in Bresson, everything is fraught with meaning.) The film is unusual in Bresson's canon in that the protagonist is not simply allowed to outlive the end of the film, but is permitted a redemptive release from his obsession, through the love he discovers for a woman. The film was inspired by Dostoevsky's *Crime and Punishment*, but is in no sense a film of that book. Both narrative and dialogue are stripped to the barest essentials, which has the typically Bressonian effect of magnifying the smallest visual gestures, and weighing them down with significance. It is as idiosyncratic and single-minded as any of his works.

PICNIC AT HANGING ROCK

Peter Weir (Australia 1975)
Rachel Roberts, Dominic Guard, Helen Morse, Jacki Weaver.
110 mins. col.

The release of *Picnic At Hanging Rock* not only signalled the arrival of a major new talent in world cinema, director ▷ Peter Weir, but also alerted audiences around the world to the remarkable upsurge then taking place in Australian film-making, drawing on distinctively Australian subjects and themes. Weir's atmospheric, haunting account of the strange, unexplained disappearance of a school-teacher and three of her pupils on an expedition to an ancient Aboriginal site captivated audiences with its eerie, sinister beauty and otherwordly atmosphere, augmented by gorgeous visuals and a hugely effective soundtrack. All these elements served to fill in the rather sparse story-line, given that the real fascination of the film lies not in what we see happen on screen, but in what is not revealed to us about the mysterious disappearances. Like many of the other films emerging from what was quickly if unimaginatively dubbed the New Australian Cinema, *Picnic At Hanging Rock* looked back into Australia's hitherto unexplored (on film at least) past for a setting, and to the ambiguous relationship between the incoming modern residents of the continent and the mysterious and mystical ways of its ancient populace. The attention to period detail (it is set in 1900), fresh and imaginative visual treatment, unconventional manipulation of the thriller genre, and unresolved air of mystery proved a potent combination, although Weir's debt to Gheorghe Zamphir's music, an evocative re-working of Beethoven on pan pipes, remains an enormous one.

PLATER, Alan Frederick

Born 15 April 1935, Jarrow-on-Tyne, County Durham, UK.

A qualified architect, Plater was first published in *Punch* (1958) and, since 1960, has built up an enormous body of work, both originals and adaptations, that reflect his working-class origins, political beliefs, abiding interest in jazz and fascination with the dignity in labour and the defining of 'man's relationship to his work'.

His first television play was *The Referees* (1961) and his earliest work includes placing some of the grit and authenticity into eighteen episodes of ▷ *Z Cars* between 1963 and 1965 and a further stint of thirty episodes on *Softly, Softly* between 1966 and 1976.

His many subsequent television plays include *Ted's Cathedral* (1964), *Close the Coalhouse Door* (1968) and *The Land of Green Ginger* (1974).

He has also been responsible for such literate and skilled small screen translations as *The Good Companions* (1980), *The Barchester Chronicles* (1981), *Fortunes of War* (1987) and *A Very British Coup* (1988).

Equally prolific in other media, he has contributed to *The Guardian* and written such film screenplays as the D.H. Lawrence adaptation *The Virgin and the Gypsy* (1969) and the Lawrence biography *Priest of Love* (1980) as well as such populist entertainments as the shipboard thriller *Juggernaut* (1974) and the mild bucolic antics of *It Shouldn't Happen To A Vet* (1974).

His novels include *Misterioso* (1987) and *The Beidebecke Affair* (1985) and *The Beiderbecke Tapes* (1986) from his television series of the same names.

POITIER, Sidney

Born 20 February 1924, Miami, Florida, USA.

Raised in the Bahamas, Poitier served as a physiotherapist in the United States Army (1942–45) before commencing his acting career as a member of the American Negro Theater.

He made his Broadway debut in an all-black production of *Lysistrata* (1946), toured in *Anna Lucasta* (1948) and made his film debut in the US Army Signal Corps documentary *From Whom Cometh My Help* (1949).

He made his Hollywood debut in *No Way Out* (1950) as a young doctor faced with the virulent bigotry of Richard Widmark (1914–). In his ensuing career of supporting roles, he did much to alter the film world's stereotypical view of black actors and he was especially notable as a student in *The Blackboard Jungle* (1955) and in *The Defiant Ones* (1958) as an escaped convict chained to white racist Tony Curtis (1925–). Although obvious in its plea for racial harmony, the film was groundbreaking at the time and earned him a Best Actor Oscar nomination.

A wholesome figure with a potent screen presence, he won the Oscar for the charm and warmth of his performance as a handyman assisting in the construction of an Arizona chapel in *Lillies of the Field* (1963) and was soon established as the cinema's first black superstar.

The screen image that accrued may have been unusually clean-cut and impossibly noble, but he brought a persuasive personality to such films as *Guess Who's Coming to Dinner* (1967) and *To Sir, With Love* (1967) and a compelling sense of controlled anger to his role of the Northern police detective fighting crime and ingrained racism in the deep South in *In the Heat of the Night* (1967). He was voted the top box-office attraction in America in 1968 and repeated his characterisation of that policeman in *They Call Me Mister Tibbs!* (1970) and *The Organisation* (1971).

He made his directorial debut with *Buck and the Preacher* (1972) and has increasingly placed the emphasis in his career on his work behind the camera with a series of blandly escapist, anonymously handled comedies like *Uptown Saturday Night* (1977), *Stir Crazy*

(1980) and *Ghost Dad* (1990).

When he appeared in the adventure *Shoot to Kill* (1988) and the thriller *Little Nikita* (1988) after more than a decade's absence from the screen his return passed virtually unnoticed.

He published an autobiography, *This Life*, in 1980.

POLANSKI, Roman
Born 18 August 1933, Paris, France.

A traumatic life that includes his internment in a German concentration camp, the early death of his mother at Auschwitz and the horrifying murder of his second wife, actress Sharon Tate (1943–69), has been reflected in a body of work by Polanski the director that seems preoccupied with alienation, individual isolation and the understanding of evil.

An actor on radio and in the theatre, he had appeared in ▷ Andrzej Wajda's *Pokolenie* (A Generation) (1954) before attending the State Film School in Todz (1954–59) where he made a number of blackly humorous, ▷ Surrealistic short films including the uncompleted *Rower* (The Bicycle) (1955).

He made his feature-length debut with *Noz W Wodzie* (Knife in the Water) (1962) (Best Foreign Film Oscar nomination) a spare, absurdist tale of a boating trip that escalates into a romantic war as two men vie for the affection of one woman.

He made his English-language debut with *Repulsion* (1965) which uses unsettling visual imagery to unsparingly convey the anguished state of a Belgian girl undergoing a mental breakdown in a London flat. He followed this with *Cul-De-Sac* (1966) and the juvenile horror comedy *Dance of the Vampires* (*also known as The Fearless Vampire Killers*) (1967).

He then moved into mainstream American filmmaking with the chilling tale of a modern day witches coven *Rosemary's Baby* (1968) and the utterly compelling detective story *Chinatown* (1974) (Best Director Oscar nomination) which is flawless in its handling of plot, period recreation and performance.

He had made a bold, full-blooded version of *Macbeth* (1971) and proved equally at ease with the langour and landscape of Hardy's *Tess* (1979) (Oscar nomination).

Charged with the rape of a thirteen year-old girl in 1977, he fled from America to avoid the remainder of a jail sentence and has been unable to work or set foot there since that date. Relocated in France, his subsequent cinema work has been uncharacteristically lacking in flair or passion and includes only the lightweight swashbuckler *Pirates* (1986), one of the biggest financial disasters in cinema history, and the derivative ▷ Hitchcockian thriller *Frantic* (1988).

On stage, he has directed *Lulu* (1974) and *Rigoletto* (1976) and acted in *Amadeus* (1981) and *Metamorphosis* (1988). His candid autobiography, *Roman*, was published in 1984.

PONTECORVO, Gillo
Real Name Gilberto Pontecorvo
Born 19 November 1919, Pisa, Italy.

A student of chemistry at the University of Pisa, Pontecorvo fought as a partisan during World War II and later served as a Youth Secretary for the Italian Communist Party, worked as a foreign correspondent for Italian newspapers and acted in the film *Il Sole Sorge Ancora* (1946).

His own film career began as an assistant to Yves Allegret (1907–) on *Les Miracles N'ont Lieu Qu'une Fois* (1951) and he made a number of short films before contributing a segment to the East German production *Die Windrose* (1956). He left the Communist Party after the Soviet invasion of Hungary in 1956 and made his feature-length debut with *La Lunga Strada Azzurra* (The Long Blue Road) (1957), followed by *Kapo* (1960) a melodramatic depiction of concentration camp horrors through the life of a teenage Jewess who is saved from death by a kindly doctor and forced to operate as a prison guard.

▷ *La Battaglia di Algeria* (The Battle of Algiers) (1965) proved to be his one undisputed masterpiece; a newsreel-like account of the guerrilla warfare waged against the French between 1954 and 1962 it conveyed the events with the political complexity and cinematic immediacy of a journalistic documentary as he adopted a ▷ cinema-vérité approach of using actual locations, non-professional actors and hand held camerawork to burn the heat of the moment into an audience's conscience.

This was followed by *Quiemada* (Burn!) (1969) a less coherent but still compelling mixture of absorbing entertainment and balanced political analysis starring ▷ Marlon Brando as a tyrannical 19th century adventurer dispatched from Britain to stir up revolt among the slaves on a Portuguese island in the Caribbean.

Sadly, he has completed only one feature since then, *Ogro* (Operation Ogre) (1979).

PORTER, Edwin S(tratton)
Born 21 April 1869, Connellsville, Pennsylvania, USA.
Died 30 April 1941, New York City, USA.

This early filmmaking pioneer served in the navy and worked at a variety of occupations before joining a company marketing the ▷ Edison Vitascope and arranging what is acknowledged as the first screening of motion pictures in New York on 23 April 1896.

He devised various forms of projector and began making a film record of newsworthy items and events like *The America's Cup* (1899) and *A Day at the Circus* (1901). He joined the Edison Company in 1900 as a designer of cameras and progressed to direct *The Life Of An American Fireman* (1902) and ▷ *The Great Train Robbery* (1903) which established primitive storytelling techniques through editing and the ability to reflect diverse perspectives on events by the cross-cutting of images.

He remained a prolific if less innovative director over the next decade on films like *Life Of An American Policeman* (1905), *Rescued from An Eagle's Nest* (1907), *The Face on the Barroom Floor* (1908) and *His Neighbour's Wife* (1913).

In 1912, he joined with ▷ Adolph Zukor in the creation of the Famous Players company but swiftly curtailed his active filmmaking career and did not direct again after *Lydia Gilmore* (1916).

In 1915, he sold his interest in Famous Players and invested in Precision Machine Corporation, the manufacturer of the Simplex projector which he had developed. He remained President of the company for many years.

POST-SYNCHRONISATION Also known as dubbing, post-synchronisation allows for the additional recording of sound after a film has been photographed and edited. A useful means of replacing dialogue that has been poorly recorded on a difficult location, it also allows foreign-language dialogue to be substituted for the language of the film's country of origin. It has also allowed the voice of non-singers to be replaced by professionals with notable examples including Marni Nixon's (1929–) soundtrack vocalising accompanying the on-screen emoting of stars like ▷ Deborah Kerr in *The King and I* (1956) and ▷ Audrey Hepburn in *My Fair Lady* (1964).

Dialogue is recorded during the shooting of the film as a guide to what will be required and subsequently divided into segments that are looped together and played repeatedly in conjunction with the relevant pictures to allow performers to rehearse and match their previously recorded on-screen efforts. When ▷ David Lean supervised the recent revision and re-release of ▷ *Lawrence of Arabia* (1962) actors like ▷ Peter O'Toole were called upon to recreate their vocal patterns of quarter-of-a-century earlier as the fine tuning led to sound re-recording.

POTTER, Dennis Christopher George
Born 17 May 1935, Joyford Hill, Coleford, Gloucestershire, UK.

A graduate in philosophy, politics and economics from New College, Oxford, Potter worked as a member of the BBC's current affairs department from 1959 to 1961 and subsequently as a widely employed journalist for *The Sun* and *Daily Herald*; he also stood as a Labour Party Candidate for East Hertfordshire in 1964.

He began a prolific career as a television dramatist with *The Confidence Course* (1965) and quickly gained attention with *Stand Up Nigel Barton* (1965) and *Vote, Vote, Vote for Nigel Barton* (1965) which went behind the scenes of contemporary British politics to produce some depressing insights into the lack of principle and air of compromise that prevails.

Throughout his work he has utilised autobiographical elements from his life as a starting point for dramas that explore the artist's relationship with his work, the dilemmas of an educated Christian who cannot survive on blind faith and the physical and psychological pain of lives lived in despair or hope. *Son of Man* (1969) was the first television screenplay that depicted Christ as a man who struggled as much with his own doubts as with those opposed to his teachings and he also aroused controversy with the banned *Brimstone and Treacle* (1978) in which a satanic stranger's rape of a comatose girl proves cathartic and results in her return to consciousness.

The purveyor of bleakly acerbic humour and grandly conceived characterisations he has also been acutely aware of the visual strengths within the television medium as well as the power of the spoken word, using elliptical editing and the ability to play with time to his dramatic advantage as well as adopting such technically audacious devices as requiring actors to mime to popular songs of the 1920s and 30s that intercut the action in *Pennies from Heaven* (1978) and *The Singing Detective* (1986) or using adult actors to impersonate children in his memory play *Blue Remembered Hills* (1979).

He has also adapted such works as *The Mayor of Casterbridge* (1976) and *Tender Is the Night* (1985) for television as well as writing the screenplays for such cinema productions as the film versions of *Pennies from Heaven* (1981) and *Brimstone and Treacle* (1982) and other commissions like *Gorky Park* (1983), *Dreamchild* (1985) and *Track 29* (1988).

He has also written novels and journalism; he moved behind the camera to direct the flashy television version of his novel *Blackeyes* (1989) and 1991 saw his debut as a film director with *Secret Friends*.

POWELL, Michael
Born 30 September 1905,
Bekesbourne, Kent, UK.
Died 19 February 1990, Avening,
Gloucestershire, UK.

Frustrated by the mundane routine of his job in a bank, Powell took work with director ▷ Rex Ingram and underwent an exacting apprenticeship on a number of silent films made at his Victorine Studios in Nice.

Back in Britain, his directorial debut *Two Crowded Hours* (1931) was one of almost two dozen ▷ quota quickies he made over six years.

His talents were more gainfully employed on *The Edge of the World* (1937) a stark depiction of the evacuation of a barren Shetland island. His enduring partnership with Hungarian writer Emeric Pressburger (1902–88) began with the fast-moving wartime thrillers *The Spy in Black* (1939) and *Contraband* (1940) and flourished with overtly propagandistic but rattingly good yarns like *49th Parallel* (1941) and *One Of Our Aircraft Is Missing* (1941).

In 1943, they cemented the partnership by forming The Archers and collaborated on films like ▷ *The Life and Death of Colonel Blimp* (1943), *I Know Where I'm Going* (1945), *Black Narcissus* (1947) and ▷ *Red Shoes* (1948) that resulted in a body of work without equal in its seductive use of a heady palette of colours, ▷ expressionist lighting and a very unBritish-like readiness to embrace the cinema as a visual language and emotional force; it aimed to express grand emotions, poetry, fantasy and sensuality rather than attempting to attain some form of earthbound social realism.

The partnership was dissolved after the standard tale of wartime heroics *Ill Met By Moonlight* (1957). Working alone, Powell made the controversial ▷ *Peeping Tom* (1959) which was viciously attacked for its 'bad taste' and 'sadism' but reclaimed by posterity as a disturbing and masterly commentary on the voyeurism of film.

The critical assault on his reputation at the time proved profoundly damaging, and his career moved into low gear with the Australian productions *They're A Weird Mob* (1966), and *Age of Consent* (1968) and the Children's Film Foundation *The Boy Who Turned Yellow* (1972) (which Pressburger scripted).

Active as a consultant at ▷ Francis Coppola's ill-fated Zoetrope studios, he remained ever hopeful of securing the funds for another project until the time of his death. (A version of *The Tempest* to star ▷ James Mason remained a favourite project). He did not direct again after *Return to The Edge of the World* (1978) but he did live to see his often scandalously underrated work finally given its due by a younger generation of critics and filmmakers like ▷ Martin Scorsese who rightly hailed The Archers as one of the most daring and distinctive forces ever seen in world cinema.

The first volume of his autobiography, *A Life In Movies*, was published in 1986 and the continuation awaits posthumous publication under the guidance of his widow, Oscar-winning film ▷ editor Thelma Schoonmaker (1945–).

PREMINGER, Otto

Born 5 December 1906, Vienna, Austria.
Died 23 April 1986, Manhattan, USA.

A student of law at Vienna University, Preminger acted with Max Reinhardt, joining the Theater in der Josefstadt in 1928 and becoming its director in 1933.

He directed his first film *Die Grosse Liebe* in 1931. Moving to the USA in 1935 he directed several Broadway productions including *Libel!* (1935) and *Outward Bound* (1938) before journeying to Hollywood, first as an actor in films like *The Pied Piper* (1942) and *They Got Me Covered* (1943) and then as a director under contract to Twentieth Century Fox, where he made a number of stuffy costume dramas and a more intriguing selection of atmospheric ▷ film noir thrillers including the classic *Laura* (1944) and *Fallen Angel* (1945).

An independent filmmaker from 1952, he boldly tackled controversial themes such as drug-addiction in *The Man With the Golden Arm* (1955), rape in *Anatomy Of A Murder* (1959), Jewish repatriation in *Exodus* (1960), homosexuality in *Advise and Consent* (1962) and racism in *Hurry Sundown* (1966).

To the public he maintained the stereotype of the old-style film director, autocratic, bullying and a strict disci-

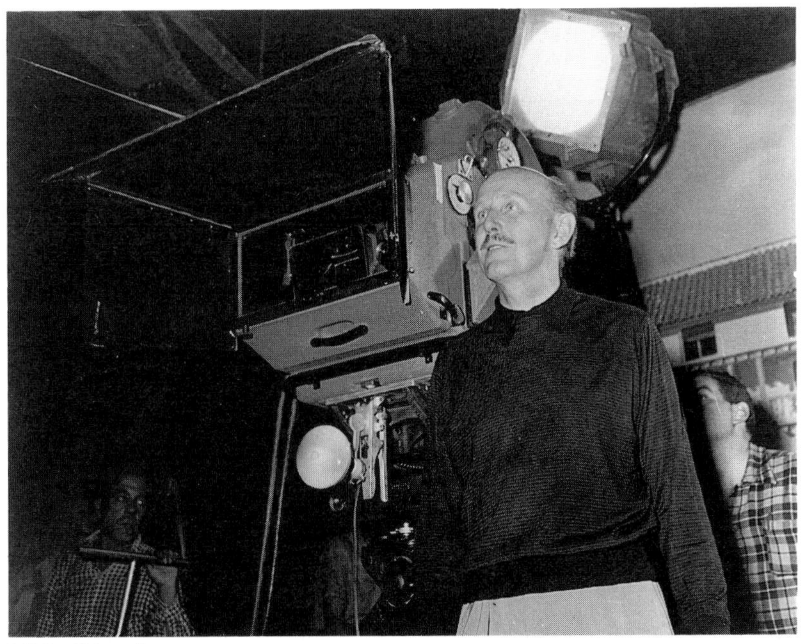

Michael Powell

plinarian known as 'Otto the Ogre'; but he was also a showman, a craftsman, a talent-spotter (▷ Faye Dunaway and Jean Seberg (1938–79) were among the performers he encouraged) and a man who took the fight against antiquated notions of censorship to the Supreme Court.

His best film remains the engrossing, excellently acted courtroom pyrotechnics of *Anatomy of A Murder* and his later career was marked by some crass misjudgements, including the desperately unfunny hippy comedy *Skidoo* (1968) and the soporific terrorist thriller *Rosebud* (1975).

An American citizen from 1943, his other theatrical work includes the *Trial* (1953) and *Full Circle* (1973). He also acted in the film *Stalag 17* (1953) and the television series ▷ *Batman* and published an autobiography, *Preminger*, in 1977. His last film was the unremarkable espionage thriller *The Human Factor* (1979).

THE PRISONER
UK 1967–68
Patrick McGoohan

The show which gave existential angst to the spy genre grew from Patrick McGoohan's (1928–) dissatisfaction with *Danger Man* (1959–62), the highly successful action adventure series of which he was the star. Accounts vary, but *Danger Man* script editor George Markstein makes the strongest claim to conceiving *The Prisoner*. He wrote a treatment which borrowed from his own experience with British Intelligence – specifically, knowledge of disorientation techniques and institutions for people who 'knew too much'. The idea was simple: if *Danger Man* had to end, why not let its central character John Drake try to retire, and subject him to these pressures?

Markstein's idea placed the ex-spy in a Kafkaesque dreamworld, where a pleasant surface concealed blanket oppression. McGoohan chose the setting, the Welsh village Portmeirion, an early *Danger Man* location, designed in a curious European style by Sir Clough Williams-Ellis. He also added piped blazers and the Village's penny-farthing logo, making it like a holiday camp from Hell. People were known by numbers. McGoohan played Number Six (the name John Drake was under copyright), and not knowing how he came to be there, engaged in a struggle to discover the identity of the faceless controller, Number One. A succession of Number Twos, meanwhile, used hallucinogenic drugs and brainwashing to try and discover why he had resigned.

Two series were commissioned, but as shooting of the first progressed, a power struggle between Markstein and McGoohan (who ensured bankability) saw the influence of the latter increase. Markstein quit, leaving McGoohan to conclude with a chaotic episode in which Number Six finally corners Number One and peels off a series of masks to reveal first an ape, then his own face. The Prisoner is his own jailer. Having programmed the destruction of the Village, he returns to London, and the street scenes which opened the series. The cyclical ending outraged viewers, who felt betrayed, but secured the mystique of the show. A documentary, *Six Into One – The Prisoner File*, was shown on Channel 4 in 1984. Despite the theme of individual responsibility and McGoohan's defiant catchphrase – 'I am not a number' – the blazer-wearing fans of The Six Of One appreciation society continue to remind Portmeirion of its murky past.

PRODUCER In contemporary film-making the producer is recognised as the individual who will see a project through from inception to completion and distribution. He or she will often have optioned a literary property (e.g. best-selling novel, Broadway play etc) or been instrumental in the encouragement and development of an original script. A prime task will then be securing the requisite funding to underwrite a fiscally responsible budget for the film's production and the gradual selection of a team of workers from the ▷ director, if one is not already attached to the project, to the ▷ cinematographer, ▷ art director and all other members of the technical crew.

Whilst the best producers are also actively involved in the creative and artistic side of the filmmaking process, their responsibilities are administrative and financial, ensuring that a shooting schedule is maintained, a budget is adhered to and any unforeseen crises can be dealt with. Trouble-shooting and problem-solving are the main endeavours during the day-to-day filming.

The producer will also work with the director and ▷ editor on a film's post-production phase whilst closely plotting the release of the film, a process that can extend from arranging distribution deals to planning publicity campaigns and approving all promotional activities. He or she will also be concerned with the full exploitation of the film in such ancillary markets as video, television, satellite and cable.

Television producers tend to specialise in a particular category of programme-making (e.g. light entertainment, drama, current affairs) or work in charge of a series of programmes with several different directors or presenters.

Notable British producers of recent vintage include ▷ David Puttnam and Jeremy Thomas (1947–) whose credits include *The Shout* (1977), *The Great Rock 'n' Roll Swindle* (1979), *Eureka* (1982), *Merry Christmas, Mr Lawrence* (1983) and the Oscar-winning *The Last Emperor* (1987). Maverick independent American producer Ed Pressman (1944–) has masterminded a bracingly eccentric group of work from *Conan The Barbarian* (1981) to *Good Morning, Babylon* (1986), *Wall Street* (1987) and *Blue Steel* (1990).

PRODUCTION DESIGNER *see* **ART DIRECTOR**

PRODUCTION MANAGER Sometimes known as a unit manager, the production manager is the individual most closely attuned to the daily practical requirements of a film during the period of its production. A breakdown of each day's needs in terms of actors, locations, props, catering and numerous other concerns will allow for proper provision to be made and a smooth shoot to ensue. The production manager will then oversee daily arrangements for transportation, accommodation, meals and every other essential element.

PSYCHO

Alfred Hitchcock (US 1960)
Anthony Perkins, Janet Leigh, Vera Miles, John Gavin.
109 mins. b/w.

The celebrated scene in which Janet Leigh (1927–) is repeatedly stabbed in the shower by a deranged Anthony Perkins (1932–) must be a contender for the most parodied moment in Hollywood movies, but ▷ Hitchcock's most overtly shocking film still retains its power to manipulate audience responses, although never more so than on first viewing. Leigh has fled with her employer's money, intending to meet up with

Janet Leigh in *Psycho* (1960)

her lover afterwards, but makes the mistake of stopping at the Bates Motel, an eerie citadel presided over by the twitchy Norman (Anthony Perkins' finest moment, and one he has been reprising less convincingly ever since) and his 'mother'. Her lover appears on the scene, and begins to try to piece together the story of her disappearance, leading in due course to the shocking moment when the truth about Norman's mother is finally revealed. The murder takes place in the opening third of the film, and that infamous shower scene has received a disproportionate amount of attention from critics, who have been conspicuously divided on the deeper merits of the film itself, and Hitchcock's work in general. Whatever the implications of that debate regarding the director's seemingly callous and even trivial manipulation of the dark or morbid themes – robbery, sex, murder, death – which run through the film, *Psycho* remains the classic example of a suspense thriller, taut, unpredictable, macabre, and genuinely shocking even on repeated viewings. Perkins starred in two sequels, the second of which *Psycho III* (1986), he also directed; Richard Franklin's *Psycho II* (1983) is by far the better of the two, but neither came close to the original.

The woefully unnecessary *Psycho IV* (1990), delved into Norman's childhood; Hitchcock's film clearly has a lot to answer for in terms of the slasher genre that the film appears to have instigated, the *Friday the 13th* series being only the most notorious example.

THE PUBLIC ENEMY

William Wellman (US 1931)
James Cagney, Jean Harlow, Edward Woods, Joan Blondell, Beryl Mercer.
96 mins. b/w.

Much of *The Pubic Enemy* seems rather dated and old-fashioned now, but the film was significant in being the first Hollywood gangster picture to tackle the social roots of crime as part of a genre film. The opening linkage between environment and the stirrings of criminality in children is rather crudely achieved by contemporary standards, but makes the point firmly enough. The subsequent account of the rise and fall of a Prohibition Era gangster set the classic Hollywood mould, but remains eminently watchable now mainly because of the power and charisma of its star, ▷ James Cagney, and that of co-star ▷ Jean Harlow, albeit to a lesser extent. Cagney plays a mean gangster named Tom Powers, snarling his way through the film in a style which would establish him as the quintessential Hollywood image of the gang-land denizen for decades to come. The moment when he squashes a grapefruit into the face of actress Mae Clarke (1910–) at breakfast has taken on iconic status in film lore. Wellman (1896–1975) directs the film as a morality tale, and underplays the graphic nature of the violence, allowing many of the film's nastier moments to take place off-camera, although suitably chilling sound-effects mean there is no real loss of impact. The explicit linking of poor living conditions with crime comes full circle when Cagney dies in the same gutters of the city which has nurtured his activities, murmuring 'I ain't so tough' in an ending which prefigures a similarly socially-motivated renouncing of his tough-guy status as he goes to the electric chair in *Angels With Dirty Faces* (1938).

PUTTNAM, David Terence

Born 25 February 1941, Southgate, London, UK.

The son of a Fleet Street news photographer, Puttnam was a highly successful advertising agency account executive and photographer's agent before founding the film company Engima in 1968.

His first feature film as a producer, the youthful romance *S.W.A.L.K.* (subsequently known as *Melody*) (1969), was written by ▷ Alan Parker and its unexpected commercial success in Japan allowed the company to keep afloat and prosper.

Documentaries like *Brother Can You Spare A Dime?* (1974) and the box-office popularity of *That'll Be the Day* (1973) and its sequel *Stardust* (1974) testified to his acumen and he gained a well-earned reputation for encouraging young directorial talents to spread their wings in such critically acclaimed low-budget efforts as *Bugsy Malone* (1976), *The Duellists* (1977) and *Midnight Express* (1978) (Best Picture Oscar nomination).

He won a Best Picture Oscar with ▷ *Chariots of Fire* (1981), an archetypal project that shows his intention to produce intelligent, humanist and often factually-based dramas that celebrate individual integrity, acts of principle and various forms of male friendship.

Its international commercial appeal allowed him to progress to larger scale explorations of human and moral dilemmas in the melancholy comedy *Local Hero* (1983) and the more epic *The Killing Fields* (1984) (Oscar nomination) and *The Mission* (1986) (Oscar nomination).

A tireless spokesman and figurehead of the British film industry in the early 1980s, he had been outspoken in his criticisms of Hollywood's financial profligacy and artistic cynicism and it therefore surprised many when he accepted a contract to become chairman and chief executive of Columbia Pictures in September 1986. Aiming to bring fiscal responsibility in to filmmaking and encourage a sense of decent moral values, he fell foul of the Hollywood establishment and his own lack of diplomacy, leaving Columbia after only twelve months of his projected three and a half year tenure. The few works of any note from his disappointing output include *Housekeeping* (1987) and *School Daze* (1988).

Returning to Britain and independent production, he secured international funding for a portfolio of pictures. The first fruits of his labours were the moderately successful saga of World War II heroism *Memphis Belle* (1990), and *Meeting Venus* (1991).

QUAID, Dennis

Born 9 April 1954, Houston, Texas, USA.

A drama major at the University of Houston, Quaid worked as part of a nightclub act whilst still a teenager and later moved to Los Angeles, encouraged by the success of his brother Randy (1953–) who had appeared in *The Last Picture Show* (1971) and *The Last Detail* (1973).

He made an inauspicious film debut in *Crazy Mama* (1975) and was seen in a number of modest ventures including *I Never Promised You A Rose Garden* (1977) and *Seniors* (1978) before more noteworthy appearances as the Indiana jock railing against his rich college rivals in *Breaking Away* (1979) and one of the Miller brothers in the western *The Long Riders* (1980).

An ingratiating charmer with a wide, wolfish grin, he made a personable and diligent supporting performer in such lacklustre settings as *Gorp* (1980) and *Caveman* (1981) but finally lit up the screen as cocky astronaut Gordon Cooper in *The Right Stuff* (1983).

Established as a leading man of irresistible dynamism, charisma and a leanly sculpted torso, he graduated to leading roles in such large-budget adventures as *Jaws 3-D* (1983), *Enemy Mine* (1985) and *Innerspace* (1987) and showed evidence of a more versatile talent as the slyly sexy Cajun cop in *The Big Easy* (1987) and the smooth Washington lobbyist in *Suspect* (1987).

Claimed as a young star of the first order, he has described the disappointing commercial fate of his recent films as evidence of his status as a box-office jinx and still awaits a major success despite such diverse efforts as *Everybody's All-American* (UK: When I Fall In Love) (1988) as a football hero living in the afterglow of his college triumphs, an over-exuberant attempt to portray extrovert rock 'n' roll legend Jerry Lee Lewis in *Great Balls of Fire* (1989) and as the union activist and soldier in the soft-centred romance *Come See The Paradise* (1990).

More recently he was seen as a sexually voracious Hollywood producer in *Postcards from the Edge* (1990) and had been announced as the star of ▷ David Lean's long postponed *Nostromo* (1991).

His stage work includes a production of *True West* (1984) opposite brother Randy and his television credits include the highly rated *Bill* (1981) and its sequel *Bill: On His Own* (1983).

QUATERMASS

UK 1953–59, 1979

Creator Nigel Kneale

Though dismissive of the science fiction genre, writer Nigel Kneale (1922–) irrevocably changed audience expectations of TV drama with three six-part series based around the doings of rocket scientist Professor Bernard Quatermass (*The Quatermass Experiment* (1953), *Quatermass II* (1955), *Quatermass And The Pit* (1958–59)).

Kneale had won the Somerset Maugham prize for literature in 1950, and in 1953 the BBC spent their entire drama budget on his idea for a 'mystifying, rather than horrific serial'. He wanted to develop characters more than had been done in US Sci-fi movies, adding humour, while exploiting the fear raised by the beginning of space exploration, the growth of nuclear technology and the Cold War.

Directed by Rudolph Cartier, the first

series marked a clear stylistic break with the stiff theatricality of previous TV productions. Because of the demands of live transmission, the special effects consisted of a rocket and a glove covered in foliage, yet the story kept the nation gripped. It followed the efforts of Quatermass to track down the sole survivor of an experimental space probe (Duncan Lamont (1918–)). Infected with an alien virus, the crewman turned into a 100 foot tall vegetable. *Quatermass II*, (with John Robinson (1908–79) in title role) told of a covert Martian invasion, and was technically superior, while the third series made good use of bleak location shots, and was the first major work of the BBC Radiophonic Workshop.

The three series were reprised in film form by Hammer. Kneale was only happy with the third movie. After *Quatermass* he had a fluctuating career, adapting a successful version of Orwell's *1984* (also with Cartier directing) which caused questions about the BBC to be asked in the House – Prince Philip defended it.

Kneale's tele-play *The Creature* (1955) was filmed by Hammer as *The Abominable Snowman* (1957), and he wrote the film scripts for *Look Back In Anger* (1959) and the *Entertainer* (1960). He later wrote Euston Films's 1979 TV revival of *Quatermass* (Sir John Mills (1908–) played the professor) but was unhappy, despite a £1·25m budget. This series was made into a TV movie for the US market. Kneale also worked on a shelved 3D version of *Creature From The Black Lagoon*, with original director Jack Arnold (1916–), and a script for *Halloween 3* (1983), which was adapted so much he asked for his name to be removed.

QUOTA QUICKIE In response to the British film industry's fears of commercial dominance by Hollywood cinema releases, an Act of Parliament once offered supposed protection to the indigenous industry by setting legal requirements for the proportion of British-made films that a cinema was obliged to show. A somewhat useless measure, this resulted in the 1920s and 1930s in quota quickies: quickly made, low-budget factory-line fare that used British crews and actors but were financed by American studios eager to subvert and exploit these legal requirements and guaranteed a wide release for the film purely on the grounds of its nationality. There was no evident concern for quality or standards and the only thing to be said in their favour is that they did at least provide employment and experience for some of the young talent then emerging in British cinema. ▷ James Mason made his film debut in a quota quickie and director ▷ Michael Powell cut his teeth on several dozen of the things in the early 1930s.

RAFELSON, Bob

Born 1935, New York City, USA.

A teenage rodeo rider in Arizona, Rafelson later played drums and bass with a jazz combo in Acapulco. A philosophy graduate from Dartmouth College, he was a radio disc jockey with the Far East Network in Japan and won the Frost National Playwriting Competition in 1953.

He later became a writer and associate producer on the DuPont Show of the Month and subsequently adapted 34 stage plays for television in the acclaimed Play of the Month slot. In 1966, he moved to California and helped form the independent production company BBS which created the rock group and television series *The Monkees* (1966–67).

He made his feature debut directing the anarchic Monkees movie *Head* (1968) which he co-wrote with actor ▷ Jack Nicholson. Nicholson starred in his next film *Five Easy Pieces* (1970), an incisive attempt to convey the mood and confusions of a particular middle-class generation through the character study of a promising bourgeois musician who has dropped out and now works as an oilrigger. The film received four Oscar nominations, including one for Best Picture. Equally impressive was *The King of Marvin Gardens* (1972) in which the tragic travails of two brothers (Nicholson and Bruce Dern (1936–)) again underline the hollowness of the universal promise inherent in the American Dream.

Through BBS, he co-produced such films as ▷ *Easy Rider* (1969), *The Last Picture Show* (1971) and *Drive, He Said* (1972), although he was absent from the director's chair until *Stay Hungry* (1976), a quirky celebration of non-conformity.

After another long absence, during which he was sacked from *Brubaker* (1980), he returned with the sexually explicit version of the classic thriller *The Postman Always Rings Twice* (1981) and the lighter ▷ film noir *Black Widow* (1987).

An inveterate traveller and adventurer, his lifelong interest in exploration surfaced in *Mountains of the Moon* (1990), a reasonably engrossing epic of the 19th century quest for the source of the Nile. He is currently scheduled to make the comedy thriller *Man Trouble* (1991).

RAGING BULL
Martin Scorsese (USA 1980)
Robert De Niro, Cathy Moriarty, Joe Pesci, Frank Vincent.
129 mins. b/w.
Academy Awards: *Best Actor* Robert De Niro; *Best Editing* Thelma Schoonmaker.

In US film magazine *Premiere*'s 1989 poll of critics and filmmakers ▷ Martin Scorsese's *Raging Bull* topped the Film of the Decade ranking with nearly double the number of votes garnered by its nearest rival, the acclaim almost unanimous for this most ferociously visceral and stylistically audacious work of its director's always challenging career. Based on the life and times of 1949 world middleweight boxing champ Jake La Motta, played here with unforgettably violent energy by Oscar-winning ▷ Robert De Niro (who put on several stones for some of the later scenes), the film chronicles his success in the ring and fairly ignominious retirement as a nightclub entertainer. Among his monologues is ▷ Marlon Brando's famous 'I coulda been a contender' speech from ▷ Elia Kazan's ▷ *On The Waterfront*. During the film Scorsese balances the victim-hero's relentlessly punishing and self-punishing brutality inside and outside the ring against his developing sense of spiritual worth. Yet if, like almost all Scorsese's work, *Raging Bull* is on one level a reflective outpouring of typically Catholic body/soul anguish, on another level it is also a harsh portrayal of straitjacketing masculine values, for as blood spurts from distorted faces in the shockingly stylised fight sequences, or terrifying family conflicts grow out of La Motta's irrationally paranoid jealousy, we see the driving impulses of machismo in auto-destruct mode. Brilliantly shot in expressionist black and white, and displaying Scorsese's almost nonchalant formal virtuosity to staggering effect, the end result is a troubling explosion of unleashed physicality which no amount of pat schematising can fully categorise.

RANK, Joseph Arthur (Baron)
Born 22 December 1888, Hull, Yorkshire, UK.
Died 29 March 1972, Sutton Scotney, Hampshire, UK.

The youngest of seven children, Rank was an apprentice in his father's flour-milling business before serving as a captain in the Royal Field Artillery during World War I.

Heavily involved in the diversification of the family business in to grain trading, the ownership of bakeries, and the eventual purchase of Hovis McDougall, he first became interested in the cinema industry whilst pursuing his religious beliefs. A devout Christian, who recognised the power of film to influence and spread the good word, he founded the Religious Film Society in 1933.

He subsequently sponsored *The Turn of the Tide* (1935) and his inability to secure a general release for the film led him to purchase a West End cinema. Soon he was on the board of the film company British National and he was also one of the founders of Pinewood Studios in 1936.

Fired with a mission to promote the health of the British film industry, he was particularly keen to support efforts directly competitive with the might of Hollywood. In 1941, he acquired control of Gaumont British Picture Corporation Limited and, the following year, Odeon Theatres Limited.

Now active in film production, exhibition and distribution, with a Rank film heralded by the appearance of a muscled strongman striking a large gong, he owned a chain of 600 cinemas and encouraged the efforts of such independent outfits as The Archers unit of ▷ Michael Powell and Emeric Pressburger (1902–88), CineGuild, and the Two Cities productions of ▷ Laurence

Olivier's Shakespeare films ▷ *Henry V* (1944) and *Hamlet* (1948).

Financial difficulties in the post-war years caused a retrenchment of his more ambitious global plans, but the Rank Charm School of promising young performers flourished and he again diversified into such areas as film stock and processing, remaining Chairman until 1962.

Rank Leisure today has a vast network of operations from hotels to leisure complexes but still includes among those film distribution, exhibition and even the occasional foray into the production of British films; *Educating Rita* (1983), *Defence of the Realm* (1985) and *The Fourth Protocol* (1987) have all received some portion of their funding, however modest, from the coffers of Rank.

RANTZEN, Esther Louise

Born 22 June 1940, Berkhamstead, UK.

Educated at Somerville College, Oxford, Rantzen joined the BBC in 1963, making sound effects for radio drama.

Shifting into research for *Man Alive* (1965–67), she joined *Braden's Week* (1968–72) as a reporter. Since 1973, she has produced, written and presented *That's Life*, a populist, campaigning consumer programme that combines responsible investigative journalism with a potpourri of comical and trivial items usually involving pets with uncanny abilities in the singing or skateboarding department.

A toothy, socially-conscious national watchdog, she has also produced the talent show *The Big Time* (1986), the chat show *Esther Interviews* (1988) and *Hearts of Gold* (1988–) a heartwarming series saluting the selfless good deeds and bravery of plucky individuals.

She has also used her position of influence to campaign on issues of child abuse and drug-addiction in a variety of documentaries, including *Childwatch* (1987) and was instrumental in setting up the national child helpline telephone service.

Married to broadcaster Desmond Wilcox (1931–) since 1977, their joint publications include *Kill the Chocolate Biscuit* (1981) and *Baby Love* (1985).

In 1988 she received the ▷ Richard Dimbleby Award for her many contributions to factual television.

RASHOMON

Akira Kurosawa (Japan 1950)
Toshiro Mifune, Masayuki Mori, Machiko Kyo, Takashi Shimura.
88 mins. b/w.
Academy Award: *Honorary Oscar* as most outstanding foreign film.
Venice Film Festival: *Golden Lion.*

Although production company Daiei were rather reluctant to submit it, fearing incomprehension if not ridicule, the stir created by the prize-winning appearance of ▷ Akira Kurosawa's *Rashomon* at the 1950 Venice Film Festival effectively opened up the Japanese cinema to appreciation and acclaim in the West. While Japan's insularity during the later Thirties and the wartime era had prevented the export of much of her cinematic output, the very exoticism of Kurosawa's work certainly contributed to *Rashomon*'s impact across Europe and America, although the film's bravura camerawork, strong performances and, above all, innovative narrative construction, made clear that here was rich artistic integrity and not just the latest cinematic novelty item. Contradictory versions of the same central event, in which a bandit comes across a couple in a forest, raping the lady and murdering the samurai husband, are related through the differing perspectives of its three participants, and the complications pile up – the killer claims his victim died a coward, but a medium informs us of the samurai's claims that he committed honourable ritual suicide, while the woman disputes the villain's report of her compliance – until a passing woodcutter offers his definitive (?) testimony while sheltering from the rain under the Rashomon gate. Perhaps less of a philosophic inquiry into the nature of truth than some have affirmed, Kurosawa's own reflection that it is about the sort of people 'who cannot survive without lies to make them feel better than they are' is perhaps more in keeping with the typically bitter humanism voiced throughout a distinguished filmography.

RAY, Nicholas
Real Name Raymond Nicholas Kienzie
Born 7 August 1911, La Crosse, Wisconsin, USA.
Died 16 June 1979, New York, USA.

A student of architecture and theatre at the University of Chicago, where his mentors included Frank Lloyd Wright (1869–1959), Ray worked extensively in the theatre and for radio in the 1930s. In 1938, he joined the Phoenix Theater of John Houseman (1902–88) and it was Houseman who appointed him War Information Radio Programme Director in 1942. The following year, he directed *Back Where I Came From* and *Lute Song* on Broadway before moving to Hollywood as an assistant to ▷ Elia Kazan on *A Tree Grows in Brooklyn* (1945).

He made his directorial debut with *They Live By Night* (1948), a sensitive account of young lovers who are fugitives from justice that proved an early indication of his sympathy for those deemed outcasts by the conventional conformist world.

In the 1950s, he established his versatility with such films as the brutal ▷ film noir *In A Lonely Place* (1950), the rodeo drama *The Lusty Men* (1952), the baroque western *Johnny Guitar* (1954), ▷ *Rebel Without A Cause* (1955) exploring the traumas of a misunderstood teenage generation and *Bigger Than Life* (1956) with ▷ James Mason on commanding form as a schoolteacher whose personality changes dramatically as he grows addicted to cortisone.

At his best, Ray brought an incisive and intense style to psychological tales that restlessly explored the neuroses of ordinary people driven to extraordinary actions through alienation from or disaffection with the rest of society.

Turning towards increasingly epic production he made *King of Kings* (1961), an intelligent life of Christ but, often at odds with the filmmaking establishment, he walked off the set of *55 Days at Peking* in 1962 never to resume commercial direction.

He worked on *The Chicago Seven* (1970), an uncompleted documentary on the violence unleashed at the Chicago Democratic Convention of 1968. He then taught at the State University of New York, making the experimental feature *We Can't Go Home Again* (1973) with his students and contributing a segment to *Dreams of Thirteen* (1974).

He acted in *The American Friend* (1977) and *Hair* (1979) and collaborated with ▷ Wim Wenders on the documentary *Lightning Over Water* (1980), a painful last testimony filmed as he was dying of cancer.

RAY, Satyajit
Born 2 May 1921, Calcutta, India.

A graduate of Santiniketan University, Ray worked as a commercial artist in an advertising agency while writing screenplays and looking for the finance to make his first film.

With government support he eventually completed ▷ *Pather Panchali* (1955) which, together with *Aparajito* (The Unvanquished) (1956) and *Apu Sansar* (The World of Apu) (1959), formed the Apu trilogy. These works of strikingly harsh beauty and originality eschew the traditional Indian cinema subjects of colourful musicals and adventures, to bring a sense of ▷ neo-realism to bear on an understated and affectionate portrait of social changes in rural life related through a young boy's painful journey from childhood to maturity.

He later worked within the documentary format and filmed a number of tales from Indian folklore, gaining international attention for his ability to render the Indian character in cinematic terms and a view of the world that could see the good and bad in all men however virtuous or malevolent. His notable early work includes *Mahanagar* (The Big City) (1963), a portrait of a housewife liberated from her domestic environment by the financial necessity of her contributing to the family income, and *Charulata* (The Lonely Wife) (1964) which delicately delineates the break-up of a middle-class marriage.

The urban settings of his later work revealed a strong interest in the complex issues of corruption and political chicanery facing his country, and he brought a gentle irony to the depiction of such problems as famine in *Ashanti Sanket* (Distant Thunder) (1973) and business ethics in *Jana-Arnaya* (The Middleman) (1975).

He worked in Hindi for the first time

with the modest parable *Shatranj Ke Khilari* (The Chess Players) (1977) and his most notable work of a late vintage is *Ghare Baire* (The Home and the World) (1982) set in the early years of this century and gracefully exploring the changes in a woman who is exposed to the personality and revolutionary beliefs of a friend of her husband's.

Poor health has restricted his recent work, but he returned to make a plodding, studio-bound version of *An Enemy of the People* entitled *Ganashatru* (1989) and *Shakha Proshakha* (Branches of the Tree) (1990), a poignant study of the tensions within four generations of one family that rise to the surface during a week's reunion.

REBEL WITHOUT A CAUSE

Nicholas Ray (USA 1955)
James Dean, Natalie Wood, Jim Backus, Sal Mineo.
111 mins.

The mid-Fifties apotheosis of the 'teenager' was probably down to the creation of a new youth market as much as any of the social forces involved, for if the nation's youngsters were questioning their relationship to the status quo then their new found rebelliousness sought to solidify itself around rampant consumption of records, clothes, movies etc to shape the new cultural identity that would mark the 'teenager' apart from the older generation. The 1955 release of both ▷ Elia Kazan's *East of Eden* and ▷ Nicholas Ray's *Rebel Without A Cause* established a nervy young actor named ▷ James Dean as the voice of and icon for frustrated youth everywhere, a cult that still flourishes as a style accessory more than three and a half decades after the ill-fated star's premature death in an auto accident. That the Dean phenomenon still has meaning today is in some part due to the effectiveness of *Rebel Without A Cause* in capturing the timeless pattern of conflict as restless adolescents seek their own values in a process of anxious self-determination, and the pained intimacy of Dean's performance still rings true as his Jim Stark faces the various rites of the new teenagerdom – drinking, tussling with father, facing a knife fight, surviving a 'chicken run', pairing off with his girlfriend and discovering police injustice for himself. In some ways it is still a problem picture in the old Warner Brothers' Thirties manner, but the unrelenting focus on the central character (Dean, according to reports from the set, virtually co-directed the picture) codified a self-consciousness that was a definitive step forward from the docile construct of Andy Hardy or the screen's usual colourless young folks.

THE RED SHOES

Michael Powell & Emeric Pressburger (UK 1948)
Anton Walbrook, Moira Shearer, Marius Goring, Leonid Massine.
133 mins. col.
Academy Awards: *Best Art Direction* Arthur Lawson; *Best Drama Music Score* Brian Easdale.

Revered by ballet afficionados as perhaps the greatest dance film ever made (yet a surprising success at the US box office on its initial release), *The Red Shoes* represents ▷ Michael Powell's commitment to a total cinema – deliriously fusing music, movement, colour, performance and camerawork – at its most impassioned and fully achieved, arguably the high water mark of the Powell/Pressburger partnership (The Archers) peak period of creativity from the mid-Forties to mid-Fifties. Originally drafted some ten years previously as a vehicle for actress Merle Oberon (1911–1979), the core of the narrative comes from Hans Christian Andersen's fairy story of the girl who is punished for succumbing to the lure of the red shoes; in this version the ballerina Vicky (Moira Shearer, 1926–) creating the role on stage is fatally driven on by the domineering influence of the Diaghilev-like impresario Lermontov (Anton Walbrook, (1900–67)) and the overwhelming music written by her jealous composer husband (Marius Goring, 1912–) to dance and dance until she cannot stop. Working with a brilliant production team which doubled the original budget in their quest for perfection, Powell and Pressburger elevated the look of the film to exquisite heights of theatricality without ever losing sight of the nightmare element of the material (the film has been criticised for the harshness of its climactic tragedy). The high romantic agony of a dedication to art that transcends even life itself has rarely been expressed on celluloid with such disturbingly zealous conviction.

REDFORD, (Charles) Robert

Born 18 August 1937, Santa Monica, California, USA.

Receiving a baseball scholarship to the University of Colorado, Redford later studied art in Europe before enrolling at the American Academy of Dramatic Arts. He made his stage debut in *Tall Story* (1959) and had built a successful theatrical and television career as a character actor before making his film debut in *War Hunt* (1962).

The long-running Broadway show *Barefoot in the Park* (1963) established his leading man potential and cinema stardom was guaranteed by the 1967 film version which revealed his pleasing way with light, domestic comedy.

Golden-haired and boyishly handsome, his good looks and image of integrity made him a treasured sex symbol opposite ▷ Paul Newman in *Butch Cassidy and the Sundance Kid* (1969) and *The Sting* (1973) (Best Actor Oscar nomination) and in the popular *The Way We Were* (1973).

A romantic icon and minimalist actor, he proved disappointing as Fitzgerald's enigmatic hero in *The Great Gatsby* (1974) but carried such liberally-minded and attractive populist entertainments as *Three Days of the Condor* (1975), *The Electric Horseman* (1979) and *Brubaker* (1980). Meanwhile, more personal projects like *Jeremiah Johnson* (1972), *The Candidate* (1972) and *All The President's Men* (1976) have reflected his interest in the American West, ecology, and the working of the American political system.

Drawn towards direction, he won an Oscar for his understated handling of mood and characterisation in *Ordinary People* (1980) a portrait of the emotional repression tearing apart a middle-class American family and has also directed the magical fable *The Milagro Beanfield War* (1987).

Seen less frequently on screen, he was well cast as the baseball player of almost legendary prowess in *The Natural* (1984) but received critical disdain for his miscast performance as English adventurer Denys Finch-Hatton in *Out of Africa* (1985) and his shallow portrayal of an aging card-sharp seeking a last chance for love in the sluggish *Havana* (1990).

The author of *The Outlaw Trail* (1978), he established The Sundance Institute in 1981 to encourage new talent and support the creation of independent feature films. A noted activist on environmental issues, he produced *The Solar Film* (1981) and has also served as a ▷ producer or executive producer on such low-budget films as *The Promised Land* (1988), *Sisters* (1989) and *Dark Wind* (1991).

In 1991 he is scheduled to appear in the comedy *The President Elopes* and direct *A River Runs Through It* from the novel by Norman McLean.

Among the many successful films he is known to have turned down are ▷ *The Graduate* (1967), *Rosemary's Baby* (1968), the title role in *Superman* (1978), *The Verdict* (1982) and *Cape Fear* (1991).

REDGRAVE, Vanessa

Born 30 January 1937, London, UK.

The eldest daughter of actors Michael Redgrave (1908–85) and Rachel Kempson (1910–), Redgrave trained at the Central School of Speech and Drama (1954–57) and made her professional debut at the Frinton Summer Theatre in 1957.

She made her London stage debut appearing opposite her father in *A Touch of the Sun* (1958) and her film debut with him in *Behind the Mask* (1958).

A respected classical actress with the Royal Shakespeare Company, she seemed wary of the cinema but attained stardom as the wealthy wife of *Morgan!* (1966) (Best Actress Oscar nomination) and as an enigmatic woman in *Blow Up* (1966) an enduring ▷ Antonioni reflection on Swinging Sixties London.

Working in all media, she has proved herself one of the most distinguished performers of her generation with a luminous grace, conviction and integrity that have served a vast emotional spread of characterisations. However, her choice of cinema material has not always seemed wise. Her best work in the medium include the eccentric, free-spirited dancer Isadora Duncan in *Isadora* (1968) (Oscar nomination), the valiant title character in *Julia* (1977) (Best Supporting Actress Oscar) whose fearless opposition to 1930s Fascism closely paralleled her off-screen principles, and the ardent feminist in *The Bostonians* (1984) (Oscar nomination).

Other notable roles include the deformed Mother Superior in the hysterical ▷ Ken Russell opus *The Devils* (1971), the popular historical travesty *Mary, Queen of Scots* (1971) (Oscar nomination) and the wartime romance *Yanks* (1979) although the sheer variability of her screen work is illustrated by the contrast in such recent films as *Prick Up Your Ears* (1987) which features a warm, precise and perceptive portrait of literary agent Peggy Ramsey, and the execrable comedy *Consuming Passions* (1988) in which her exuberant turn as a lascivious Maltese widow was, to say the least, misguided.

Winner of an Emmy for her staggering portrayal of an Auschwitz inmate in *Playing for Time* (1980), her other work for television includes *My Body, My Child* (1982), *Wetherby* (1985), and the mini-series *Three Sovereigns for Sarah* (1986).

Highlights of her stage career include *The Prime of Miss Jean Brodie* (1966), *The Lady from the Sea* (1976–77) and *Orpheus Descending* (1988–89).

She will shortly be seen in the film *Ballad of the Sad Café* (1991) and *Whatever Happened to Baby Jane?* (1991).

REED, Carol (Sir)
Born 30 December 1906, London, UK.
Died 25 April 1976, London, UK.

The illegitimate son of actor Sir Herbert Beerbohm Tree (1853–1917), Reed briefly tackled the task of farming in America before, perhaps inevitably, following in his father's illustrious footsteps and making his stage debut at the Holborn Empire in *Heraclius* (1924).

An association with author Edgar Wallace (1875–1932) on the latter's popular West End thrillers gained him experience as a stage director and Wallace's involvement with British Lion Film Corporation brought him employment as an assistant on such films as *The Valley of Ghosts* (1929) and *The Flying Squad* (1929).

He spent a number of years as a dialogue director and co-director before making his fully-fledged directorial debut with the adventure *Midshipman Easy* (1935). He first attracted critical attention and commercial success with *Bank Holiday* (1937), a kaleidoscopic portrait of British working-class life.

The mining melodrama *The Stars Look Down* (1939) won him a reputation for his skilful handling of actors and interest in social issues although the vigorous spy thriller *Night Train to Munich* (1940) has more firmly stood the test of time.

Working on wartime propaganda, his documentary *The True Glory* (1944) (co-directed with Garson Kanin) won an Oscar and he brought a similar veneer of accuracy and realism to *The Way Ahead* (1944).

A storyteller and highly competent technician, Reed firmly believed that the director's function was to serve the intentions of a script's original author and to illustrate an aptitude for versatility by working in many genres. Fond of tilted camera angles and stories involving children bewildered by the machinations of the adult world, his career lacks any element of personal perspective or consistency, although his collaborations with Graham Greene (1904–91) on *The Fallen Idol* (1948) (Best Director Oscar nomination), ▷ *The Third Man* (1949) (Best Director Oscar nomination) and *Our Man in Havana* (1959) were among the more felicitous and good-mannered of unions between writer and director.

▷ *Odd Man Out* (1946), an evocative account of a hunted IRA gunman played by ▷ James Mason, completes the list of his finest post-war achievements.

In June 1952 he became the first British film director to receive a knighthood. Latterly a talent for hire on impersonal internationally-funded projects, his reputation suffered, although *Trapeze* (1956), at least, was a box-office success. His dismissal from the ▷ Marlon Brando *Mutiny on the Bounty* in 1960 was said to have irreparably damaged his self-confidence although he continued to work on major projects and won an Oscar for the old-fashioned virtues displayed in the family musical *Oliver!* (1968) which featured his nephew Oliver Reed (1938–) as Bill Sykes. His last film was *Private Eye* (1972).

REEL (1) A flanged spool on which long lengths of film or tape may be wound, or (2) the roll of motion picture film that forms a convenient length for handling a section of a programme during editing and printing. A typical feature film comprises 5 or 6 reels, each between 500 and 600 metres in length.

LA REGLE DU JEU (The Rules of the Game)

Jean Renoir (France 1939)
Marcel Dalio, Nora Gregor, Roland Toutain, Jean Renoir.
Premiered at 115 mins then cut to 85 mins; reassembled version restored original length, first screened at 1959 Venice Film Festival. b/w.

Greeted with much public approbation on its initial three-week run in Paris because it satirised the French ruling class in the run-up to war, heavily cut then banned altogether, not seen in the director's original version for the next twenty years, *La Règle du Jeu* has survived its inhospitable early reception to reach the point where it is regarded as ▷ Jean Renoir's supreme masterpiece and remains a regular feature of critics' all-time ten best lists. Rarely quoted in full, the film's emblematic line 'There's one thing that's terrible and that's that everyone has his reasons' is some indication of the richly ambiguous currents of drama, irony and social observation it brings to the basic material of an updated Beaumarchais boulevard comedy. The plotting brings assorted upper class types to the country home of Marcel Dalio's (1900–83) Marquis de la Chesnaye over one ill-fated weekend when the plethora of sexual jealousies that erupt between masters and servants are to climax in a tragic case of mistaken identity, the gamekeeper (Gaston Modot (1887–1970)) spotting dashing aviator André (Roland Toutain (1905–77)) embracing a woman he wrongly supposes to be his wife and shooting him. Renoir's purpose however is in examining the state of flux of the nation's moral values, so it follows that the other guests quickly rationalise the event to keep up appearances, their judgement one of justifiable homicide on the part of the Marquis to protect the honour of his spouse Christine (Nora Gregor (c.1890–1949)), for it is she who is the woman in question. With the camera fluidly marshalling the attention first towards one character then another, the eternal intrigue of the film lies in the tension between its mocking disapproval and the generosity of spirit it extends to these various dilettantes, the final product one of those rare masterpieces that seem to present a different facet on every subsequent repeat viewing.

REITH, John Charles Walsham (Lord)

Born 20 July 1889, Stonehaven, UK.
Died 16 June 1971, Edinburgh, UK.

Educated at Glasgow Academy and the city's Royal Technical College, Reith worked for the North British Locomotive Company and at the Royal Albert Docks before seeing service with the 5th Scottish Rifles during World War I.

Invalided out, he worked for the Ministry of Munitions and later moved to London, envisaging some kind of career in politics. In 1922 he secured the position of general manager at the newly-formed British Broadcasting Company and single-mindedly strove to establish radio as a medium of national importance and significance. When the company received a Royal charter and became the British Broadcasting Corporation in 1926, he became the first Director General and laid down the guiding principles of a monopolistic public service broadcasting network that sought to inform and entertain whilst earning international respect for its authority and accuracy.

In 1936 he inaugurated British television, but he left the BBC in 1938 when it no longer seemed to represent the challenge that his uncompromising and often dictatorial talents required.

His subsequent, varied career included a spell as an M.P., a return to the Royal Navy, and the use of his administrative skills as the chairman of such bodies as the Commonwealth Telecommunications Board (1946–50), the National Film Finance Corporation (1949–51) and the Colonial Development Corporation (1950–59).

His autobiographical works include *Into the Wind* (1949) and *Wearing Spurs* (1966). Knighted in 1927, he became the first Baron Reith in 1941. In 1947, the BBC established an annual lecture series in his name which continues to this day.

RENOIR, Jean
Born 15 September 1894, Paris, France.
Died 12 February 1979, Beverly Hills, California, USA.

The son of impressionist painter Pierre Renoir, Renoir graduated from the University of Aix-en-Provence with a degree in mathematics and philosophy. After World War I service with the cavalry and in the French Flying Corps, he worked as a potter and ceramicist before a viewing of *Foolish Wives* (1924) is reputed to have influenced his decision to become a filmmaker.

He made his directorial debut with *La Fille de l'Eau* (1924) and created a diverse body of silent work from a version of Zola's *Nana* (1926) to the fantasy *Sur un Air de Charleston* (Charleston) (1927) and the panoramic historical pageant *Le Tournoi* (The Tournament) (1928).

With the advent of sound, his talents flourished on the Feydeau farce *On Purge Bébé* (1931), the thrillers *La Chienne* (1931) and *La Nuit du Carrefour* (1931) and *Boudu Sauvé des Eaux* (Boudu Saved from Drowning) (1932) which uses ▷ deep focus photography and direct sound to enhance the story of a tramp who is saved from drowning and wreaks havoc on the bourgeois household of his benefactor.

Jean Renoir

He followed this with a succession of ever more impressive films that highlight the unobtrusive elegance and discretion of his direction, his deep feeling for nature and a warm humanist appreciation of mankind's foibles and failings. His greatest work from this period includes the uncompleted *Une Partie de Campagne* (A Day in the Country) (1936), a beautiful ode to the countryside and the flow of a nearby river, and his two masterpieces ▷ *La Grande Illusion* (1937) which illustrates how common humanity can transcend petty personal, class and nationalistic differences, and ▷ *La Règle du Jeu* (The Rules of the Game) (1939), revealing a microcosm of pre-war French society during the course of a weekend party in a country chateau.

In 1940, he left France for America and worked in Hollywood on such films as *Swamp Water* (1941) and *The Southerner* (1945) (Best Director Oscar nomination) which show his affinity for ▷ neo-realist tales of the land, the black comedy *The Diary of A Chambermaid* (1946) and the ▷ film noir *Woman on the Beach* (1949).

Returning to Europe, he subsequently ventured to India for *The River* (1951), a serene, documentary-like drama of an English family's experiences of the Indian continent and of mortality. He then experimented with a rich palette of colours in a loosely linked trilogy of 19th century romances that explored his fascination with the theatre and the distinctions between artifice and reality: *The Golden Coach* (1952), *French Can Can* (1955) and *Elena et les Hommes* (Paris Does Strange Things) (1956).

His last films were *Le Caporal Epinglé* (The Vanishing Corporal) (1962) a gently amusing echo of ▷ *La Grande Illusion* as a young French P.O.W. proves indefatigable in his efforts to escape from a German camp, and *Le Petit Théâtre de Jean Renoir* (The Little Theatre of Jean Renoir) (1969) a four-part potpourri of dramatic reflections on the themes from his lifetime's work: his humanity, perceptiveness on the compromises and exigencies in human relationships and fondness for La Belle Epoque.

He made an acting appearance in *The Christian Licorice Store* (1971), published *My Life and My Films* (1974) and received an honorary Oscar in April 1975 to confirm his status as 'a genius who, with grace, responsibility and enviable devotion has won the world's admiration'.

RESNAIS, Alain

Born 3 June, 1922, Vannes, Brittany, France.

An amateur filmmaker from his teenage days, Resnais was a student at the IDHEC (Institut des Hautes Etudes Cinématographiques) in Paris, before serving in the armies of occupation in Germany.

After the War, he worked as an editor and cameraman and made his feature-length directorial debut with the 16mm endeavour *Ouvert Pour Cause d'Inventaire* (1947). Over the next decade, he continued to edit other people's films and directed a number of documentaries, of which *Van Gogh* (1948) and *Gaugin* (1950) remain the best known.

Associated with the first flourishings of the ▷ nouvelle vague, he made *Hiroshima Mon Amour* (1959) in which a French actress's affair with a Japanese architect brings memories flooding back of her wartime liaison with a German soldier. Making extensive use of complex flashbacks and soundtrack devices, the film becomes a meditation on the relationship of the past and present and the role that memory plays in dictating the future actions of any person.

He followed this with the even more challenging and cryptic ▷ *L'Année Dernière A Marienbad* (Last Year At Marienbad) (1961) in which a man persists in his assertion of having had an affair the previous year with a female guest at a vast mansion. Using a sense of geometric design in the visual patterns on screen, allied to long tracking shots down limitless corridors and a general air of enigma, he had created another Chinese box exploring the complex ties between past, present and future, memory and imagination.

Subsequent films like *Muriel* (1963) and *La Guerre Est Finie* (The War Is Over) (1966) provided further variations on his preoccupations with highly stylised, elliptical narratives on the theme of lives haunted by memories of the past.

He worked in New York for a while, returning to mainstream cinema with the more accessible entertainments of *Stavisky* (1974), recreating a notorious French scandal from the 1930s, and *Providence* (1977) an elegantly acted ▷ David Mercer script in which an aging author jumbles the characters and themes of his latest novel with his embittered perceptions of his nearest and dearest.

Mon Oncle d'Amérique (My American Uncle) (1980) was a more characteristic piece that intertwined three lives that are analysed in terms of the theories of an animal behaviourist. His more recent work, often criticised for a lack of warmth and an obsessive emphasis on ambiguity, includes *La Vie Est Un Roman* (Life Is A Bed of Roses) (1983), *Mélo* (1986) and the heavyhanded comic chaos of *Je Veux Rentrer A La Maison* (I Want To Go Home) (1989).

RIEFENSTAHL, Leni

Full Name Helene Berta Amalie Riefenstahl.
Born 22 August 1902, Berlin, Germany.

A student of fine art and ballet, Riefenstahl was a lithe professional dancer before making her film début as a performer in *Der Heilige Berg* (1926). She appeared in a number of the mountaineering films made by Arnold Fanck (1889–1974) including *Der Grosse Sprung* (1927) and *Stürme Uber Dem Montblanc* (1931) and he has been credited with teaching her most of the basics of filmmaking.

In 1931, she formed Riefenstahl Films and made her directorial debut with *Das Blaue Licht* (The Blue Light) (1932). Appointed 'film expert to the Nationalist Socialist Party' by Hitler, she subsequently made *Sieg Des Glaubens* (Victory of the Faith) (1933). ▷ *Triumph Des Willens* (Triumph of the Will) (1934) was a compelling record of a Nazi rally at Nuremberg that used over 40 cameramen to create a record of the event that uses expert editing and photography to glorify Hitler's magnetism and the future he offered Germany. *Olympische Spiele 1936* (Olympiad) (1938) was another propagandistic documentary of impressive technique that celebrates the Berlin Olympic Games as a 'song of praise for the ideals of National socialism' and was given a gala première on Hitler's 49th birthday.

Whilst it is impossible to deny the beauty, grace and filmmaking artistry of her work, it is difficult to stomach claims that Riefenstahl was merely a documentarist creating an objective record of events and equally impossible to divorce judgement of her work from the cause it served.

The war years were littered with unrealised projects including *Van Gogh* (1943) and afterwards she was interned by the Allies and held on charges of pro-Nazi activity that were subsequently dropped. Unsurprisingly she was never able to resume her filmmaking career. *Tiefland* (Lowland) (1944), a version of Eugene D'Albert's opera in which she acted and served as co-director with ▷ G.W. Pabst, was finally released in 1954. A 1956 film *Schwartze Fracht* (Black Cargo) on the slave trade was never completed and a 1977 documentary *Nuba* was never released.

However she did enjoy a further career under the name of Helen Jacobs as a photojournalist, covering the 1972 Olympic Games for the *Sunday Times* and building a portfolio of work on the East African tribe of Mesakin Nuba.

RIGGER A rigger is the stagehand responsible for erecting the scaffolding and catwalks that support sets, the placement of lighting equipment and his fellow workers as they fulfil their tasks within a film unit.

RIO BRAVO

Howard Hawks (USA 1959)
John Wayne, Dean Martin, Ricky Nelson, Angie Dickinson, Walter Brennan.
141 mins. col.

Recovering from the failure of his Egyptian adventure *Land of the Pharaohs* some four years previously, with *Rio Bravo* ▷ Howard Hawks produced perhaps the most cogent exposition of the themes that had interested him throughout his widely varied career and a film that was, arguably, one of the greatest of all Hollywood westerns. Unlike the historicist approach favoured by fellow giant ▷ John Ford, Hawks's Texas town of Rio Bravo is no real moment in the development of the American frontier but an almost abstract background in which the director can play off his favourite archetypal characters against each other, so it is hardly surprising that in its examination of the integrity of the male grouping it bears a strong resemblance to earlier Hawks action fare, most notably 1939's Andean flying thriller *Only Angels Have Wings* and the Martinique-set wartime smuggling drama *To Have and Have Not* (1944). Here the relationships rather than plotting are emphasised, the central focus is on the bond between ▷ John Wayne's sheriff John T. Chance and his drunken buddy Dude, a well-cast Dean Martin (1917–), as the latter fights to beat the bottle and win back his self-esteem, and the pair's interaction with Walter Brennan's (1894–1974) plucky old jailer Stumpy, Ricky Nelson's (1940–85) singing upstart gunslinger Colorado, and Angie Dickinson's (1931–) alluring showgirl Feathers – each of whom plays their own part in defending the town jail from the villains' efforts to spring a prisoner from its confines. A leisurely and marvelously executed exercise in genre conventions, the film inspired Hawks himself to more or less repeat the same form (to lesser effect, it must be said) in the later *El Dorado* (1967) and *Rio Lobo* (1970), while John Carpenter (1948–) virtually remade it in an urban setting as *Assault on Precinct 13*, acknowledging his source by giving his own editor credit to a certain John T. Chance.

ROACH, Hal

Full Name Harriett Eugene Roach
Born 14 January 1892, Elmira, New York, USA.

After an adventurous life as a mule-skinner and gold prospector in Alaska, Roach entered the film industry in 1911 as a stuntman and ▷ extra.

Three years later, he began producing short comedy films featuring ▷ Harold Lloyd as the character Willie Work and later as Lonesome Luke. Titles include *Willie* (1914), *Luke the Candy Cut-Up* (1916) and *Lonesome Luke-Lawyer* (1918).

Responsible for scores of short films, he became an expert in the mechanics of screen humour and slapstick, helping to foster the careers of Charlie Chase (1893–1940), Will Rogers (1879–1934) and, most successfully, the partnership of ▷ Laurel and Hardy. He also provided opportunities for the enhancement of the directorial talents of such future comedy masters as ▷ George Stevens and Leo McCarey (1898–1969).

He also devised the popular and long-running series of 'Our Gang' films featuring the antics of precocious ragamuffins and won Oscars for *The Music Box* (1932) and *Bored of Education* (1936).

Concentrating on feature-length productions, his greatest successes include *Bonnie Scotland* (1935), *Way Out West* (1937), *Of Mice and Men* (1939) and the prehistoric fantasy *One Million Years B.C.* (1940) which he co-directed.

During World War II, he made a number of propaganda and training films and subsequently diversified into television production. His production company was dissolved in 1962 and his final involvement with the cinema came as the co-producer of the Hammer version of *One Million Years B.C.* (1966) and the compilation feature *The Crazy World of Laurel and Hardy* (1967). He received an honorary Oscar in April 1984 in recognition of his 'unparalleled record of contributions to the motion picture art form'.

ROBERTS, Julia

Born 28 October 1967, Smyrna, Georgia, USA.

The sister of actor Eric Roberts (1956–) (seen in such films as *Star 80* (1983) and *Runaway Train* (1985)), Roberts made her film debut in *Baja Oklahoma* (1988) and subsequently appeared with her brother in *Blood Red* (1988) and in *Satisfaction* (1988) before catching the eye as one of three young women undergoing a series of class-ridden romantic tribulations in the well-observed smalltown character study *Mystic Pizza* (1988).

Impressive amidst a welter of maudlin sentimentality as the diabetic daughter in *Steel Magnolias* (1989) (Best Supporting Actress Oscar nomination) she became a star overnight in *Pretty Woman* (1990) (Best Actress Oscar nomination). A warmhearted variation on Pygmalion with a hint of Cinderella, she brought a freshness and vitality to the cliché-ridden character of a gauche Los Angeles prostitute who enters a different world as the strictly-business consort of visiting East Coast hotshot ▷ Richard Gere.

At the time of writing, she seems set to become one of the few female stars of recent vintage to also carry a box-office magnetism that brings customers over the door on her name alone. Cast as one of the medical students dabbling with the afterlife in *Flatliners* (1990) her presence made the film an unexpected commercial hit and she followed this as a brutalised wife escaping a tyrannical husband in *Sleeping With The Enemy* (1991) and the tragic romance *Dying Young* (1991).

In demand for almost every female role available she has been announced to play Tinkerbell in ▷ Steven Spielberg's *Hook* (1991) and has also been mooted as a possible female lead for the forthcoming film version of *Phantom of the Opera*.

ROBINSON, Edward G.

Real Name Emanuel Goldenberg
Born 12 December 1893, Bucharest, Romania.
Died 26 January 1973, Los Angeles, California, USA.

Emigrating to America around the turn of the century, Robinson later studied at the American Academy of Dramatic Arts in New York before making his stage debut in *Paid in Full* (1913).

He made his Broadway debut in *Under Fire* (1915) and served in the US Navy during World War I. A prolific stage performer early in his career, his work includes *Banco* (1922), *Androcles and the Lion* (1925) and *The Racket* (1927).

He made his film debut in *The Bright Shawl* (1923) and remained a stage performer until *The Hole in the Wall* (1929). However it was his electrifying portrayal of vulgar, merciless, megalomaniac gangster 'Rico' Bandello (a libel-conscious imitation of Al Capone) in *Little Caesar* (1930) that brought him overnight stardom.

A short, squat figure with thick features and a wide mouth, he brought magnetism and refreshing vestiges of humanity to a rogue's gallery of larcenous hoodlums in films like *The Whole Town's Talking* (1935), *The Last Gangster* (1937) and *Key Largo* (1948), and also made fun of this image in lighter fare like *A Slight Case of Murder* (1938) and *Brother Orchid* (1940).

A cultured man who spoke eight languages and was a noted art connoisseur in real life, his screen versatility was displayed throughout the 1940s in a succession of vital, detailed characterisations that reveal the gentle humanity of a dedicated scientist in *Dr Ehrlich's Magic Bullet* (1940), the grim paranoia of the captain in *The Sea Wolf* (1941), the weary doggedness of an insurance investigator in ▷ *Double Indemnity* (1944), a hen-pecked husband in *Scarlet Street* (1945) and the patriarch in *All My Sons* (1948).

The hysteria of the McCarthy witch-hunts harmed his career, but he reasserted his reputation as a powerful character actor on stage in *Middle of the Night* (1956–58). Transcending through sheer force of personality the formulaic nature of many of the 1960s scripts he accepted, he was also memorable as the veteran poker player in *The Cincinnati Kid* (1965) and the nostalgic old man in *Soylent Green* (1973), his last work.

His autobiography, *All My Yesterdays*, was published in 1973 and his widow accepted a posthumous Oscar in March of that year dedicated to a 'Renaissance man. From his friends in the industry he loves'.

ROEG, Nicolas Jack

Born 15 August 1928, London, UK.

Beginning on the lowest rung of the filmmaking ladder, Roeg worked as a tea boy at the Marylebone Studio in 1947 and was a clapper boy on *The Miniver Story* (1950). He pursued a slow rise through the ranks, acting as camera operator on such films as *Passport to Shame* (1958) and *The Sundowners* (1960).

A second-unit photographer on ▷ *Lawrence of Arabia* (1962), he subsequently became a noted ▷ cinematographer making valuable contributions to such diverse films as *The Masque of Red Death* (1964), *Fahrenheit 541* (1966), *Far From the Madding Crowd* (1967) and *Petulia* (1968).

He began his directorial career in collaboration with Donald Cammell on *Performance* (1970), an original if pretentious exploration of the shifting power struggles between a rock star and the London gangster who takes refuge in his home. He followed this with the solo directorial effort *Walkabout* (1971) which captures the awesome beauty of the Australian outback as two children go walkabout and are saved from starvation and thirst by an Aborigine. Even more accomplished was *Don't Look Now* (1973), a chilling psychological thriller set in Venice which fragments time and narrative to striking effect as it recounts the story of a man haunted by the drowning of his daughter and the foreknowledge of his own death.

His interest in outsiders, obsessive behaviour and an unsettling ability to juggle time was apparent in works that were complex in structure and horrifying in their conclusions. The kaleidoscopic *The Man Who Fell to Earth* (1976) follows an extra-terrestrial visitor who is destroyed by his attempts to assimilate with our culture. *Bad Timing* (1979) is a powerful study in sexual obsession and psychopathic behaviour, *Eureka* (1982) a grandiose portrait of a rugged gold prospector whose life is led in the afterglow of his greatest discoveries and *Insignificance* (1985) a playful examination of celebrity, the theory of relativity and the horror of nuclear annihilation.

Gaining a reputation as an original artist struggling against the commercial drive of modern mainstream filmmaking, he grew increasingly prolific as the 1980s wore on but the results were usually of a less idiosyncratic and conventional nature like *Castaway* (1987), a television remake of *Sweet Bird of Youth* (1989) starring ▷ Elizabeth Taylor, and the children's adventure *The Witches* (1990). Even more typical projects like the provocative Oedipal drama *Track 29* (1988) seemed like a mannered echo of earlier and richer works.

ROHMER, Eric

Real Name Jean-Marie Maurice Scherer

Born 4 April 1920, Tulle, France.

A teacher of literature at a lycée in Nancy, Rohmer had published a novel, *Elizabeth* (1946), under the pseudonym of Gilbert Cordier, before moving to Paris and establishing his reputation as a film critic for such publications as *La Revue du Cinema* and *Les Temps Modernes*.

In 1950, he founded *La Gazette du Cinema* and began to make short films like *Journal d'Un Scelerat* (1950) and *Presentation ou Charlotte et son Steak*

(1951). Linked with the members of the ▷ nouvelle vague who strove to turn their intellectual attacks on established notions of European cinema into groundbreaking feature films, his own feature debut *La Signe du Lion* (The Sign of Leo) (1959), a portrait of an irresponsible American composer surviving on the Left Bank, failed to generate the same kind of enthusiasm as their early efforts.

Editor-in-chief of *Cahiers du Cinema* (1957–63) and co-author of ▷ *Hitchcock* (1957), he returned to direction with *La Boulangerie de Monceau* (1963) the first in a series of six 'Moral Tales'. Struggling to maintain a cinema career, he worked extensively for French educational television before finding critical acclaim for *La Collectioneuse* (1967) in which an artist and an antique dealer attempt to resist the bikini-clad charms of a girl who shares a friend's villa in St Tropez.

He followed this with *Ma Nuit Chez Maud* (My Night With Maud) (1968), *Le Genou de Claire* (Claire's Knee) (1970) and *L'Amour l'Après-Midi* (Love in the Afternoon) (1972) which allowed free rein to his talent for a microscopic scrutiny of the moral quandaries in love and the tricky business of maintaining a steady course on the path of righteousness.

After the stylised historical drama *Die Marquise Von O* (The Marquise of O) (1976) and *Perceval Le Gallois* (Perceval) (1978), he embarked on a series of six 'Comedies and Proverbs' that included *La Femme de l'Aviateur* (The Aviator's Wife) (1980), *Pauline à la Plage* (Pauline at the Beach) (1983), *Les Nuits de la Pleine Lune* (Full Moon in Paris) (1984) and *L'Ami de mon Amie* (My Girlfriend's Boyfriend) (1987).

In this series his young, affluent bourgeois characters are usually played by unknowns, his style is seriously lacking in flamboyance and his scripts are dense with verbal and intellectual debate, yet he speaks eloquent, cinematic truths about pain, loneliness and emotional manipulation.

He then began a series based around the four seasons with *Conte de Printemps* (A Tale of Springtime) (1990) which found him delicately attuned to the murmurs of the heart and all the petty jealousies, misunderstandings, deceits and desires that are woven around the chance friendship of a philosophy teacher and a piano student and their respective emotional entanglements.

ROME, OPEN CITY (Roma, Città Aperta)

Roberto Rossellini (Italy 1945)
Anna Magnani, Aldo Fabrizi, Marcello Pagliero, Harry Feist.
100 mins. b/w.
Cannes Film Festival: *Palme D'Or.*

Critics continue to debate whether ▷ Luchino Visconti's 1942 James M. Cain adaptation *Ossessione* or ▷ Roberto Rossellini's 1945 *Roma, Città Aperta* (Rome, Open City) might be termed the first flowering of Italian ▷ neo-realism, but while the earlier offering was consciously designed to cohere with a set of realist artistic principles, Rossellini's chronicle of life under the Nazi occupation was more concerned with preserving on screen real historical incidents – filmed in the real locales wherever possible, and utilising a largely non-professional cast to gain a greater feeling of authenticity. Shot within two months of the Allied liberation of Rome, the project had been in development since the previous year but the director's wealthy sponsor soon ran out of money and Rossellini was left to sell his clothes and even his furniture to scrape together the funds required to purchase leftover lengths of newsreel stock. The adverse conditions under which the film was evidently put together however lend its grainy images, rough performances and downbeat locations an immediacy that speaks emotively of a certain historical experience. Still moving today when its documentary-influenced techniques have become a part of the film language, *Roma, Città Aperta*'s raw vigour aroused a considerable stir on its initial release. Analytically speaking, its melodramatic plotting, manipulative music and sentimentalised use of children hardly render it a model of objectivity along doctrinaire neo-realist theoretical lines, but the humanism that suffuses so much of Rossellini's work here transcends such narrow representational nitpicking, while the piece's international success was to rejuvenate the home industry and offer much encouragement to the subsequent careers of his Italian peers, ▷ Visconti and ▷ Vittorio De Sica.

ROMERO, George A(ndrew)

Born 5 February 1940, New York, USA.

A student of art, design and theatre, Romero began making his own short films on 8mm and created such titles as *The Man From the Meteor*, *Gorilla* and *Earthbottoms* between 1954 and 1956.

Working in television and as an actor and director in Pittsburgh, he made his feature-length debut with *Night of the Living Dead* (1968), a chilling and visceral low-budget horror story of traditional American values and the family unit under threat from flesh-eating zombies.

He continued his gory re-invigoration of the horror genre with *The Crazies* (1973) in which a fearsome military elite develops when a small town is struck by a biological plague. He returned to earlier themes with *Zombies – Dawn of the Dead* (1978), an emotional rollercoaster ride whose horrific content is underpinned by a sharply satirical subtext that attacks contemporary consumerist impulses and human greed.

He also directed *Jack's Wife* (1973) in which a bored housewife dabbles in witchcraft, and *Martin* (1977) in which a disturbed and insecure teenager suffers under the delusion that he may be a vampire.

A defiantly independent filmmaker who has maintained his base in Pittsburgh, he ventured into more mainstream waters with *Knightriders* (1981) in which contemporary bikers re-enact medieval jousts, and the undistinguished comicbook horror omnibus *Creepshow* (1982).

He returned to the zombie series with *Day of the Dead* (1985) but parsimonious budgets and poor acting have proved a hindrance rather than an inspiration to the development of his artistic vision and this entry lacked the complexity and undercurrents of his earlier work.

Monkey Shines (1988) in which a capuchin monkey suffers malevolent personality changes when injected with human cranial fluid, failed to overcome the lack of credibility in its plot and his decision to collaborate on a script for a poorly received remake of *Night of the Living Dead* (1990) does not appear to have been a wise move.

His latest film is *The Dark Half* (1991).

ROOM AT THE TOP

Jack Clayton (UK 1959)
Laurence Harvey, Simone Signoret, Heather Sears, Donald Houston.
115 mins. b/w.
Acadmy Awards: *Best Actress* Simone Signoret; *Best Adapted Screenplay* Neil Paterson.
Cannes Film Festival: *Best Actress* Simone Signoret.

Britain in the late Fifties saw a burgeoning resistance to the apparent changelessness of its society during the austere post-war period, a sense of youth revolt founded on a hostility to the Establishment and/or working-class consciousness and culturally defined by rock 'n' roll and the so-called 'angry young men' of the theatre (John Osborne, Arnold Wesker) and the literary scene (John Braine, Allan Stillitoe, Stan Barstow). In film terms, the screen adaptations of Osborne's *Look Back in Anger* (1959) and Braine's *Room at the Top* were to provide a new impetus for the British film industry, with the substantial international success of the latter film in particular a key component of the growing confidence that was to launch the careers of a number of innovative new talents – notably the loosely-aligned ▷ Free Cinema grouping including directors Karel Reisz (1926–), with ▷ *Saturday Night and Sunday Morning* (1960), and Lindsay Anderson (1923–), with *This Sporting Life* (1963). Looking at these later offerings contextualises Jack Clayton's (1921–) *Room at the Top* as a transitional work, its much-touted sexual openness less significant perhaps than its grim Northern setting and harsh assessment of the means necessary to surmount the ossification of the British class system. Shot on location in Bradford, it follows the fortunes of Laurence Harvey's (1928–73) Joe Lampton, a working class aspirant who's sharp enough to realise that ability is no guaranteed passport to social mobility and who instead doggedly courts the daughter (Heather Sears (1935–)) of a local business magnate. However, in his single-minded drive towards wealth and social ranking he sacrifices his personal happiness, for agreeing not to see his lover, older woman Simone Signoret (1921–85), is the price he must pay for a marriage of arid respectability. In the final analysis, freedom from the constriction of social codes is shown to be not a matter of financial security but one of self-respect.

Followed in 1965 by a sequel, *Man at the Top* (director Ted Kotcheff (1931–)), also starring Laurence Harvey, while the Joe Lampton character, as played by Kenneth Haigh (1929–), was also revived for a British television series, *Man at the Top*, which spawned a movie spin-off under the same title in 1973, the third year of its run.

ROOTS

USA 1977
Writer Alex Haley

The original series of *Roots* was a 12-hour epic journey, tracing the family tree of ex-coastguard Haley through seven generations and the American experience of slavery. In the US it ran over eight consecutive nights and was promoted as the 'biggest event in TV history'. The result of the hype, for what was an above average mini-series – albeit one based on 12 years of research – was that 85% of all homes with a television tuned in.

Haley claimed to have traced his roots back to one Kunta Kinte, who was born in the Gambia in 1750. Kinte (played by LeVar Burton and John Amos) was captured by slave traders and auctioned in Savannah, Georgia. His progress was followed, through to his grandson, Chicken George (Ben Vereen (1946–)), who was emancipated in 1870. The story was picked up in *Roots: The Next Generation* (1979), which concluded with Haley (played by James Earl Jones (1931–)) arriving in the Gambia to meet members of his ancestral tribe. ▷ Marlon Brando contributed a nine minute cameo as American Nazi, George Lincoln Rockwell, and won an Emmy.

While the historical veracity of Haley's story was disputed, the symbolic importance of a mainstream television drama showing black history as a heroic struggle was immense. That it achieved record ratings was even more remarkable.

ROSEANNE

USA 1988–
Roseanne Barr

As developed by *Cosby Show* writer Matt Williams, the original idea for *Roseanne* was a show based around the lives of working mothers. With the introduction of Barr (1954–), Williams (and Cosby producers Marcy Carsey and Tom Werner), focused on a single woman in a warts-and-all, blue collar family setting.

Barr, a popular stand-up comedienne, had won an Ace Award (a cable Emmy) for 1987's *Roseanne Barr Show*, which included a sketch with her as the maternal head of a household – a loose prototype for her own series. The plots are not highly developed, the family rolls with the punches of everyday life. Though Barr is the focus – controlling her three children with sarcastic put-downs – husband Dan (John Goodman) (1952–) often threatens to steal the show.

An antidote to the superficial gloss of the 1980s, *Roseanne* was an immediate success, but was soon marred by production difficulties. Barr complained that she and Dan spent too much time eating, and resisted ideas she thought might undermine the dignity of the characters. Then the star began to rewrite the scripts. By the thirteenth episode Barr threatened to quit unless writer Williams did so first. Barr stayed and won full creative control. Not the most fluid of actresses, she made her film debut in the 1989 flop *She Devil*. Goodman continues to carve a film career as a solid supporting actor and was the star of *King Ralph* (1991).

ROSENTHAL, Jack Morris

Born 8 September 1931, Manchester, UK.

Educated at Sheffield University, Rosenthal joined the promotions department of Granada television in 1956 and made his professional writing debut with over 150 episodes of ▷ *Coronation Street* between 1961 and 1969.

He also contributed to the innovative satirical show ▷ *That Was The Week That Was* (1963) and later created the comedy series *The Lovers!* (1970) which spawned a feature-film in 1972.

His many individual plays have won acclaim for their ability to mine a rich vein of humour in everyday situations and the eccentricities of behaviour revealed under the strain of awkward social occasions. *The Evacuees* (1974) was a nostalgic tale of two Jewish boys evacuated to Lytham St Anne's and striving, by hook or by crook, to escape back to Manchester, *Ready When You Are Mr McGill* (1976) concerned an ▷ extra unable to handle his moment in the limelight when entrusted with a speaking role, the award-winning *Barmitzvah Boy* (1976) humorously explored the domestic pressures on a young boy approaching his barmitzvah, whilst *Spend! Spend! Spend!* (1977) revealed one way to cope with a massive pools win.

His rare film scripts include *Lucky Star* (1980), *P'Tang, Yang Kipperbang* (1982), *Yentl* (1983) in collaboration with ▷ Barbra Streisand and *The Chain* (1985), a multi-character piece using the seven deadly sins as a dramatic basis for investigating the emotional traumas of a series of interlinked house removals.

His stage work includes *Smash!* (1981) and *Our Gracie* (1983) whilst his more recent television plays include *The Knowledge* (1979), *London's Burning* (1986) which led to a long-running series, and the simple nostalgia of *And A Nightingale Sang* (1989).

He has been married since 1973 to actress Maureen Lipman (1946–).

ROSI, Francesco

Born 15 November 1922, Naples, Italy.

A radio journalist and theatre actor, set designer and assistant director, Rosi was hired by ▷ Luchino Visconti as an assistant on *La Terra Trema* (1947) and worked in this capacity on a number of features including *Bellissima* (1951) and *Senso* (1954).

After co-directing *Kean* (1956), he made his solo directorial debut with *La Sfida* (The Challenge) (1958) which won a special jury prize at the Venice Film Festival. Concerned with contemporary social issues and the poverty and injustices he witnessed in Italian society, he quickly made his mark with *Salvatore Giuliano* (1961), a mixture of drama and documentary that amasses a mountain of information on the Sicilian Robin Hood-style bandit whose bullet-ridden corpse was discovered in 1950.

Seeking realism through the use of non-professional actors, he nevertheless employed ▷ Rod Steiger (1925–) as a political boss in *Le Mani Sulla Citta* (Hands Over the City) (1963), an angry indictment of property speculation in the poorest neighbourhoods of Naples.

Straying from his accepted interests, he made the bullfighting story *Il Momento Della Verita* (The Moment of Truth) (1964) and the charming fairytale confection *C'Era Una Volta* (Cinderella – Italian Style) (1967).

In the 1970s, he returned to incisive examinations of political corruption with *Il Caso Mattei* (The Mattei Affair) (1972), a documentary-like speculation on the demise of the socialist oil magnate, the gangster chronicle *Lucky Luciano* (1973) and *Cadaveri Eccellenti* (Illustrious Corpses) (1976) an elegantly visualised and gripping revelation of a right-wing conspiracy to dupe the public.

His subsequent films have lacked the consistency of theme and rigour of his earlier work, often substituting pictorial flair and an artistic langour for narrative drive. However, the leisurely *Cristo Si E Fermato A Eboli* (Christ Stopped At Eboli) (1979) features an excellent performance from Gian Maria Volonté (1930–) as a man exiled to southern Italy, *Tre Fratelli* (Three Brothers) (1981) is a thoughtful portrayal of Italy's North-South divide, *Carmen* (1983) a rousing, sun-drenched version of the opera and *Cronaca Di Una Morte Annunciata* (Chronicle of a Death Foretold) (1986) a sumptuous if stately adaptation of the novel by Gabriel Garcia Marquez.

Most recently he has returned to the political thriller with *Di Menticare Palermo* (To Forget Palermo) (1990), the labyrinthine story of an Italo-American political candidate inexorably drawn into Mafia chicanery.

ROSSELLINI, Roberto

Born 8 May 1906, Rome, Italy.
Died 3 June 1977, Rome, Italy.

A sound technician and ▷ editor, Rossellini first made his mark in the cinema with experimental short films like *Daphne* (1936) and *Prelude à l'Après-Midi* (1938) before making his feature-length directorial debut on *La Nave Bianca* (The White Ship) (1940).

Immediately after the war, his trilogy

of ▷ *Roma, Citta Apperta* (Rome – Open City) (1945), *Paisa* (Paisan) (1946) and *Germania, Anno Zero* (Germany, Year Zero) (1947) were fundamental to the establishment of the ▷ neo-realist movement, conveying the painful aftermath of the conflict and the hardships of war-torn Italy with stories which combined drama with factual accuracy and were told with raw naturalism, non-professional actors and true life settings.

An adulterous liaison with ▷ Ingrid Bergman provoked worldwide condemnation, and the five films they made together were often banned. Critically undervalued at the time they are still under-appreciated and provide a compelling focus on the problems of the individual adjusting to changes in post-war Europe and portraits of a marriage in crisis. They include *Stromboli Terra Di Dio* (Stromboli) (1950) with Bergman as a Lithuanian refugee escaping internment by marriage to a simple fisherman, *Viaggio In Itali* (Voyage to Italy) (1953), in which an estranged English couple are reconciled during a visit to Naples and *Angst* (Fear) (1954) in which a wife is blackmailed over her adulterous activities.

Although he married Bergman in 1950 (they were divorced in 1957), his reputation had been irreparably damaged in international circles. However, he did enjoy popular successes once more with the wartime stories *Il Generale Della Rovere* (General Della Rovere) (1959) starring ▷ Vittorio De Sica as a conman disguised as an Italian general and *Era Notte A Roma* (It Was Night In Rome) (1960) detailing the price paid for providing sanctuary for three escaped P.O.W.s.

He spent his later years creating intellectually stimulating and unorthodox television documentaries on historical figures, including *The Rise To Power of Louis XIV* (1965), *Socrates* (1970) and *The Messiah* (1977).

His daughter by Ingrid Bergman, Isabella Rossellini (1952–) has been seen in such films as *Blue Velvet* (1986), *Tough Guys Don't Dance* (1987) and *Cousins* (1989).

ROSSITER, Leonard

Born 21 October 1926, Liverpool, UK.
Died 5 October 1984, London, UK.

An insurance clerk until he was 27, Rossiter subsequently joined Preston Repertory company making his stage debut in *The Gay Dog* (1954) and his London debut in *Free As Air* (1957–58).

His hawk-like features, predatory mouth and lizard-like tongue, combined with expert timing and energetic attack, could portray the furtively sinister or the manically comic. He made his film debut in *A Kind of Loving* (1962) and subsequently played supporting roles in such major British productions as *Billy Liar* (1963), *King Rat* (1965), ▷ *2001: A Space Odyssey* (1968) and *Barry Lyndon* (1975).

Following his Broadway debut in *Semi-Detached* (1953) his notable theatre work included *The Resistible Rise of Arturo Ui* (1968–69), *Richard III* (1971) and *Banana Box* (1973). The latter was transformed into the television series *Rising Damp* (1974–78) where his leering, lecherous landlord Rigsby, presiding over the seediest of boarding houses, remains a supreme comic achievement. A film version followed in 1980 and his lugubrious talents were further employed on television as the businessman who masterminds his own demise in the equally popular situation comedy *The Fall and Rise of Reginald Perrin* (1976–80).

Later film work includes *Britannia Hospital* (1982) and *Water* (1984); he died during the interval of a West End production of *Loot* (1984).

ROUGH CUT The first or 'rough' version of a finished film, the rough cut is prepared by the ▷ editor from the various takes of scenes printed during a film's production. Selecting the best of the individual takes he or she will cut these together in the order dictated by the script thus providing those concerned with an approximation of the completed film and an indication of what fine tuning may still be required.

RUDOLPH, Alan

Born 18 December 1943, Los Angeles, California, USA.

The son of film and television director Oscar Rudolph, Rudolph made a brief childhood appearance in his father's film *Rocket Man* (1954) written by Lenny Bruce (1926–66). A graduate in accountancy, he later joined the Director's Guild Training Program and worked on numerous television productions and the film *Riot* (1968).

He directed several short films and made his feature-length debut with the independent, low-budget efforts *Premonition* (1970) and *Terror Circus* (1973). Long-associated with ▷ Robert Altman, he worked as an assistant-director to him on *The Long Goodbye* (1973), *California Split* (1974) and *Nashville* (1975) and co-wrote the screenplay for his *Buffalo Bill and the Indians* (1976).

When his directorial career began in earnest, Altman produced both *Welcome to L.A.* (1976) and *Remember My Name* (1978). Something of a maverick, his most personal films create an elegantly stylish private world of heightened reality to showcase Pirandellian ruminations on dangerous liaisons, broken hearts and confusing coincidences provoked by that old devil called love. Films like *Choose Me* (1984), *Trouble in Mind* (1985) and *Love at Large* (1990) use stylised performances, plaintive soundtracks, artfully composed lighting and set design to provide almost surreal settings for his mosaic-like portraits of the sexually and socially maladjusted.

He has alternated these personal endeavours with more mainstream assignments like the broad comedy *Roadie* (1980) and the ecological thriller *Endangered Species* (1982), although the peregrinations of his career have also produced such diversionary delights as the political documentary *Return Engagement* (1983) and the charming romantic fantasy *Made in Heaven* (1987). In 1988, he realised a long-cherished project *The Moderns*, a beautifully mounted if somewhat arid depiction of Paris in the 1920s and the ragtag of artistes and charlatans that it attracted.

His latest film is *Mortal Thoughts* (1991).

THE RULES OF THE GAME *see* **LA RÈGLE DU JEU**

RUNNING SHOT A shot in which the camera is rendered mobile to keep pace with a moving object, whether an actor or vehicle. This can be achieved by also mounting the camera on a vehicle or by the exertions of an energetic cameraman utilising a hand-held portable camera.

RUSHES (or DAILIES) During any day of filming a particular scene may run to any number of ▷ takes from which the ▷ director will select the ones that he or she wishes to consider for use in the final editing of the film. Positive prints will then be struck from the motion picture negative the same night of shooting and sent unedited to the director, ▷ producer and others most closely involved, for viewing the next day. The prints offer some indication of the progress of filming, whether more footage may be required to cover certain scenes and will, to some extent, shape future plans. Rushes are also referred to as dailies.

RUSSELL, Ken

Full Name Henry Kenneth Alfred Russell

Born 3 July 1927, Southampton, UK.

After service in the Merchant Navy (1945) and Royal Air Force (1946–49), Russell pursued a variety of careers as a ballet dancer, actor and photographer before such amateur film efforts as *Amelia and the Angel* (1957) and *Peep Show* (1958) secured him employment at the BBC.

His numerous television documentaries include *Portrait of A Goon* (1959), and *Lotte Lenya Sings Kurt Weill* (1962) and he made his feature film debut with the comedy *French Dressing* (1963).

This was followed by the idiosyncratic film concluding the Harry Palmer spy trilogy, *Billion Dollar Brain* (1967), which is notable for a lavish climactic confrontation on ice that evoked memories of ▷ *Alexander Nevsky* (1938).

However, he attracted wide attention with his television biographies for the arts programme *Omnibus* which increasingly attempted to break the conventions of portraying great lives by seeking some dramatic interpretation of the inner conflicts and personality that illuminated the artist's work. His successes include *Isadora Duncan, The Biggest Dancer in the World* (1966), *Song of Summer* (1968) on Delius and *The Dance of the Seven Veils* (1970) on Richard Strauss.

International success followed with his vibrant adaptation of *Women In Love* (1969) (Best Director Oscar nomination) which displayed his expressive use of landscape, sensitive eye and ability to spot previously untapped potential in actors like Oliver Reed (1938–) and ▷ Glenda Jackson.

He then embarked on his most prolific and ambitious period of filmmaking, with *The Music Lovers* (1970), a brazenly overstated biography of Tchaikovsky, followed by the grotesque power of *The Devils* (1971), which conveys an outbreak of sexual hysteria in a 17th century monastery, his technically proficient but rather charmless version of *The Boyfriend* (1971) and *The Savage Messiah* (1972) a more subdued portrait of sculptor Henri Gaudier-Brzeska.

A controversial figure, who seemed to court the label of enfant terrible, his flamboyant style and unorthodox approach have divided viewers into outraged observers or staunch followers. However, what at first seemed like a bracing, barricade-storming assault on the plodding conventions of social realist British cinema with a barrage of stylish images and iconoclastic sexual attitudes appeared more like flatulent self-indulgence by the time of such lesser works as *Lisztomania* (1975) and *Valentino* (1977).

Allied to the more conventional yokes of *Altered States* (1980) and *Crimes of Passion* (1984) he produced some of his best work with the latter proving a highly effective satire on American mores and an outlandish plea for the redemptive power of love.

However, a subsequent return to Britain resulted in such banal and fatuous low-budget efforts as *Gothic* (1987) and *Salome's Last Dance* (1989).

He has also staged opera and directed pop videos for such singers as Elton John (1947–) and Cliff Richard (1940–), published an autobiography entitled *A British Picture* (1989) and made an acting appearance in *Russia House* (1990). His latest film as a director is *Whore* (1991).

RUTHERFORD, Margaret (Dame)

Born 11 May 1892, London, UK.
Died 22 May 1972, Chalfont St Peter, Buckinghamshire, UK.

A teacher of speech and piano, Rutherford studied acting at the Old Vic Theatre making her professional stage debut in their 1925 pantomime *Little Jack Horner*.

She worked in a variety of regional repertory companies and eventually made her film debut in the ▷ quota quickie *Dusty Ermine* (1936). A useful supporting performer, her star rose on the stage with a series of triumphant characterisations both comic and sinister in such roles as Miss Prism in *The Importance of Being Earnest* (1939), the housekeeper Mrs Danvers in *Rebecca* (1940) and medium Madame Arcati in *Blithe Spirit* (1941).

Regularly described as a shapeless form with the jowls of a bloodhound and the manner of any number of querulous birds, she continued to add vignettes of eccentric humour to such films as *Quiet Wedding* (1941) and *The Demi-Paradise* (1943).

A warm-hearted performer, she brought a torrent of enthusiasm and a cherishable comic sensibility to her stage role of the irrepressible Arcati in the film of *Blithe Spirit* (1945). A top character actress, she was equally memorable as the bluff school headmistress sublimely teamed with ▷ Alistair Sim in *The Happiest Days of Your Life* (1950), as Miss Prism in *The Importance of Being Earnest* (1952) and the usherette in *The Smallest Show On Earth* (1958).

British screen comedy of the 1950s was not the most sparkling showcase for her abilities, but she gamely ploughed through a mire of mirthless films in support of such unequal comedians as Norman Wisdom (1920–) in *Trouble in Store* (1954) and Ronald Shiner (1903–66) in *Aunt Clara* (1955).

Late in life, her popularity soared even further as Agatha Christie's sleuth Miss Marple in a series of British thrillers that transcended their low budgets and modest intentions largely thanks to her joyous professionalism. The titles include *Murder She Said* (1962), *Murder at the Gallop* (1963) and *Murder Ahoy* (1965).

The charm of her fussy Duchess of Brighton offered welcome comic relief in the all-star *The V.I.P.s* (1963) and brought her a Best Supporting Actress Oscar. Her subsequent career brought a late flourish of differentiated and well-handled roles as a moving Mistress Quickly in ▷ Orson Welles's *Chimes At Midnight* (1966), a greedy passenger in ▷ Chaplin's *A Countess From Hong Kong* (1967) and the Princess Ilaria in the Italian comedy *Arabella* (1968), her last film.

Her many stage successes include *Ring Round the Moon* (1951), *Farewell, Farewell Eugene* (1959–60) and *The School for Scandal* (1962).

In 1945, she married actor Stringer Davis (1896–1973) who played small roles in many of her subsequent films and was her sleuthing partner in the Miss Marple films.

Margaret Rutherford: An Autobiography was published in 1972.

SAFETY FILM Also termed non-flam, and contrasted with the highly dangerous cellulose nitrate used before 1950, safety film is film with its emulsion coated on a celluloid triacetate or polyester base and is thus slow burning and of low inflammability.

SAFETY LAST

Fred Newmeyer, Sam Taylor (US 1923)
Harold Lloyd, Mildred Davis, Bill Strothers, Noah Young.
70 mins. b/w.

Safety Last is a typical vehicle for the comic genius of ▷ Harold Lloyd, combining his famous penchant for hair-raisingly dangerous stunts with a degree of psychologically poignant character-building as subtle as anything in ▷ Charlie Chaplin's films. Just as Chaplin made an instantly recognisable character out of the woebegone little tramp, so Lloyd took a bespectacled, accident-prone country hick and set him loose in the hostile city. On this occasion, he is due to begin a new job in a large store, but realises that he is late in setting out. The misadventures which follow are the stuff of everyday anxieties, but his solutions are essentially cinematic. Unable to find even a handhold on the crowded street-car, for example, he finally lies in the street as if injured, and is whisked away to the department store in an ambulance. Audiences are invited to both sympathise with the motivating panic and celebrate the ingenuity of this awkward everyman who is simultaneously victim and unlikely hero. That strange synthesis is even more evident in the long scene which occupies the final quarter of the film, arguably the most famous in all of Lloyd's work, in which he finds himself taking over the job of his friend the human fly, who daily climbs an enormous skyscraper, but who is temporarily prevented from doing so by a pursuing policeman. He sets out on the long climb, expecting his friend's imminent return. The audience is well able to guess that no such rescue lies at hand, and the painstaking and increasingly dangerous climb turns into a real *tour de force* for the comic, culminating in his hanging from the hands of a clock over the dizzying drop. Once again, Lloyd is the unwitting victim of circumstance (his most hazardous stunts are always undertaken at the behest of, or to help, someone else), and once again emerges as a gormless but triumphant hero from his trial, a universally recognisable character who strikes a chord with audiences everywhere. The version now shown on television includes an episode from another Lloyd film, *Hot Water*, adding eight minutes to the running time.

THE SAINT

(UK 1963–68)
Roger Moore

The character of The Saint was created in 1928 by novelist Leslie Charteris, who wrote over two dozen books featuring a suave criminal in constant battle with his brothers in crime. Charteris was not slow to see the potential of his hero, and soon adapted his adventures for radio, comic strips and films. In the 1950s he formed a television production company with the aim of shooting a series with ▷ David Niven in the title role. It was not to be, and it took ▷ Lew Grade to steer the character into his most famous incarnation. By now the hero had been updated, and was a sophisticated operator in the world of espionage. Or as Charteris put it 'a roaring adventurer who loves a fight . . . a dashing daredevil, imperturbable, debonair, preposterously handsome, a pirate or a philanthropist as the occasion demands.' He was – who else? – Roger Moore (1927–), though he had to diet before taking the part. Moore was vital to Grade's strategy. Already known to American audiences from *Ivanhoe* and *Maverick*, he boosted the export potential. *The Saint*, after all, was the most expensive British series to be made without American assistance. The show was smooth, enjoyable nonsense, with Moore visiting exotic locations and falling for a new woman (among them Julie Christie (1940–)) in each episode. His two-seater Volvo sports car became a cult object. The switch to colour filming in 1966 strained the budget, and led to cutbacks elsewhere, so later episodes tend to be more formulaic. The 1978 revival, *The Return Of The Saint*, starring Ian Ogilvy (1943–) and a further revival in 1990, were both ill-timed and forgettable.

SATURDAY NIGHT AND SUNDAY MORNING

Karel Reisz (UK 1960)
Albert Finney, Shirley Ann Field, Rachel Roberts, Hylda Baker.
British Academy Awards: *Best British Film*; *Best British Actress* Rachel Roberts; *Most Promising Newcomer* Albert Finney.
89 mins. b/w.

Saturday Night and Sunday Morning is the best of the stream of social realist films which dominated British cinema in the late 1950s and early 1960s. Tagged 'kitchen-sink dramas' because of their focus on the domestic details of working class life, which had not previously been envisaged as a suitable subject for serious filmmakers, they characteristically drew on the work of the new generation of novelists and playwrights then chronicling the mores of the working class, including such writers as Stan Barstow, John Osborne, and Alan Sillitoe, whose novel provided the basis of this film. Albert Finney (1936–) made a powerful and highly convincing debut, and much of the enduring appeal of the film is down to his remarkable performance in the central role. His character, Arthur Seaton, is centrally concerned with his relationships with two women, his fiancée Doreen (Shirley Ann Field) (1938–) and a married woman, Brenda (Rachel Roberts) (1927–80), with whom he has an affair. It is one of the great strengths of the film that the women are depicted as strongly individualised characters in their own right, rather than simply pawns in the male game. Arthur himself is neither angry young man, like Osborne's Jimmy Porter in *Look Back In Anger*, nor avid social climber on the make, like Joe Lampton in John Braine's ▷ *Room At The Top*. His philosophy is simple; enjoy life as best you can, and 'don't let the bastards grind you down.' That rather self-defeating limitation, along with director Karel Reisz's (1926–) gritty, semi-documentary approach, ultimately gives the film a dour, almost despairing air, but it remains a landmark in British cinema, particularly in its frank treatment of sexuality.

SAURA, Carlos

Born 4 January 1932, Huesca, Spain.

Trained as an engineer, Saura was a professional photographer before studying at the Instituto de Investigaciones Y Experiencias Cinematograficos in Madrid from 1952 to 1957 and making his directorial debut with the short *La Tarde del Domingo* (Sunday Afternoon) (1957) and the documentary *Cuenca* (1958).

He made his feature-length debut with *Los Golfos* (The Hooligans) (1959), a harsh study of the slum-dwelling youth on the outskirts of Madrid. One of the generation keen to explore the emotional and psychological damage of the Franco years to the national psyche, he was able to subtly circumvent the country's repressive cinematic guardians with material that used allusion and allegory to make its points. *La Caza* (The Hunt) (1965), for instance, is a lean psychological thriller following four participants in a day's rabbit-hunting, that also reveals much about their perspective on the Civil War and its bitterly divisive legacy.

Consistently drawn to the interplay between past and present, reality and fantasy, his films comment on the Franco era and the consequences of sexual and political repression. Notable work includes *Peppermint Frappe* (1968), *Ana y los Lobos* (Ana and the Wolves) (1972), and *Cria Cuervos* (Raise Ravens) (1975).

There is a noticeable shift in his career after the death of Franco as he was able to be less oblique in his reportage of modern Spain. Early post-Franco work includes *Elisa, Vida Mia* (Alisa My Love) (1977) an ambiguous, personal drama of the relationship between a father and daughter, *Los Ojos Vendados* (Blindfold) (1978) about terrorism and *Deprisa, Deprisa* (Fast, Fast) (1981) a gritty, energetic account of Madrid's criminally-orientated punks using non-professional actors.

More recently, he found an international audience for a trio of stunning flamenco dance features; *Bodas de Sangre* (Blood Wedding) (1981), *Carmen* (1983) and *El Amor Brujo* (A Love Bewitched) (1985).

His lavish conquistador epic *El Dorado* (1988) proved an acute disappointment in its sluggish treatment of material memorably covered in *Aguirre, Wrath of God* (1972) but he returned to form with *Ay, Carmela* (1990) a tragi-comedy of two-travelling actors and the troubles they encounter by straying on to the wrong side of the enemy lines during the Civil War.

SAYLES, John Thomas

Born 28 September 1950, Schenectady, New York, USA.

A graduate of Williams College, Massachusetts, Sayles worked as an actor and had published two novels, *Pride of the Bimbos* (1975) and *Union Dues* (1977), before securing work as a scriptwriter for ▷ Roger Corman's New World Pictures.

Employed exclusively on exploitation subjects made on parsimonious budgets, he wrote scripts noted for their invention, wit and ability to both satisfy the requirements of any genre whilst displaying a knowing humour at their ludicrous excesses. His many scripts include *Piranha* (1978), *Battle Beyond the Stars* (1980), a space age variation on ▷ *The Seven Samurai* and *The Howling* (1980) a nicely judged werewolf shocker with a welcome tongue-in-cheek tone that inspired several inferior sequels.

Using his earnings he financed independent projects of his own, making his directorial debut with *The Return of the Secaucus Seven* (1980) in which a reunion of 1960s college radicals serves to illustrate the extreme social and political changes that have befallen an entire generation of Americans.

Lianna (1982) told of a young woman painfully freeing herself from a bleak marriage and growing to accept her lesbianism. Bereft of vast financial resources, it treated its subject with frankness and integrity offering realistic dialogue and sensitive performances.

The Brother from Another Planet (1984) was a typically sly, individualistic piece of science fiction in which a mute black alien crashlands in Harlem, whilst *Matewan* (1987), luminously photographed by Haskell Wexler (1926–), represented a more ambitious scale of production as it conveyed a bitter coal-mining strike from the 1920s with the sweep of a Western and the intimacy of a more personal drama.

Still a scriptwriter for hire on films like *The Clan of the Cave Bear* (1986) and *Breaking-In* (1989), he continues to write plays and short stories, has directed pop promos for the music of Bruce Springsteen and still acts on stage and screen often playing small roles in his own films.

His interest in politics and American history unravelling at a level of ordinary human emotions was evident once more when he directed *Eight Men Out* (1988), the story of the infamous 1919 baseball World Series when the Chicago White Sox team accepted bribes to lose matches.

SCHLESINGER, John Richard

Born 16 February 1926, London, UK.

As a student at Oxford University from 1945 to 1950, Schlesinger was a member of the dramatic society and directed his first short film *Black Legend* (1948).

Thereafter, he worked as a small part actor, making his film acting debut in *Singlehanded* (1952) and appearing in the likes of *Oh, Rosalinda* (1955) and *Brothers-in-Law* (1957). At the BBC from 1956 to 1961, he directed documentaries for the *Tonight* and *Monitor* series and won praise for his short film *Terminus* (1961) which covered a day in the life of Waterloo Station.

He made his feature-length debut with *A Kind of Loving* (1962) and gained a reputation for illuminating complex human relationships and his sensitive handling of actors in such subsequent films as the comedy *Billy Liar* (1963), the modish *Darling* (1965) and *Far From The Madding Crowd* (1967).

His transatlantic career began with promise when he won a Best Director Oscar for *Midnight Cowboy* (1969), a well-acted character study of the relationship between a hick Texas stud and a consumptive con man that was entirely calculated but benefited from the incisive playing of its stars and the director's eye for the authentic blend of smells, sights and seediness that constitute mid-Manhattan.

He subsequently tackled the troubled bisexual triangle in *Sunday, Bloody Sunday* (1971) and made a vastly overblown version of *The Day of the Locust* (1975). His more recent work has shown some alarming variations in quality and ranges from the reasonably exciting thriller *Marathon Man* (1976), to the elephantine comedy *Honky Tonk Freeway* (1981) and such unadulterated potboilers as *The Believers* (1987).

After the staid and highly theatrical character study of *Madame Sousatzka* (1988) he made his best film for some time with the derivative but well-crafted and entertaining thriller *Pacific Heights* (1990).

He has also staged opera, including *The Tales of Hoffman* in London, and occasionally directs for television, notably *An Englishman Abroad* (1982) and *Separate Tables* (1983).

SCHRADER, Paul

Born 22 July 1946, Grand Rapids, Michigan, USA.

Raised in a strict Calvinist household, Schrader did not see his first film until late in his teenage years. Fascinated by the medium, he became a combative film critic, some of whose philosophies are explored in his book *Transcendental Style in Film: Ozu, Bresson, Dreyer* (1972).

He subsequently enrolled at UCLA (University of California at Los Angeles) Film School and became a scriptwriter whose earliest efforts to make it to the screen includes *The Yakuza* (1974), *Obsession* (1975) a latterday variation on ▷ Hitchcock's ▷ *Vertigo* (1958), and ▷ *Taxi Driver* (1976) which was brilliantly realised by director ▷ Martin Scorsese as a portrait of a contemporary urban hell and the redemption through violence of an alienated individual.

He made his directorial debut with the gritty thriller *Blue Collar* (1978), and quickly followed this with *Hardcore* (UK: The Hardcore Life) (1979) in which a deeply religious father searches for his runaway daughter through a living hell of the seedy Los Angeles porn industry.

His best work came with *American Gigolo* (1980) which made highly effective use of a sleek visual style and the narcissistic image of star ▷ Richard Gere for a ▷ Bresson-like examination of the redemption through love experienced by a high-class male prostitute whose chic world crumbles when he is arrested on a murder charge.

Thereafter, a highly seductive visual style and an emotional coldness have tended to dominate his work, placing barriers between the subject and the possibility of audience involvement. His subsequent directorial efforts include the remake of *Cat People* (1982), a stylistically ambitious and beautiful attempt to convey the life and work of Japanese writer *Mishima* (1985), the little seen rock 'n' roll family drama *Light of Day* (1987) and a dispassionate and alienating recreation of controversial events in the life of *Patty Hearst* (1988).

He has continued to work on occasional scripts like *The Mosquito Coast* (1986) and *The Last Temptation of Christ* (1988) and he recently directed *Comfort of Strangers* (1990) which reflected his recurring problems of a seductive surface gloss and impeccable technique concealing a lack of dramatic conviction and weak characterisation.

SCHWARZENEGGER, Arnold

Born 30 July 1947, Thal, near Graz, Austria.

A dedicated bodybuilder from the age of 14, Schwarzenegger was named Mr Germany in 1966 and went on to become the youngest Mr Universe in history. Winning the title on seven occasions, he changed the face of bodybuilding as a popular sport and with his charm, chutzpah and business-sense built a lucrative empire of bodybuilding books, videos and personal endorsements.

In America from 1968, his physique

Arnold Schwarzenegger in *The Terminator* (1984)

secured him a role in *Hercules Goes To New York* (1969) where he appeared under the name of Arnold Strong. He had a small role in *The Long Goodbye* (1973) and made a number of television appearances before being cast by ▷ Bob Rafelson in *Stay Hungry* (1976) for which he received a ▷ Golden Globe as best newcomer.

The bodybuilding documentary *Pumping Iron* (1977) showed his charisma, mischievous sense of humour and competitiveness and led to further roles in *The Villain* (1979) and the television biography *The Jayne Mansfield Story* (1980).

His physical prowess was used to popular effect as sword wielding supermen in such epics as *Conan, the Barbarian* (1981), *Conan the Destroyer* (1984) and *Red Sonja* (1985). However, his appearance as the impassive, cyborg killer in *The Terminator* (1984) suggested the potential for acting growth that had not always been apparent in his rather monolithic and inexpressive roles to date.

Shafts of blunt humour and leadenly delivered comic one-liners helped to lighten and popularise his image as a cartoon-like superhero in such violent, macho thrillers as *Commando* (1985), *Predator* (1987) and *Red Heat* (1988) and he won much praise for his endearing performance as the naive brother in the comedy *Twins* (1988).

Now one of the world's most potent box-office attractions, he enjoyed another major success with the psychotic science-fiction adventure *Total Recall* (1990) before continuing to develop his comic skills and a lighter touch in *Kindergarten Cop* (1990). *T2 – Terminator 2: Judgement Day* is scheduled for release in 1991.

An American citizen since 1983, and known as 'Conan the Republican' because of his very public support for that party, he was named Chairman of the President's Council on Physical Fitness and Sports in 1990.

SCORSESE, Martin

Born 17 November 1942, Queens, New York, USA.

A film buff from childhood, Scorsese abandoned his original intention to study for the priesthood and instead attended New York University where he made a number of short films including *What's A Nice Girl Like You Doing In A Place Like This?* (1963).

He then worked towards his first feature *Who's That Knocking At My Door?* (1969) and subsequently lectured, made commercials and served as an editor on such films as *Woodstock* (1969), before an association with ▷ Roger Corman brought him the opportunity to direct *Boxcar Bertha* (1972), a vivid Depression-era story in the manner of ▷ *Bonnie and Clyde*.

More characteristic was *Mean Streets* (1973), an electrifying portrayal of life on the streets of New York's Little Italy that began his lengthy association with ▷ Robert De Niro and a recurring fascination with such issues as violence, guilt, comradeship and the mobster's lifestyle.

He brought a sympathetic eye to bear on the plight of a widow struggling to pick up the pieces of her life in *Alice Doesn't Live Here Anymore* (1974), which won an Oscar for Ellen Burstyn (1932–).

He firmly established himself as one of the foremost directors of his generation with ▷ *Taxi Driver* (1976), a compelling insight into the repressed inner life and corrosive inadequacies of a New York cabbie and the catharsis experienced in his climactic explosion of violence. Making exemplary use of the music of Bernard Herrmann (1911–75), his urban setting and fluid camerawork, he created a classic of contemporary urban alienation.

His subsequent work, including the lavish 1950s style musical *New York, New York* (1977) and the masterful ▷ *Raging Bull* (1980) (Best Director Oscar nomination) sought to illuminate the cause of masculine aggression and sexual inadequacy whilst questioning some of the more traditionally honoured American values.

His versatility was again underlined by his chillingly sarcastic view of the contemporary obsession with celebrity and the power of the media in *King of Comedy* (1982), in which a prospective stand-up comedian kidnaps a veteran chat-show host in return for his shot at fifteen minutes of television fame.

Its lack of commercial success and the cancellation of a cherished project to film *The Last Temptation of Christ* in 1983 led him to a more economic and less dramatically intense period of filmmaking that resulted in the financial successes of the yuppie-in-peril comedy *After Hours* (1985) and *The Color of Money* (1986).

He then went to Morocco for the controversial *Last Temptation of Christ* (1988) (Oscar nomination), a moving contemporary vision of the Messiah and a profound expression of the director's much troubled faith, and followed this with a bravura return to the peak of his form with *GoodFellas* (1990) (Oscar nomination), a blood-drenched saga of thirty years in the everyday life of a mobster.

He has recently completed a remake of *Cape Fear* (1991) that teams De Niro with Nick Nolte (1940–).

He has also directed pop videos, including Michael Jackson's *Bad* (1987), documentaries, including *The Last Waltz* (1978), commercials and the Broadway show *The Act* (1978–79) and has acted in his own and other films including *'Round Midnight* (1986), *Dreams* (1990) and *Guilty By Suspicion* (1991).

SCREWBALL COMEDY Largely a product of the Depression-era in America, screwball comedy offered a particularly zany antidote to the gloom of the real world, as attractive characters cavorted through a series of illogical escapades that combined a verbal sophistication with an obvious sense of physical humour (pratfalls, slapstick, etc) and a genuinely infectious anarchic spirit. Noted for their liberated, strong-willed heroines who easily outwitted the nominal male hero and an unusual edge of satire and class-consciousness, the genre more or less began in 1934 with the release of *It Happened One Night* and *Twentieth Century* and was a relatively shortlived phenomenon that did not survive the onset of World War II. Notable examples include *My Man Godfrey* (1936), *Nothing Sacred* (1937), *Bringing Up Baby* (1938) and *It's A Wonderful World* (1939) and the most skilled practitioners of the art include ▷ Cary Grant, ▷ Katharine Hepburn, ▷ Carole Lombard and William Powell (1892–1984).

Echoes of the genre's most mirthful moments can be heard in the glorious knockabout character romps of ▷ Preston Sturges in the 1940s and ▷ Peter Bogdanovich paid slavish homage in *What's Up Doc?* (1972).

THE SEARCHERS

John Ford (US 1956)
John Wayne, Jeffrey Hunter, Vera Miles, Ward Bond, Natalie Wood.
119 mins. col.

▷ John Ford reached the summit of his achievement in the Western genre with this dark, complex tale of revenge and

John Wayne in *The Searchers* (1956)

self-discovery. The great strength of *The Searchers* lies in the way in which it takes one of the most basic of all Western plotlines, and transforms it into a psychologically and imagistically rich reflection on the genre itself, and the myths it reflects. ▷ John Wayne, in arguably his finest film performance, plays Ethan Edwards, a virulently Indian-hating hero who is himself more than halfway to the savage state. Edwards sets out to find a band of Indians who have murdered his brother's family (Ford skilfully suggests an unfulfilled relationship between Edwards and his dead sister-in-law) and kidnapped his niece, whom he considers to be contaminated by her association, and intends to kill. The search, ultimately in the company of a young part-Indian, is framed against the savagely beautiful Monument Valley backdrop so beloved of the director. As the search progresses, so the complexities of Edwards's character, and the moral dilemmas which he poses, multiply. His own behaviour, including the adoption of Indian customs like scalping, is as savage as any atrocity perpetrated by the Indians. It becomes increasingly clear that Edwards himself is caught in a strange no-man's land which is both psychological and territorial, a point made clear in the symbolic closing of the door as he hesitates outside in the final frames. He is trapped between the two poles of the film's central divide, excluded by custom and character from the civilisation which the settlements represent, but unable to acknowledge his much closer connections with the Indians (and by extension the wilderness) he so detests. The dilemma of this ambiguous character takes on added resonance from being played by Wayne, the quintessential embodiment of the classic Western hero. The enduring appeal of *The Searchers* lies in the fact that it is a superb adventure story, but one which expands to contain a revelatory probing of the values inherent in the genre itself.

SECOND-UNIT A small unit of film technicians who work in addition to the main film unit on a major production and are under the distant supervision of the ▷ director, but who are entrusted with the shooting of locations, action sequences or vast crowd scenes that do not require the presence of the director or any of the principal performers involved in the film. Used as a means of cutting costs and saving time, the operation will work under a second-unit director and has included some notable work with Andrew Marton (1904–) carrying out the chariot race sequence from ▷ *Ben-Hur* (1959), Yakima Canutt (1895–1986) staging the cable-car fight in *Where Eagles Dare* (1968) and such luminaries as actor James Coburn (1928–) shooting second-unit on his friend ▷ Sam Peckinpah's *Convoy* (1978).

SEIDELMAN, Susan

Born 11 December 1952, Abington, Pennsylvania, USA.

Never a movie buff as a child, Seidelman studied graphic design at Drexel University and only shifted her major to film out of a sense of boredom and the impression that 'it seemed like an easy way to earn a degree'.

After graduating with a B.A., she worked for her local television channel 48 on their news programme Community Bulletin Board before being accepted for New York University Graduate School of Film and TV in 1974. Her satirical short film *And You Act Like One, Too* (1976) won a student Oscar, whilst *Yours Truly, Andrea G. Stern* (c. 1978) garnered prizes at several international festivals, including Athens and Chicago.

Her first feature, the self-financed *Smithereens* (1982) told of a rootless young girl from the suburbs and her false dreams of finding success in the punk rock scene of New York. It became the first independent American feature to be accepted in the main competition at the Cannes Film Festival.

Moving more into the mainstream, she directed *Desperately Seeking Susan* (1985) a fizzy, contemporary ▷ screwball comedy that made excellent use of its New York locale and the popular image of emerging music sensation Madonna (1958–).

Signed to a three-film contract with Orion Pictures, she then made the underrated *Making Mr Right* (1987), a role-reversal Pygmalion in which a Florida image consultant's assignment of promoting a wayward android serves as a metaphor for the difficulties in contemporary relationships.

Cookie (1989), a Mafia spoof, and *She-Devil* (1989), adapted from the Fay Weldon novel, were not commercial successes and revealed less attractive aspects of the director's talent including a taste for garish set-design, broad comic zaniness and a surprisingly toothless approach to some insubstantial scripts.

The most consistently employed woman director in contemporary American cinema, Seidelman's best work shows a strong visual sensibility, appreciation of pop culture and a flair for creating appealing feminist heroines from the unlikeliest of characters.

SELLERS, Peter

Full Name Peter Richard Henry Sellers
Born 8 September 1925, Southsea, Hampshire, UK.
Died 24 July 1980, London, UK.

A child actor in the revue *Splash Me* (1931), Sellers persevered as a stand-up comic and impressionist, first with ENSA (Entertainments National Services Association) during his Army Service and then at the Windmill Theatre. He subsequently moved into radio. A meeting with Spike Milligan (1914–) resulted in *The Crazy People* (1951) which was the forerunner of the popular and innovative *The Goon Show* (1951–59) a landmark in British comedy for its imaginative use of sound effects, an exotic mixture of surrealism and slapstick, and a humour that respected no conventions, revelling in the unpredictable and anarchic.

He made his film debut in *Penny Points to Paradise* (1951) and put his facility with mimicry at the disposal of a pungent series of characterisations that made him one of the stalwarts of British film comedy in the 1950s and 1960s. His most notable roles include the spiv member of *The Ladykillers* (1955), the elderly cinema projectionist in *The Smallest Show On Earth* (1958), the bumptious union leader in *I'm Alright Jack* (1959) and the amorous Welsh librarian in *Only Two Can Play* (1961).

Two films with director ▷ Stanley Kubrick established his international reputation; *Lolita* (1962) as Quilty and *Dr Strangelove* (1964) (Best Actor Oscar nomination) in which he enjoyed one of his finest hours as the American President, an RAF officer and a deranged scientist with Nazi tendencies.

His popularity was sustained as the bumbling, incompetent French detective Inspector Clouseau who mangled the English language with merciless abandon and was usually the butt of an elaborate pratfall or slapstick routine. The long-running series began with *The Pink Panther* (1963) and extended beyond his death to include the tasteless *Trail of the Pink Panther* (1982), which shamelessly constructed a flimsy new adventure using previously unseen out-takes of Sellers material and a substitute disguised with bandages.

Mister Topaze (1961) was the only film which he directed and he was often not the best judge of material, particularly in his search for serious roles that overstretched his abilities, and during the last decade of his career when he appeared in some of the most witless comedies of the day. His more successful ventures include *The World of Henry Orient* (1964) as an idolised concert pianist, *There's A Girl In My Soup* (1970) as a lecherous big-headed television star and *The Optimists Of Nine Elms* (1973) as a busker, and he gave one of his finest and most disciplined performances as the simpleton gardener whose inane pronouncements are hailed as the wisdom of a messianic sage in *Being There* (1979) (Oscar nomination).

He made his final film appearance in *The Fiendish Plot of Dr Fu Manchu* (1980).

SELZNICK, David O(liver)

Born 10 May 1902, Pittsburgh, Ohio, USA.
Died 22 June 1965, Los Angeles, California, USA.

The son of Russian-born film distributor Lewis J. Selznick (1870–1933) and younger brother of the producer and agent Myron (1898–1944), Selznick began his career working for his father and became a producer of such intriguing-sounding short films as *Will He Conquer Dempsey?* (1923) and *Rudolph Valentino and His 88 American Beauties* (1923).

An assistant story editor and associate producer at M-G-M in 1926, he subsequently worked as an associate director at Paramount before becoming Vice-President in charge of production at RKO where his most notable endeavours include *A Bill of Divorcement* (1932), *What Price Hollywood?* (1932) and ▷ *King Kong* (1933).

Back at M-G-M between 1933 and 1936, when he worked on *David Copperfield* (1935), *Anna Karenina* (1935) and *A Tale of Two Cities* (1935), among many others, he became an independent producer in 1936, forming Selznick International.

A stickler for detail, he was renowned for scrupulously supervising every aspect of a production and for making his every wish and random thought known to his associates through a voluminous correspondence contained in memos.

He made an auspicious start to his independent career with the ▷ screwball comedy *Nothing Sacred* (1937), the rousing swashbuckling adventure *The Prisoner of Zenda* (1937) and the first version of ▷ *A Star Is Born* (1937) (Best Picture Oscar nomination) but his crowning achievement was as the man who masterminded ▷ *Gone With the Wind* (1939) (Best Picture Oscar), searching the globe for the correct actress to play Scarlett O'Hara, convincing ▷ Clark Gable that he was the only man to play Rhett Butler and sparing no expense to lavishly re-create the power and the glory of Margaret Mitchell's novel and launch the film amidst unprecedented ballyhoo.

His other work pales by comparison but includes an association with ▷ Alfred Hitchcock that led to *Rebecca* (1940) (Oscar), *Spellbound* (1945) (Oscar nomination) and *The Paradine Case* (1948), as well as the wartime melodrama *Since You Went Away* (1944) (Oscar nomination), the gloriously overwrought western *Duel in the Sun* (1946), ▷ *The Third Man* (1949) and a poor version of *Farewell to Arms* (1957) that proved to be his last work for the cinema.

Many of his later films starred actress Jennifer Jones (1919–) who was his wife from 1949 until his death. The book *Memo From David O. Selznick* (1972) by Rudy Behlmer gives a strong flavour of the man.

SEMBENE, Ousmane

Born 8 January 1923, Ziguinchor, Senegal.

A mechanic and manual labourer, Sembene joined the Free French forces fighting in Africa during the War. Demobilised at Marseilles in 1945, he made only a brief return to Senegal before settling in France as a dock worker, trade union activist and, ultimately, a writer.

His first novel, *Le Docker Noir* (1956), was followed by a succession of texts including *O pays, mon beau peuple* (1957) and *Les Bouts de Bois de Dieu* (1960). Returning to Senegal, he was drawn to film as a more accessible form of communicating his concerns and studied in Moscow for a year (with ▷ Mark Donskoi among others) before making his directorial debut with an unreleased 16mm documentary on Mali entitled *Songhays* (L'Empire Sonrai) (1963).

Borom Sarret (1963) which followed uses a poor cart-driver's journey to juxtapose the extremes of poverty and wealth that exist in Dakar.

A keen observer of colonialism and its legacy for the political and cultural life of Africa, he was responsible for the first feature-film to come from sub-Saharan Africa – *La Noire de . . .* (The Black Girl From . . .) (1966) in which a young girl is driven to suicide by the callous insensitivity of her French employers.

Whilst utilising conventional narrative techniques and non-professional actors, his work has been groundbreaking in bringing sub-Saharan filmmaking to a wider audiences and placing on record the struggle for dignity and survival that he sees around him. He has also shown an acute awareness of women's roles in contemporary Africa.

Mandabi (The Money Order) (1968), a wry, pointed comedy, was also the first film to use a native African language – Wolof. He later used the Diola language in *Emitai* (1972), which tells of a village's defiant protection of its traditional ways.

Overcoming the technical difficulties of filming in Africa, his work grew more assured and he won growing critical attention for *Xala* (The Curse), (1974) an adaptation of his satirical novel that finds a metaphor for contemporary Senegal in the figure of a businessman cursed with impotence when his selection of a young third wife meets the disapproval of his other spouses and a militant daughter.

Ceddo (1977), banned in Senegal, controversially addressed issues of religious faith as a princess is both victim and vanquisher when an Imam attempts to forcibly convert a nation to Islam.

Still active as a novelist, his most recent film work has been as the co-director of *Camp Thiaroye* (1987), an episodic account of the stirring of nationalistic feelings among the African soldiers in a 1944 repatriation camp.

SENNETT, Mack

Real Name Mikall Sinnott
Born 17 January 1880, Danville, Quebec, Canada.
Died 5 November 1960, Hollywood, California, USA.

A child singing prodigy, Sennett hoped to pursue a career in opera and appeared in minor roles on Broadway and in burlesque between 1902 and 1908.

He joined Biograph Studios in 1908, making his film debut that year in *Baked in the Altar*. Under the tutelage of ▷ D.W. Griffith he became a leading man and turned to direction with *The Lucky Toothache* (1910).

By 1912, he had formed his own Keystone Company in Los Angeles, and set about altering and defining the conventions of American screen comedy. An ability to spot comic talent led to his discovery of ▷ Charles Chaplin, Fatty Arbuckle (1887–1933) and numerous others, whilst scores of short comedies illustrated his skill for staging breathless chases, custard pie fights and the pricking of pomposity. He also established the popular, bumbling boys in blue the Keystone Kops and the Sennett Bathing Beauties. His fuller length films include *Tillie's Punctured Romance* (1914), *The Goodbye Kiss* (1928) and *Way Up There* (1935).

Semi-retired from 1935, he made cameo appearances in *Hollywood Cavalcade* (1939) and *Down Memory Lane* (1949) and ▷ W.C. Fields presented him with a special ▷ Oscar in March 1938 dedicated to a 'master of fun, discoverer of stars, sympathetic, kindly, understanding comedy genius'.

Mack Sennett: King of Comedy, an as-told-to autobiography, was published in 1954.

SENSURROUND Perfected for use with the film *Earthquake* (1974), Sensurround was a ▷ special-effect that sought to replicate the high-decibel sound levels and tremors of an earthquake with a feeling like a low rumbling coursing under the floor of the cinema auditorium. The process involved adding air vibrations during the sound dubbing of the film which were then boosted in the auditorium by a system involving anywhere between ten and twenty speakers. A short-lived gimmick it was also used to enhance sequences in *Midway* (1976) and *Rollercoaster* (1977).

SESAME STREET

USA 1969–

Under the guidance of producer Joan Ganz Cooney, *Sesame Street* applied the techniques of advertising to the serious matter of educational television. A product of New York's non-profit-making Children's Television Workshop, and set in Harlem, it was specifically aimed at the infant children of low-income families, and avoided ethnic and class stereotyping – and condescension – by creative use of animation, puppets and film inserts. Few items lasted longer than 30 seconds, and any one show might include up to fifty items. Undoubtedly the most popular characters were ▷ Jim Henson's Muppets (a mix of marionettes and puppets), among them Kermit the Frog, who was created in 1954 from an old coat, a ping pong ball and a piece of cardboard. Also on hand to help with arithmetic, geometry and the fundamentals of spelling, were Big Bird (an eight foot canary), the Cookie Monster and Ernie and Bert Snuffleupagus, who were one third anteater to two thirds elephant, as well as being invisible to adults. The Muppets had previously appeared in the five-minute Sam And Friends, and in 1976 – after a spell on *The Julie Andrews Show*, would graduate to their own, exceptionally successful, spin-off series.

SET DESIGNER The individual who, quite literally, designs the sets for use in a film, making the requisite drawings and specifications from the sketches

Kermit the Frog, a regular on *Sesame Street*

supplied by the ▷ art director and sometimes constructing three-dimensional models of what the finished article will look like.

SEVEN SAMURAI (Shichinin No Samurai)

Akira Kurosawa (Japan 1954)
Takashi Shimura, Yoshio Onaba, Osao Kimura, Seiji Miyaguchi, Toshiro Mifune.
Venice Film Festival: *Silver Prize*
200 mins. b/w.

▷ Akira Kurosawa's reputation as arguably the leading director of large-scale action in the world was built on the back of this epic adventure story. It is the most popular of all Kurosawa's films in the West, and that may be in large part because it so closely (and consciously) parallels the familiar conventions of the western genre itself. The director has acknowledged the influence of the western (and especially the classic westerns of ▷ John Ford) on this film, a correspondence underlined by its subsequent metamorphosis into John Sturges's (1911–) Hollywood western *The Magnificent Seven* (1960), although its influence is equally evident in many other films of the period. Set in a 16th century Japan ravaged by civil war, the film revolves around the adventures of a group of Samurai warriors who are hired to protect a remote but impoverished village from the annual harvest-time incursion by plundering bandits hidden in the surrounding woods. That battle between the civilised settlers and the savage raiders, with the tarnished defenders in the middle, has been fought out in countless westerns, and has recurred again in Kurosawa's own work. The choreography of the ultimate violent clashes is breath-taking, and prefigures the kind of highly complex battle sequences which he created on a vaster scale for *Kagemusha* (1980) and *Ran* (1985) at a later stage of his career. The film's high standing in world cinema is due in part to its excellence as a totally absorbing adventure story, but is equally attributable to the depth and intensity of emotions which the director succeeds in building around the framework of that adventure. The ultimate confrontation is carefully postponed while an intricate complex of relationships between the villagers and the Samurai, at base two inimical groups, is developed. The climax, when it arrives amid the pouring rain so characteristic of Kurosawa, is all the more effective as a consequence.

THE SEVENTH SEAL (Det Sjunde Inseglet)

Ingmar Bergman (Sweden 1957)
Max Von Sydow, Gunnar Bjornstrand, Bengt Ekerot, Nils Poppe, Bibi Andersson.
Cannes Film Festival: *Special Jury Prize.*
96 mins. b/w.

This enigmatic and highly allusive allegory about man's relationship with God and Death established ▷ Ingmar Bergman as a major artist in world cinema, although he later turned against many of the films he made in this period. *The Seventh Seal* is a remarkable piece of work, both in terms of the audacity of the subject, and in the starkly etched visual style, the clean austerity of which is emphasised by the heavily stylised dialogue and a mournful music track. ▷ Max von Sydow plays a 14th-century knight, Antonius Blok, who returns from the Crusades to find his native land ravaged by plague and a strange cult of self-

The Seventh Seal (1957)

flagellation, inspired by a demented monk. The basic theme of the film is encapsulated in Blok's early encounter with the hooded figure of Death, who has come to tell him of his inevitable end. In an attempt to forestall that end, and also to reconcile himself with the workings of God and the shattered remnants of his faith, he challenged Death to a game of chess, earning a brief reprieve. During that time, he encounters the happier, innocent side of religious experience as represented by the travelling player Joff and his wife Mia, whose joyous visions are in stark contrast to the knight's dreadful doubts. Ultimately, the persecuted Joff is rescued from the famous dance of Death across the hilltop horizon which ends the film by the sacrifice of the knight, his lady, and the unfortunate squire Jons, who takes a more pragmatic view of life than his master, but must share his fate. Bergman described the film as being about 'the fear of death', and it reflects something of his own troubled attitude to religious belief and a malevolent God at this time in his life.

SHADOWS

John Cassavetes (US 1959)
Lelia Goldoni, Ben Carruthers, Hugh Hurd.
Venice Film Festival: *Critics Award.*
87 mins. b/w.

Director ▷ John Cassavetes's debut film is probably more important for its ground-breaking role in establishing the credibility of the American Independent cinema sector than for its own intrinsic merits. The film is not Cassevetes's best, but it did allow him to demonstrate the deeply personal and highly improvisational cinema which he envisaged could be translated into significant films. Made on a very low budget as an offshoot from a method-acting workshop which Cassavetes was then running, it was partly financed by his acting role in the television series *Johnny Staccato.* Shot in a rigorously striking ▷ cinéma vérité manner, and set in a curious underworld redolent of the Beat counterculture of the late 1950s, the film is given real atmosphere by Charles Mingus's jazz music score. *Shadows* was minimally scripted, and utilised a very loose plot line about a light-skinned black woman who falls in love with a white boy, but is ultimately rejected when he discovers her racial origin. Within that loose framework, Cassavetes allows the actors to explore the various human relationships and the issues of personal and racial identity which arise in the film, while simultaneously permitting them to explore the processes of acting itself through the considerable improvisational latitude allowed to them. It is inevitably a little rambling and even incoherent at times, but remains both a significant landmark and a fascinating achievement. The director would go on to refine the techniques he pioneered here in more fully realised fashion in films like *Husbands* (1970) and *A Woman Under The Influence* (1974), but the germ of his art is already fully present in this painfully realistic experiment in personal expression.

SHANE

George Stevens (US 1953)
Alan Ladd, Jean Arthur, Van Heflin, Brian De Wilde, Jack Palance.
Academy Awards: *Best Cinematography* Loyal Griggs.
118 mins. col.

Shane is the archetypal western; it exemplifies the western's mythicising function more completely than any other single film in the genre. The plot is simple, but contains multiple resonances. The Indian Wars waged for control of the plains are over (Hollywood had not yet learned to see them as genocide), but a new conflict is emerging in the wide open spaces, this time between the ranchers, greedy for vast open ranges to breed their cattle, and the settling farmers, eager to fence in and contain. Shane, a gunman whose time is passing, and who wearies of the morality of his trade, rides into the valley and attempts to align himself with the farmers, not as a gunman but as one of them. The pressures from the hired killer set loose by the ranchers, however, will allow no such remedy for his ills. Shane is fated to be a loner, and his actions in confronting the evil, vanquishing it, and disappearing from a settled life which cannot safely contain him, has the stark simplicity and tragic resonance of Greek myth. If it lacks the ambiguity and questioning self-doubt which pervades ▷ John Ford's ▷ *The Searchers*, *Shane* remains among the handful of very great westerns, and therefore among the handful of greatest American films, since no genre captured the American psyche and its values so vividly or so nakedly. Alan Ladd (1913–64) may have lacked the literal stature of a great romantic hero, but carried off the role for which he will be longest remembered with distinction. Shane is far from being the first solitary figure to stalk through a western landscape, nor the last, but he stands at the centre of the most starkly delineated example of the power and significance of that figure in American cinema, and in the American mind. *Shane*, more than any other western, strips down the fundamentals of the genre to its innermost, essential ideal, and gives that ideal memorable form.

SHICHININ NO SUMARAI *see* SEVEN SAMURAI

SHINDO, Kaneto

Born 22 April 1912, Hiroshima Prefecture, Japan.

Employed in the art department at the Shinko-Kinema Kyoto Studio from 1927, Shindo won prizes for a number of his unfilmed screenplays and worked as an assistant art director on *Aizo Toye* (The Mountain Pass of Love and Hate) (1934) directed by ▷ Kenji Mizoguchi.

A member of the studio's screenwriting team from 1939, he wrote such scripts as *Nanshin Josei* (South Advancing Woman) (1939), *Anjo-ke no Butokai* (The Ball of the Anjo Family) (1947) and *Shitto* (Jealousy) (1949). Heavily influenced by his intermittent collaborations with Mizoguchi on such films as *Josei no Shori* (The Victory of Women) (1947) and *Waga Koi wa Moenu* (My Love Burns) (1949), he made his directorial debut with *Aisai Monogatari* (Story of My Loving Wife) (1951), a tribute to his first wife who had died in her twenties.

Throughout his career, he has dramatised those issues closest to his own experiences: the penury of the rural life his father had known and the horrors of the atomic age. *Genbaku-no-No* (Children of Hiroshima) (1952) follows a schoolteacher's return to Hiroshima seven years after the dropping of the first atom bomb, whilst *Haha* (Mother) (1963) and *Honno* (Instinct) (1966) deal with the social and sexual consequences of the bombing on a humanistic level. Expressing revulsion at the power of nuclear weapons, his films eschew bitterness to find strength in survival and hope that future generations will learn from the mistakes of their forefathers.

His best known films in the West are *Hadaka no Shima* (The Island) (1961) and *Onibaba* (1964). The former is a wordless allegory on the dignity to be found in man's daily grind whilst the latter displays his talent for conjuring up mood and pictorial beauty with a chilling tale of two women who rob and kill passing samurai warriors.

Adept at many genres from sensationalistic horror to sexually graphic dramas, his style adapts to suit the tenor of his scenario. In 1975, he created the affectionate documentary *Aru Eiga-Kantoku no Shogai: Mizoguchi Kenji no Kiroku* (Life Of A Film Director: Record of Kenji Mizoguchi). His most recent films include the semi-documentary *Chikuzan Hitoi Tabi* (The Life Of Chikuzan) (1977) the story of a wandering minstrel, and *Hokusai Manga* (Hokusai, Ukiyoe Master) (1982).

SHOOTING RATIO Every scene in the script of a feature film is photographed in a variable number of ▷ takes and the shooting ratio is the average ratio of takes shot to scenes in the picture. A low-budget film, where financial concerns are paramount, may hope to operate on a ratio of 5:1, a larger-budget production may allow the director a ratio of 10:1 or even higher. Because of the nature of the beast, the shooting ratio of a documentary (footage shot to that actually used) is generally much higher.

SIGNORET, Simone

Real Name Simone-Henriette-Charlotte Kaminker
Born 25 March 1921, Wiesbaden, Germany.
Died 29 September 1985, Normandy, France.

A teacher and a typist for the newspaper *Le Nouveau Temps*, Signoret abandoned thoughts of any other career to enter the cinema as an extra in *Le Prince Charmant* (1942) and *Les Visiteurs du Soir* (1942).

An attractive, languorous blonde she soon graduated to leading roles and was frequently cast as a prostitute or courtesan bringing an alluring warmth and heavy-lidded sensuality to such films as *La Ronde* (1950), *Casque d'Or* (1952) and the thriller *Les Diaboliques* (1954).

A rare participant in English-language productions she won a Best Actress Oscar as the older woman ruthlessly sacrificed in the rise of Laurence Harvey (1928–73) in *Room At The Top* (1959) and gained further distinction for her performances as the drug-addicted countess in *The Ship of Fools* (1965), a concentration camp survivor in *The Deadly Game* (1966) and Arkadina in *The Seagull* (1968).

Unafraid to show her age, she matured very quickly into one of France's most distinguished character actresses, seen at her best as the Resistance leader in *L'Armée des Ombres* (Army in the Shadows) (1969), the dowdy housewife in *Le Chat* (1971), the aged prostitute in *Madame Rosa* (1977) and the grandmother in *L'Adolescente* (1979).

She made her last film appearances in *L'Etoile du Nord* (1982) and *Guy De Maupassant* (1982).

Her theatre work includes a production of *The Crucible* (1954) and a much criticised *Macbeth* (1966) in London with ▷ Alec Guinness.

Married to actor ▷ Yves Montand from 1951, she later turned to writing, completing an intelligent volume of autobiography *La Nostalgie N'Est Plus Ce Qu'Elle Etait* (Nostalgia Isn't What It Used To Be) (1976) and a novel *Adieu Volodia* (1985).

SILVERS, Phil

Real Name Philip Silver
Born 11 May 1912, Brownsville, Brooklyn, New York, USA.
Died 1 November 1985, Los Angeles, California, USA.

A stagestruck youngster, Silvers was a singer and boy tenor who made his professional bow as part of *The Gus Edwards Revue* (1925) in Philadelphia.

He worked in vaudeville and with the *Minsky Burlesque Troupe* (1934–39) before his Broadway debut in *Yokel Boy* (1939).

Signed to a contract with M-G-M, he made his film debut in *Hit Parade of 1941* (1940) and subsequently added sparkle to a succession of escapist wartime comedies and musicals as bald, bespectacled hapless suitors, friends of the leading man and general comic relief in films like *Tom, Dick and Harry* (1941), *You're In The Army Now* (1941) and *Cover Girl* (1944).

After World War II, he enjoyed notable Broadway hits with *High Button Shoes* (1947) and *Top Banana* (1951) for which he received a Tony Award.

However, he won lasting fame in television as the star of *You'll Never Get Rich* (1955–59), later known as *The Phil Silvers Show,* which earned him three ▷ Emmy Awards for his perfectly realised portrayal of Sergeant Bilko 'a Machiavellian clown in uniform' forever pursuing get-rich-quick schemes with explosive, fast-talking braggadocio.

Irrevocably associated with the character, *The New Phil Silvers Show* (1964) did not find favour, but he enjoyed further Broadway successes with *Do Re Me* (1960) and *A Funny Thing Happened On the Way to the Forum* (1972) (Tony Award) and brought some bright moments to films like *It's A Mad, Mad, Mad, Mad World* (1963), *Carry On ... Follow That Camel* (1967), *Buona Sera Mrs Campbell* (1968) and *The Cheap Detective* (1978).

Latterly in poor health, he continued to make guest appearances on television and grace increasingly inferior films. His autobiography, *The Laugh Is On Me,* was published in 1973.

SIM, Alistair

Born 9 October 1900, Edinburgh, UK.
Died 19 August 1976, London, UK.

Destined to make his career in the family tailoring business, Sim found his theatrical interests pulling him in a different direction. A lecturer in elocution at Edinburgh University from 1925 to 1930, he later left his job to make a professional stage debut in the London production of *Othello* (1930) starring Paul Robeson (1898–1976) and Peggy Ashcroft (1907–91).

Further stage work, including a season with the Old Vic, led to his film debut in *Riverside Murder* (1935).

His lugubrious manner, distinctive physiognomy, balding pate and range of rollercoaster vocalisations made him a cherished performer equally adept at mirthful or menacing characterisations.

A reliable supporting performer in many ▷ quota quickies of the 1930s, his more memorable roles include the genie of the lamp in *Alf's Button Afloat* (1938), the excitable newspaper editor in *This Man Is News* (1938) and *This Man In Paris* (1938) and Sergeant Bingham in the series of *Inspector Hornleigh* thrillers that included *Inspector Hornleigh On Holiday* (1939) and *Inspector Hornleigh Goes To It* (1941).

Although far from the conventional image of a leading man, the virtuosity and assurance of his eccentric characterisations brought him a level of audience affection that had to be satisfied with leading roles and he became a star as the police detective in the superior thriller *Green for Danger* (1946) and the Ealing comedy *Hue and Cry* (1946).

His many comedy successes include *The Happiest Days of Your Life* (1950), *Laughter in Paradise* (1951), *The Belles of St Trinians* (1954) in which he offered a hilarious portrait of disreputable schoolmistress Miss Frinton, and *The Green Man* (1956) in which he played a professional assassin. Dramatically he made a perfect *Scrooge* (1951), capturing the full range of the misanthrope's redemption and offered gentle probity as the title character in *An Inspector Calls* (1954).

After supporting roles in *School for Scoundrels* (1960) and *The Millionairess* (1961) he was absent from the screen until providing some ribtickling moments as a befuddled cleric in *The Ruling Class* (1972) and made further welcome contributions to *Royal Flash* (1975) and the ▷ Walt Disney children's film *Escape From the Dark* (1976).

His television work over the years includes the long-running comedy series *Misleading Cases,* several plays and the film *Rogue Male* (1976).

On stage he enjoyed a long association with playwright James Bridie and also appeared in *The Tempest* (1962), *Too True To Be Good* (1965), *The Magistrate* (1969) and *Dandy Dick* (1973) among many others.

SINGIN' IN THE RAIN

Gene Kelly, Stanley Donen (US 1952)

Gene Kelly, Donald O'Conner, Debbie Reynolds, Jean Hagen, Millard Mitchell.
103 mins. col.

Singin' In The Rain is regularly declared the greatest of all Hollywood musicals, and is regarded with the greatest affection by almost everyone who values the genre. The film takes some of the modernising influences which the production team of producer Arthur Freed (1894–1973) and co-directors Stanley Donen (1924–) and ▷ Gene Kelly had pioneered in ▷ *On The Town*, but sets them in a more conventional Hollywood studio setting. The plot revolves around the attempts of a film studio to convert their two major stars, Don Lockwood, played by Kelly, and his on-stage romantic partner Lina Lamont, from the silent era into the new talkies. Sadly, though, Lina's voice – cruelly travestied by Jean Hagen (1924–77) in the film – proves to be a hopeless case. Lockwood devises a plan to recruit a promising young singer and actress (Debbie Reynolds (1932–) in her first major film role) to provide Lamont's voice, but inevitably their romantic involvement is threatened by the subsequent demands made on her. Most of the song and dance routines (with the exception of the fantasy ballet involving Cyd Charisse (1921–), which is as charmingly out of place as anything concocted in the 1930s) are incorporated into advancing the plot, as in *On The Town*, but the final effect is less (relatively) naturalistic, doubtless a consequence of the stylised sets and showbiz subject matter. The film has at least two sequences to rate with the very best routines ever to come out of Hollywood, in Kelly's celebrated performance of the title song as he splashes through the rain, and the undervalued Donald O'Conner's (1925–) brilliant comic rendition of 'Make 'Em Laugh', which is as virtuoso and athletic as anything we see from Kelly himself. The film remains completely irresistible even after many viewings.

Gene Kelly and Cyd Charisse in *Singin' in the Rain* (1952)

SIRK, Douglas

Real Name Claus (later Hans) Detlef Sierck
Born 26 April 1900, Hamburg, Germany.
Died 14 January 1987, Lugano, Switzerland.

A student of law, philosophy and art history, Sirk began writing for the *Neue Hamburger Zeitung* in 1920 and commenced his lengthy association with the theatre the following year when he was chosen as assistant dramaturge for the *Deutsches Schauspiele* in Hamburg.

Artistic director of the *Bremen Schauspielhaus* (1923–29), he was subsequently appointed director of the Leipzig Altes Theater in 1929 and directed his first film *April, April* in 1935. He fled Germany in 1937, working throughout Europe before leaving for Hollywood in 1939. He made his American directorial debut with *Hitler's Madman* (1943) and enjoyed a modicum of success with the thrillers *Lured* (1947) and *Sleep My Love* (1948).

Employed at Universal Studios from 1950, he showed a flair for colour composition and a nostalgic appreciation of America's past in films like *Has Anybody Seen My Girl?* (1952), *Meet Me At The Fair* (1952) and *Take Me To Town* (1953).

However, he seemed most at ease utilising major studio stars like Rock Hudson (1925–85) in well-heeled, chintzy melodramas that depicted the emotional repression and sexual traumas of the middle class whilst also serving as critical assaults on conservative and patriarchal American values. His many popular films in this vein include *Magnificent Obsession* (1954), *All That Heaven Allows* (1955), *Written on the Wind* (1956) and *Imitation of Life* (1959).

Citing ill-health, he retired from filmmaking in 1959 but worked for the German theatre in the 1960s.

DET SJUNDE INSEGLET *see* **THE SEVENTH SEAL**

SJÖSTRÖM, Victor David
Born 20 September 1879, Silbodal, Sweden.
Died 3 January 1960, Stockholm, Sweden.

One of the founding fathers of Swedish cinema, Sjöström was raised in New York but returned to his homeland and was established as a leading actor and director for the stage before joining the expanding film company Svenska Bio in 1912. He made his film debut in *Vampyren* (1912) directed by Mauritz Stiller (1883–1928) and directed for the first time with *Tradgarrdsmaastaren* (The Gardener) (1912).

His many notable successes as actor/director include *Ingeborg Holm* (1913), a social drama, *Terje Vigen* (1918) based on Ibsen's poem, and, above all, *Korklaren* (The Phantom Carriage) (1921) where his psychological restraint in the portrayal of the leading character and sophisticated use of such techniques as multi-exposure and flashbacks were landmarks in cinematic acting and philosophy.

Noted for his 'sombre artistic mind', his work shows a preoccupation with revenge and redemption, a highly dramatic use of landscape to mirror the emotional state of his characters, a pantheistic sense of nature and an approach to acting that displays a subtlety and restraint far ahead of its time.

In 1923, he moved to Hollywood and used the name Seastrom to direct around nine films, few of which are still known to exist. However, his name endures as the director responsible for the ▷ Lon Chaney film *He Who Gets Slapped* (1924) and two masterpieces with ▷ Lillian Gish: a lyrical adaptation of Nathanial Hawthorne's 17th century *The Scarlet Letter* (1926) and the remarkable *The Wind* (1928) in which a young Virginian girl's journey to the wilds of Texas becomes a descent into madness starkly paralleled in the hostile landscapes of the Mojave desert and the howling winds that charge across the bleakest of terrains.

He made his sound debut with *A Lady to Love* (1930) but did not adapt well to the demands of the new era and his directing career ended with a return to Sweden for *Markurells I Wadkoping* (1931) and the ▷ Alexander Korda costume drama *Under The Red Robe* (1937) made in Britain.

As an actor however he created several memorable parts, as Knut Borg in Molander's *Ordet* (The Word) (1943) and above all Professor Borg in *Smiltronstallet* (Wild Strawberries) (1957), an egocentric, academically successful old doctor forced to admit his failure in human relationships. It crowned his career and was said to be a tribute from one great artist, ▷ Ingmar Bergman, to another.

SLEEPER A film industry term for a film which appears with little publicity or kudos and suddenly becomes a major financial or artistic success on the basis of its own merits and enthusiastic word-of-mouth. Two of the major hits of 1990, *Ghost* and *Home Alone* could be termed sleepers on the basis of the modest expectations that they soon spectacularly outpaced, whilst on the art-house circuit *Metropolitan* (1990) arrived relatively unheralded to become a popular attraction.

SLOW MOTION A process achieved by running the camera at a speed faster than the standard rate of projection (twenty-four frames per second) so that when the film is projected at the standard rate the action is apparently slowed down. Each movement is slower because it occupies more frames, i.e. an action shot at forty-eight frames a second and projected at twenty-four will take twice as long as it otherwise would have. Now something of a cliché in the horror and action genres, one of its greatest proponents was director ▷ Sam Peckinpah whose films like ▷ *The Wild Bunch* (1969) and *The Getaway* (1972) added a beauty and balletic quality to their depiction of blood-letting by the use of slow motion.

SMITH, Maggie

Born 28 December 1934, Ilford, Essex, UK.

A student at the Oxford Playhouse, Smith made her stage debut with the Oxford University Dramatic Society in a production of *Twelfth Night* (1952) and, after revue experience, appeared in New York as one of the *New Faces of '56*.

Her inimitable vocal range, mastery of stagecraft and precise timing have enabled her to portray vulnerability to both dramatic and comic effect. Gaining increasing critical esteem for her performances in *The Rehearsal* (1961) and *Mary, Mary* (1963), she joined the National Theatre to play in *Othello* (1963), *Hay Fever* (1966) and *The Three Sisters* (1970) among others.

She made her film debut as a calculating socialite in the minor thriller *Nowhere To Go* (1958) and later contributed scene-stealing performances to such films as *The V.I.P.s* (1963) and *The Pumpkin Eater* (1964) and was nominated for a Best Supporting Actress Oscar as Desdemona in the film version of *Othello* (1965).

Revealing a delightful flair for comedy in *Hot Millions* (1968), she gave a tour de force performance of charged theatricality as the nonconformist 1930s Scottish schoolteacher with Fascist leanings in *The Prime of Miss Jean Brodie* (1969) (Oscar), forever fussing over the hearts and minds of her beloved 'gels'.

Her subsequent selection of film roles shows a penchant for eccentric comedy and acidulous spinsters. Her most notable work includes the Auntie Mame-style character in *Travels With My Aunt* (1972) (Oscar nomination), a razor-sharp portrait of an English actress feeling the pressures of an American ▷ Oscar night in *California Suite* (1978) (Best Supporting Actress Oscar), a bitchy companion to ▷ Bette Davis in *Death on the Nile* (1978), the social-climbing chiropodist's wife in *A Private Function* (1984), the lady's companion in *A Room With A View* (1985) (Oscar nomination) and the lovelorn Irish woman in *The Lonely Passion of Judith Hearne* (1987).

Her more recent stage work includes *Virginia* (1980) and *Lettice and Lovage* (1988–90) and she will next be seen in the film *Hook* (1991).

SNOW WHITE AND THE SEVEN DWARFS

Walt Disney (US 1937)
Voices of Adriana Caselotti, Harry Stockwell, Lucille Laverne, Moroni Olsen.
Academy Award: *Special Award* Walt Disney.
Venice Film Festival: *Great Art Trophy.*
New York Film Critics Awards: *Special Award.*
83 mins. col.

▷ Walt Disney acted as producer rather than supervising director on the first full-length cartoon produced by his studio, but the project is very definitely his nonetheless. It was Disney who saw the writing on the wall for the eight-minute cartoon which had previously been the staple of animated filmmakers, and who had the ambition to launch an animated feature film, widely viewed as an act of supreme folly by his competitors.

In order to realise the project, the studio were required to develop many technical changes in both animation and camera techniques, and to deal with the particular problems of a 'realistic' depiction of human figures, previously almost unknown in the cartoon medium. Even the drawing boards used in the studio had to be changed in order to accommodate the larger, more detailed images required, while the development of the ▷ multi-plane camera helped to give a greater illusion of depth, augmented by a much refined attention to small but telling effects like the movement of smoke or rain, and other similar 'special effects' which had not been considered worthwhile in the shorter format, but greatly enhanced the overall quality of the animation. Equally unprecedented attention was paid to developing characterisation, both in the 'realistic' human figures like Snow White, the Huntsman, or the Prince, and in the dwarves, each of whom was given a distinctive speech or action which the audience could recognise, a strategy successfully employed for minor characters in live action films. *Snow White* set the standard for subsequent feature-length animation films, and if it has been technically surpassed since, it remains a classic of the form.

SOAP
USA 1977–80

An intentionally absurd tale of two sisters and their families, *Soap* spoofed daytime tele-dramas so successfully that many of its critics – among them the Catholic Church – failed to see it as satire, and launched a campaign to have what they took to be its immoral excesses removed from the screens. Created, written and produced by Susan Harris, *Soap* followed the increasingly bizarre lives of two sisters and their families in Dunns River, Connecticut. The amiable, but dim, Jessica Tate (Katherine Helmond (1933–)) was married to the super-rich Chester (Robert Mandan), who was generous with his extra-marital favours. Her ex-army father was living under the delusion that he was still at war. Her black butler Benson (Robert Guillaume (1930–), later given his own show) was irritable and rude. Less rich, but no more blessed was sister, Mary (Cathryn Damon (1933–87)) who was hitched to the oft-crazed Burt Campbell (Richard Mulligan (1932–)) – he killed her first husband, and later was kidnapped by extra-terrestrials. The sisters had four sons between them. One was a mobster, and one was a ventriloquist who believed his dummy, Bob, was alive. A third was gay, and planning a sex change, while the fourth, a tennis coach, was murdered early on (Jessica was charged, Chester later confessed). As with the soaps it aped, the plots grew ever more convoluted. A baby was possessed by the devil, and Jessica had a fling with a South American revolutionary. Such developments were treated with suitably comic disdain – though the show's critics didn't get the joke. Having broken new ground by centring a sitcom on two women, Harris went one better and created *The Golden Girls*, which all but dispensed with men.

SOAP OPERA Now used as a general term for melodrama that involves a good deal of superficial sex, shopping and emotional torment, soap opera was originally used to describe the afternoon serials on radio and television that involved some form of continuing domestic drama and were sponsored by the large soap and detergent manufacturing companies in a none-too subtle ploy to influence householders towards the purchase of their product. Although now more of a social phenomenon than anything else, ▷ *Coronation Street* has been British television's longest-running soap opera and ▷ *Peyton Place* is acknowledged as American primetime's first soap opera.

SOFT FOCUS A visual effect in which the image outline displays a softness or haziness and the focus is not clearly defined. This can be achieved by the simple means of filming slightly out of focus or by shooting through a special lens, filter or gauze, or a substance like petroleum jelly. Generally used to create a mood of nostalgia or romance it has also proved efficacious in softening the features of aging thespians when it comes to close-ups and portraiture.

SOME LIKE IT HOT
Billy Wilder (US 1959)
Marilyn Monroe, Tony Curtis, Jack Lemmon, George Raft, Pat O'Brien.
Academy Award: *Costume Design.*
120 mins. b/w.

Like many of the films which have come to be regarded as the classic Hollywood movies, *Some Like It Hot* was largely passed over at the ▷ Academy Awards, possibly because its rather risqué sub-

Tony Curtis and Jack Lemmon in *Some Like It Hot* (1959)

ject-matter discomfited the notoriously conservative members of that hidebound organisation. Certainly, the gender confusions thrown up in ▷ Billy Wilder's superbly funny ▷ screwball comedy, harmless though they are, might have been seen as threatening to a status quo not yet rocked by the social and sexual revolutions of the 1960s. Tony Curtis (1925–) and ▷ Jack Lemmon are excellent as Joe and Jerry, the hapless witnesses to a gangland killing in Chicago who make their escape by dressing up as women and joining an all-girl band bound for Florida. The chase is continued when gangland killer Spats Colombo (George Raft (1895–1980), parodying his famous *Scarface* (1932) role), arrives at the hotel where the band are working, but the real comedy is generated by a complex and very funny examination of sexual identity. Joe (Curtis) falls in love with Sugar Kane, memorably played by ▷ Marilyn Monroe, but is caught between his urgent desire to declare himself, and Jerry's insistence that he maintain the pretence of being a woman to fend off the pursuing hoods. Jerry himself, meanwhile, becomes the focus of the amorous attractions of the smitten Osgood Fielding III (Joe E. Brown (1892–1973)), who provides one of the most engaging punch-lines in Hollywood when he greets the discovery of Jerry's real gender with a besotted 'Nobody's perfect!', adding a hilariously inappropriate final twist to the theme of sexual identity. The film's combination of a razor-sharp script (by Wilder and his regular writing partner I.A.L. Diamond (1915–88)), outstanding performances, and vibrant action makes it one of the most consistently funny, and certainly most enduring, of all Hollywood comedies.

THE SOUND OF MUSIC

Robert Wise (US 1965)
Julie Andrews, Christopher Plummer, Eleanor Parker, Peggy Wood, Richard Haydn.
Academy Awards: *Best Picture*, *Best Director*, *Best Score*, *Best Adaptation* Irwing Kostal, *Best Editing* William Reynolds, *Best Sound* James P. Corcoran and Fred Hynes.
174 mins. col.

The Sound of Music has never enjoyed even a respectful critical reception, far less an admiring one, but as one of the most popular films ever to emerge from Hollywood, it has unquestionably had the last laugh. Based on the true story of the Von Trapp family's escape from Austria into neutral Switzerland in 1938 to avoid the Nazi occupation, the film version was developed from Rodgers and Hammerstein's rather weak stage musical. The film, though, had the incalculable advantage of the staggeringly beautiful Swiss scenery as a backdrop to its sentimental, shamelessly tear-jerking account of the adventures and tribulations of the fugitives, punctuated by hugely popular songs like 'Do Re Mi' and 'My Favourite Things'. ▷ Julie Andrews turned in a winning performance as the saintly 'mother', Maria Trapp, ushering her charges to safety, and no opportunity was missed to ensure that there would not be a dry eye in the house. Artistically, it is easy to dismiss the film for its blatant manipulation of sentiment, but to do so would be to ignore the considerable brilliance of both director and performers in realising those specific aims. *The Sound of Music* succeeds admirably in what it sets out to do, and no amount of critical disdain seems capable of denting its astonishing popularity. A shorter 147 minute version of the film was prepared subsequently for television.

SOUTH BANK SHOW

UK 1978–

Since it replaced the pioneering *Aquarius*, the *South Bank Show* has been ITV's flagship arts programme. Edited and presented by Melvyn Bragg (1939–), and running hour-long shows for 26 weeks of the year, it has maintained a wide brief, reflecting Bragg's desire to popularise the arts. Novelist Bragg, who has been head of Arts at London Weekend Television since 1982, earned his spurs on the BBC's *Monitor*, and later developed the popular paperback book review show *Read All About It*. His programmes reflect his six years at the BBC, being informative and sometimes staid. He was awarded the Royal Television Society's gold medal for achievements in 1990, by which time the *South Bank Show* was coming under fire for being uncritical of its subjects, like singer George Michael – who was given an hour

Julie Andrews in *The Sound of Music* (1965)

to treat his songs more seriously than they deserved.

SPAGHETTI WESTERN Although Hollywood had already remade the ▷ Akira Kurosawa film ▷ *Shichinin No Samurai* (Seven Samurai) (1954) as the western *The Magnificent Seven* (1960), when Italian director ▷ Sergio Leone turned the same director's *Yojimbo* (1961) into ▷ *Per un Pugno di Dollari* (A Fistful of Dollars) (1964), he completely revitalised the western and made a star of television cowboy ▷ Clint Eastwood.

Leone's approach of accentuating the mythic elements of the western through the character of the 'Man With No Name' and stressing the savagery with highly stylised bursts of violence and the screeching sounds of ▷ Ennio Morricone's music, spawned a thousand imitators and soon every cheaply-made western from Spain and Italy bore the generic identification of 'spaghetti western'.

The genre reached its apotheosis with Leone's own operatic *C'Era Una Volta Il West* (Once Upon A Time in the West) (1968) but other popular examples include *Escondido, Un Minuto Per Pregare, Un Instante Per Morire* (A Minute to Pray, A Second To Die) (1968), *Sabata* (1970), *Companeros!* (1971) and *My Name Is Trinity* (1973).

SPECIAL EFFECTS A general term for scenes in motion pictures that involve some kind of trick photography either achieved entirely in front of the camera or in which the recorded image is substantially modified by subsequent technical operations.

Among the former are the many ways in which the area photographed is extended, by scenes painted on glass or miniature sets built in perspective; scale models shot in ▷ slow-motion represent trains, liners and spaceships while animated models using ▷ stop-motion photography bring fantastic creatures to life. A frequent need for image combination is to show live actors performing against a different background which may be an actual distant location, a model, or even a drawing. An early technique was ▷ back projection. In another method the actors are shot against a large uniform ▷ blue-screen to be combined with the background later, using ▷ travelling mattes for film and chromakey for video.

Two or more camera images may be combined in ▷ dissolves, ▷ wipes and ▷ split-screen effects; in film all these need optical printing at the laboratory after photography, but in video they can be made directly, either during shooting or in videotape editing. Vision mixers provide vast scope for image manipulation effects which are completely beyond those possible on film.

SPIELBERG, Steven

Born 18 December 1947, Cincinatti, Ohio, USA.

An ambitious, award-winning amateur filmmaker whose earliest efforts include *Escape to Nowhere* (1960) and *Firelight* (1963), Spielberg's *Amblin'* (1968) so impressed executives at Universal Studios that he became one of the youngest television directors in the history of the medium when he made his debut on a segment of *Night Gallery* (1969) starring ▷ Joan Crawford.

The imagination and verve displayed in subsequent television work like the chilling contemporary nightmare of a motorist terrorised by the relentless pursuit of an apparently driverless truck in *Duel* (1971), the haunted house nail-biter *Something Evil* (1972) and the thriller *Savage* (1973) boosted his reputation, particularly in Europe where the first of these was accorded a cinema release.

He made his cinema debut with *The Sugarland Express* (1974) a picaresque, modest tale of an escaped convict and his wife whose flight from the law makes them media folk heroes. Thereafter his faultless technique and sure storytelling touch quickly made him the most commercially successful film director of all time as he focused on man's involvement with the extraordinary and inexplicable, exploring our primeval fear of the shark in the superbly orchestrated *Jaws* (1975) and expressing child-like wonder at the marvels of this world and beyond in *Close Encounters of the Third Kind* (1977, and Special Edition in 1980) (Best Director

Steven Spielberg directs *Jaws* (1975)

Oscar nomination) and the phenomenally popular ▷ *E.T. – The Extra-Terrestrial* (1982) (Oscar nomination).

His only real failure came with the heavy-handed anarchy of the wartime comedy *1941* (1979). More recently he has alternated the breathtaking adventures of his daredevil archaeologist hero Indiana Jones in the expert escapism of *Raiders of the Lost Ark* (1981) (Oscar nomination), *Indiana Jones and the Temple of Doom* (1984) and *Indiana Jones and the Last Crusade* (1989) with grand literary adaptations that display his mastery of the filmmaking art but are debilitated by his tendency towards sentimentality and a lack of depth. These include *The Color Purple* (1985) (Oscar nomination) and *Empire of the Sun* (1987).

His most recent works include the charming, easygoing romance of *Always* (1989), (a remake of *A Guy Named Joe* (1943)), and the forthcoming *Hook* (1991), his long-planned version of Peter Pan.

His company Amblin' has been responsible for many other films which he has not directed, the most successful including *Poltergeist* (1982), *Back to the Future* (1985) and its two sequels, *An American Tail* (1986), *Who Framed Roger Rabbit* (1988) and *Arachnophobia* (1990).

SPITTING IMAGE
UK 1984–

The latex caricatures of Peter Fluck and Roger Law quickly gained *Spitting Image* a reputation for biting satire, not consistent with its scripts, which were patchy at best. In the main, the controversies the programme generated came from its irreverent approach to the Royal Family, particularly the portrayal of the Queen Mother as a beer-guzzling horse racing enthusiast. Produced by John Lloyd (of *Not The Nine O'Clock News*), the humour was a cruder version of that practiced by Oxbridge comedy groups and *Private Eye* (indeed, *The Eye*'s editor Ian Hislop was a writer for the show). The Reagan Presidency was something of a Godsend – the search for his missing brain was a long running theme. Ex-Prime Minister Margaret Thatcher was shown as a stern, be-suited man, with her voice provided by Steve Nallon. As the show became established its targets were drawn more from the world of entertainment and television. Fluck and Law's achievement was celebrated in a 1990 ▷ *South Bank Show*, in which several real life politicians engaged in interviews with their likenesses. Not a pretty sight, it was further evidence of the show's political impotence. *Spitting Image* puppets are now displayed at a museum in London's Covent Garden.

SPLIT SCREEN Often created by the use of ▷ matte techniques, split screen denotes the appearance of two images in the same picture which have been separately filmed and now appear together, with the boundary between them sometimes made invisible. The shot is regularly used to show two people in a telephone conversation, as in *When Harry Met Sally* (1989). 'Split screen' has also been used to refer to multiple images recorded separately and used in the same picture but this is more correctly known as 'multi-image' and can be seen in such works as *Grand Prix* (1966) and *The Boston Strangler* (1968).

STAGECOACH
John Ford (US 1939)
John Wayne, Andy Devine, Thomas Mitchell, Claire Trevor.
Academy Awards: *Best Supporting Actor* Thomas Mitchell, *Best Music Score.*
New York Film Critics Award: *Best Direction.*
105 mins. b/w.

Stagecoach marked the emergence of the western from its role as a simplistic action vehicle into a genre capable of carrying more serious themes within its adventure narrative. The theme is a classic and much repeated one, throwing together a disparate (but far from random) group of characters, and then placing them under great stress, in this instance from Indian attack in hostile territory. The quality and tension of the action drama itself marked it out as a major development in the genre, notably the famous chase scene in which the stagecoach is besieged by galloping Indians against the broad sweep of ▷ Ford's quintessential location, Utah's statuesque Monument Valley. While it may look like old hat to modern audiences, it was a significant advance at the time. Beyond that action level, however, the film suggested a more complex ambition within the relationships which grew up between the various characters, who were to be models for many westerns to come. The characterisation has more depth and complexity than anything which preceded it, and touched on such issues as racism and social and moral values, which had previously found no place in the unquestioning action mores of the genre. The emphasis on the group rather than the lone hero (even ▷ John Wayne as The Ringo Kid is not isolated from the concerns and inter-relationships of the group) is a particularly Fordian refinement, and if he would go on to greater achievements in the genre, *Stagecoach* remains a landmark film in the development of the director, and of the western itself.

Neither the 1966 remake, nor the 1986 television version can hold a candle to the original.

STANWYCK, Barbara

Real Name Ruby Stevens.
Born 16 July 1907, Brooklyn, New York, USA.
Died 20 January 1990, Santa Monica, California, USA.

A career girl from the age of 13, Stanwyck became a dancer, appearing in *The Ziegfeld Follies of 1923*, and made her dramatic stage debut in *The Noose* (1926).

She made her film debut in *Broadway Nights* (1927). An important featured role in *The Locked Door* (1930) was followed by a succession of roles that established her versatility and screen presence, including the evangelist in *The Miracle Woman* (1931), *Night Nurse* (1931) and *Forbidden* (1932) in which she is a librarian who becomes pregnant out of wedlock.

Best remembered as gutsy, pioneering women in westerns like *Annie Oakley* (1935) and *Union Pacific* (1939), her professionalism and good nature were a byword among Hollywood directors and technicians. Frequently seen as tough-talking, cynical, bossy women, often struggling to escape from the wrong side of the tracks, her range also extended to a poignant performance as the self-sacrificing mother in the classic tearjerker *Stella Dallas* (1937) (Best Actress Oscar nomination), deft comic turns in the likes of *The Lady Eve* (1941) as a cunning con trickster and *Ball of Fire* (1941) (Oscar nomination) as a gum-chewing showgirl invading the halls of academe, and the sultry, archetypal femme fatale in the ▷ film noir ▷ *Double Indemnity* (1944) (Oscar nomination).

A durable leading lady, her later work of note includes *Sorry, Wrong Number* (1948) (Oscar nomination), as the terrorised invalid, and the all-star boardroom melodrama *Executive Suite* (1954).

Keen to work but faced with material of, at best, lacklustre quality, her film career ended with performances as a lesbian madame in *Walk on The Wild Side* (1962), supporting Elvis Presley (1935–77) in *Roustabout* (1964) and in the standard chiller *The Night Walker* (1965) which reunited her with her former husband Robert Taylor (1911–69).

However, she turned to television and, after the short-lived series *The Barbara Stanwyck Theatre* (1960), found fresh popularity as the star of the western series *The Big Valley* (1965–69). She was subsequently seen in such television films as *The House That Would Not Die* (1970), *A Taste of Evil* (1971) and *The Letters* (1973) and enlivened the mini-series *The Thorn Birds* (1982) and the short-lived ▷ soap-opera *The Colbys* (1986–87).

She received an honorary Oscar in March 1982 for her 'superlative creativity and unique contribution to the art of screen acting' and was the 1987 recipient of the ▷ American Film Institute Life Achievement Award.

A STAR IS BORN

William Wellman (US 1937)
Fredric March, Janet Gaynor, Adolphe Menjou, Lionel Stander.
Academy Awards: *Best Original Story* Robert Carson and William Wellman; *Best Cinematography*.
111 mins. col.

George Cukor (US 1954)
Judy Garland, James Mason, Charles Bickford, Jack Carson.
154 mins. col.

Frank Pierson (US 1976)
Barbra Streisand, Kris Kristofferson, Gary Busey.
Academy Awards: *Best Song* 'Evergreen' (Paul Williams and Barbra Streisand)
140 mins. col.

Hollywood has always been fascinated by its own carefully nurtured myths, a preoccupation which is nowhere more obviously manifest than in the heavily sentimental melodrama of *A Star Is Born*. William Wellman's (1896–1975) version was inspired by *What Price Hollywood?* (1932) and atmospherically shot in early Technicolour. This story of a fading matinée idol whose star wanes in direct opposition to the rise of his young actress wife is the most delicately handled of the three versions, while ▷ Frederic March and Janet Gaynor (1906–84) succeed in striking the right balance between sentiment and sentimentality. ▷ George Cukor's version is the most famous, largely because it signalled the return to screen popularity of ▷ Judy Garland, cast in this version as an aspiring young singer who marries a hopelessly alcoholic actor, portrayed with distinction by ▷ James Mason. The relationship between these two characters is less convincingly defined this time around, and the film relies much more on the sympathetic appeal of Garland's brittle vulnerability, her singing, and a strong supporting cast. The film originally ran 181 minutes, but was heavily cut after the premiere; some of the lost material was restored in a 1983 version, which runs at 170 minutes. The musical connection was taken to further extremes in Frank Pierson's (1945–) version, which switched the setting from Hollywood to the world of 1970s Rock, but was essentially a vehicle for ▷ Barbra Streisand to emote expansively. It lacks any of the virtues of the earlier versions, and its lack of critical success may have laid this particular Hollywood warhorse to permanent rest. Since it is now unlikely to be bettered, that would be no bad thing.

STAR WARS

George Lucas (US 1977)
Mark Hamill, Harrison Ford, Carrie Fisher, Alec Guinness, Peter Cushing.
Academy Awards: *Best Art Direction/Set Direction* Norman Reynolds, Leslie Dilley, Roger Christian; *Best Sound* Don MacDougall, Ray West, Bob Minkler, Derek Ball; *Best Original Score* John Williams; *Best Film Editing* Paul Hirsch, Marcia Lucas, Richard Chew; *Best Costume Design* John Mollo; *Best Visual Effects* John Sears, John Dykstra, Richard Edlund, Grant McCune, Robert Blalack.
Special Academy Award: Ben Burtt, Jr (Sound Effects)
121 mins. col.

Star Wars opened the floodgates for the series of high-budget, effects-laden action adventure spectacles which dominated the box-office returns in the decade after its explosive arrival on screen. George Lucas (1945–) and ▷ Steven Spielberg were to dominate the field as directors and producers, and if both came from film school backgrounds, their own movies looked to classic Hollywood epics rather than European art cinema for inspiration. A rolling credit establishes the setting as 'a long time ago in a galaxy far, far away', the backdrop to a classic conflict between good and evil, personified on the one hand by the youthful hero, Luke Skywalker (played by the unknown Mark Hamill (1952–)), and on the other by his nemesis Darth Vader, and the Empire he represents. Skywalker, however, is no simplistic Flash Gordon type. His battle with the evil legions is also a battle with the dark forces in himself (made even more clear in the subsequent *Return of the Jedi* (1983), where Vader is revealed to be his father), and the helpers he recruits, notably Han Solo (▷ Harrison Ford), are not exactly unambiguously clean-cut either, although the robots R2D2 and C3PO, and a variety of odd, furry extraterrestrials, gave an extra boost to the appeal of the film, and foreshadowed the later success of Spielberg's ▷ *E.T. – The Extraterrestrial*. If there is more to the film than just spectacular action, that action is undoubtedly the root of its enormous box-office appeal. *Star Wars* raised the stakes in the art of ▷ special effects by several notches, as the half-dozen technical ▷ Oscars suggest, and also sent budgets spiralling through the roof. Two sequels, *The Empire Strikes Back* (1980) and *Return of the Jedi* (1983), proved equally popular, although neither had quite the gut-wrenching impact of *Star Wars*, itself. Lucas spoke of the trilogy as being the middle trio of a nine part epic cycle, but no further additions to the mythology of the Empire have appeared as yet.

STAR TREK
USA 1966–69
Creator Gene Roddenberry

In order to comment on broader political issues like the Cold War and Vietnam, former World War II aviator and LA cop Roddenberry (1931–) conceived an adult science fiction series based in a 'new world with new rules'. He had previously written for *Highway Patrol*, *Naked City* and *Dr Kildare*, but *Star Trek* was more ambitious. The working title was *Wagon Train To The Stars*, with the Western pioneers of the 1840s swapped for those of the 23rd century, where space, as the titles put it, was 'the final frontier'.

Leonard Nimoy in *Star Trek*

It was set aboard the USS Enterprise (originally to be called the Yorktown). The crew was headed by the courageous Captain James T. (for Tiberius) Kirk (William Shatner (1931–)), his half-Vulcan first officer Mr Spock (Leonard Nimoy (1931–)), Chief Engineer Montgomery Scott (James Doohan) (1920–), and the ship's doctor Leonard 'Bones' McCoy (DeForest Kelley (1919–)). The crew was the United Nations in miniature, though the Russian character Chekov (Walter Koenig) was only added in the second series, after critical jibes in *Pravda*.

Inspired by a famous split infinitive, the Enterprise promised 'to boldly go where no man has gone before', thus confronting allegorical inter-planetary disputes. The themes were the conflicts between scientific discovery and morality – symbolised in the relationship between the logical, pointy-eared Spock and the flawed, human Kirk. The special effects were, of necessity, inventive. The Enterprise was a 12′ × 2′ model, and the ship's transporter system (which spawned the catchline 'Beam me up Scottie') was done with a beam of light and some aluminium dust. Thus no expensive sequences showing the ship landing were required.

Though axed due to poor ratings its strong cult appeal ensured frequent repeats. A cartoon with the original actors doing the voices ran from 1973–75. A full-scale revival was planned from 1972, but none of the cast were available. The huge success of the *Star Trek* films brought the idea back to life. Set 70 years later, *Star Trek: The Next Generation*, (1987) had Roddenberry as executive producer and Shakespearian actor Patrick Stewart as Captain Jean-Luc Picard. Women were in stronger roles, and the ship had children on board. But with values still rooted in the optimism of Kennedy and Johnson's 1960s, its allegorical clout had diminished.

The original team however have resurfaced in a series of cinema adventures begun with *Star Trek: The Motion Picture* (1979) and continuing until the present, with number six currently in the pipeline.

STEADICAM Used to provide a steadying influence for the camera during the filming of particularly tricky mobile hand-held shots, the Steadicam is an articulated arm which extends from a harness that fits around the body of the camera operator and counterbalances the weight of the camera it supports by means of spring force. A free-floating gimbal (or support) attaches the arm to the camera mounting. Allowing the fluidity of the camera movement to remain unaffected regardless of the terrain or cramped conditions in which the operator may find himself, the Steadicam was made striking use of in *The Shining*

(1980) and the La Donna E Mobile section of *Aria* (1987).

STEPTOE AND SON

UK 1962–74

There was nothing very humorous about the situation of two rag and bone men, father and son, trapped in argumentative squalor. And in many ways, the strength of *Steptoe And Son* was that it reversed many of the rules of sitcom writing and was more tragic than comic. The work of writers Ray Galton (1930–) and Alan Simpson (1929–), who also provided many of the classic ▷ Tony Hancock scripts, Steptoe was first seen in *The Offer*, a series of one-off plays run under the banner *Comedy Playhouse*. The two central characters were the lazy, dirty, often downright malicious Albert (Wilfrid Brambell (1912–85)), and his middle-aged son Harold (Harry H. Corbett (1925–82)) who dreamed of bettering himself, only to find his schemes constantly undermined by his father's emotional blackmail. The relationship between the two was not unlike a loveless marriage, with the added pathos that though the two men were tied by blood they were both reluctant to admit to their interdependence. Harold was forever plotting to escape from his situation, eternally hoping for a relationship with a woman, but his efforts were always scuppered by the jealous Albert (who punctured his water bed in one memorable episode). The outlook of the show was not optimistic, and the humour borne of the trapped relationship was dark – with Albert feigning a heart attack to stop his son leaving home, or Harold sleepwalking and trying to cut his father's head off with a meat cleaver. Some have ascribed its popularity to the notion that it tied in with the prevailing mood of decline in Britain, yet it was successfully adapted for Holland as *Stiefbeen En Zoon*, Sweden as *Albert Og Herbert* and the USA as *Sandford And Son* (with black actors Redd Foxx and Desmond Wilson in the lead roles). When repeated in 1988, *Steptoe* still made the UK ratings Top Ten, and it spawned the feature films *Steptoe and Son* (1972) and *Steptoe and Son Ride Again (1973)*.

STEVENS, George Cooper

Born 18 December 1904, Oakland, California, USA.
Died 9 March 1975, Paris, France.

Raised in a showbusiness family, Stevens first appeared on stage in 1909 and later became an actor and stage manager with his father's theatre company. His interest in photography took him to Hollywood in 1921 where he worked as an assistant cameraman. He later joined ▷ Hal Roach and was the cameraman on such ▷ Laurel and Hardy comedies as *Leave 'Em Laughing* (1928) and *Big Business* (1929).

Roach allowed him to direct a number of two-reel comedies and he made his feature-length debut with *The Cohens and The Kellys in Trouble* (1933). ▷ Katharine Hepburn then requested him as the director of *Alice Adams* (1935), a pleasing slice of small town Americana. He worked with Hepburn again on *Quality Street* (1937) and *Woman of the Year* (1942), her first film with ▷ Spencer Tracy.

He then established his niche as a proficient storyteller, craftsman and versatile purveyor of light, lively escapism with a string of popular entertainments including the ▷ Astaire-Rogers musical *Swing Time* (1936), the comedy *Vivacious Lady* (1938), the hearty adventure *Gunga Din* (1939) and the farcical *The More The Merrier* (1943) (Best Director Oscar nomination).

During the war, he headed the Special Motion Picture Unit attached to the US Army Signal Corps and captured documentary footage of the D-Day landings and the liberation of the Dachau concentration camp.

He returned to commercial filmmaking with the sentimental *I Remember Mama* (1948). Over the next decade, he won Best Director Oscars for *A Place in the Sun* (1951) and *Giant* (1956) and created the classic western ▷ *Shane* (1953) (Oscar nomination). Whilst still interested in American mores and issues of class and wealth, the spriteliness of his earlier work was increasingly replaced with a painstaking ponderousness that manifested itself in excessive running times, grandiose notions of good taste and a certain pomposity.

His last films include *The Diary of Anne Frank* (1959) (Oscar nomination), the Biblical epic *The Greatest Story Ever*

Told (1965) and the overblown romance *The Only Game In Town* (1969).

His son, George Stevens Jnr (1932–), has been head of the American Film Institute since 1977 and directed the documentary *George Stevens: A Filmmaker's Journey* (1984).

STEWART, James Maitland

Born 20 May, 1908, Indiana, Pennsylvania, USA.

An architecture student at Princeton University, Stewart's work in summer stock prompted him to make a career in acting and he made his New York stage debut in *Carrie Nation* (1932).

He made his film debut in the short *Important News* (1935) and was incongruously cast as a reporter called Shorty in his feature-length debut *Murder Man* (1935). Tall, gangly and innately decent, with a distinctive, hesitant drawl, he was initially cast as gauche suitors and unwordly country boys in films like *Wife vs. Secretary* (1936) and *The Gorgeous Hussy* (1936) and was memorably miscast as lovelorn French sewer worker Chico in *Seventh Heaven* (1937). However, his engaging manner with comedy became evident in *Vivacious Lady* (1938) and *You Can't Take It With You* (1938) and he was expertly cast as naive, all-American idealists fighting corruption in films like ▷ *Mr Smith Goes To Washington* (1939) (Best Actor Oscar nomination) and *Destry Rides Again* (1939).

Possessed of a simplicity of technique, naturalness of manner and effortless ability to imbue any situation with conviction and emotional truth, his pre-war stardom was crowned as the cynical reporter in ▷ *The Philadelphia Story* (1940) (Oscar).

After distinguished service as a Colonel in the United States Air Force, he returned to acting as the quintessential small town man in the enduring Christmas story ▷ *It's A Wonderful Life* (1946) and developed a more mature image as flinty, psychologically complex westerners in films like *The Naked Spur* (1953) and *The Man from Laramie* (1955), and as the resourceful, sometimes perversely romantic hero of ▷ Hitchcock thrillers like *Rear Window* (1954), *The Man Who Knew Too Much* (1956) and the classic ▷ *Vertigo* (1958).

His benign comic talents were seen to good effect as the inebriated chum of invisible rabbit *Harvey* (1950) and he also offered a warm impersonation of the wartime bandleader in the sentimental biography *The Glenn Miller Story* (1953) and gave one of his best dramatic performances as the cagy, country lawyer whose folksy manner conceals an incisive legal mind in *Anatomy of A Murder* (1959) (Oscar nomination).

In the 1960s, his career settled into less challenging material of mild domestic comedies and routine westerns, although his credits do include *The Man Who Shot Liberty Valance* (1962), a grizzled character performance as the veteran pilot in *The Flight of the Phoenix* (1965) and an excellent account of the weary paterfamilias in the Civil War story *Shenandoah* (1965).

More recently, he has been seen as a supporting actor in such films as *The Shootist* (1976), *Airport '77* (1977) and *The Big Sleep* (1978) and made his last cinema appearance to date in *A Story of Africa* (1981).

On television, he has appeared in the situation comedy *The Jimmy Stewart Show* (1971–72), the critically admired courtroom dramas *Hawkins on Murder* (1973), the maudlin euthanasia drama *Right of Way* (1983) and the mini-series *North and South Book II* (1987).

The recipient of the ▷ American Film Institute Life Achievement Award in 1980, he received an honorary Oscar in March 1985 for 'fifty years of meaningful performances, for his high ideals, both on and off the screen, with the respect and affection of his colleagues'.

His most recent involvement with the cinema has been as one of the vocal talents behind the animated characters in *An American Tail II* (1991).

STILL MAN The individual assigned to a film during its production who is responsible for taking still photographs that will subsequently be used for publicity and promotional activities. Attempting to capture action from the film, he or she will sometimes take specially posed photographs and also attempt to capture some of the less formal behind-the-scenes activities.

STOCK SHOT *see* **LIBRARY SHOT**

STONE, Oliver

Born 15 September 1946, New York City, USA.

As a volunteer for the 25th Infantry Division of the US Marine Corps, Stone saw active service in Vietnam in 1967, receiving the Bronze Star for Valor and the Purple Heart with First Oak Leaf Cluster. The intensity of his experiences under fire have fuelled much of the rage and force in his subsequent writing as well as the bravura primitivism of the technique he would display as a director.

A student of film at New York University, where his lecturers included ▷ Martin Scorsese, he wrote numerous unfilmed screenplays before making his directorial debut with the Canadian horror film *Seizure* (1973). He won an Oscar for his screenplay of *Midnight Express* (1978) and his other scripts include *Conan the Barbarian* (1982), *Scarface* (1982) and *Year of the Dragon* (1985).

He returned to direction with the underrated psychological thriller *The Hand* (1981) but first made his mark with *Salvador* (1986), a muscular true story of a journalist's experiences in strife-ridden El Salvador.

His own experiences in Vietnam were finally distilled in to *Platoon* (1986) in which a teenage soldier is confronted with good and evil, as personified by two sergeants, when he undertakes an odyssey into the heart of darkness. Striking an emotional and intellectual nerve with audiences, the film won Stone a Best Director Oscar.

He then turned to the battlefront of American materialism in *Wall Street* (1987) (Oscar nomination) and accepted the technical challenge of bringing the Eric Bogosian play *Talk Radio* (1988) to the screen.

A director of liberal sympathies with a love of language, raw emotion and a penchant for restless, agitated camera movement, he returned to the impact of Vietnam on the American psyche with the much lauded epic *Born on the 4th of July* (1989) which earned him a further Best Director Oscar. He then turned his attention to rock star Jim Morrison with the film biography *The Doors* (1991) and was then set to examine the Kennedy assassination in *JFK* (1991).

STOP MOTION A technique most often used in trick photography and animation, stop motion is based on a filming process of exposing the film one frame at a time and thus allowing inanimate objects to be rearranged between shots and an illusion of movement created when the end result is processed and projected. Much of the motion in the monster from the original ▷ *King Kong* (1933) was created this way and it forms an important element of the Dynamation process created by Ray Harryhausen (1920–) and seen in such fantasy films as *The Seventh Voyage of Sinbad* (1958) and *Jason and the Argonatus* (1963).

See also **TIME LAPSE PHOTOGRAPHY**

STORYBOARD A form of visual blueprint for a film or a particularly intricate scene within a film, a storyboard is a series of sketches and drawings illustrating the sequence of shots that will comprise the finished production. Part of the preparation that many ▷ directors undertake before filming begins, they are a means of anticipating problems and allowing forward planning with such individuals as the ▷ cinematographer. Some directors make no use of the device at all, others, like ▷ Hitchcock and all animators, are renowned for using this process to meticulously pre-plan their movements and camera angles.

STREEP, Meryl

Full Name Mary Louise Streep
Born 22 June 1949, Summit, New Jersey, USA.

A graduate of Vassar College and a student at Yale Drama School, Streep made her New York stage debut in *The Playboy of Seville* (1969) and subsequently appeared in summer stock, off-Broadway and in *Trelawny of the Wells* (1975) before moving on to such television work as *The Deadliest Season* (1977) and *Holocaust* (1977) and making her film debut with a small role in *Julia* (1977).

Receiving a Best Supporting Actress Oscar nomination for *The Deerhunter* (1978), she subsequently revealed the range and depth of her versatility with varied supporting assignments in such films as *Manhattan* (1979), *The Seduction of Joe Tynan* (1979) and as the estranged wife in *Kramer Vs Kramer* (1979) (Oscar).

A fair-haired woman with angular features and an alabaster-like complexion, she became established as a first rank star, and has constructed a gallery of sensitively handled characterisations that show her facility with languages and ability to render a gamut of emotions with finesse, naturalistic conviction and subtle use of body language. Her numerous roles of distinction include the crazy, visionary Victorian heroine and her contemporary counterpart in *The French Lieutenant's Woman* (1981) (Oscar nomination), the emaciated, emotionally distraught concentration camp survivor with a guilty secret in *Sophie's Choice* (1982) (Oscar) and the impassioned and politicised working woman in *Silkwood* (1983) (Oscar nomination).

As proof that there was indeed some limits to the roles she could attempt, she was less convincing as the ice-cool blonde at the centre of the ▷ Hitchcock-style thriller *Still of the Night* (1982) but impressed anew as Karen Blixen in *Out of Africa* (1985) (Oscar nomination), with her transformation into a dipsomaniac Depression-era derelict in *Ironweed* (1987) (Oscar nomination) and for her frank account of the unsympathetic Australian mother accused of murdering her child in *Cry in the Dark* (1989) (Oscar nomination).

She revealed a flair for comedy as a romantic novelist in the otherwise unappealing *She-Devil* (1989) and gave one of her less flamboyant performances as the daughter trying to fight drug addiction and a smothering relationship with her mother in *Postcards from the Edge* (1990).

Her latest film is *Defending Your Life* (1991).

A STREETCAR NAMED DESIRE

Elia Kazan (US 1951)
Vivien Leigh, Marlon Brando, Kim Hunter, Karl Malden.
Academy Awards: *Best Actress* Vivien Leigh, *Best Supporting Actor* Karl Malden; *Best Supporting Actress* Kim Hunter; *Best Art Direction/Set Decoration* Richard Day, George James Hopkins.
Venice Film Festival: *Best Actress* Vivien Leigh; *Special Jury Prize.*
New York Film Critics Awards: *Best Motion Picture, Best Actress* Vivien Leigh; *Best Direction.*
125 mins. b/w.

▷ Elia Kazan's film adaptation of the Tennessee Williams play which he had previously directed on Broadway brought a new maturity to Hollywood's treatment of sexuality. Although one or two of the more controversial elements of the play were dropped from the film screenplay, notably the homosexual theme and Blanche's nymphomania, the filmmakers insisted on a greater level of explicit realism than anything which had previously been attempted under the primly restrictive Production Code. In particular, they insisted on keeping the crucial scene in which the brutish Stanley Kowalski rapes his sister-in-law Blanche, who has come to stay in their home. Eventually, they were allowed to retain the scene, but not to show the actual physical action, leaving the audience to work out what had happened from Stella's subsequent rejection of her husband. ▷ Marlon Brando turned in a magnificent performance in his first great starring role, although he lost out to ▷ Humphrey Bogart in ▷ *The African Queen* at the Oscars. The sharp contrast between ▷ Vivien Leigh's winsome, classically English acting style in the role of Blanche, and Brando's whole-hearted adoption of the new and largely unknown Method style, derived from Stanislavsky and propagated in America by Lee Strasberg (1899–1982), is not the least of the film's many attractions. The poignant, hot-house sexual atmosphere of the play is convincingly captured even under the restrictions of the Production Code, and the film paved the way for a more honestly adult Hollywood cinema to emerge.

STREISAND, Barbra

Full Name Barbara Joan Streisand
Born 24 April 1942, Brooklyn, New York, USA.

A performer in Greenwich Village amateur talent contests and a nightclub singer, Streisand made her professional New York debut in *Another Evening With Harry Stones* (1961) and was soon the toast of Broadway in *I Can Get It For You Wholesale* (1963) and *Funny Girl*

(1964), a musical biography of Ziegfeld Follies star Fanny Brice (1891–1951).

Her television special *My Name Is Barbra* (1965) won five Emmy Awards and she had received consecutive music industry Grammy awards as female vocalist of the year in 1964 and 1965 before making her film debut in the screen version of *Funny Girl* (1968) which won her a Best Actress Oscar.

A prodigiously talented entertainer with a show-stopping vocal range and an attraction towards ugly duckling roles that exploited her lack of conventional beauty, Jewishness and aggressively kooky sense of comedy, she became the last great star of the traditional Hollywood musical in such lavish, big-budget affairs as *Hello, Dolly!* (1969) and *On A Clear Day You Can See Forever* (1970).

Her brittle comic skills were evidenced in *The Owl and the Pussycat* (1970) and *What's Up Doc?* (1972) and she gave one of her most impressive dramatic performances as the left-wing activist torn between principles and romance in the exceptionally popular *The Way We Were* (1973).

Playing safe with such unadventurous material as *For Pete's Sake* (1974) and *Funny Lady* (1975) she then appeared in an over-inflated rock industry remake of ▷ *A Star Is Born* (1976) which earned her a further Oscar as the co-writer of the love song 'Evergreen'.

Letting an increasing length of time pass between projects, she was then seen in the bland comedy *The Main Event* (1979) and was miscast as the daffy blonde housewife seeking romance in *All Night Long* (1981), but disarmed criticism with an underrated performance of considerable low-key charm.

In 1983 she became the complete ▷ auteur as the producer, director and co-writer of the musical *Yentl* in which she also acted and sang. A warmly engaging, old-fashioned affair it revealed further talents in her directorial abilities.

Maintaining parallel careers as a top-selling recording artist (she won a further Grammy in 1978), she now rarely performs in public. Her recent film work comprises *Nuts* (1987) in which she gave a strident, tour de force as a prostitute attempting to prove her sanity, and *Prince of Tides* (1991) which she has also co-produced and directed.

STURGES, Preston

Real Name Edmund Preston Biden
Born 29 August 1898, Chicago, Illinois, USA.
Died 6 August 1959, New York, USA.

Educated in America and Europe, Sturges enlisted in the Air Corps in 1917 and later worked in the cosmetics industry, inventing a 'kiss-proof' lipstick.

A dramatist from 1927, he later moved to Hollywood making his debut as a scriptwriter with *The Big Pond* (1930) and subsequently creating such highly regarded scripts as *The Power and The Glory* (1933) and *The Good Fairy* (1935).

He made his directorial debut with *The Great McGinty* (1940) and won an Oscar for the script of this sharp political satire. He then enjoyed a brief but glorious run of successes with inventive, freewheeling comedies that combined wit, slapstick, social satire and characters bearing the unlikeliest of monikers. Fragrant examples include Woodrow Lafayette Pershing Truesmith, The Princess Centimillia, Judge Alfalfa O'Toole and Trudy Kockenlocker.

His enduring hits include the romance of predatory trickster and virginal victim in *The Lady Eve* (1941), *Sullivan's Travels* (1942), in which a pretentious film director discovers that the world wants to ease its cares with comedies rather than social realism, *Hail, the Conquering Hero* (1944) (Oscar nomination) in which an army medical reject is nominated for mayor on account of his supposed gallantry and the outrageous *Miracle of Morgan's Creek* (1944) in which the aforementioned Trudy finds herself pregnant by an unknown G.I. and winds up giving birth to sextuplets. ('Hitler Demands Recount' blazes from the screen).

A move to work under contract to ▷ Howard Hughes proved the beginning of the end for his career and after the failures of the poor ▷ Harold Lloyd comedy *Mad Wednesday* (1947), *Unfaithfully Yours* (1948) in which a symphony conductor mentally plots the demise of his unfaithful wife, and *The Beautiful Blonde From Bashful Bend* (1949) about a hotheaded lady sharpshooter, he retired to France making one sad final film *Le Carnet Du Major Thompson* (The Diary of Major Thompson) (1957).

He received a posthumous Laurel Award for Achievement in 1974 from the Writer's Guild of America.

Preston Sturges, The Rise and Fall of An American Dreamer (1990) was a documentary film portrait of the director with contributions from such collaborators as Eddie Bracken (1920–), Betty Hutton (1921–) and Joel McCrea (1905–90).

SUNRISE

F.W. Murnau (US 1927)
George O'Brien, Janet Gaynor, Bodil Rosing, Margaret Livingstone.
Academy Awards: *Best Actress* Janet Gaynor; *Best Cinematography* Charles Rosher, Karl Struss; *Best Artistic Quality of Production.*
117 mins. b/w.

▷ F.W. Murnau made the transition from Germany to Hollywood with this classic silent drama, but brought much of the visual language of German ▷ expressionism with him. The plot is a straight-forward one, involving the story of a young farmer who is seduced by another woman in the city they visit, and plans to murder his wife in order to be with her. Murnau and his screenwriter, Carl Mayer (1894–1944), opted for a happy ending rather than the tragic dénoument of the original book by Hermann Sudermann, but that does not seriously deflect the power of the dramatic material. The film is most notable, though, for its direction, and the power of its visual images. Murnau abandoned the static conventions of silent cinema in favour of a camera which is in almost constant motion, to the point where the action is moulded to fit the needs of a fluid camera, rather than the other way around. The film jumps around between scenes which demand highly stylised, ▷ expressionistic sets and acting, and much more conventionally naturalistic ones, following a rhythm perceived in the material by the director (although not always by the viewer). The use of lighting effects in establishing the moods appropriate to these stylistic and thematic shifts is extraordinary even by contemporary standards, while Murnau's camera-work would be highly influential on subsequent generations of film-makers.

SUNSET BOULEVARD

Billy Wilder (US 1950)
William Holden, Buster Keaton, Gloria Swanson, Erich von Stroheim, Nancy Olson.
Academy Awards: *Best Screenplay* Charles Brackett, Billy Wilder, D.M. Marshman Jr; *Best Score for a Dramatic or Comedy Picture* Franz Waxman; *Best Art Direction/Set Decoration* Hans Dreier, John Meehan, Sam Comer, Ray Moyer.
110 mins. b/w.

Like ▷ *A Star Is Born* and ▷ *Singin' In The Rain, Sunset Boulevard* is a film about Hollywood itself, but it is a darker creature by far than anything envisaged by either the melodrama or the musical. While brilliantly directed by ▷ Billy Wilder, it is hard to avoid the feeling that there is something almost barbarically cruel in his treatment of the film's star, ▷ Gloria Swanson. She plays a former screen goddess of the silent era, Norma Desmond, whose career languished with the coming of sound, much as Swanson's own did. She enlists a young screen-writer, Joe Gillis, played by William Holden (1918–81), to help her with a hopeless comeback script, and moves him as her kept lover into the mansion she shares with her chauffeur, himself a former film director. In a particularly black touch, the chauffeur/director is played by film director ▷ Erich von Stroheim; at one point, Norma plays Joe one of her old films, and the film we see on screen is von Stroheim's *Queen Kelly*, an epic disaster which effectively ended Swanson's career. Whatever the ethics, Wilder undeniably obtains a chillingly convincing performance from his star, and the picture reveals the workings of the Hollywood system in a light which makes even the crassest moments of despair and hypocrisy in *A Star Is Born* look gentle. Although the film has a distinctly ▷ film noir feel and look, and ends with the jealous Norma murdering Joe because he has met another woman (and one, moreover, with a script which might actually succeed), it is not played as a suspense thriller, since the viewer has been primed by the opening shots of Joe's body floating in the pool at Desmond's mansion to expect tragedy. Instead, Wilder probes with a merciless and unrelenting gaze at the weaknesses of his protagonists, and the evils of the system which produced them.

SURREALISM A cultural movement in France that manifested itself in the art, literature and cinema worlds of the 1920s, Surrealism was influenced by the theories of Karl Marx and Sigmund Freud and sought to give a full, poetic expression to the thoughts and images more readily encountered in unconscious states of dream and hallucination. Iconoclastic, irrational and absurdist, it produced work of shocking images, unexpected savagery and the grotesque that removed film from the tyranny of narrative or logic to concentrate on emotion and symbolism. Among the more notable films to embody such free-association, artistic daring were *L'Etoile de Mer* (1928) directed by Man Ray (1890–1976) and the collaboration of ▷ Luis Bunuel and Salvador Dali (1904–89) on *Un Chien Andalou* (1928) in which an eyeball is memorably sliced by an open-blade razor.

ŜVANJKMAĴER, Jan

Born 3 September 1934, Prague, Czechoslovakia.

A student at the Institute of Artistic Industry (1950–54) and the Marionette Faculty of the Academy of Fine Arts (1954–58), Ŝvanjkmajer worked extensively as a director with the Theatre of Masks and the Lanterna Magika Theatre in Prague before beginning to explore the boundaries of film as a means of illustrating his morbid preoccupations.

His first work, the short *Posledn Trik Pana Schwarcewalldea A Pana Edgara* (The Last Trick of Mr Schwarzwald and Mr Edgar) (1964) features two rival magicians who indulge in a competitive display of disgorging increasingly unlikely objects before dismembering each other and leaving only two disconnected arms to unite in reluctant conciliation.

His subsequent films are a bizarre collection of the macabre and the surreal, that use puppets, trick photography, animation and live action to depict his mordantly comical obsession with the human body, cannibalism and man's darkest fears.

Titles include the blackly humorous *Byt* (The Flat) (1968) in which a man finds himself locked in a room and at the mercy of inanimate objects, *Don Sajn* (Don Juan) (1970) a more traditional marionette narrative, *Leonarduv Denik* (Leonardo's Diary) (1972), a mixture of animation and found footage that comments on the sketches of Leonardo Da Vinci and *Moznosh Dialogu* (Dimensions of Dialogue) (1982), a harshly unsettling trio of chapters that display man's intolerance of nonconformity through a juxtaposition of grotesque imagery, brutal editing and violent emotions.

Influenced by the work of Lewis Carroll and Edgar Allan Poe, he has made short versions of *Jabberwocky* (1971), *Zanik Domu Usheru* (The Fall of the House of Usher) (1981) and *Jama, Kivadlo A Nadeje* (The Pit, The Pendulum and Hope) (1983) and made his feature-length debut with *Neco Z Alenky* (Alice) (1988), an individualistic interpretation of *Alice in Wonderland* that typically highlights the obsessions and anxieties of a troubled child.

SWANSON, Gloria

Real Name Gloria May Josephine Svensson
Born 27 March 1897, Chicago, Illinois, USA.
Died 4 April 1983, New York City, USA.

A clerk who was studying to be a singer in Chicago, Swanson entered the nascent film industry as an ▷ extra and subsequently worked as a bit part player, receiving her first mention in the credits of *The Fable of Elvira and Farina and The Meal Ticket* (1915).

One of ▷ Mack Sennett's decorative Bathing Beauties, a professional association with ▷ director ▷ Cecil B. De Mille brought her leading roles as chic sophisticates in the front line of the battle of the sexes. A stylishly predatory female, her vivacity and poise were crucial to such glamorously decadent silent features as *Male and Female* (1919), *Why Change Your Wife?* (1919) and *The Affairs of Anatol* (1921).

Now a fully-fledged star whose life was lived in the full blaze of publicity, her significant later roles include Napoleon's laundress in *Madame Sans-Gene* (1925) and the title role in *Sadie Thompson* (1928) (Best Actress Oscar nomination), a version of Somerset Maugham's *Rain*.

Despite the ill-fated extravagances of *Queen Kelly* (1928) which was uncompleted and left her in debt for years to come, she did survive the transition to sound, and sang in her talkie debut *The Trespasser* (1929) (Oscar nomination).

Perhaps too closely identified with the glamour and artifice of a certain type of silent film, her career then gradually drifted away and she left the screen after playing a prima donna in *Tonight or Never* (1931) and a further musical appearance in the operetta *Music in the Air* (1934).

She made an unremarked-upon comeback in the feeble comedy *Father Takes A Wife* (1941) and was absent again until her sensational performance as reclusive silent screen idol Norma Desmond in > *Sunset Boulevard* (1950) (Oscar nomination). Her subsequent work in *Three for Bedroom C* (1952) and *Nero's Weekend* (1956) was undistinguished.

Never relinquishing her glamorous star status she continued to appear on stage and television and also found time to sculpt, design clothes and actively promote the benefits of health foods. She made her last screen appearances in the television movie *Killer Bees* (1974) and the all-star disaster film *Airport '75* (1974) in a cameo performance as herself.

Married six times, she published her autobiography *Swanson on Swanson* in 1980.

THE SWEET LIFE *see* LA DOLCE VITA

SWEET SMELL OF SUCCESS

Alexander Mackendrick (US 1957)
Burt Lancaster, Tony Curtis, Susan Harrison, Sam Levene.
96 mins. b/w.

The Sweet Smell of Success is another of those Hollywood films which broke the rules a little too much for its merits to be recognised fully at the time of its release, leaving it to a later critical re-evaluation to establish its rightful standing. While James Wong Howe's (1899–1976) celebrated camera work, shot on the night streets of New York in evocatively moody fashion, align it most closely with > film noir, it falls neatly into no specific genre category, either in manner or subject matter. Tony Curtis (1925–) gained credibility as a major actor playing Sidney Falco, a charming, ambitious and unprincipled publicity agent obsessed with earning the approval of the smoothly destructive gossip columnist J. J. Hunsecker, a strangely out of character role for > Burt Lancaster. In order to find favour, Falco sets about destroying the reputation of a young jazz musician who has taken up, much to the columnist's (implicitly sexual) chagrin, with Hunsecker's sister. > Alexander Mackendrick brings a genuine sense of suspense and largely psychological menace to this powerful, seamy revelation of the underside of the glamorous New York media world and its night life, while the script, by radical left-wing playwright Clifford Odets (1903–63) and Ernest Lehman (1920–), is completely uncompromising in its attack on the corrupt values at the heart of their abuse of media power. The stylish realisation of the film has lost none of its appeal, while the issues it raises are arguably even more pertinent in today's unrelenting wash of media-glare than they were when it was made.

SZABÓ, István

Born 18 February 1938, Budapest, Hungary.

Educated at the Academy of Theatre and Film Art in Budapest, Szabo gained immediate attention with his diploma film *Koncert* (Concert) (1961) and made two further shorts *Variációk Egy Témára* (Variations On A Theme) (1961) and *Te* (You) (1963) before his feature-length directorial debut on *Álmodozások Kora* (The Age of Daydreaming) (1964), a poetic tale of a young engineer's infatuation with a female colleague.

In films like *Apa* (Father) (1966) and *Tüzoltó Utca 25* (25 Fireman's Street) (1973), he created intimate well-observed dramas of everyday joys and woes that also served as an unofficial record of the national mood in post-war Hungary.

He first attracted wider international interest with *Bizalom* (Confidence) (1979), a subtle two-hander in which fugitives from the Nazis are forced to pose as man and wife in the Budapest of 1944. His next project *Mephisto* (1981) earned an Oscar as Best Foreign Film and featured a bravura performance from ▷ Klaus Maria Brandauer as an ambitious actor who betrays his highest principles and dearest loves to secure renown and glory in Nazi Germany. A sleek, well-mounted production it glitteringly portrayed the motivations behind such a betrayal and the wider lessons in power's ability to seduce and corrupt.

He worked with Brandauer again on *Colonel Redl* (1984) and *Hanussen* (1988), rich, complex portraits of charismatic men of destiny and the forces that cause their fall from grace.

Moving further into the international mainstream he directed *Meeting Venus* (1991) which chronicles the travails of an opera diva played by ▷ Glenn Close.

He has also acted in *Túsztörténet* (Stand Off) (1989).

TAKE A single, uninterrupted record of the action of any scene or part of a scene in the filmmaking process. Sometimes one take may be sufficient to satisfy the requirements of the ▷ director but attempts to improve performance, sound recording or any of the other disparate elements involved may see the number rise into dozens. Information on the ▷ clapperboard identifies the scene and take number with the clapperboard boy reading the number for recording on the soundtrack. The best individual takes are used to edit together for the final film.

TANNER, Alain

Born 6 December 1929, Geneva, Switzerland.

A graduate in maritime administration, Tanner travelled extensively before arriving in London in 1955 and finding work with the ▷ British Film Institute. Influenced by the ▷ free cinema movement, he joined forces with Claude Goretta (1929–) to make the documentary *Nice Time* (1957).

Resident in Paris from 1958, he returned to Switzerland in 1960 and worked on a number of biographical documentaries and experimental short films like *Ramuz, Passage d'un Poète* (Ramuz, A Poet's Way) (1961), *L'École* (The School) (1962) and *Une Ville à Chandigarh* (A City at Chandigarh) (1966).

Active in television documentaries between 1964 and 1969, he was a key figure in establishing a feature film industry in Switzerland, making his debut with *Charles, Mort ou Vif* (Charles Dead or Alive) (1969). The success of that film allowed him to pursue a career marked by recurring autobiographical themes of rootlessness, revolt and crises of identity in films like *La Salamandre* (The Salamander) (1971), *Le Retour d'Afrique* (Return from Africa) (1973) and *Jonas, Qui Aura 25 Ans en l'An 2000* (Jonah, Who Will Be 25 in the Year 2000) (1975), a cinematically stylish mosaic of four couples and their efforts to maintain the Marxist idealism that flourished in May of 1968.

His films grew more pessimistic and elliptical with *Messidor* (1979) and thereafter he sought projects abroad, reinforcing his own feelings of exile and being at odds with the traditions of his homeland. The mystical *Light Years Away* (1981) was set in Ireland and *Dans La Ville Blanche* (In The White City) (1983) features a sun-drenched Lisbon where a ship's mechanic films his mournful, alienated life for the wife who awaits him in Switzerland.

More recently, he has reflected his own disillusionment with the perceived retreat from political idealism in *No Man's Land* (1985) and stirred controversy with his unflinching stories of sexual desire in *Une Flamme dans Mon Coeur* (A Flame in My Heart) (1987) and frank appraisal of Swiss attitudes to Third World immigration in *La Femme de Rose Hill* (The Woman From Rose Hill) (1989).

TARKOVSKY, Andrei Alexandrovich

Born 4 April 1932, Zavroshne, USSR.
Died 29 December 1986, Paris, France.

The son of a poet, Tarkovsky was a student of Oriental languages and worked as a geological prospector in Siberia before studying at the State Film School and directing the short film *Segodnya Otpuska Nye Budyet*

(There Will Be No Leave Here Today) (1959).

He made his feature-length debut with *Ivanovo Detstvo* (Ivan's Childhood) (1962), an uncharacteristically conventional, somewhat lyrical account of an orphan boy who gains renown as a daring watime spy before his death at the hands of the Germans.

He followed this with ▷ *Andrei Rubelev* (1966), the sweeping parable of a 15th century icon painter whose faith is systematically destroyed by the sights and sounds he witnesses on an epic journey through feudal Russia.

His bleak science-fiction epic *Solaris* (1972) was followed by *Zerkalo* (Mirror) (1974) which made use of film as a personal expression of his inner thoughts, dreams and memories and contributed most significantly to his recognition as one of the cinema's true poets with a distinctive, slow-moving style incorporating elliptical imagery and lengthy, enigmatic and often impenetrable subject matter.

Stalker (1979) proved another production of epic length in which the title character guides a writer and a scientist through a forbidden Zone to the Room where it is rumoured that all prayers will be answered. An impressive technical feat displaying some startling imagery and inventive use of colour, it was otherwise a somewhat stilted and obscure drama.

Nostalghia (Nostalgia) (1983), made in Italy, incurred official displeasure with its melancholy story of a Russian musicologist in Tuscany who is asked to cross an ancient sulphur pool carrying a lighted torch as an act of faith. Filled with arresting Christian symbols its austere visual elegance and measured pace worked against wide understanding or sympathy for his artistic preoccupation.

Now in exile from his homeland and suffering from failing health, he made *Offret* (The Sacrifice) (1986) in which a man is willing to relinquish his own life and possessions to prevent a forthcoming apocalypse. Beautifully composed and characteristic of his worries about the future of the planet and advocacy of peace, it remains, like most of his work, emotionally cold and requiring a good deal of patient indulgence on the part of the viewer.

TASHLIN, Frank

Born 19 February 1913, Weekhawken, New Jersey, USA.
Died 5 May 1972, Hollywood, USA.

Whilst still a teenager, Tashlin began his career as an errand boy for animator ▷ Max Fleischer. Gradually following in his mentor's footsteps, he contributed to the *Merrie Melodies* and *Looney Tunes* series and used the pseudonym of Tish-Tash to sell cartoons to various periodicals; his comic strip Van Boring was syndicated from 1934 to 1936.

A gag writer for ▷ Hal Roach, he was also employed at the Disney studios on Mickey Mouse and Donald Duck cartoons. During the War, he directed Private Snafu cartoons for the Army Signal Unit of ▷ Frank Capra and made his first attempt at a non-animated feature as the co-writer of *Delightfully Dangerous* (1944).

In 1946, he published his first cartoon book *The Bear That Wasn't* whilst gainfully employed as a deviser of comic situations and jokes for the likes of ▷ Bob Hope, the ▷ Marx Brothers and Eddie Bracken (1920–). Bidding a permanent farewell to the world of cartoons, he became a scriptwriter on such comedies as *The Paleface* (1948) and *Miss Grant Takes Richmond* (1949) before making an uncredited directorial debut on *The Lemon Drop Kid* (1951) at the behest of its star Bob Hope.

He moved into his stride as a director with films like *Son of Paleface* (1952), *The Girl Can't Help It* (1956) and *Will Success Spoil Rock Hunter?* (1957) in which he managed to transfer his animator's sensibilities to a surreal and anarchic live-action landscape. His films may lack a coherency in their structure or sophistication of characterisations, but they offer clever visual jokes and verbal innuendo alongside a garish, satirical view of American pop culture and a sense of comic invention achieved through wild overstatement.

His many films with ▷ Jerry Lewis resulted in some of the madcap star's more palatable comedies like *Who's Minding the Store?* (1963) and *The Disorderly Orderly* (1964) but the inspiration he drew from the advent of rock and roll and the spread of saturation television in the 1950s seemed to desert him in the following decade and his career ended with some poor ▷ Doris Day comedies and the weak Bob Hope vehicle *The Private Navy of Sergeant O'Farrell* (1968).

TATI, Jacques

Real Name Jacques Tatischeff.
Born 9 October 1908, Le Pecq, France.
Died 4 November 1982, Paris, France.

A skilled rugby player in his youth, Tati began his entertaining career with a wordless cabaret act in which he mimicked various sporting personalities of the day, an amusing talent that he captured in the short film *Oscar, Champion de Tennis* (1932).

He travelled extensively throughout European music halls and circuses and served with the French Army during the Second World War. Afterwards, he appeared in the films *Sylvie et Le Fantôme* (1945) and *Le Diable au Corps* (1946) before writing, directing and appearing in *L'École des Facteurs* (1947).

He expanded the latter short film into the feature *Jour de Fête* (1949) in which he plays a village postman who enthusiastically embraces modern American methods of efficiency.

Continually satirising the cold, impersonal nature of mechanical advances and the drive for efficiency, he created his most famous character in *Monsieur Hulot's Holiday* (1952). The pipe-smoking, lugubrious Hulot, who reappeared in *Mon Oncle* (1958), displays the gait of a man permanently walking into the fury of a hurricane and is forever beset by physical mishaps and confrontations with unyielding modern technology.

A graceful pantomimist and exacting perfectionist, he created inventive visual humour that was presented and framed within the camera in such a way that the viewer is never drawn into a situation or forced towards a punchline by editing or close-ups or any similar device. Instead, the action unfurls at a relaxed pace and a reasonable distance that makes the viewer come to it.

Nine years passed before *Playtime* (1967) but the now painfully elaborate structuring of his visual humour, the rendering of Hulot as one figure in an ant-hill of confused humanity and an increasing aloofness in his satirising of modern society made this less of a joy and left him close to bankruptcy.

The last films in a very modest body of work were *Traffic* (1971) and *Parade* (1973), a sad, hour-long film for Swedish television in which he recreated parts of a circus mime performance.

TAVERNIER, Bertrand

Born 25 April 1941, Lyons, France.

Tavernier first felt that he wanted to make films at the age of 14 and had already noted the names of ▷ John Ford, Henry Hathaway (1898–1985) and William Wellman (1896–1975) as directors whose work he admired. The abundant cinemas of Paris and the Cinématheque Française provided his education and he later formed the film club Le Nickel-Odeon which opened with ▷ Vincente Minnelli presenting *The Bandwagon* (1953) and could boast ▷ King Vidor and Delmar Daves (1904–77) as its honorary chairmen.

Briefly studying law, he became a writer and critic for publications like *Positif* and *Cahiers du Cinema* before working as an assistant to ▷ Jean-Pierre Melville on *Leon Morin, Prêtre* (Leon Morin, Priest) (1961) and directing segments of the multi-story features *Les Baisers* (1963) and *La Chance et L'Amour* (1964).

He then spent almost a decade as a freelance press agent and scriptwriter before making his feature-length debut with *L'Horloger de Saint-Paul* (The Watchmaker of St. Paul) (1973), a quietly affecting character study with ▷ Philippe Noiret as a widower forced to re-examine his life when his son is accused of murder.

His frequent collaborations with Noiret include *Que La Fête Commence!* (Let Joy Reign Supreme!) (1975) concerning intrigue at the 18th century court of Phillippe D'Orleans, and *Le Juge et L'Assassin* (The Judge and the Assassin) (1976), a rich political parable in which the judge must decide if a child murderer is responsible for his actions.

Eschewing any form of flashy technique in favour of well-told narratives with fully-rounded characters, he illustrated the diversity of his interests with *Des Enfants Gâtes* (Spoiled on Children) (1977) about a film director's creative block, and the imaginative science-fiction story *La Mort En Direct* (Deathwatch) (1979).

Throughout the last decade, he has produced an enviable body of work that utilises elegant cinematic storytelling in celebration of non-conformity and the myriad emotions experienced in daily living and loving. Notable successes include the lyrical, ▷ Renoir-like *Dimanche A La Campagne* (Sunday in the Country) (1984) and *'Round Midnight* (1986), his moving salute to the bebop legends of the 1950s.

A darker more disturbing side of his talent was shown in the thriller *Coup de Torchon* (Clean Slate) (1981) and the medieval drama *La Passion Beatrice* (1988).

More recently, he has provided some sombre reflections on mortality and mature love in *La Vie Est Rien D'Autre* (Life and Nothing But) (1989) and *Daddy Nostalgia* (These Foolish Things) (1990).

Still a writer and documentarist, he has also produced such films as *La Question* (1978) and *Le Mors aux Dents* (1979).

TAVIANI BROTHERS

Vittorio *Born* 20 September 1929, San Miniato, Italy.
Paolo *Born* 8 November 1931, San Miniato, Italy.

Educated at the University of Pisa, Vittorio in law and Paolo in liberal arts, the Tavianis wrote criticism and formed a film club before collaborating with Cesare Zavattini (1902–) on *San Miniato, Luglio 44* (San Miniato, July 44) (1954) a documentary about a Nazi massacre in their native village.

Continuing to work in the documentary format for the next eight years, they also served as assistants to such distinguished directors as ▷ Joris Ivens and ▷ Roberto Rossellini and their work betrays the influence of the ▷ neo-realists.

They made their directorial debut with *Un Uomo da Bruciare* (A Man for Burning) (1962) in which a man's exhortation to the Sicilian peasantry to adopt collective action against the stranglehold of the Mafia is met by his assassination.

Their early solo work, including *Sovversivi* (Subversives) (1967) and *Sotto Il Segno dello Scorpione* (Under the Sign of Scorpio) (1969) offered pessimistic perspectives on the process of revolution and its ideological adherents. Sympathetic to the travails of working people, their films began to veer away from social realism to a more fantastical depiction of their struggles taking cognizance of the power of dreams and the effectiveness of symbolism and metaphor in conveying this to a wider audience.

Allonsanfan (1974) was an ornately eyecatching but somewhat ambiguous portrait of disillusioned nobleman's treacherous intervention in a 19th century peasant revolt but *Padre, Pardone* (1977), winner of the Cannes Palme D'Or, was a much more clear-sighted and effective tale of personal growth and commitment as a young shepherd boy is inspired by his military service to break from his domineering father.

Their international reputation secured, they won continued praise for the beauty and power of *La Notte di San Lorenzo* (The Night of San Lorenzo) (1982) a bravura evocation of Tuscan resistance to Nazi aggression, and *Kaos* (1984) four tales of peasant life in turn of the century Sicily that once again illustrated their ability to find hope for the future in the lessons of the past.

However, their mastery of dazzling imagery and sometimes flamboyant theatricality has been squandered of late on the sentimental and novelettish *Good Morning Babylon* (1987), in which two Italian brothers journey to America and employment on ▷ D.W. Griffith's ▷ *Intolerance*, whilst *Il Sole Anche di Notte* (Night Sun) (1990) was a very dull plod indeed through the search for spiritual redemption by a broken-hearted suitor turned hermit and healer.

TAXI DRIVER

Martin Scorsese (USA 1976)
Robert De Niro, Cybill Shepherd, Jodie Foster, Harvey Keitel, Peter Boyle.
113 mins. col.
Cannes Film Festival: *Palme D'Or.*

It perhaps took a collaboration between avowedly Catholic filmmaker ▷ Martin Scorsese and the sternly Calvinist-influenced screenwriter ▷ Paul Schrader to create the quintessential picture of contemporary urban life as hell on earth. Vietnam veteran and New York cabbie Travis Bickle (as played by ▷ Robert De Niro in an immesurably perceptive investigation of mania) burns with disgust at the moral decay around him, particularly represented by ▷ Jodie Foster's child prostitute. This almost existential sense of frustration, fed by his feelings of sexual inadequacy and paranoid disdain for other people, leads him to take up arms in an effort to effect his own form of change, failing to assassinate the President before slaughtering the young girl's pimp (Harvey Keitel, 1947–) and several other miscreants in a psychopathically violent bloodbath. In genre terms, Travis Bickle's actions fuse those of the moralising wild west hero (Scorsese has compared him to ▷ John Wayne's Ethan Edwards character in ▷ John Ford's ▷ *The Searchers*) now transposed to the modern metropolis with the relentless capacity for carnage of the post-68 horror film's numerous mad slayers, but on a deeper level the dismaying enthusiasm with which some audiences cheered on De Niro's crazed quest envinces the degree to which *Taxi Driver* resonates with a widespread sense of ideological confusion, the feeling of powerlessness experienced by those still clinging to the crumbling value systems of moral rectitude and the American Way in the face of an entropic social formation's increasingly variable standards of conduct. Depending on the viewer's outlook, the final confirmation of Bickle as folk hero is either a neo-Fascist endorsement of the vigilante ethos or a heavily ironic, angst-drenched admission that the democratic or spiritual ideal has lost contact with those dark mean streets out there.

TAYLOR, Elizabeth Rosemond

Born 27 February 1932, London, UK.

Returning to America with her family at the onset of World War II, Taylor was living in Beverly Hills at a time when America's love affair with juvenile performers was at its height. Possessed of a youthful grace, sparkling violet eyes and an eyecatching beauty she was soon signed to Universal (home of ▷ Deanna Durbin) and made her film debut in *There's One Born Every Minute* (1942).

Moving to M-G-M, she appeared in such films as *Lassie Come Home* (1943) and *The White Cliffs of Dover* (1944) and forced herself to gain the extra three inches required to play the young girl who dreams of success with horse *National Velvet* (1944).

Allowed to mature before the cameras, she made a smooth progress to adulthood as spoiled daughters and petulant teenagers in the likes of *A Date With Judy* (1948) and *Father of the Bride* (1951) and gave the first evidence of maturer skills in *A Place in the Sun* (1951), as a rich society girl.

A number of routine roles followed before she firmly established her credentials as a leading lady and a performer of some modest skill and fire. Her run of more than competent performances includes the wife of a Texas rancher in *Giant* (1956), a conniving Southern belle in *Raintree County* (1957) (Best Actress Oscar nomination), a scorching interpretation of Maggie the Cat in *Cat On a Hot Tin Roof* (1958) (Oscar nomination) and the disturbed young woman in *Suddenly, Last Summer* (1959) (Oscar nomination).

She won the Oscar for her performance as a New York call girl in *Butterfield 8* (1960), a film she had disliked, and then became involved in the lengthy production of the tedious and costly *Cleopatra* (1963) which, at least, brought her the love of co-star ▷ Richard Burton.

Always in the public eye, she was now part of the most famous showbusiness couple on the planet whose every tiff, lavish party or latest diamond purchase could command world headlines. Her screen career continued, although somewhat subordinate to Burton's, and the best of their films together include the lavish melodrama *The V.I.P.s* (1963), *Who's Afraid of Virginia Woolf?* (1966) (Oscar) in which she unleashed a startling ferocity as the blowsy, embittered Martha and *The Taming of the Shrew* (1967) in which she was a fiery Katherine.

Many of her subsequent roles seemed but variations on her Martha, glamorous vipers whose foul-mouthed malevolence she played to the hilt in films like *Zee and Co.* (1971).

Less active on screen in the latter half of the 1970s, few of her recent films, from the disastrous American-Soviet co-production *The Blue Bird* (1976) to the poor musical *A Little Night Music* (1977) and the Agatha Christie mystery *The Mirror Crack'd* (1980), have found any degree of box-office popularity.

She made her Broadway debut in *The Little Foxes* (1981) and returned there for a heavily criticised production of *Private Lives* (1983), a professional reunion with Burton.

In the 1980s she emerged from The Betty Ford Clinic where she has undergone treatment for alcohol and drug addiction and soon recaptured her acting career with a series of alarmingly variable performances and, over the last decade, has worked mostly in television. The better films include *Between Friends* (1983), *Malice in Wonderland* (1985) and *Sweet Bird of Youth* (1989).

For the record, her list of husbands comprises Nicky Hilton Jnr (1950–51), Michael Wilding (1952–57), Mike Todd (1957–58), Eddie Fisher (1959–64), Richard Burton (1964–74 & 1975–76) and Senator John W. Warner (1976–82).

Continuing to capture headlines for her private life, ill-health, battles of the bulge and other trivia, she has also become a noted campaigner and fund-raiser for AIDS charities.

TECHNICOLOR The trademark for a number of colour cinematography processes developed by the Technicolor Motion Picture Corporation, the earliest form of Technicolor exposed two negatives in the camera by the use of a beam splitter on the red and green components of light and then projected the separate prints through the red and green filter in a single projector. The results were seen in *The Gulf Between* (1917) and were further refined in films like *The Black Pirate* (1926).

Internationally dominant between 1932 and 1955, the next system involved special three-strip cameras and multiple release prints by photo-mechanical dye-transfer. ▷ Walt Disney first used the process for his cartoon *Flowers and Trees* (1932) and the first feature-film to use this richer version of Technicolor was ▷ *Becky Sharp* (1935). Even after the introduction of colour negatives in 1953, dye-transfer printing, or 'imbibation', continued until 1978.

TEMPLE, Shirley Jane
Born 23 April 1928, Santa Monica, California, USA.

A precociously talented child, Temple made her first film appearance in a series of short *Baby Burlesks* spoofs including *War Babies* (1932), *The Runt Page* (1932) and *Polly-Tix in Washington* (1932).

She also began to appear in feature films and her rendition of the song 'Baby Take A Bow' in *Stand Up and Cheer* (1933) plus her performance as the adorable orphan in *Little Miss Marker* (1934) made her a star. Very swiftly, she became a phenomenon who inspired an unending series of dolls, books, games and variegated merchandising that would put today's promoters to shame. On her eighth birthday she received a staggering 135,000 gifts from her fans across the world.

Voted America's top box-office draw between 1935 and 1938 (she would be supplanted by the equally precocious Mickey Rooney (1920–)), she was an unspoilt personality of undeniable cuteness and good cheer who sang, danced and did impressions. Her sweetness captivated Depression-era audiences and she was the world's favourite golden-haired moppet in films like *Curly Top* (1935), *Dimples* (1936), *Wee Willie Winkie* (1937), *Heidi* (1937) and *Rebecca of Sunnybrook Farm* (1938).

Inevitably her appeal waned, and she made few films as a teenager, the best of which is probably *Since You Went Away* (1944). Subsequent comeback attempts revealed neither the material nor the talent to sustain a distinctive adult career and she retired from the screen after such forgotten 1949 ventures as *A Kiss for Corliss* and *The Story of Seabiscuit* although she did present and occasionally act in *The Shirley Temple Storybook* (1958–60) for television.

One of the few individuals not to bear any scars from a childhood spent in the employ of Hollywood, she was later involved in Republican Party politics as Mrs Shirley Temple Black, serving as America's representative to the United Nations General Assembly in 1969 and later Ambassador in Ghana (1974–76) and White House Chief of Protocol (1976–77).

She received a special juvenile Oscar in February 1935 for her 'outstanding contribution to screen entertainment' and has published the autobiography *Child Star* (1988).

THALBERG, Irving Grant

Born 30 May 1899, Brooklyn, New York, USA.
Died 14 September 1936, Hollywood, California, USA.

A clerk in his grandfather's department store, Thalberg attended night school classes in shorthand and Spanish before embarking on a secretarial career that brought him to Universal Studios.

A personal secretary to studio boss Carl Laemmle (1867–1939), his efficiency and organisational flair secured his eventual promotion to general manager of the company. He then joined forces with independent producer ▷ Louis B. Mayer and remained with him to play a key role in the formation of Metro-Goldwyn-Mayer.

An ambitious, dedicated head of production, his early successes included *The Merry Widow* (1924), *The Big Parade* (1925) and ▷ *Ben-Hur* (1926).

Overseeing every aspect of a production, insisting on re-writing and re-editing until the finished film was shaped into his vision of popular, cultured entertainment, he was renowned for his meticulous attention to detail and helped the studio weather the transition from the silent to sound era with the major box-office attraction *Broadway Melody of 1929*, among others.

Drawn to literary adaptations and prestige drama, he helped foster M-G-M's reputation as the studio with 'more stars than there are in heaven' and was responsible for putting into production such films as *Anna Christie* (1930), *Grand Hotel* (1932), *The Barretts of Wimpole Street* (1934), *Mutiny on the Bounty* (1935), *A Night at the Opera* (1935), *Camille* (1936) and *The Good Earth* (1937).

A notorious workaholic, frequently in poor health, his obsessive devotion to the creation of motion pictures and wonder boy status are said to have inspired the character of Monroe Stahr in *The Last Tycoon* by F. Scott Fitzgerald (1896–1940). He was married to actress Norma Shearer (1900–1983) from 1928 until his death.

THAT WAS THE WEEK THAT WAS

UK 1962–63

Frequently lamented and much-imitated, this satirical revue show was irreverent, topical, and a sign that the BBC was beginning to loosen its corporate tie. Devised by members of the *Tonight* team – including Alasdair Milne, Antony Jay, and Ned Sherrin (who then produced it), *TW3* was made by the BBC's Current Affairs department, the idea being to discuss anything that people might talk about on a Saturday night. Fronted by newcomer ▷ David Frost, the regular team included Roy Kinnear (1934–88), Kenneth Cope (1931–), Lance Percival (1933–), Willie Rushton (1937–) and Millicent Martin (1934–) (who sang a song incorporating the week's events). A visible innovation was the use of the TV studio and its audience, which added to the feeling of immediacy. Spoofish news reports were mixed with sketches and Bernard Levin's interviews, which went beyond polite chat. Though its subversive impact has been exaggerated, *TW3* did take individuals to task, including the then Home Secretary Henry Brooke who was savaged for his position over the regulations governing the granting of political asylum. The BBC's liberalism did not stretch to running the show in an election year, and with a poll imminent in 1964, it was axed. Frost then took the format to the USA. Although it tried to hitch itself to the beginnings of 1960s radicalism, the show ran for just one season.

THE THIRD MAN

Carol Reed (UK 1949)
Joseph Cotten, Trevor Howard, Orson Welles, Alida Valli.
British version at 104 mins, narrated by Carol Reed; American version at 93 mins, narrated by Joseph Cotten. b/w.
Academy Award: *Best Cinematography* Robert Krasker.
Cannes Film Festival: *Best Film.*

Having worked together on 1948's *The Fallen Idol*, producer ▷ Alexander Korda was anxious to maintain the partnership between novelist and screenwriter Graham Greene (1904–91) and director ▷ Carol Reed, and indeed it was with *The*

Third Man that their work together was to result in one of the most enduring of all British films. Greene it was who turned out a richly suggestive script, redolent of the themes of corruption and betrayal that mark much of his best work. It details the efforts in post-World War II occupied Vienna of Joseph Cotten's (1905–) American pulp novelist Holly Martin in tracking down the body of his late friend Harry Lime – only to at length discover that his old buddy has faked his own death to cover for his nefarious smuggling of lethally-diluted black market penicillin. While Reed's marvellously atmospheric location work throughout the narrow streets and shadowy sewers of the bombed-out city lends the film much of its visual panache, and he also takes credit for implementing zither virtuoso Anton Karas' insistently effective scoring, the influence of ▷ Orson Welles imposes its distinctive stamp over much of the action. Certainly the angular camerawork and Robert Krasker's (1913–81) stylized chiaroscuro cinematography evoke Welles's own directorial personality, but it is his mesmerisingly charismatic performance as disturbingly cynical individualist Harry Lime that sparks true screen electricity, for as writer of his own dialogue he created one of the cinema's most delicious rationalisations of villainy. 'In Italy for thirty years under the Borgias, they had warfare, terror, murder, bloodshed. They produced Michelangelo, Leonardo Da Vinci and the Renaissance,' he purrs from the aerial perspective of a fairground ferris wheel. 'In Switzerland they had brotherly love, five hundred years of democracy and peace. And what did that produce? The cuckoo clock.'

THIRTYSOMETHING
USA 1988–

On its launch, *thirtysomething* (the lower-case initial letter is obligatory) was quickly dubbed a yuppie drama. The label was misleading, for though the characters were all under forty (hence the title) none – in the first series, anyway – represented the fast-buck amorality of the Reagan-era. Based around seven ex-college friends and the central couple of Michael and Hope Steadman (Ken Olin, Mel Harris), it was a post-*Big Chill* tale of middle-class folk in Philadelphia, which achieved its impact by treating angst, professional and emotional, with therapeutic vigour.

The show's creators, Marshall Herskovitz and Ed Zwick met in 1975 at the Los Angeles Film Institute. Their film company is named Bedford Falls, after the town in ▷ Frank Capra's ▷ *It's A Wonderful Life* – a favourite movie, reflecting their concern with the balance between joy and sadness, a *thirtysomething* staple. Early episodes were blighted by over-frequent use of fantasy sequences, which detracted from the strength of the characters. So intense was the drama that many viewers took an instant dislike to the show, only to be slowly won round. The second series saw relationships more strained, and Nancy (Patricia Wettig) developed cancer – previously a taboo subject for US drama. Perhaps the most sympathetic character, aspiring photographer Melissa (Melanie Mayron) has also been given a bigger role, with talk of a spin-off series. The show's cohesion may be challenged by the fact that the actors now direct. Whatever happens, the term 'thirtysomething' has entered common parlance.

3-D Three-dimensional filmmaking using a stereoscopy process to create the illusion of depth and perspective in screen imagery flourished briefly in the 1950s as one of the many weapons that Hollywood resorted to in an attempt to combat falling cinema attendances and the growth of its television rival. The boom era for the process began in 1953 with such films as *Bwana Devil*, *House of Wax* and *Fort Ti*.

The impracticalities of distributing and collecting the special glasses required to view the effect, the technical difficulties of projection and the generally poor quality of the films involved ensured that it was a short-lived fad although more prestigious works like *Kiss Me Kate* (1953) and *Dial M for Murder* (1954) were originally shot with 3-D images but released flat.

It has briefly resurfaced over the years and enjoyed an unexpected and modest resurgence in the 1980s on such films as *Jaws 3-D* (1983) and *Emmanuelle IV* (1984).

TILL DEATH US DO PART

UK 1966–74
Writer Johnny Speight

The unfocused rage of the West Ham supporting, black-hating bigot Alf Garnett (Warren Mitchell (1926–)) was first seen in a *Comedy Playhouse* production (though the family were at that time known as the Ramseys). Also considered for the lead role were ▷ Peter Sellers and Leo McKern (1920–), while Alf's long suffering wife Else – Dandy Nichols' (1907–86) 'silly old moo' – was first played by Gretchen Franklin (1911–) (later Ethel in ▷ *Eastenders*). Also in the Garnett household were daughter Rita (Una Stubbs (1937–)) and son-in-law Mike (Anthony Booth (1937–)). Though loosely based on Frank Muir (1920–) and Denis Norden's (1922–) comic family *The Glums*, Speight's Garnett brought a rare complexity to TV comedy. He was loudmouthed and unapologetic, ranting with equal force against immigrants, Labour politicians, and the permissive society. Mike (variously dismissed as a 'randy Scouse git' and 'Shirley Temple', because of his over-collar-length hair) represented everything Alf hated – he was a Trotskyite, had no respect for the Royal Family, and was untouched by the work ethic. Garnett's profanity upset some – clean-up campaigner Mary Whitehouse counted 78 'bloodys' in a single episode. There were fears, too, that the ironic intent of the writing was missed by some viewers, to whom Garnett seemed to be articulating real fears and concerns.

Till Death Us Do Part was successfully transplanted to the USA (as ▷ *All In The Family*), and Germany (where the Garnetts became the Tetzlaffs). When revived in 1985 as *In Sickness And In Health*, Garnett's racism was toned down, and with a Conservative government in power, his outrage was harder to sustain. Though Mitchell's performances remained outstanding, the show has lost its social context. The death of Dandy Nichols after seven shows, was a further blow.

TIME-LAPSE PHOTOGRAPHY

An accelerated means of conveying slow changes in a given situation or process, time-lapse photography involves a camera taking a series of photographs from the same viewpoint at regular intervals. When these single frames are projected at a normal speed rapid changes are seen to occur. Often used to record the growth and bloom of a flower, it can also capture the development of cloud formation, metallic corrosion or traffic flow, among many other examples.

TOP OF THE POPS

UK 1964–

Born in a boom-time for British pop music, *Top Of The Pops* first show came from a converted Manchester church, and had Jimmy Savile (1926–) spinning the discs of the Dave Clark Five, the Rolling Stones and Dusty Springfield. Other early DJs were Pete Murray (1928–) and Alan Freeman (1930–). Though *Top Of The Pops* has undergone cosmetic changes, its democratic format – the country's best-selling records are featured – has enabled it to survive the trends which have tripped up competitors like Mike Mansfield's *Supersonic* (in the early 1970s), or ITV's live effort *The Roxy* (in the mid-1980s). Though generally presented by Radio One disc jockeys, *TOTP*'s practice of playing the records was replaced early on by largely mimed performances from the artists. Additionally, from 1967, songs were subject to artistic interpretations by the female dance troupe Pans People (and later Legs and Co). The show has had problems with the content of certain songs, like Jane Birkin's *Je T'Aime*, Frankie Goes to Hollywood's *Relax* and the Sex Pistol's *God Save The Queen*, though by ignoring them, it fuelled their popularity. The latest competitor, *The Chart Show*, is a slickly executed collage of video-clips, with no presenter.

TORNATORE, Giuseppe

Born 1956, Bagheria, Sicily.

After a career in television and as the director of such short films as *Il Caretto* (1972) and *Diario Di Guttuso* (1980), Tornatore made his feature-film debut with *Il Camorrista* (1987) which starred Ben Gazzara (1930–) in the story of a Mafia boss's rise to power.

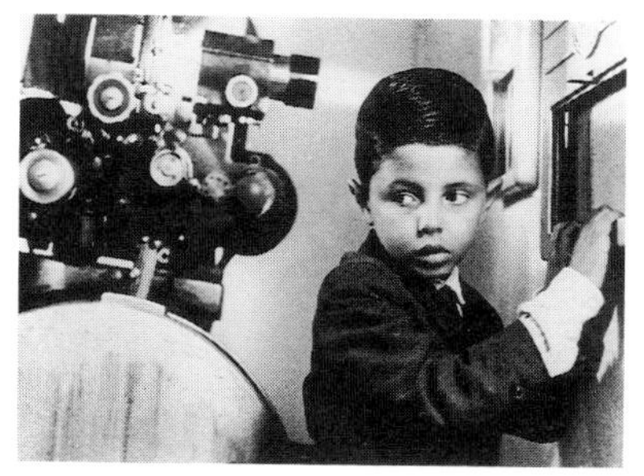

Salvatore Cascio in *Cinema Paradiso* (1988) directed by Giuseppe Tornatore

He followed this with *Nuovo Cinema Paradiso* (Cinema Paradiso) (1988) which captured the hearts of world audiences with its sentimental tale of a famous film director reflecting on his life-long love affair with the cinema and his affectionate memories of the aged projectionist who introduced him to the delights and wonders of the cinema world.

A winner of the Best Foreign Film Oscar it is currently scheduled for an American remake.

His subsequent work *Stanno Tutti Bene* (Everybody's Fine) (1990) was another persuasive assault on the heartstrings with a winning performance from ▷ Marcello Mastroianni as a Sicilian pensioner who decides to visit his far flung offspring and receives a rather different impression of their lifestyles than the illusions they had presented to him.

A director with a refreshingly innocent and emotional view of the world, he then proceeded to make *Especially on Sunday* (1991), a reunion with his *Cinema Paradiso* star ▷ Philippe Noiret.

TORRE-NILSSON, Leopoldo

Born 5 May 1924, Buenos Aires, Argentina.
Died 8 September 1978, Buenos Aires, Argentina.

The son of Argentine director Leopoldo Torres Rios (1899–1960), Torre-Nilsson made his directorial debut as a teenager with the short film *El Muro* (The Wall) (1947) and collaborated with his father as a co-director on the features *El Crimen De Oribe* (Oribe's Crime) (1949) and *El Hijo Del Crack* (Son of the 'Star') (1953) before striking out on his own with *Dia De Odio* (Days of Hatred) (1953).

A prolific director with his own production company, his work in particular and Argenine cinema in general was more closely scrutinised after *La Casa Del Angel* (The House of the Angel) (1957) was screened at the Cannes Film Festival. A baroque, claustrophobic tale of a virginal young girl crushed under the weight of a sheltered, repressive Catholic upbringing it established the style and content of much of his future endeavours.

Working in close collaboration with his wife, novelist and scriptwriter Beatriz Guido (1924–88) he would explore the hypocrises and pretensions of middle-class morality in a style that was heavily influenced by European films of the period. His work has also been compared to that of ▷ Luis Bunuel in its macabre sense of humour and penchant for the grotesque.

Among the more notable of the more than two dozen features he made are *La Caida* (The Fall) (1959), in which a virginal young girl awakens to life's rich possibilities during her sojourn with an eccentric family, *La Mano En La Trampa* (The Hand in the Trap) (1961) a vivid depiction of the relationship between a convent schoolgirl and a ageing, lovelorn aunt, and *Boquitas Pintadas* (Painted Lips) (1974). His final film was *Piedra Libre* (Free Stone) (1976).

TOWNE, Robert Burton

Born 1934 or 1936, Los Angeles, California, USA.

A writer from a very young age, Towne is said to have completed his first short story at the age of six. Later a student of English and philosophy at Pomona College in Claremont, he studied acting with Jeff Corey (1914–) and found himself sharing a room with ▷ Jack Nicholson.

Encouraged by ▷ Roger Corman, he wrote a screenplay for *The Last Woman on Earth* (1960) and worked on *Creature from the Haunted Sea* (1961) and *The Tomb of Ligeia* (1964) before moving into television and contributing to such series as ▷ *The Man from U.N.C.L.E.* and *Outer Limits*.

He was hired by ▷ Warren Beatty to revise the script for ▷ *Bonnie and Clyde* (1967) and decided to emphasise the inevitability of the final, bloodspattered outcome. Soon in demand as a 'script doctor', he has gained legendary renown as a writer capable of working his magic on a poor script by rewriting, rearranging, or revising substandard material. His work in this area has often been uncredited but he received public acknowledgement of his skills in ▷ Francis Coppola's Oscar acceptance speech for ▷ *The Godfather* (1972), on which Towne had worked.

A painstaking craftsman, he received a Best Screenplay Oscar nomination for *The Last Detail* (1973) in which the story of two roughneck sailors escorting a callow colleague to the brig becomes a statement on the grinding compromises and iniquities of daily living and doing a job.

He won the Oscar for his brilliant script of *Chinatown* (1974) a sleek ▷ film noir in which a private detective unravels a startling web of corruption and incest in a lovingly recreated and suitably seedy 1930s Los Angeles.

A co-writer with Warren Beatty of the social satire *Shampoo* (1975), he also worked as a 'doctor' on such varied films as *Marathon Man* (1976), *The Missouri Breaks* (1976) and *Orca, The Killer Whale* (1977).

In anticipation of realising his dream project to make an authentic version of Edgar Rice Burrough's Tarzan novels, he turned director on *Personal Best* (1982), a sensitive and well-handled account of a lesbian relationship that develops between two competitors in the 1980 Olympic Games.

Production difficulties on the latter film forced him to relinquish the right to direct *Greystoke, The Legend of Tarzan, Lord of the Apes* (1984) and his displeasure at the final film made him transfer his screen credit to P. H. Vazak, his recently deceased Hungarian sheepdog, on the basis that 'If he could've written, he would have done it better'. Thus the dead mutt received Towne's Best Screenplay Oscar nomination.

He returned to direction with the muddled-morality romance of *Tequila Sunrise* (1988) and has contributed, publically or privately, to such recent scripts as *8 Million Ways to Die* (1986), *Frantic* (1988), *Days of Thunder* (1990) and the much-troubled Chinatown sequel *Two Jakes* (1990). The disastrous commercial reception for the latter film may place in doubt the viability of the third-part of the trilogy which is set in 1959 and finds private eye Gittes being sued by his own wife.

TOP HAT

Mark Sandrich (USA 1935)
Fred Astaire, Ginger Rogers, Edward Everett Horton, Helen Broderick.
105 mins. b/w.

Although RKO's 1933 frothy musical release *Flying Down To Rio* was ostensibly a vehicle for box office attraction Dolores Del Rio (1905–83), it was actually the pairing of fifth-billed former Broadway hoofer ▷ Fred Astaire and ex-vaudevillian supporting blonde Ginger Rogers (1911–) in their featured number 'The Carioca' which caught the attention of the moviegoing public to such an extent that a starring movie of their own together was only a matter of time. In the next two years the studio swiftly adapted two established theatrical properties *The Gay Divorcee* (1934) and *Roberta* (1935) for the new dance sensations, but it was with the first film created specifically for their talents, 1935's *Top Hat*, that the particular screen magic of the coupling was to achieve its most fully realised expression. Here the featherweight scenario as Astaire's musical star Jerry Travers and Rogers's no-nonsense dame Dale Tremont surmount mistaken identity to dance their way to romance admist Van Nest Polglase's (1898–1968) ravishingly stylised Art Deco version of Venice matters less than the way in which Irving Berlin's (1888–1989) marvellous tunes and the choreography – for which Hermes Pan (1905–90) and Astaire himself take credit – allow the emotional thrust of the material to be expressed through the production numbers. While the title song, synonymous with the Astaire legend, allows Fred to strut his solo stuff most admirably, the witty love-hate duet 'Isn't This A Lovely Day?' and the later highly-charged 'Cheek To Cheek' routine chart the fortunes of two people whose superficial hostility inevitably gives way to the most touching affection. Typifying sophisticated Thirties cool as their bodies swirled with precise grace across the screen, here was a professional chemistry that would stand the test of time and be good for another six movies together.

TRACKING SHOT A shot in which the camera, and its operators, are moved to follow the chosen part of the action. Usually a forwards or backwards motion, the process often involves the use of a set of tracks, hence the term.

TRACY, Spencer

Born 5 April 1900, Milwaukee, Wisconsin, USA.
Died 10 June 1967, Los Angeles, California, USA.

A student at the American Academy of Dramatic Arts in New York, Tracy made his New York debut in *R.U.R.* (1922) and was a noted theatrical performer in such Broadway productions as *Yellow* (1926), *Baby Cyclone* (1927) and *The Last Mile* (1929) before making his film debut in the shorts *Taxi Talks* (1930) and *The Hard Guy* (1930).

Part of the creative influx of Broadway actors to Hollywood during the early talkie era, he made his feature-length debut in the film version of *Up The River* (1930) and signed a contract with Twentieth Century-Fox.

His burly physique and plain looks found him typecast as tough guys, working men and gangsters but he began to acquire a critical cachet as the hardened criminal in *20,000 Years in Sing Sing* (1932) and the callous railroad President in *The Power and the Glory* (1933), which was later cited as an influence on the storytelling techniques in ▷ *Citizen Kane* (1941).

Under contract to M-G-M from 1935, he developed in to a leading man of rock-like reliability who became a major star with his performances as a pugnacious priest in *San Francisco* (1936) (Best Actor Oscar nomination), the sea dog in *Captains Courageous* (1937) (Oscar) and the kindly Father Flanagan in *Boys Town* (1938) (Oscar).

The romantic comedy *Woman of the Year* (1942) began a long professional and personal relationship with ▷ Katharine Hepburn that resulted in a number

Spencer Tracy, Judy Holliday and Katharine Hepburn in *Adam's Rib* (1949)

of sharply-written and performed battle-of-the-sexes comedies, among them *Adam's Rib* (1949), *Pat and Mike* (1952) and *The Desk Set* (1957).

A master of understatement and effortless honesty in his acting, Tracy brought an eloquent truth and down-to-earth integrity to his work that won him a reputation as one of the finest of all cinema actors. Among the many distinguished performances later in his career are the title role in *Father of the Bride* (1950) (Oscar nomination), the disapproving father in *The Actress* (1953), the one-armed man of mystery in *Bad Day At Black Rock* (1955) (Oscar nomination) and the Clarence Darrow-like lawyer in *Inherit the Wind* (1960) (Oscar nomination). He received further Oscar nominations for *The Old Man and The Sea* (1958) and *Judgement at Nuremberg* (1961).

Suffering from poor health during the last decade of his life, and known to be an alcohol-loving and cantankerous colleague, he was unable to make a number of projects including ▷ *The Leopard* (1963), *Cheyenne Autumn* (1964) and *The Cincinnati Kid* (1965), but he returned to the screen after a four year absence to give a performance of heartfelt sincerity that transcended the sentimentality and dated subject matter of the comedy *Guess Who's Coming To Dinner* (1967) (Oscar nomination). Completed just ten days before his death, it provided a fitting swansong to his career.

TRAVELLING MATTE-SHOT A useful ▷ special-effect shot in many lavish contemporary fantasy adventures, this is an image combination process in cinematography for superimposing foreground action on a separately photographed background scene by printing at the laboratory.

A ▷ matte, a strip of film with opaque silhouettes of the foreground, is used to reserve this area when printing the background, and the foreground action is inserted into this space at a second printing using a complementary matte. In its simplest form, the process deals with stationary mattes but more sophisticated systems can employ mattes that change shape from frame to frame, hence the term 'travelling'. This allows moving action to be combined with other elements. An example of the process is ▷ blue-screen process photography.

TREE OF WOODEN CLOGS (L'Albero Degli Zoccolli)
Ermano Olmi (Italy 1978)
Luigi Ornagli, Francesca Moriggi, Omer Brignoli, Antonio Ferrari.
178 mins. col.
Cannes Film Festival: *Palme D'Or*.

Although citing the poetic film ethnology of America's ▷ Robert Flaherty as a major influence, regional filmmaker Ermano Olmi's (1931–) lengthy study of rural Lombardy is something of a return to the Italian ▷ neo-realist tradition in its concentration and insistence on the nobility of the everyday experience of its non-professional cast of Lombardian peasants. Eschewing the posturing spectacle and fabulist tragicomedy of contemporary offerings drawing on similar material, namely ▷ Bernardo Bertolucci's *Novencento* (1900) (1976) and the ▷ Taviani Brothers' *Padre Padrone* (1977), Olmi's unhurried chronicling of the lives of four farming families across the seasons on the surface at least appears closer to screenwriter and theorist Cesare Zavattini's (1902–89) original uncompromising neo-realist ideal. The relaxed unfolding of the main narrative events – birth and marriage, the slaughter of a pig, the discovery of a gold coin – almost render the film a humanist documentary were it not for more obviously fictive moments like the miracle of the cow, allowing Olmi's optimistic Catholicism to shine through when prayers bring about the sudden recovery of a stricken animal. Indeed, perhaps because Olmi's authorial imprint is so delicately manifested it is all too easy not to realise that the viewer is being delicately guided towards definite spiritual and ideological positions. A careful consideration of the section from which the title is derived, wherein a young peasant boy breaks his shoe on the way home from school, resulting in his father cutting the landowner's tree to make him a new sandal, and in turn precipitating the eviction of the entire family, reveals a Marxian overview of class exploitation all the more potent for the undemonstrative lucidity of its expression.

TRINTIGNANT, Jean-Louis

Born 11 December 1930, Piolenc, Vaucluse, France.

Interested in the theatre, Trintignant abandoned his legal studies to become an actor and made his Paris stage debut in *A Chacun Selon Sa Faim* (1951).

His first major stage role in *Responsibilité Limitée* (Limited Responsibility) (1954) led to his film debut in the short *Peauchinef* (1955) and he was subsequently cast as innocent, unassuming young men in films like *La Loi des Rues* (The Law of the Streets) (1956) and the controversial *Et Dieu Créa La Femme* (And God Created Woman) (1956).

His pale-skinned impassivity and sensitive, butterscotch eyes have lent themselves to the portrayal of romantic vulnerability and the illumination of the interior life of the psychologically disturbed.

He enjoyed success with the uncharacteristic role of a Fascist assassin in *Le Combat Dans L'Île* (Island Battle) (1962) and gained an international following as the racing driver in the sentimental romance *Un Homme et une Femme* (A Man and a Woman) (1966).

He followed this with a string of distinguished performances as the playboy in *Les Biches* (The Does) (1968), the obsessive, publicity-seeking psychotic in *Le Voleur des Crimes* (Thief of Crime) (1968), the resolute magistrate in the gripping *Z* (1968), the narrator of *Ma Nuit Chez Maud* (My Night At Maud's) (1969) and the chillingly repressed assassin in *Il Conformista* (The Conformist) (1970).

He subsequently turned to direction, making his debut behind the camera with *Une Journée Bien Remplie* (A Full Day's Work) (1972), in which a smalltown baker aims to murder the nine jurors who have sentenced his son to death.

Passionately interested in motor racing, he has often taken time off from acting and publically toyed with the idea of retirement but has resumed his career with batteries recharged even if he has an increasing preference for supporting or secondary assignments rather than starring roles.

He directed again with *Le Maitre Naguer* (The Master Swimmer) (1979) and his notable screen work in recent years includes the film within a film *Les Violons du Ball* (1974), his edgy portrayal of a bank executive implicated in financial misedmeanours in *L'Argent des Autres* (1978), the popular black and white thriller *Vivement Dimanche!* (Finally Sunday) (1983), *Un Homme et une Femme: Vingt Ans Déjà* (A Man and a Woman Twenty Years Later) (1986) and *La Vallée Fantôme* (The Phantom Valley) (1987) in which he portrayed a filmmaker in search of renewed inspiration.

Resistant to such British and American offers as *The Servant* (1963), *Close Encounters of the Third Kind* (1977) and ▷ *Apocalypse Now* (1979), he made a rare foray into an English-language production as a smoothly sinister and nonchalantly cynical businessman in *Under Fire* (1983).

Following another one of his periodic absences, he returned to the screen in *Merci, La Vie* (1991) and was said to be preparing another directorial project.

THE TRIUMPH OF THE WILL (Triumph des Willens)

Leni Riefenstahl (Germany 1934)
b/w. 120 mins.

From the outset, Hitler and Minister of Information, Josef Goebbels, recognised that the film medium could play a significant role in communicating Nazi power and achievement to the widest of possible audiences across Germany: newsreels, shorts and documentary features soon became the tools of direct propaganda. Such is the context in which the work of film artist ▷ Leni Riefenstahl must be assessed, her critical reputation irrevocably complicated by her acknowledged status as Hitler's favourite filmmaker – the passing of a 'clean bill of ideological health' by Allied de-Nazification units after the war notwithstanding – with *Triumph des Willens* commissioned by Hitler himself as a filmed record of the 1934 National Socialist Party congress at Nuremberg. In the most basic of terms it is thus a documentary, but Riefenstahl's control of composition and montage is so complete that the basic material of actuality is at length transformed into the overpowering stuff of myth. From the messianic overtones of the Fuhrer's initial aeroplane descent from the clouds, to the constant motion of the 18-minute parade section or the massive architectural vistas of the wreath laying sequence, Riefenstahl's masterly editing creates impressive abstract patterns of images and sounds. Constantly cutting from panoramic long shot (a vast array of Nazi flags) to extreme close-up (one flag in one man's hand) she both dislodges the viewer's sense of perspective and creates an aura of visual harmony implicitly rhyming with the political message being put forward. It can be argued that such extraordinary formal skill transcends the content on view thus compelling us to judge it as absolute film, but by concentrating on the glories of Riefenstahl's form are we then tacitly approving her work as a Nazi propagandist? When the director visited Paris to accept the film's Grand Prix at the 1937 Exposition Internationale des Arts et des Techniques her appearance met with protests from French workers.

TRUFFAUT, François

Born 6 February 1932, Paris, France.
Died 21 October 1984, Paris, France.

The survivor of an unhappy, near-delinquent childhood, and an army deserter, Truffaut developed a fanatical passion for the cinema and became a ruthlessly critical writer on film for *Cahiers du Cinema* where he was an influential voice among those positing the ▷ auteur theory.

His contempt for much of the conservatism he perceived in the French cinema of the 1950s led him to direction, firstly in such short films as *Une Visite* (1955) and then in the feature *Les Quatre Cents Coups* (The Four Hundred Blows) (1959), a haunting autobiographical study of a deprived childhood that was influenced by ▷ neo-realism and introduced his enduring cinematic alter ego of Antoine Doinel, played by Jean-Pierre Leaud (1944–) here and in a succession of films that followed the amorous complications of the character's life from adolescent to married man in films like *Baisers Volés* (Stolen Kisses) (1968), *Domicile Conjugale* (Bed and Board) (1970) and *L'Amour en Fuite* (Love on the Run) (1979).

One of the key figures in the ▷ nouvelle vague, he worked in all genres bringing a refreshing cinematic fluidity and spontaneity to such subject matter as the gangster story *Tirez le Pianist* (Shoot the Pianist) (1960), the bittersweet ménage-à-trois ▷ *Jules et Jim* (Jules and Jim) (1961) and the Hitchcock-style revenge thriller *La Mariée Etait en Noire* (The Bride Wore Black) (1967).

Capable of infusing his work with the most effortless charm and artistry, his recurring themes were the many pains and pleasures of love and the tensions between life and art, artifice and reality. The range of his accomplishments from the 1970s along include the delightful film-within-a-film *La Nuit Americaine* (Day for Night) (1973), the brooding gothic romance of *L'Histoire d'Adèle H* (The Story of Adele H) (1975), and the utterly beguiling portrait of childhood in *L'Argent de Poche* (Small Change) (1976).

Towards the end of his life, he enjoyed a major commercial success with *Le Dernier Métro* (The Last Metro) (1980), an opulent tale of a wartime theatrical troupe's struggle to survive. His last films were the modest romantic drama *La Femme A Côté* (The Woman Next Door) (1981) and the black and white pastiche of Hollywood thrillers *Vivement, Dimanche!* (Finally, Sunday) (1983).

An actor in his own films, he was also seen in *Close Encounters of the Third Kind* (1977). His books include *Les Films De Ma Vie* (1975) and a collection of his letters *Francois Truffaut Correspondance* (1988) both of which have been translated into English.

TURNER, Kathleen

Born 19 June 1954, Springfield, Missouri, USA.

The daughter of an American diplomat, Turner spent her childhood in Canada, Cuba, Washington D.C., Venezuela and London where she began to study at the Central School of Speech and Drama.

Returning to the United States, she later graduated with a Bachelor of Fine Arts degree from the University of Maryland and moved to New York. After waiting on tables, appearing in commercials and doing the rounds of casting agencies she was hired as villainess Nola Dancy Aldrich for the television ▷ soap-opera *The Doctors* (1978–80).

During the run of the show, she also appeared on stage and subsequently left to test for a role in the film *All the Marbles*

Kathleen Turner with Michael Douglas in *Romancing the Stone* (1984)

(1981). She did not win the part, but was cast as the sultry, conniving wife in the stylish, contemporary ▷ film noir *Body Heat* (1981).

Honey-blonde, with sensuous lips and an ebullient, no-nonsense manner, her deep, velvety tones and sexy manner made her an exceptional femme fatale but she quickly avoided stereotyping by eschewing similar roles and providing a delightful pastiche of the character in the madcap comedy *The Man With Two Brains* (1983).

Her stardom was consolidated with the wildly popular cliffhanging adventure *Romancing the Stone* (1984) in which she gave a gutsy and winning performance as a dowdy romantic novelist who finds her life transformed by her unwitting involvement with real-life adventure and the roguish Jack Colton, played by ▷ Michael Douglas.

Her subsequent accomplishments include a bravura performance in the controversial *Crimes of Passion* (1984), in which she stars as a frigid fashion designer who finds a form of liberation as hooker China Blue, *Prizzi's Honor* (1985), in which her female assassin is more than a match for ▷ Jack Nicholson, and *Peggy Sue Got Married* (1986) (Best Actress Oscar nomination) in which she gave a fully rounded and poignant perspective on a middle-aged woman who is afforded the opportunity to revisit her teenage years from the vantage point of wise maturity.

Her lesser ventures include the disappointing *Jewel of the Nile* (1985) (a sequel to *Romancing the Stone*), and the redundant psychological thriller *Julia and Julia* (1987) and the lacklustre comedy *Switching Channels* (1987), but a mark of her professionalism has been her decision to accept supporting roles and display a willingness to expand the scope of her talents. Thus, she played the emotionally distraught wife in *The Accidental Tourist* (1988) and provided the husky vocal accompaniment to animated bombshell Jessica Rabbit in *Who Framed Roger Rabbit* (1988) and the short cartoons that followed.

More recently, she was on fine form as the sympathetic wife in the black battle-of-the-sexes comedy *The War of the Roses* (1989). Her most recent film is *Warshawski* (1991) in which she plays a private-eye and she is next scheduled to appear in *House of Cards* (1991) as the mother of an autistic child.

She has continued to work in the theatre over the years, most recently as Maggie in a Broadway revival of *Cat On A Hot Tin Roof* (1990).

TWILIGHT ZONE

USA 1959–64
Creator Rod Serling

An anthology of bizarre journeys to the sixth dimension, ('... the middle ground between light and shadow, between science and superstition, between the pit of a man's fears and the sunlight of his knowledge ...') *The Twilight Zone* was the work of Purple Heart-winning army paratrooper Rod Serling (1924–75), who began writing to help distance himself from the war. Having graduated from Antioch College he won three Emmys for weighty TV plays, but turned to fantasy after several battles with sponsors over programme content. His first effort was *The Time Element* (1957), in which a pilot foretells of Pearl Harbor, but is dismissed as mad by the Army. A warm response to this led CBS to commission *Where Is Everybody?* in which a man finds himself in a deserted town, only to discover that the experience was the hallucinatory result of a simulated space journey. Serling remained the central creative force during the show's five-year run, writing the majority of the stories, though his whimsical, sometimes sentimental style was complemented by the more disturbing approach of Charles Beaumont (1929–87), and the suspenseful tales of Richard Matheson (1926–). Popular themes in the series were dopplegangers, corrupted notions of time, and characters who found that they were acting in some kind of larger drama. Many distinguished actors appeared, among them ▷ *Star Trek*'s Leonard Nimoy (1931–), George Takei and William Shatner (1931–) (the latter in the memorable *Nightmare At 20,000 Feet*, as a former mental patient who sees the wing of his aircraft being destroyed, but is not believed by his fellow passengers). There were roles, too, for Telly Savalas (1924–), Burt Reynolds (1936–), Roddy McDowall (1928–), ▷ Robert Redford, Lee Marvin (1924–87), Charles Bronson (1922–), and Peter Falk (1927–), while ▷ Dennis Hopper was involved in *He's Alive*, where Hitler's ghost returned to taunt America.

In 1983, *The Twilight Zone – The Movie* put old scripts in the hands of directors John Landis (1950–), ▷ Steven Spielberg, Joe Dante (1946–) and George Miller (1945–). Miller's segment, a reworking of *Nightmare At 20,000 Feet*, was the most successful. In 1985 a new series began on television, with variable results.

TWIN PEAKS

USA 1990–
David Lynch/Mark Frost

The collaboration of left-field filmmaker ▷ Lynch and ex-▷ *Hill Street Blues* story editor Frost was expected to throw some strange shapes, but its success led some to suggest that the grammar of television had been changed. The Twin Peaks of the title is a picturesque lumber town in Washington state, which has been rocked by the murder of 17-year-old cheerleader, Laura Palmer. FBI agent Cooper (Kyle MacLachlan (1959–)) – Frost based his quirkiness and attention to detail on Lynch – is drafted in to assist local sheriff Harry S. Truman (Michael Ontkean (1946–)) in the search for the killer, and their investigations reveal a seamy underside to the town.

The show borrowed heavily from several genres, and had an upended sense of narrative. Essentially a ▷ soap, it nodded to ▷ *Gunsmoke* to the extent of featuring a hillbilly deputy and an obsession with a 'Damn fine cup of coffee'. Cooper's habit of recording the minutiae of the day on a micro-cassette was an FBI version of Kirk's Captain's log in ▷ *Star Trek*. The plot, though convoluted, was as plausible as the mainstream *Dynasty* had been, and the catchline 'Who Killed Laura Palmer?' echoed the fuss which surrounded ▷ *Dallas* when JR was shot. Indeed, at the end of the first series the killer had still not been uncovered, and agent Cooper found himself on the wrong end of a gun. Lynch's influence, doughnuts and cherry pie aside, was obvious in the few episodes he directed, whereupon inspiration would come from ghostly giants or games of pitch and toss. Also notable was the music by Lynch and Angelo Badalamenti – a pastiche of soundtrack work, with repeated themes to suggest danger or romance. As ratings fell in the show's second season, the claims that *Twin Peaks* had broken the mould looked over-optimistic.

2001: A SPACE ODYSSEY

Stanley Kubrick (USA/UK 1968)
Keir Dullea, Gary Lockwood, the voice of Douglas Rain, William Sylvester.
Premiered at 160 mins, cut by Kubrick to release length of 141 mins. col.
Academy Award: *Special Visual Effects* Wally Veevers, Douglas Trumbull.

The deliberate Homeric connotations of the title offer some indication of the thematic ambition of ▷ Stanley Kubrick's *2001: A Space Odyssey*, which takes on nothing less than the history and future development of all mankind and is exemplified in the celebrated cut from prehistoric man's bone-tool flying through the air to the similar outline of a huge spacecraft floating in orbit. Ever the determinist, Kubrick's narrative development is marked by the appearance of mysterious black monoliths at 'The Dawn of Man', on the moon millions of years later, and finally in space near Jupiter. Each of these propels man further along an evolutionary path controlled by some higher alien intelligence (if not the concept of God). Humankind, runs the general interpretation of a film described by its director as a 'non-verbal experience', has transcended the human state through technology but must move beyond that technology (as envinced by the deadly clash between the two central astronauts and their manipulative on-board computer HAL 9000) before rebirth as astral superman. Keir Dullea's (1936–) surviving space traveller journeys through the dimensions to undergo the cycle of old age and childhood on a new plane of existence. For all its would-be grand Nietzchean theorising on 'humanity' however, *2001*, like much of Kubrick's later work, is perhaps flawed by its simple lack of warmth for the human race. Asserting man's dehumanisation by his technology, the film's visual stress is on the magnificently conceived and executed future hardware, or the bravura imprint of its own directorial styling (most obviously signalled by the eccentric virtuosity of its choice of classical score), rather that its cast of characters. Still, the final irony remains that this most cerebral of superproductions was championed by contemporary conoisseurs of psychedelia as 'the ultimate trip', a line later used on the poster for the film's reissue.

It was followed in 1984 by an almost inevitably disappointing sequel *2010*, directed by Peter Hymans (1943–), in which American space agency official Roy Scheider (1935–) attempts to discover just what did happen to the previous film's mission to Jupiter.

ULLMANN, Liv Johanne

Born 16 December 1939, Tokyo, Japan.

An acting student at the Weber-Douglas School in London, Ullmann began her stage career with a repertory company in Stavanger and made her film debut in *Fjols Til Fjells* (Fools in the Mountains) (1957).

Small roles in a number of little-seen Scandinavian films followed and her stage reputation grew through a body of work with the National Theatre in Oslo. However, her greatest screen roles came from a long personal and professional association with director ▷ Ingmar Bergman that began with ▷ *Persona* (1966), in which she sensitively delineates the anguish of a former actress suffering from a psychosomatic illness that has deprived her of speech.

In such films as *Skammen* (The Shame) (1968), *En Passion* (A Passion) (1969) and *Visknigar Od Rop* (Cries and Whispers) (1972) she laid bare, with astonishing simplicity and directness, the inner turmoil of women experiencing various emotional and sexual crises. Ruthlessly scrutinised by the camera and dispensing with any flamboyance of action or manner, her face is rendered a map of pain and puzzlement with every flicker of the eye or movement of a muscle signifying volumes of emotion.

Work away from the special atmosphere of her relationship with Bergman

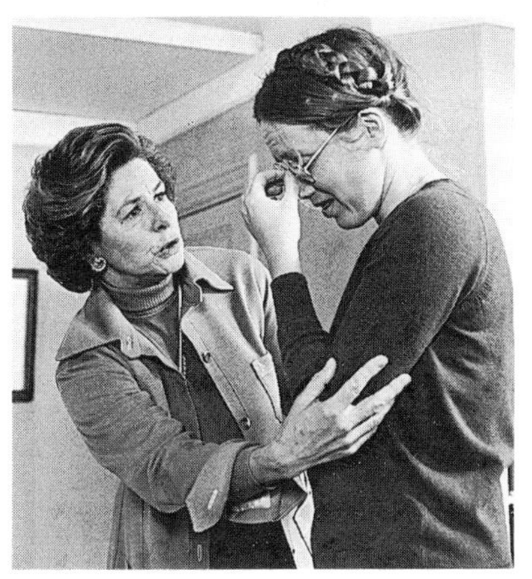

Liv Ullmann with Ingrid Bergman in *Autumn Sonata* (1978)

has generally been less rewarding and his disastrous English-language credits include *Pope Joan* (1972), the ill-conceived musical version of *Lost Horizon* (1973) and the heavyhanded romantic trifle *Forty Carats* (1973). However, she did receive a Best Actress Oscar nomination as a Swedish woman settling in the American mid-West of the 19th century in *The Emigrants* (1972).

She triumphed for Bergman again in *Ansikte Mot Ansikte* (Face to Face) (1975) (Oscar nomination), as a psychiatrist seeking to make sense of the world around her and the many anxieties that undermine her confidence. Their final collaborations included *Scenes from*

A Marriage (1976), *Das Schlangerei* (The Serpent's Egg) (1977) and *Hostsonaten* (Autumn Sonata) (1978) an examination of the tensions, jealousies and regrets in a mother-daughter relationship where she was eloquently partnered by ▷ Ingrid Bergman.

Wisely eschewing most of the English-language offers that have come her way, her international career over the past decade includes the Swiss Cold War drama *La Diagonale du Fou* (Dangerous Moves) (1983), the Italian comedy *Speriamo Che Sia Femmina* (Let's Hope It's A Girl) (1985) and *Gaby – A True Story* (1987), as the mother of a paraplegic daughter.

She made her Broadway debut in *A Doll's House* (1975) and her regular theatre appearances include the musical *I Remember Mama* (1979), *Ghosts* (1982) and *Old Times* (1985) in London.

She has worked extensively for the UNICEF organisation and written the autobiographical volumes *Changing* (1977) and *Choices* (1984). At the time of writing she is said to be preparing to make her debut as a feature film director.

ULTIMO TANGO A PARIGI *see* LAST TANGO IN PARIS

THE UNTOUCHABLES
USA 1959–63

After surveying the early efforts of his scriptwriters, producer Quinn Martin is reported to have issued the command 'More action!' He was thus in no position to complain when *The Untouchables* was dubbed 'the weekly bloodbath' by critics, stunned by the violence on display. The show was a dramatisation of the efforts of real-life G-Man Eliot Ness, who headed a Treasury Department squad which aimed to crack down on mob activities and corruption in prohibition-era Chicago. Ness died in 1957, just as work was beginning on the TV production. He was played by Robert Stack (1919–), and Neville Brand (1921–) was Al Capone, in a show which was watched by one in three American households at its peak. Though the frequency and brutality of the violence shocked some, it appealed to many. *The Untouchables* posed moral questions, showing – in a stark, near-documentary style – that crime was a product of individual weakness. Staccato voiceovers by radio regular Walter Winchell, and musical scores by Nelson Riddle, underlined the need for constant vigilance. Nevertheless, the assassination of John F. Kennedy gave fresh impetus to the anti-violence campaigners, and the show was axed in 1963. It remains a popular repeat, aided in 1987 by ▷ Brian De Palma's stylish re-working of the story for cinema.

UPSTAIRS, DOWNSTAIRS
UK 1971–75

A period drama with a soapy storyline which gave London Weekend Television a worldwide hit, *Upstairs, Downstairs* was conceived beside a swimming pool in the South of France by resting actresses Jean Marsh (1935–) and Eileen Atkins (1934–). The parents of both had been in service, and the idea was to centre a drama on the lives of two maids, (Marsh played Rose in the series, the part of Sarah, though taken by Pauline Collins (1940–), was meant for Atkins). After refinements by script-editor Alfred Shaughnessy and producer John Hawkesworth, the drama was located in the household of Lord Bellamy (David Langton) just prior to the outbreak of World War I. Bellamy was a politician, so the scripts were able to address events like the war, the suffragette movement and the General Strike. Much of the drama, though, concerned the relationships between the Bellamy household and their servants. Gordon Jackson's (1923–90) butler, Hudson, was particularly memorable, as was the wheezy cook Mrs Bridges (Angela Baddeley (1904–76)). As the show developed it became more melodramatic – Lady Marjorie (Rachel Gurney) was drowned on the Titanic, allowing Bellamy to marry Virginia Hamilton (Hannah Gordon (1941–)). The younger generation, Captain James (Simon Williams (1946–)) and Georgina (Lesley-Anne Down (1954–)), then came more to the fore. With 300 million viewers in 50 countries, *Upstairs, Downstairs* was the most financially successful show of all time. It was particularly popular in the USA, where it was twice voted Best Series by the National Academy of TV Arts.

VALENTINO, Rudolph
Real Name Rodolpho Alfonzo Raffaelo Pierre Filibert Gugliemi di Valentina D'Antonguolla.
Born 6 May 1895, Castellaneta, Italy.
Died 23 August 1926, New York, USA.

The son of an army vet, Valentino briefly studied agriculture before abandoning his home country for New York where he made a living of sorts as a dancer and bisexual gigolo.

He began a stage career as a dancer and moved to Hollywood as a $5 a day extra in films like *My Official Wife* (1914) and *Alimony* (1918). His career made little progress and he soon found himself typecast as villains in the likes of *Out of Luck* (1919) and *Once To Every Woman* (1920).

Cast as a ne'er-do-well who makes good as a World War I hero in *The Four Horsemen of the Apocalypse* (1921), public response to his smouldering manner and erotic tango made him an overnight sensation. His newfound magnetism was confirmed by the healthy audiences for *The Conquering Power* (1921) and *The Sheik* (1921) in which his lusty desert chieftain sent females swooning. Soon a craze for all things Arabian swept the American nation.

A romantic star of the first order, he had good looks, a athletic bearing and a graceful gait to conceal a grasp of dramatics that seems to have stretched to flaring his nostrils as a warning of amorous intent and flashing his eyes to convey menace or bravado. Dominated in his private life and his career decisions by actress Natacha Rambova (1897–1966), he enjoyed a string of successes in ever more exotic and outré material, including the bullfighter in *Blood and Sand* (1922), the title character in *The Young Rajah* (1922), the foppish swashbuckler in *Monsieur Beaucaire* (1924) and the more manly womaniser in *Cobra* (1925).

He made his last appearances in *The Eagle* (1925) and *The Son of the Sheik* (1926) in which he at least displayed a sense of humour in sending up some of the more senseless elements in his previous work.

His early death from peritonitis provoked international mourning and rendered a man of unassuming talent and modest accomplishment a cinematic legend: the silent screen's greatest lover.

Over the years, he has been impersonated, always unsuccessfully, by a number of actors including Anthony Dexter (1919–) in *Valentino* (1951), and Rudolf Nureyev (1938–) in *Valentino* (1977).

Rudolph Valentino

VAN DYKE, Dick
Born 13 December 1925, West Plains, Missouri, USA.

After performing in school plays and amateur dramatics, Van Dyke became a radio announcer in the US Air Force during World War II. He later toured as part of the nightclub act The Merry Mutes as half of 'Eric and Van'.

Moving into television, he acted as master of ceremonies on such programmes as *The Morning Show* (1955), *The Cartoon Show* (1956) and *Flair* (1960). His Broadway debut in *The Boys Against the Girls* (1959) was followed by *Bye, Bye Birdie* (1960–61) a role which won him a Tony Award and which he repeated in his 1963 film debut.

A gangling, breezy fellow with a ready smile, strong sense of visual humour and a talent for clowning, singing and dancing, his television sit-com ▷ *The Dick Van Dyke Show* (1961–66) was one of the most popular in the history of the medium and won him ▷ Emmys in 1962, 1964 and 1965 for his genial demeanour and expert sense of timing.

His film career flourished with the spectacular success of *Mary Poppins* (1964) in which he made an unlikely Cockney chimneysweep and he was soon established in such family entertainment as *Fitzwilly* (1967), *Never A Dull Moment* (1967) and *Chitty, Chitty, Bang Bang* (1968).

His other film work of note includes *The Comic* (1969) in which he touchingly portrays the rise and fall of a beloved silent screen comedian, and *Cold Turkey* (1971), a caustic satire on American mores in which he plays the Reverend Clayton Brooks.

On television *The New Dick Van Dyke Show* (1971–72) failed to emulate the popularity of his earlier work and he was then absent for some time in an ultimately triumphant struggle with alcoholism.

He displayed his dramatic worth with a notable performance as a corporate executive facing his alcoholism in the television film *The Morning After* (1974) but was overtly earnest as a priest accused of murder in *The Runner Stumbles* (1979).

Happily slipping into semi-retirement, he has worked selectively on stage, in television variety specials and such comic television movies as *Dropout Father* (1982) and *Found Money* (1983) and the mini-series *Strong Medicine* (1986).

More recently, he has been seen as a one-off guest star in *The Golden Girls* (1990) and made a welcome return to cinema screens as D. A. Fletcher in *Dick Tracy* (1990).

In 1970, he published *Faith, Hope and Hilarity*.

VARDA, Agnès

Born 30 May 1928, Brussels, Belgium.

Educated in literature and psychology at the Sorbonne, Varda was a student of art history at the Ecole du Louvre when she began taking night classes in photography and pursuing a professional career in this subject.

She made her directorial debut with *La Pointe Courte* (1954) which is often cited as an early influence on the evolution of the ▷ nouvelle vague. Melding two stories, of a couple's attempt to save their marriage and of independent fishermen struggling to survive against major competition, it established her basic approach of seeing personal issues in a broader social context and examining the dichotomy between inner subjectivity and social objectivity.

She followed this with a number of documentaries and the feature *Cleo de Sept à Cinq* (Cleo From 5 to 7) (1961); scenes from the life of a nightclub singer as she awaits the results of test to determine whether she has terminal cancer. *Le Bonheur* (Happiness) (1965), chided for its amorality, tells of a young carpenter who readily compensates for the death of his wife by welcoming his mistress as a substitute life partner and mother for his children.

Throughout her career, she has maintained a commitment to the documentary form and explored a curiosity about international political situations in films like *Salut Les Cubains* (Salute to Cuba) (1963), *Loin du Vietnam* (Far From Vietnam) (1967) and *Black Panthers* (1968).

Resident in America during the late 1960s, she made *Lion's Love* (1969), a mosaic of reality and fiction exploring violence on a personal and wider level.

She co-wrote ▷ *Ultimo Tango A Parigi* (Last Tango in Paris) (1972) and began an involvement with the burgeoning women's movement that bore fruit most significantly in *L'Une Chante, L'Autre Pas* (One Sings, The Other Doesn't) (1977) which charts a friendship over two decades and illustrates how the women are affected and politicised by their diverse experiences of life.

More recently, she won the Golden Lion at the Venice Film Festival for *Sans Toit Ni Loi* (Vagabonde) (1985), the story of a young drifter's bleak and aimless existence which becomes an investigation of loneliness and a disillusioned generation.

Married to director Jacques Demy (1931–90), she had completed filming *Jacquot de Nantes* (1991) at the time of his death. An autobiographical piece, it tells of a young boy's desire to become a filmmaker.

VERHOEVEN, Paul

Born 1938, Amsterdam, Holland.

Vivid childhood memories of the devastating wartime bombing of The Hague have undoubtedly scarred Verhoeven, whose films generally display a disturbing vision of a world rent by brutality, violence and man's basest desires.

A student at the University of Leyden, he received a PhD in maths and physics and worked as a teacher before serving in the Dutch Navy as a documentary filmmaker. He directed such short films as *Een Hagedis Teveel* (A Lizard Too Much) (1960), *Feest* (Let's Have A Party) (1963) and *Hets Korps Mariniers* (The Dutch Marine Corps) (1966) before moving into television with *Mussert* (1968), a controversial portrait of the Dutch SS leader, and the series *Floris* (1969) which he has described as a 'Dutch Ivanhoe' and which began his frequent collaborations with actor Rutger Hauer (1944–).

He has dismissed his first feature film, *Wat Zien Ik* (Business Is Business) (1971) as 'a parochial comedy about prostitution in Amsterdam', but his second film *Turks Fruit* (Turkish Delight) (1973) received an Oscar nomination as Best Foreign Film. Based on a bestselling novel, its emphatic sexual content won an international audience for this story of a marriage torn apart by class differences.

His interest in class issues continued with the period Cinderella-story *Keetje Tippel* (Cathy Tippel) (1975) and his interest in the war resurfaced with *Soldaat Van Oranje* (Soldier of Orange) (1977) which graphically captures the experiences of six University students as they face the Nazi onslaught.

The virtuosity of his camerawork and unflinching depiction of graphic acts of sex and violence found a comfortable narrative home in *Spetters* (1980), a flashy tale of teenage alienation and he brought a delirious visual sensibility and baroque touch to the homoerotic thriller *De Vierde Man* (The Fourth Man) (1983).

Attracted to working within the Hollywood system, he toiled on a variety of unrealised projects from a seafaring epic to adaptations of Agatha Christie and H. P. Lovecraft before making his American debut with *Flesh & Blood* (1985), an unremittingly brutal 16th-century adventure.

However, he has enjoyed major box-office successes with the big-budget science-fiction stories *Robocop* (1987) and *Total Recall* (1990) which have allowed him to add layers of psychological depth to spectacular action-adventures whilst retaining his penchant for bone-crushing, gratuitous violence and a darkly satirical view of the future.

At the time of writing he has vowed to extend the boundaries of screen sexuality with the erotic thriller *Basic Instinct* (1991) which will star ▷ Michael Douglas.

VERTIGO

Alfred Hitchcock (USA 1958)
James Stewart, Kim Novak, Barbara Bel Geddes, Tom Helmore.
127 mins. col.

Arguably ▷ Alfred Hitchcock's supreme masterpiece, where his technical gifts most completely mesh with the richness of his material, *Vertigo* is also one of the bleakest, most perverse offerings to come out of the mainstream American cinema of the Fifties. ▷ James Stewart, whose presence assures us of the normality of the subsequently disturbing goings-on, plays retired detective Scottie Ferguson, hired by a friend to trail his apparently unstable wife Madeleine Elster (an extraordinary Kim Novak, (1933–)) with whom he falls passionately in love; he is deeply distraught when his vertigo renders him unable to prevent her falling to her death from a church steeple. Some time later he meets a shopgirl, Judy Barton, who bears an uncanny resemblance to the previous suicide victim, and, typically flaunting all the rules of suspense, Hitchcock uncovers the central twist to the audience halfway though the film – Madeleine and Judy are the same woman, the earlier 'fatality' a purposely staged spectacle cooked up by her husband and mistress to cover for the murder of the real wife. The revelation however, only deepens the film's emotional and psychological impact, with perhaps the key scene the one in which Scottie fervently remodels (clothes, hairstyle, the lot) the new Judy in the image of his former love. The director has described Stewart's role here as a man who 'wants to go to bed with a woman who's dead, he's indulging in a form of necrophilia', but it is also that the filmmaker is turning the everyday insecurities of human relationships into a cynical game of manipulation. Given the master of suspense's own voyeuristic proclivities and distinctive sexual fantasies, it is highly indicative of the peculiarly personal obsessions with which he loaded the film that he has his protagonist transform the female lead into the paradigm Hitchcockian glacial blonde, a purely visual object of gratification.

VICTIM

Basil Dearden (UK 1961)
Dirk Bogarde, Sylvia Sims, Dennis Price, John Barrie.
100 mins. col.

Basil Dearden (1911–71) is widely regarded as a classic journeyman director in British cinema. His long professional relationship with producer Michael Relph (1915–) and the Ealing Studios is reminiscent of the old Hollywood studio contract system, and the films which he turned out are generally seen as skillful but rather safe. There were a number of notable exceptions to that rule, however, of which *Victim* is not only the best, but also the most controversial. As in the earlier *Sapphire* (1959), which used the murder of a prostitute as a vehicle for a study of racial prejudice, or the subsequent *Life For Ruth* (1962), which dealt with religious prejudice, *Victim* uses a tense thriller framework to investigate the previously taboo subject of homosexuality. If homosexuals figured at all in British film, it was almost certainly as comic parodies derived from the music hall or Nöel Coward (1899–1973), but *Victim* took a very different line. ▷ Dirk Bogarde played a successful and socially admired lawyer who finds events from his past returning to haunt him when he becomes involved in a case with homosexual implications, and attempts to stand up to a gang of blackmailers who have murdered his former lover. It was one of the earliest films anywhere to deal seriously with homosexual subject matter, and does so in a thoughtful and entirely non-sensationalist fashion, while simultaneously succeeding in being a tense and highly watchable drama. Bogarde is particularly good in the central role. Films like *Victim* and *Sapphire*, while clearly in the Ealing tradition of well-intentioned, socially-conscious drama, reflect great credit on the daring and conviction of an allegedly cautious and predictable director, and for that reason tend to stand out amid the solid achievement of his work.

VIDOR, King Wallis

Born 8 February 1894, Galveston, Texas, USA.
Died 1 November 1982, Paso Robles, California, USA.

A cinema projectionist and freelance newsreel cameraman, Vidor made his debut as the director of the documentary *Hurricane in Galveston* (1913).

In Hollywood from 1915, he worked as a writer and ▷ extra before directing a series of films on juvenile crime, and a feature, *The Turn of the Road* (1919).

A successful mounting of *Peg O' My Heart* (1922) brought him a long-term contract with M-G-M where his interest in social issues and the everyday struggles of the average American were manifested in his 'wheat, steel and war' trilogy. *The Big Parade* (1925) focused on a young soldier's emotional journey from patriotic exuberance to cowed and philosophical veteran, *The Crowd* (1928) (Best Director Oscar nomination) charts the trials and tribulations of a face in the crowd, his marriage and struggle to survive the economic hardships of the Depression years, whilst *Our Daily Bread* (1934) offered a celebration of the virtues of a life on the land.

His work ranged across almost all of the most popular genres and includes the witty Hollywood comedy *Show People* (1928), the all-black musical *Hallelujah* (1929) (Oscar nomination), the western *Billy the Kid* (1930) and the tearjerker *The Champ* (1931) (Oscar nomination).

An adept and conscientious craftsman, with a skill in handling actors, dealing in unashamed emotions and conveying a strong humanist message, his later work includes *Stella Dallas* (1937), *The Citadel* (1938) (Oscar nomination) and the large-scale adventure *Northwest Passage* (1940). He subsequently concentrated on savagely overripe melodramas of sexual greed and jealousy told in florid style and lurid colours; notable titles include *Duel in the Sun* (1946),

Beyond the Forest (1949) and *Ruby Gentry* (1952).

His last feature films were a dull version of *War and Peace* (1956) (Oscar nomination) and the average Biblical epic *Solomon and Sheba* (1959), although he did direct two short documentaries in the 1960s.

His autobiography *A Tree Is A Tree* was published in 1953. Towards the end of his life, he received an honorary ▷ Oscar in April 1979 for his 'incomparable achievements as a cinematic creator and innovator' and acted in the film *Love and Money* (1982).

VIEWFINDER An optical or video device forming part of a camera or an accessory to it, showing an image of the scene being recorded, with indication of the exact limits of the field of view involved. The scene is viewed through the camera's lens by means of a mirrored shutter, or a rotating glass prism or membrane called a pellicle.

VIGO, Jean

Born 25th April 1905, Paris, France.
Died 5th October 1934, Paris, France.

Vigo's entire legacy may only amount to a few titles and several hours of film, but his position as one of the most inventive and influential of celluloid artists is undisputed.

His father, the anarchist Miguel Almereyda (a pseudonym for Eugene Bonaventure de Vigo), was apparently murdered in his cell at the Fresnes prison during the summer of 1917. Left in the care of his grandparents, Vigo attended school in Millau, studied in Chartres from 1922 to 1924 and followed courses in ethics, sociology and psychology at the Sorbonne.

Perennially in poor health, he was stricken with tuberculosis and settled with his wife in Nice where he purchased a camera and began working as an assistant at Franco-Film.

His first film, *A Propos de Nice* (1929), was influenced by Dziga Vertov's (1896–1954) *Man With A Movie Camera* (1929) and was photographed by the latter's brother Boris Kaufman (1906–80). A restless exploration of the contrast between the idle rich who visit Nice and the vitality of the working-class residents, the documentary displays Vigo's love of experimentation and ▷ surrealism as he juxtaposes apparently random images to heighten his social critique.

His second film, *Taris* (1931), was an eleven-minute celebration of the champion French swimmer, offering some playful underwater scenes and further exploration of slow-motion and reverse shots.

Zéro de Conduite (1932) reflects his own years at provincial boarding schools as Vigo conveys the anarchic spirit of youth rebelling against the drudgery and regimentation inflicted by their authoritarian bourgeois teachers. Banned in France until 1945, the film inspired directors of the ▷ nouvelle vague and was the model for Lindsay Anderson's (1923–) *If* (1967).

His final film, ▷ *L'Atalante* (1934), follows the relationship of a Seine barge skipper and his new country bride as they painfully accommodate the changes needed to keep their love alive. Creating a palpable mood of tenderness, the film tested the boundaries of film technique and successfully utilises fast-editing, ▷ Surrealism and some arresting pyrotechnics to create a poetic paean to romance.

Vigo died of septicaemia shortly after the film's poorly received release. However, its reputation has grown and a major restoration and re-release in 1990 served to underline the lyricism, innovation and exuberance of a work now regarded as a masterpiece.

VISCONTI, Luchino

Real Name Count Don Luchino Visconti Di Morone
Born 2 November 1906, Milan, Italy.
Died 17 March 1976, Rome, Italy.

An aristocrat by birth and a Marxist by conviction, these twin facets of Visconti's life are evident throughout the style and substance of his lengthy film career where a sumptuous, overstated surface dazzle distracts from epic sagas of the decline of the aristocracy and the betrayal of the peasantry.

A member of a cavalry regiment, he began to act and design for the theatre before moving to France and working as an assistant to ▷ Jean Renoir on such films as *Las Bas-Fonds* (1936) and *La Tosca* (1940). He made his directorial debut with *Ossessione* (1942), an adaptation of James M. Cain's tale of grand passion and murder *The Postman Always Rings Twice* that he relocated to the Po Delta and imbued with a naturalistic feel for character and location that made it a precursor to the ▷ neo-realist movement.

A similar realistic approach was evident in *La Terra Trema* (1948), which details the struggles of Sicilian fishermen against poverty and the devious machinations of the bourgeois classes. His subsequent work includes the comedy *Bellissima* (1951) starring ▷ Anna Magnani, and *Senso* (1954) which views the tangled emotional ties of a 19th-century Italian noblewoman and an Austrian officer as symbolic of the way the peasant classes betrayed revolutionary ideals. A characteristic work, set during the Risorgimento, it offers stunningly seductive photography, set design and period detail.

In many respects, his most melodramatic films embody the qualities of screen opera with their often overbearing sense of production values and explosive, larger-than-life emotions. These characteristics are certainly evident in his best work from the 1960s; *Rocco E I Suoi Fratelli* (Rocco and His Brothers) (1960) a sometimes risibly over-emphatic account of the hardships faced by a family who escape from the poverty of the South to the hope of Milan; the majestic ▷ *Il Gattopardo* (The Leopard) (1963), an elegiac reflection on the passing of the aristocratic classes with a performance of grace and mellowness from ▷ Burt Lancaster, and *La Caduta Degli Dei* (The Damned) (1969) a stylish but hollow wallow in the high camp extremes of decadent behaviour that flourish as the Nazis rise to power in Germany.

There was no less of an operatic sensibility to *Morte A Venezia* (Death in Venice) (1971), an elegant transposition of Thomas Mann's novel with ▷ Dirk Bogarde as a Mahler-like composer whose dying days are spent in obsessive longing for a young boy who seems to embody the beauty, grace and harmony he has been seeking all his life.

This was a highpoint in his career, and he made only three further features: *Ludwig* (1972), the more restrained *Gruppo Di Famiglia In Un Interno* (Conversation Piece) (1974) with Lancaster as an ageing professor coaxed out of his seclusion and *L'Innocente* (The Innocent) (1976), a doom-laden tale of sexual double standards.

VON STERNBERG, Josef

Real Name Jonas Sternberg
Born 29 May 1894, Vienna, Austria.
Died 22 December 1969, Hollywood, California, USA.

Educated in both Austria and the USA, Sternberg made training films for the U.S. Army Signal Corps during World War I and later worked as a scenarist and assistant director on such films as *The Mystery of the Yellow Room* (1919) and *Vanity's Price* (1919).

He made his directorial debut with the low-budget *The Salvation Hunters* (1925) and then created *Woman of the Sea* (The Seagull) (1926) for ▷ Charles Chaplin's company. He then made a name for himself on atmospheric, smoke-filled and beautifully lit tales of decadence, love and gangsterism like *Underworld* (1927) and *The Docks of New York* (1928).

In 1930, at the request of Emil Jannings (1882–1950), he went to Germany to direct ▷ *The Blue Angel* which established the stardom of ▷ Marlene Dietrich as the provocative singer Lola-Lola. In Hollywood he made six further features with Dietrich that vie with each other in their extravagant embellishment of her screen image as the exotic, alluring femme fatale. A poetic artist in light and shadow, his boundless visual imagination was given rull rein in films like *Shanghai Express* (1932), *The Scarlet Empress* (1934) and *The Devil Is A Woman* (1935).

An autocratic figure who fulfilled the riding crop and jodhpurs caricature of the director, his abortive attempt to film *I Claudius* (1937) appears to have damaged his standing and, after the cynical melodrama *Shanghai Gesture* (1941), he worked infrequently. *Jet Pilot* (1950) and *Macao* (1952), both made for ▷ Howard Hughes, were taken out of his control and either re-edited or re-shot by other hands and his final film *Anatahan* (1953) was made in Japan.

He spent his last years teaching and published a suitably colourful autobiography entitled *Fun In A Chinese Laundry* (1965).

VON STROHEIM, Erich
Real Name Erich Oswald Stroheim
Born 22 September 1885, Vienna, Austria.
Died 12 May 1957, Paris, France.

Said to have served in the Austro-Hungarian Army, Von Stroheim emigrated to America and worked in a number of modest jobs before entering the film industry as an actor in *Captain McLean* (1914) and military advisor on *Old Heidelberg* (1915). An assistant director on such ▷ D. W. Griffith productions as ▷ *Birth Of A Nation* (1915), ▷ *Intolerance* (1916) and *Hearts of the World* (1918), he made his solo directorial debut with *Blind Husbands* (1919).

Films like *The Devil's Passkey* (1920), *Foolish Wives* (1921) and *Merry Go-Round* (1922) contributed to his reputation as a profligate chronicler of sexual mores, who explored a series of triangular relationship with a polish and sophistication that were rare for the time and spiralled into lavish perfectionist scrutinies of the lifestyles of the rich and famous.

Continually at odds with producers over his extravagance in expenditure and the increasingly expansive running times of his work, his five hour treatise on the avariciousness of man, *Greed* (1923), was badly cut by other hands and signalled the beginning of his fall from grace even if his acting career continued to flourish as the monocle wielding seducer and 'Man You Love to Hate'.

He enjoyed box-office successes with *The Merry Widow* (1925) and *The Wedding March* (1927) but when *Queen Kelly* (1928) veered alarmingly over budget, the unfinished film virtually bankrupted its star ▷ Gloria Swanson and made Von Stroheim *persona non grata* to the Hollywood financiers.

In the 1930s, he moved to France and although he never directed again his bullet-headed features and ramrod bearing made distinctive acting appearances in ▷ *La Grande Illusion* (1937), as Rommel on *Five Graves to Cairo* (1943) and, in a poignant piece of casting as the supportive butler to deranged silent screen legend Norma Desmond (Swanson) in ▷ *Sunset Boulevard* (1950).

He made his final acting appearances in *Serie Noire* (1955) and *La Madone Des Sleepings* (1955).

VON SYDOW, Max Carl Adolf
Born 10 April 1929, Lund, Sweden.

A student at the Royal Academy in Stockholm, Von Sydow made his film debut in *Bara En Mor* (Only A Mother) (1949).

A member of various theatrical companies, he began a long association with director ▷ Ingmar Bergman at the Municipal Theatre of Malmö. On film their many collaborations include ▷ *Det Sjunde Inseglet* (The Seventh Seal) (1957), *Sasom I En Spegel* (Through A Glass Darkly) (1961) and *Skammen* (The Shame) (1968) all of which make effective use of his long, lean frame and gaunt features to reflect the troubled emotional inner lives of his guilt-ridden characters.

He made his American film debut as Jesus Christ in *The Greatest Story Ever Told* (1965) and brought gruff conviction to starring roles as a fanatical missionary in *Hawaii* (1966) and a Swedish peasant venturing forth to America in *Utvandrarna* (The Emigrants) (1971) and its sequel *Nybyggarna* (The New Land) (1972).

A character actor of international standing over the past twenty-five years, his range of work includes the priest in ▷ *The Exorcist* (1973), the amiably professional assassin in *Three Days of the Condor* (1975), and the Chief Justice in *Cadaveri Eccellenti* (Illustrious Corpses) (1976).

In more recent years, he has shown a willingness to tackle any kind of colourful supporting assignment and display lighter shadings of his sometimes gloomy screen career and his ever more varied list of credits include the evil Emperor Ming in *Flash Gordon* (1980), a sympathetic Nazi commander in *Escape to Victory* (1981), a sporting participation in the crazy Canadian comedy *Strange Brew* (1983) and a gentle interpretation of the older lover in *Hannah and Her Sisters* (1986).

He has also continued to appear on stage, making his Broadway debut in *The Night of the Tribades* (1977), appearing in *Duet for One* (1981) and an Old Vic production of *The Tempest* (1988).

More recently, he won considerable acclaim for his performance as the widowed farmer enduring penury and hardship at the turn of the century in *Pelle Erobreren* (Pelle, The Conqueror)

(1988) and made his directorial debut with the lyrical period romance *Katinka* (1988). His recent work includes *Awakenings* and *Till The End of the World* (1991).

VON TROTTA, Margarethe

Born 21 February 1942, Berlin, Germany.

Educated in Romance languages and literature, Von Trotta studied acting in Munich before making a number of critically applauded stage and television appearances, including *Baal* (1969) her first collaboration with future husband Volker Schlöndorff (1939–).

Her various film roles include *Schräge Vögel* (1968), *Gotter Der Pest* (Gods of the Plague) (1969) and *Der Plötzliche Reichtum der Armen Leute von Kombach* (The Sudden Fortune of the Poor People of Kombach) (1971) which she also co-wrote.

Continuing to work as a scriptwriter on most of Schlondorff's projects, she wrote and co-directed *Die Verlorene Ehre der Katherina Blum* (The Lost Honour of Katharina Blum) (1975), a potent critique of police practices and the excesses of the gutter press as a young woman faces public disgrace after unwittingly spending a night with a terrorist suspect.

She made her solo directorial debut with *Das Zweite Erwachen der Christa Klages* (The Second Awakening of Christa Klages) (1977) in which a woman's desperate act of robbery to save her beleagured nursery has catastrophic repercussions. The film signalled her talent for addressing wider contemporary issues through the acts of committed individuals and her emphasis on female characters and their position in the social order.

Schwestern Oder Die Balance des Glücks (The German Sisters) (1981) provided acutely observed and psychologically detailed analyses of the bonds between women and the sexual and ideological politics at play in sibling relationships.

Her most ambitious film to date has been *Rosa Luxemberg* (1986), a somewhat heavy handed biography of the revolutionary, but she continued to mine a rich source of material in challenging dramas that insightfully explore the destructive and supportive sides of women's relationships. Her recent notable work includes *Heller Wahn* (Friends and Husbands) (1982) and *L'Africana* (The Return) (1990).

WAJDA, Andrzej

Born 6 May 1926, Suwalki, Poland.

The son of a cavalry officer, Wajda joined the Polish Home Army during World War II but spent most of these years as a cooper's apprentice and joiner.

In 1946, he began to study painting at the Academy of Fine Arts in Krakow and later transferred to The School of Theatre and Cinematography in Lodz. He subsequently spent a brief time as an assistant before making his directorial debut with *Pokolenie* (A Generation) (1954), the first of a starkly impressive trilogy, completed with *Kanal* (1956) and *Popiol I Diament* (Ashes and Diamonds) (1958), that dramatically conveyed the legacy of pain, confusion and fear experienced by his young countrymen during the wartime struggles. The themes of war's aftermath, the hollowness of the traditional idea of military heroism, and the predicament of individuals caught up in political events were ones to which he would return in *Lotna* (1959) and *Krajobraz Po Bitwie* (Landscape After Battle) (1970).

Throughout the 1960s he displayed a range of interests, focusing on the disaffection of modern youth in *Niewinni Czarodzieje* (The Innocent Sorcerers) (1960), creating a moving homage to star Zbigniew Cybulski (1927–67) in *Wszystko Na Sprzedaz* (Everything for Sale) (1967) and later gaining a reputation for his lyrical adaptations of 19th-century literature like *Brzezina* (The Birchwood) (1970) and *Panny Z Wilka* (The Young Ladies of Wilko) (1979).

However, it is his status as the uncompromisingly critical cinematic heart and soul of Poland that brought forth some of his best work including *Czlowiek Z Marmur* (Man of Marble) (1972) which examined the fall from grace of a revered bricklayer from the 1950s, and the stirring *Czlowiek Z Zelaza* (Man of Iron) (1981) which vividly captures the feel of history in the making and follows a radio reporter covering the Gdansk shipyard strike and the rise of Solidarity.

A noted theatre director whose many productions include *The Possessed* (1972) in London, *Hamlet* (1973) in Poland, and *The White Marriage* (1976) at Yale University, he later moved to France and cannily paralleled the events in his own country with *Danton* (1982), which captured the bloody intrigue of the French Revolution with the sweep of an epic and the intimate human moments of a finely acted chamber work.

His most recent film work includes several stories of World War II, notably *Eine Liebe In Deutschland* (A Love in Germany) (1983) and *Korczak* (1990) which sombrely saluted the heroism of the defiant, self-sacrificing Jewish doctor who befriended and defended the children of the Warsaw ghettoes.

WALSH, Raoul

Born 11 March 1887, New York, USA.
Died 31 December 1980, Hollywood, California, USA.

Walsh's early life is as colourful as many of the flamboyant characters who would appear in his adventure films. He worked on a trading ship to Cuba, was employed as a wrangler in Mexico and drifted across the West as everything from an undertaker to a surgeon's assistant. In 1910, he appeared on stage in *The Clansman* and began to play cowboy roles in silent films like *The Banker's Daughter*. In 1912, he became an assistant to ▷ D. W. Griffith and moved with his company to Hollywood.

Griffith dispatched him to Mexico to capture Pancho Villa on film and the footage was incorporated into his first feature as a director *The Life of General Villa* (1914). Also for Griffith, he was an assistant director on ▷ *Birth of A Nation* (1915), in which he played John Wilkes Booth.

Developing his career as a director, he enjoyed popular successes with the Arabian Nights fantasy *The Thief of Bagdad* (1924), the army comedy *What Price Glory?* (1926) and *Sadie Thompson* (1928).

A pioneer in the western, he made *In Old Arizona* (1929), one of the first films to record good quality sound on location, and the epic *The Big Trail* (1930) which utilised the new 70mm Grandeur process and introduced ▷ John Wayne to leading roles. During the making of the former, he lost an eye when a jackrabbit smashed through the windshield of his car.

A largely unremarkable spell of work ended in 1939 when he became a contract director at Warner Brothers. There, his career flourished as he mass-produced muscular, fast-paced examples of all the popular genres where his sense of pace, unpretentious narrative skills and vigorous use of action served the careers of such studio stalwarts as ▷ James Cagney, ▷ Errol Flynn and ▷ Humphrey Bogart. His most memorable films from this prolific decade include *The Roaring Twenties* (1939), *High Sierra* (1941), *They Died With Their Boots On* (1941) and ▷ *White Heat* (1949).

He showed a softer side in the period romance *The Strawberry Blonde* (1941) but the last years of his career were taken up with more formulaic westerns and war movies, including *The Naked and The Dead* (1958). He retired after *A Distant Trumpet* (1964).

A master storyteller whose varied career spanned the history of Hollywood, Walsh's autobiography *Each Man in His Time* was published in 1974.

WARHOL, Andy

Born 6 August 1928, Pittsburgh, Pennsylvania, USA.
Died 22 February 1987, New York, USA.

A student at the Carnegie Institute of Technology and pioneer of 'pop art', with his colourful reproductions of familiar everyday objects, such as the famous Campbell soup can label, and magazine illustrations directly reproduced by silk screen, Warhol began making experimental and ▷ avant-garde films in 1963.

A resolute minimalist, who refused to imbue camera with the viewpoint of any authorial hand, his early (frequently unwatchable) work often consisted of pointing a camera at a given object and watching the results over a lengthy period of time. Thus *Sleep* (1963) in which an unconscious, naked man is observed over a period of six hours and *Empire* (1964) which consists of eight hours of footage watching the Empire State Building.

Still eschewing the use of the most standard cinematic devices like editing or continuity, he eventually added dialogue and loosely-structured narratives of a sort to produce marginally more cohesive and accessible work that often passively utilised the camera to observe sexual and personal anecdotes from the sub-cultural activities of the individuals who formed part of his 'Factory'. *Mario Banana* (1964) features transvestite Mario Montez eating a banana, *Harlot* (1964), his first sound film, contains much the same with the thrilling addition of a woman in an evening dress to keep him company. Among the many titles that teasingly explored, both explicitly and implicitly, the sexual behaviour and fantasies of the coterie were *Blow Job* (1963) and *My Hustler* (1965) and *Bike Boy* (1967).

Underground notoriety eventually mushroomed into arthouse success with the like of the three hour *Chelsea Girls* (1966) and *Lonesome Cowboys* (1968) in which a variety of suitably attired gay men are posed around a minimal storyline.

Following the 1967 attempt on his life, much of the direction on his subsequent films was entrusted to Paul Morrissey (1939–) with Warhol merely serving as the producer. Notable titles, often starring the muscular and permanently naked torso of Joe Dallesandro (1948–), include *Flesh* (1968), *Trash* (1970) and *Heat* (1972) a modern-day variation on ▷ *Sunset Boulevard.*

Warhol himself directed *Women in Revolt* (Sex) (1972) in which a selection of transvestites depict the shattered dreams of a group of women. Thereafter, he seemed to lend merely his name to such films as *Flesh for Frankenstein* (1973), *Blood for Dracula* (1974) and *Bad* (1976).

Late in life he made an acting appearance in *The Love Boat* television series and he can be seen in the film *Tootsie* (1982).

WARNER, Jack Leonard

Real Name Jack Leonard Eichelbaum
Born 2 August 1892, London, Ontario, Canada.
Died 9 September 1978, Los Angeles, California, USA.

The son of a cobbler and part of a large and impoverished immigrant family, Warner was a boy soprano and inveterate performer who embarked upon a showbusiness career and later became a partner with his brothers Harry (1881–1958), Albert (1884–1967) and Samuel (1887–1927) in the exhibition and distribution of motion pictures in Pennsylvania, opening the Cascade Theatre in Newcastle during 1903.

Building a nationwide company, they moved into production with *Peril of the Plains* (1910) and *Raiders on the Mexican Border* (1910). Later, encouraged by the vast commercial success of *My Four Years in Germany* (1918) they built a studio of their own in Los Angeles which opened for business in 1919.

Warner Brothers came into official existence in 1923 and they enjoyed early hits with *The Gold Diggers* (1923) and *Where The North Begins* (1923), starring canine wonder Rin-Tin-Tin. Pioneering work on the synchronisation of sound and picture led to *Don Juan* (1926) and ▷ *The Jazz Singer* (1927) and the sensational popularity of the latter made them a major studio.

Specialising in gangster yarns, Depression-era musicals, respectful historical biographies and torn-from-the-headlines social dramas, the studio was also home to such stars as ▷ Humphrey Bogart, ▷ James Cagney, ▷ Bette Davis and ▷ Errol Flynn all of whom, in one way or another, rebelled against the working conditions and contracts that Warner attempted to impose causing Bogart to remark 'This studio has more suspensions that the Golden Gate Bridge'.

A tough head of production, he oversaw such films as *Yankee Doodle Dandy* (1942), *Auntie Mame* (1958), *My Fair Lady* (1964) and *Camelot* (1967) and continued to work in the industry as an independent after his retiral from the company in 1969. The last productions to bear his name were the western *Dirty Little Billy* (1972) and the musical *1776* (1972).

His autobiography, *My First Hundred Years in Hollywood* was published in 1965.

WASHINGTON, Denzel

Born 28 December 1954, Mount Vernon, New York, USA.

The son of a security guard and pentecostal preacher, Washington began studying Medicine at New York's Fordham University before changing his major to Drama and Journalism. After winning a scholarship to the American

Conservatory Theater in San Francisco, he returned to New York and worked with the Shakespeare in the Park ensemble.

He appeared in a number of off-Broadway productions and television movies like *Wilma* (1977) and *Flesh & Blood* (1979) before his feature film debut in the comedy *Carbon Copy* (1981).

His performance in the Negro Ensemble Company's off-Broadway production of *A Soldier's Play* (1981) earned him an Obie Award and brought him a starring role in the television medical drama *St. Elsewhere* (1982–88).

He has continued to work in all media: portraying *Malcolm X* (1925–65) in the play *When the Chickens Come Home to Roost* (1983), starring in the television movie *The George McKenna Story* (1986) and proving a dependable and authoritative supporting actor in films like *A Soldier's Story* (1984) and *Power* (1985).

He managed to escape the pitfalls that result from television stardom in a long-running series, receiving a Best Supporting Actor Oscar nomination for his credible, authentic-seeming depiction of Steve Biko (1946–77) in *Cry Freedom* (1987). He displayed a facility with accents and a propensity to appear in projects that explore racial issues and various aspects of the black experience, including *For Queen and Country* (1988) and *Glory* (1989) in which his stubbornly proud and defiant slave-turned-soldier earned him a Best Supporting Actor Oscar.

Also adept at expressing intelligence, integrity and charm, he has underlined his status as a charismatic romantic leading man in less worthy and more populist fare like *The Mighty Quinn* (1989), *Heart Condition* (1990) and *Mo' Better Blues* (1990).

He made his Broadway debut in *Checkmates* (1988).

WAYNE, John

Real Name Marion Michael Morrison
Born 26 May 1907, Winterset, Iowa, USA.
Died 11 June 1979, Los Angeles, California.

The son of a druggist, Wayne was raised in California and attended University on a football scholarship. A prop man at Twentieth-Century Fox, he made early unbilled appearances in a number of films including *Brown of Harvard* (1926) and *Mother Machree* (1928) on which he first encountered director ▷ John Ford, later responsible for many of his greatest successes.

After a number of bit parts, he was entrusted with the lead in the western *The Big Trail* (1930), a technically innovative early talkie shown in wide-screen 70mm. Stardom did not ensue and he found himself consigned to Saturday matinée westerns and serials, including a spell as the musically inclined cowboy Singin' Sandy and as part of The Three Mesquiteers.

It was Ford who rescued him from obscurity and cast him as the Ringo Kid in ▷ *Stagecoach* (1939) and he began to take centre stage as a reliable man of action in such larger budget fare as *Reap The Wild Wind* (1942). *The Spoilers* (1942) and *They Were Expendable* (1945).

Working with Ford again, he starred in the cavalry trilogy of *Fort Apache* (1948), *She Wore A Yellow Ribbon* (1949) (one of his best roles as ageing cavalry officer Nathan Brittles) and *Rio Grande* (1950) and began to win appreciative reviews for his acting as the single-minded pioneer cattle man Thomas Dunson in *Red River* (1948), a fearless marine sergeant in *Sands of Iwo Jima* (1949) (Best Actor Oscar nomination) and as the brawling Sean Thornton home from America in the Irish romantic whimsy *The Quiet Man* (1952).

Cherished for the decency and integrity he brought to his portrayals of the cowboy, his notable exercises within the genre include ▷ *The Searchers* (1956), ▷ *Rio Bravo* (1959) and *The Man Who Shot Liberty Valance* (1962).

He turned to direction with the long-winded but suitably spectacular *The Alamo* (1960) and his much-derided Vietnam epic *The Green Berets* (1968) which reflected his specific support of America's involvement in the war and the extreme right-wing politics that he happily espoused throughout his life.

He overcame a first brush with cancer to return to the screen in the lively westerns *The Sons of Katie Elder* (1965) and *El Dorado* (1967) and amiably spoofed his own invicinbility as fat, one-eyed law enforcer Rooster Cogburn in *True Grit* (1969) (Oscar).

He prolonged his action career with a

series of formulaic westerns and starred in a number of urban cop thrillers like *McQ* (1974) and *Brannigan* (1975) which were westerns in all but name. He gave a poignant and underrated final performance in *The Shootist* (1976) as a legendary gunfighter dying of cancer who cannot escape the violence of his past.

His distinctive swaggering gait, drawling speech and two-fisted Galahadish character had made him the quintessential man of the West and the record books reveal him to have been one of the most popular film stars in movie history, making a record-breaking twenty-five appearances in the list of American top-ten box-office draws and topping that list in 1950, 1951, 1954, 1969 and 1971.

WEIR, Peter Lindsay

Born 21 August 1944, Sydney, Australia.

The son of a real estate broker, Weir studied to be a criminal lawyer at Sydney University but left to work in his father's firm and travel extensively in Europe. On his return, he became a stagehand at the local television channel ATN-7 and used their facilities to make his first short comedy films *Count Vim's Last Exercise* (1967) and *The Life and Times of The Rev. Buck Shotte* (1968).

His imaginative flair, thoughtful camerawork and sense of the macabre were soon evident in *Michael* (1970) and *Homesdale* (1971) which won him consecutive Australian Film Institute Grand Prix.

After such documentaries as *Incredible Floridas* (1972) and *What Ever Happened to Green Valley?* (1973) he made his full-length feature debut as the director of *The Cars That Ate Paris* (1974), a bizarre black comedy in which the residents of a small town lure motorists into deliberate accidents so they can cannibalise the wreckage of their jalopies and keep the local economy bouyant.

He followed this with a carefully mounted tale of enigma and mystery in the languid period piece ▷ *Picnic at Hanging Rock* (1975) which concerns the true story of a group of schoolgirls who disappeared during an outing to a sacred aboriginal site. He was established at the forefront of the newly emerging Australian film industry renaissance and kept his position with *The Last Wave* (1977), an exploration of the tensions between modern and ancient cultures told through the story of a white lawyer defending a group of aborigines charged with murder.

Increasing the scope of his projects, he brought an attractive visual sensibility to bear on the simple anti-war statement contained within the Boy's Own adventures of two young soldiers in *Gallipoli* (1980) and to the political romance *The Year of Living Dangerously* (1982).

Subsequently pursuing an American career, he brought an old-fashioned craftsmanship to bear on the culture clash thriller *Witness* (1985) but uninspired handling and a miscast central star gave him a rare failure with *The Mosquito Coast* (1986).

However, he returned to box-office favour with the life-enhancing celebration of non-conformity in *Dead Poets Society* (1989) and the amiable comedy of a Frenchman's American marriage of convenience to gain his coveted *Green Card* (1990).

WELLES, George Orson

Born 6 May 1915, Kenosha, Wisconsin, USA.
Died 10 October 1985, Hollywood, California, USA.

A child prodigy who performed with the Gate Theatre in Dublin, stunned Broadway with his innovative theatre productions and shocked American radio listeners with his vivid dramatisation of *The War of the Worlds*, Welles made no less of an impact on the film world when he was brought to Hollywood by RKO in 1939.

He had already dabbled in film with the 16mm short *The Hearts of Age* (1934) and the now destroyed *Too Much Johnson* (1938) and toyed with a number of ideas, including a version of *Heart of Darkness* and his own script *The Smiler With The Knife*, before making his dazzling debut with ▷ *Citizen Kane* (1941). Few beginnings in the history of the cinema have been more auspicious, and this showed a boy-wonder brimming with the confidence and restless curiosity of an alert mind set loose on a spellbinding new toy. Stylistically daring, technically innovative and a compelling story, it won him an unprecedented four Oscar nominations for Best Director, Best Actor, co-writer of the script and producer of the year's Best Film.

He followed this with the no less impressive and electrifying ▷ *The Magnificent Ambersons* (1942) (Best Film Oscar nomination), a dazzling account of a wealthy American family's gradual decline. However, the troubled shooting of his unfinished Mexican documentary *It's All True* (1942) and studio interference that resulted in the savage mutilation of his version of Ambersons signalled the beginning of his lifelong battle to maintain authorial independence and secure the requisite funding for future projects. The combined experiences also saddled him with an unwarrented reputation for profligacy and lack of professionalism.

It would be some time before he would direct again and then it was on the less flamboyant political thriller *The Stranger* (1946). However, he was back on form with the perverse, often incomprehensible and frequently mesmerising ▷ film noir *The Lady from Shanghai* (1948), famed for its climactic shoot out in a hall of mirrors and wilful maltreatment of a cropped, peroxide blonde ▷ Rita Hayworth who had once been his real life wife.

He worked on a number of Shakespeare adaptations over the years, including a low-budget *Macbeth* (1948), inventive *Othello* (1952) and elegiac *Chimes at Midnight* (1966) culled from five plays.

His other directorial work of note includes the bravura pulp thriller *Touch of Evil* (1958), which he transforms through sheer technique and style into a powerful and brooding study of justice among the seedy denizens of a Mexican border town, an ▷ expressionist version of *The Trial* (1962), and *F for Fake* (1973), a witty essay on charlatanism that reflected his own interest in illusion and reality and also marked his last completed work for the cinema.

Many uncompleted works are known to exist in segments around the world, the most famous of which are his uncompleted 1958 version of *Don Quixote* and an exposé of the movie world *The Other Side of the Wind* (1975), which stars the late ▷ John Huston and ▷ Peter Bogdanovich.

An actor of some distinction with dozens of appearances to his credit, his most notable cinema roles in the hands of other directors include Rochester in *Jane Eyre* (1944), dapper blackmarketeer Harry Lime in ▷ *The Third Man* (1949), the Clarence Darrow-style lawyer in *Compulsion* (1959) and Cardinal Wolsey in *A Man for All Seasons* (1966). He made his last appearance in *Someone to Love* (1987).

Cheerfully consigning himself to inane cameo roles and using his rich, resonant tones to sell sherry on television commercials, the wastage of his talent is one of the more regrettable crimes in the filmmaking community.

He was not however unappreciated, receiving a special ▷ Oscar in April 1971 for 'superlative artistry and versatility', the 1975 ▷ American Film Institute Award and a 1984 Life Achievement Award from the Director's Guild of America.

WENDERS, Wim

Full Name Wilhelm Wenders
Born 14 August 1945, Dusseldorf, West Germany.

The son of a surgeon, Wenders was a student of medicine and then philosophy before dropping out and moving to Paris where he painted and soaked up film culture at the Cinémathèque Française. On his return to Germany in 1967 he enrolled at the Hochschüle fur Fernsehen und Film in Munich.

During his time there he worked as a film critic and made a number of short films including *Schauplätze* (Locations) (1967), *3 Amerikanische LPs* (3 American LPs) (1969) and *Polizeifilm* (Police Film) (1970).

Natassja Kinski in *Paris, Texas* (1984) directed by Wim Wenders

He made his feature-length debut with *Summer in the City* (1971) and followed this with *Die Angst Des Tormanns Beim Elfmeter* (The Goalkeeper's Fear of the Penalty) (1972), a hypnotic, slow-moving metaphor for the anxieties of modern living expressed in the story of an alienated goalkeeper who murders a cinema cashier and aimlessly wanders, awaiting his arrest by the police.

He declared himself unhappy with *Der Scharlachrote Buchstabe* (The Scarlet Letter) (1972) adapted from a Nathaniel Hawthorne novel and dealing with the puritanism of 17th-century New England but began his celebrated road movie trilogy with *Alice in Den Städten* (Alice in the Cities) (1974), a melancholy portrait of a photojournalist travelling down the east coast of America with a nine year-old girl in search of her grandmother.

Concerned with the influence of American popular culture on post-war German society, his bleak and elliptical work often dealt with isolation and alienation involving journeys of self enlightenment in films like *Falsche Bewegung* (Wrong Move) (1975) and ▷ *Im Lauf Der Zeit* (Kings of the Road) (1976).

His own love of the American thriller and the work of directors like ▷ Sam Fuller was reflected in his adaptation of Patricia Highsmith's *Ripley's Game, Der Amerikanische Freund* (The American Friend) (1977). He subsequently went to America to work for ▷ Francis Coppola on *Hammett* (1982), a stylish ▷ film noir that wove fact and fiction into a reflection on creativity and the real-life roots of the detective stories written by Dashiell Hammett.

His frustrations over the many delays and artistic wrangles on the latter production were reflected in *Der Stand der Dinge* (The State of Things) (1982), a static and gloomy account of a film production in crisis.

He then collaborated with Sam Shepard (1943–) on *Paris, Texas* (1984), a beautifully filmed evocation of America and exquisite delineation of the emotional agenda faced by a man who has wandered through the desert and now attempts to pick up the pieces of his family and married life.

He followed this with the monochrome *Der Himmel Über Berlin* (Wings of Desire) (1987), an alternatively sublime and ridiculous story of a benign angel's desire to experience the joys and woes of mortality and express his love for a circus performer.

Over the years, he has produced a few films and continues to make documentaries, most notably *Lightning Over Water* (1980) which chronicles the dying days of film director ▷ Nicholas Ray.

His youthful criticisms have been collected in the volume *Emotion Pictures* (1986) and 1991 will see the release of his long awaited *Till The End of the World.*

WERTMÜLLER, Lina

Real Name Arcangela Felice Assunta Wertmuller von Elgg Spanol von Braueich.

Born 14 August 1928, Rome, Italy.

Born into an aristocratic family of Swiss descent, Wertmuller graduated from the Academy of Theatre in Rome (1951) and spent a decade working in all aspects of the theatre before being hired by ▷ Federico Fellini as an assistant director on *Otto E Mezzo* ($8\frac{1}{2}$) (1963).

She made her directorial debut with *I Basilischi* (The Lizards) (1963), an observant account of the indolent, aimless young men in a sleepy southern Italian town. However, she encountered some difficulty in establishing a viable film career and returned to the more conducive climes of television and the theatre.

She found more favour in the 1970s, when her slyly comic and combative contributions to the battle of the sexes were hailed by critics and feminists. This most successful phase of her career began with *Mimi Metallurgico Ferito Nell' Onore* (The Seduction of Mimi) (1972), an acidic assault on the idiocies of Sicilian life illustrated in the story of a simple labourer who loses all he holds dear when caught between the conflicting demands of the Mafia and the Communist Party.

Other heady mixtures of political provocation and sexual parable include the entertaining *Film D'Amore E D'Anarchia* (Love and Anarchy) (1973) in which a farmer's mission to assassinate Mussolini is disrupted by his love for a prostitute, *Tutto A Posto* (All Screwed Up) (1973) about a Milanese commune of young people and *Travolit Da Un Insolito Destino Nell'Azzurro Mare D'Agosto* (Swept Away ... By an

Unusual Destiny in the Blue Sea of August) (1975) in which a deckhand and his boss's wife abandon their class roles and sexual inhibitions when castaway on a remote island.

Her work had secured a growing popularity in America and *Pasqualino Settebellezze* (Seven Beauties) (1975), the story of how a smalltime romeo survives the war, earned Oscar nominations for her script and regular leading actor Giancarlo Giannini (1942–). She signed a four-picture contract with Warner Brothers which was cancelled after the failure of the fantasy *The End of the World In Our Usual Bed in a Night Full of Rain* (1978).

Her work was re-evaluated in some quarters as vulgar, overstated and tending to reinforce sexual stereotypes rather than explode them but she has maintained a prolific career of such eccentrically titled films as *A Joke of Destiny Lying In Wait Round the Corner Like A Street Bandit* (1983) and *Summer Night With Greek Profiles Almond Eyes and Scent of Basil* (1986) and her version of *Saturday, Sunday, Monday* (1990) starring ▷ Sophia Loren was well-received at the opening of that year's Chicago Film Festival.

WEST, Mae

Born 17 August 1892, New York, USA.
Died 22 November 1980, Los Angeles, California, USA.

Best recalled for her invitation to 'come up and see me sometime' and a host of quotable innuendo from 'Is that a gun in your pocket or are you just glad to see me?' to 'a man in the house is worth two in the street', West's verbal witticisms made her the scourge of 1930s moralists and allowed her to push forward the boundaries of sexual humour in American cinema.

The daughter of a heavyweight boxer, she was steeped in showbusiness from an early age and her ability to sing, dance, act and mimic prominent celebrities of the day led to a vaudeville career first as 'The Baby Vamp' and later 'The Original Brinkley Girl'.

She made her legitimate theatre debut in *Sometime* (1919) and spent the next decade alternating between vaudeville and the theatre with appearances in her own plays *Sex* (1926) and *Diamond Lil* (1928) that briefly landed her in jail on charges of obscenity.

Mae West

She made her film debut with a scene-stealing supporting assignment in *Night After Night* (1932) and sashayed her way to notoriety as the buxom woman of ill-repute in such saucy fare as *She Done Him Wrong* (1933), *I'm No Angel* (1933) and *Belle of the Nineties* (1934).

An inventive writer with a talent for concocting cheerful vulgarity, double entendres and sultry sexual innuendo, she was clearly no great actress but a unique personality who appeared like a grotesque female personator and tossed off her crisp one-liners with a monotonous delivery that consisted of a swivel of the hips, puckering of the lips or suggestive flash of the eyes.

The new ▷ Hays Production Code of censorship began to affect the limits of her screen ribaldry but she remained popular, in a somewhat diluted form, throughout *Goin' to Town* (1935), *Klondike Annie* (1936), *Go West Young Man* (1937) and *Every Day's A Holiday* (1938).

However, her box-office appeal waned and she retired from the screen after *My Little Chickadee* (1940), in which she was uneasily partnered by ▷ W. C. Fields, and *The Heat's On* (1943).

She returned to the stage in *Catherine Was Great* (1947–48) and spent a number of years in the 1950s on intermittent tours with *Diamond Lil*.

Among the film offers she is said to have rejected are ▷ *Sunset Boulevard* (1950), *The Belle of New York* (1952) and *Pal Joey* (1957), but she did return to the screen, looking every inch a star and revelling in her own dialogue, as a Hollywood agent in the execrable *Myra Breckinridge* (1970) and surprised many with her starring role as a bride attempting to consummate her sixth marriage in *Sextette* (1978), a belated version of one of her plays that did not even enjoy a brief vogue among curiosity seekers.

WHALE, James

Born 22 July 1889, Dudley, UK.
Died 30 May 1957, Hollywood, California, USA.

A somewhat enigmatic figure, Whale trained as a graphic artist and worked as a cartoonist for *The Bystander* before serving in the First World War and being interned as a prisoner of war. Afterwards, he joined the Birmingham Repertory Company as an actor and set designer, making his London debut in *A Comedy of Good and Evil* (1925). His acclaimed stage direction of the anti-war play *Journey's End* (1928) by R. C. Sherriff (1896–1975) took him to New York and thence to Hollywood where he worked as a dialogue director on *The Love Doctor* (1929) and *Hell's Angels* (1930) before committing *Journey's End* to celluloid in 1930.

At Universal Studios, he directed *Frankenstein* (1931) bringing sensitivity, style and a lightness of touch to this tale of the macabre and eliciting a performance of great pathos from ▷ Boris Karloff as the monster.

The Old Dark House (1932) and *The Invisible Man* (1933) proved him a master of bizarre fantasy and established many of what were to become the conventions of the genre. ▷ *Bride of Frankenstein* (1935) was the apotheosis of his work in the area, blending atmospheric visuals, subtly suggestive horror, touches of eccentricity and just the right amount of tongue-in-cheek humour alongside performances of grotesque individuality from Elsa Lanchester (1902–86), Ernest Thesiger (1879–1961), Colin Clive (1898–1937) and Karloff.

He established his versatility with the drawing-room divorce drama *One More River* (1934) and an entertaining version of *Showboat* (1936). After studio interference with *The Road Back* (1937), his subsequent films, in a variety of genres, display little of his earlier wit or inventiveness and he retired in 1941. There has been some suggestion that his homosexuality created problems for him in Hollywood and that his enthusiasm for film waned with the death of his friend Clive.

Active in stage direction and as a painter, he made the experimental television play *Hello Out There* (1949) and had begun designing a science-fiction operatta based on works by Ray Bradbury and Max Beerbohm when he was found dead in his swimming pool.

WHICKER, Alan Donald

Born 2 August 1925, Cairo, Egypt.

Commissioned in the Devonshire Regiment during World War II, Whicker rose to the rank of major and served with the Army Film Unit and Photo Unit.

Thereafter, he was a Fleet Street war correspondent, who reported on the Inchon landings in Korea, and later joined the BBC (1957–68) where he worked on the *Tonight* (1957–65) programme and began his *Whicker's World* documentary series in 1958.

An urbane gentlemanly figure, perenially well groomed and dapper in blazer and flannels, he has become television's most travelled man using his discreet interviewing technique and good manners to allow viewers a privileged eavesdrop on the lives of the rich and famous as well as discovering the exotic and extraordinary aspects of everyday lives in all parts of the world.

Among his many specific series over the past three decades are *Whicker Down Under* (1961), *Whicker Within A Woman's World* (1972), *Whicker's World Aboard the Orient Express* (1982) and *Living With Waltzing Matilda* (1987–88).

Providing an unvaryingly high standard of television documentary that has rarely figured outside the top ten ratings, he has received numerous awards including the Royal Television Society Silver Medal (1968) and the ▷ Richard Dimbleby Award (1977).

Among his books are *Some Rise By*

Sin (1949) and the autobiography *Within Whicker's World* (1982).

WHISKY GALORE

Alexander Mackendrick (UK 1948)
Basil Radford, Joan Greenwood, James Robertson Justice, Jean Cadel, Gordon Jackson.
82 min. b/w.

▷ Alexander Mackendrick's debut feature film for Ealing Studios converted Compton MacKenzie's popular novel about the wreck of a wartime ship (based on the real wreck of the S.S. *Politician*) on a small Scottish island into a classic of British comic cinema. As in all the best comedies, there is a sharp undercurrent of something darker, and even cruel, notably in the treatment of the unfortunate but incorruptible Captain Waggett, the English head of the Home Guard (wartime civil defence) unit on the island. When the ship is wrecked with its precious cargo of whisky on an island suffering severely from wartime shortages, Waggett appoints himself as stern preserver of the law, setting himself against the plundering islanders intent on solving the shortage and having a good time in the process. There is no contest for the viewer's affections, and if Mackendrick is slightly malicious in making us side so resolutely with the lawbreakers, their crime is a highly understandable one, and of no great severity, especially since Waggett is content to let the precious cargo simply go to the bottom. The film adds a twist to the book by having the thirsty salvagers anxiously wait for the passing Sabbath, thereby heightening the tension of the race against the excise officers. Much of the appeal of the film lies in the playing off of the rather stereotyped crafty, couthy islanders against the upright bearers of authority, but, like the subsequent work of ▷ Bill Forsyth, the comic surface cloaks more subtle hues and shades. The film has remained eminently watchable, and the celebratory ceilidh (which included many of the real islanders acting as extras in the cast) is a classic moment of joyous warmth. The film was issued in America as *Tight Little Island*, and subsequently spawned an inferior sequel of sorts, *Rockets Galore* (also known as *Mad Little Island*), directed by Michael Relph (1915–).

WHITE, Pearl

Born 4 March 1889, Greenridge, Illinois, USA.
Died 4 August 1938, Paris, France.

A stage actress from the age of six, White performed in stock theatre companies and circuses before making her screen debut as a horseback rider in *The Life of Buffalo Bill* (1910).

Initially employed as a stuntwoman for her equestrian skills, she was seen as an actress in scores of westerns and comedies before winning lasting fame as the star of cliffhanging serials like *The Perils of Pauline* (1914) and *The Exploits of Elaine* (1915). She established the screen image of the damsel in distress, forever in danger from an oncoming train, rising water level or some other apparently inescapable fate.

Audiences warmed to her cheerful personality and fair-haired charm but she was unable to transfer her popularity into feature film success and efforts like *The White Moll* (1920), *The Mountain Woman* (1921) and *Know Your Men* (1921) failed to generate a lasting career.

She returned to serials with *Plunder* (1922) and her final appearance in *Terreur* (1924) was made in France which is where she spent the rest of her life.

An autobiography, *Just Me* was published in 1919.

WHITE HEAT

Raoul Walsh (US 1949)
James Cagney, Virginia Mayo, Edmond O'Brien, Margaret Wycherly.
114 mins. b/w.

▷ James Cagney had established himself as the major player in Hollywood gangster films in the 1930s, but *White Heat* cast him in a more psychologically complex variant of the role. The film is something of a bridge between the often socially conscious, almost documentary accounts of gangland existence which dominated the 1930s, and the more atmospheric, consciously artistic, and ambiguous treatment of the theme which would follow with the ascendance of ▷ film noir. Cagney plays Cody Jarrett, a criminal with a strange obsession about his mother, and a tendency to blinding migraine headaches which serve to emphasise the psychopathic elements of his character. While serving time in prison, he is first apparently befriended, and subsequently betrayed, by an undercover police officer named Fallon, played by Edmond O'Brien (1915–85) and by his mistreated wife Verna. As the world seems to gang up on him, Jarrett's spectacular demise on top of an exploding oil tank, a moment now enshrined in Hollywood lore, takes on a genuinely tragic significance, elevating the gangster from social pariah to something much closer to a defiantly doomed anti-hero taking his dramatic farewell from a changing world where the individual increasingly counts for little. The film wears this dimension lightly, however, and much of its continued appeal is down to its relentless action, pacy dialogue, and charismatic performances, directed with unfussy skill by ▷ Raoul Walsh.

WIDERBERG, Bo

Born 8 June 1930, Malmö, Sweden.

After briefly working in a mental hospital, Widerberg pursued a literary career as the author of short stories and as a newspaper journalist. In the early 1960s he became established as one of Sweden's most influential film critics and published the book *The Vision of Swedish Cinema* (1962) which was scathingly critical of the indigenous industry for its dominance by ▷ Ingmar Bergman and lack of concern for the plight of ordinary folk.

He made his directorial debut with the television drama *Pojken Och Draken* (The Boy and the Kite) (1961) and won critical approval for *Kvarteret Korpen* (Raven's End) (1963), an attempt at autobiographical social realism telling of an aspiring writer's struggles in Malmö during the Depression.

He pursued a similar course with the self-indulgent *Karlek 65* (Love 65) (1965) about the professional and personal crises of a film director.

Displaying the characteristics of an incurable romantic, he filmed *Elvira Madigan* (1967), the tale of a doomed 19th century love affair that is swathed in ▷ soft focus photography, slow-motion cavorting through the most picturesque of locales, all to the strains of Mozart's Piano Concerto No. 21. Nauseating or heartrending according to individual taste, it proved his greatest international success and spawned a thousand imitative television commercials.

Also notable from this period are *Adalen 31* (1969) which uses pastoral prettiness and Duke Ellington music to convey a bitter Swedish strike, a well-liked if soft-edged portrait of labour leader *Joe Hill* (1971) and *Fimpen* (Stubby) (1974), an engaging modern day fairy-tale about a six year-old soccer star who plays for Sweden in the World Cup.

He has also turned his attention to reasonably stylish suspense thrillers like *Mannen Pa Taket* (The Man on the Roof) (1976) about a police manhunt for a sniper and *Manen Fran Mallorca* (The Man from Majorca) (1984) in which the investigation of a modest post-office robbery reveals a vast network of political corruption.

Most recently, he made a restrained adaptation of the novel *Ormens Vag Pa Halleberget* (The Serpent's Way) (1988), a 19th-century tale of a landowner's merciless exploitation of a widowed tenant farmer.

THE WILD BUNCH

Sam Peckinpah (US 1969)
William Holden, Ernest Borgnine, Robert Ryan, Edmond O'Brein, Warren Oates.
143 mins. col.

The Wild Bunch is one of the most controversial films ever to emerge from American cinema, and opinion has long been divided on its merits. It is arguably

William Holden in *The Wild Bunch* (1969)

the most relentlessly violent western ever made, but, unlike some of the equally graphic adventure films which followed in its wake, its appalling violence is framed in a consciously and carefully constructed moral context which borders on the mythic. The plot revolves around the most elemental of movie strategies, an extended chase, which opens and closes with two horrific massacres involving the eponymous Bunch of ageing outlaws faced by a changing world (it is set in 1913, and ▷ Peckinpah continually emphasises that encroaching modernity), but holding on to their well-defined code of loyalty and honour. As the outlaws are hunted down by pursuing bounty hunters, Peckinpah sets up a highly skilful moral dichotomy, representing the outlaws as simultaneously vicious disturbers of the social order and something close to folk heroes, an ambiguity which is never quite resolved even by their willed and astonishingly bloody death in the Mexican Revolution. The film rewrote the rules on the graphic representation of violence on screen, perhaps with regrettable consequences, but did so in the context of the most remarkable ▷ montage editing in Hollywood history. Technically, it is a *tour de force* to rival anything committed to celluloid, and remains as disturbing now as it was on its release in 1969. That capacity to disturb is not simply a consequence of the violence, which has sadly become commonplace, but is rather rooted in the fusion of extreme visual violence with a genuinely dark vision of the human condition, all framed in the context of a group of characters with whom the viewer becomes inextricably involved. The formula is familiar enough, but it has seldom been realised with this degree of intensity. After its initial screenings, the studio cut four scenes explaining the relationship between the Bunch and their pursuer, Thornton, a former member, giving a new running time of 135 minutes.

WILDER, Billy

Real Name **Samuel Wilder**
Born 22 June 1906, Sucha, Austria.

Nicknamed Billy after his fascination with Buffalo Bill, Wilder's cultural addictions as a youth were jazz music, dancing, westerns and the sophisticated films of ▷ Ernst Lubitsch.

A crime and sports reporter, he is said to have ghost-written scripts for silent films and later co-directed *Menschem Am Sontag* (People On Sunday) (1929). The pleasing result showing how four strangers pass a day in Wannsee park on the outskirts of Berlin brought him employment as a scriptwriter on films like *Emile Und Die Detektive* (Emil and the Detective) (1931).

He fled Germany as the Nazis came to power, stopping in Paris to co-direct *Mauvaise Graine* (Bad Seed) (1933) en route to his final destination in Los Angeles. A Hollywood scriptwriter, he enjoyed a fruitful partnership with theatre critic Charles Brackett (1892–1969) on such spry and frothy romantic folderols as *Bluebeard's Eighth Wife* (1938), *Midnight* (1939) and *Ninotchka* (1939) which earned them their first Oscar nomination.

Subsequent scripts include *Ball of Fire* (1941) (Oscar nomination) and *Hold Back the Dawn* (1941) (Oscar nomination).

He made his directorial debut with the bright comedy *The Major and The Minor* (1942) and built a body of work unrivalled in its bittersweet romanticism, humour, insight into the human character and a distinctive brand of cynicism which Wilder has described as merely a romantic's painfully realistic view of the world around him.

His mastery of dialogue, characterisation and narrative were immediately apparent in the wartime adventure *Five Graves to Cairo* (1943), the archetypal ▷ film noir *Double Indemnity* (1944) (Best Director and Screenplay Oscar nominations) which set the standard by which all others have been judged, and *The Lost Weekend* (1945) (Direction and Screenplay Oscars) which used ▷ Expressionist techniques to explore the fall and redemption of a chronic alcoholic portrayed by Ray Milland (1905–86).

Oscar-nominated for the screenplay of *A Foreign Affair* (1948), his last collaboration with Brackett was on the mordantly macabre ▷ *Sunset Boulevard* (1950) (Screenplay Oscar, Direction Nomination) which is told in flashback by a corpse floating in a pool and beguiles with its audacious acting and storyline of a reclusive silent screen star who declares 'I am big. It's the pictures that got small.'

His virtually unbroken run of successes over the next decade includes *An Ace in the Hole* (1951) (Screenplay nomination), the gripping prisoner of war story *Stalag 17* (1953) (Direction nomination), the smart comedy *Sabrina* (1954) (Direction and Screenplay nominations), the rather flat but very popular *Seven Year Itch* (1955), the Agatha Christie mystery *Witness for the Prosecution* (1957) (Direction nomination), and the uproarious farce ▷ *Some Like It Hot* (1959) (Direction and Screenplay nominations) which began his long partnership with ▷ Jack Lemmon.

The poignant romance ▷ *The Apartment* (1960) (Direction, Screenplay and Producer Oscars) proved another major success, and the frenetic Cold War comedy *One, Two, Three* (1961) moves at a breathtaking pace dictated by the dynamic performance of its star ▷ James Cagney in his last leading role.

Thereafter, his sure touch began to falter and the dissenting voices that accused his work of bad taste grew louder. However, there are many notable works from his later years including the deliciously devious comedy *The Fortune Cookie* (UK: *Meet Whiplash Willie*) (1966) (Screenplay nomination), the underrated romance *Avanti!* (1972) and *Fedora* (1978) an excellent companion piece to ▷ *Sunset Boulevard* that unravels the mystery of a reclusive star's ever youthful demeanour.

Only his final film, the laboured, leaden and vulgar farce *Buddy, Buddy* (1981) is unworthy of his prodigous talents and awesome reputation.

A funny but affectionate critic of Hollywood and noted raconteur, he was the 1986 recipient of the ▷ American Film Institute Life Achievement Award.

WILLIAMS, Robin

Born 21 July 1952, Chicago, Illinois, USA.

After a somewhat peripatetic childhood, Williams moved with his parents to Marin County, California and entered Claremont Men's College as a political-science major. There, he discovered his love of performing and later studied acting at Julliard School in New York with John Houseman (1902–88).

Settling in San Francisco, where he was unable to secure employment as a dramatic actor, he began to develop a nightclub act based on his endless repertoire of vocal permutations, and penchant for quick-witted, free-association improvisation.

Appearing at the Los Angeles Comedy Store, he was spotted for television and appeared in such shows as *Laugh-In* and *America 2-Night* during the 1977–8 season. Producer Garry Marshall (1934–) then chose him for a guest role in the long-running series *Happy Days.*

Robin Williams in *Good Morning Vietnam* (1987)

His character of an amiable extra-terrestrial, baffled by the ways of earthlings, proved so popular that he was given a top-rated series of his own, *Mork and Mindy* (1978–81).

He made his film debut with the title role in *Popeye* (1980). Initially, cinema seemed unable to capture or contain his spontaneous wit, utter unpredictability and manic inventiveness and films like *The World According to Garp* (1982), *The Survivors* (1983) and *Club Paradise* (1986) were all box-office disappointments.

However, the role of an iconoclastic armed forces disc jockey in *Good Morning, Vietnam* (1987) finally allowed his comic creativity to blossom within an acceptable dramatic framework. He received an Oscar nomination and, after a cameo contribution to *The Adventures of Baron Munchausen* (1988), he was similarly rewarded as an inspirational schoolteacher breezing through the halls of academe in *Dead Poets Society* (1989).

Still very active as a stand-up performer, and a strong supporter of Comic Relief in America, he capitalised on his newfound comfort with the film medium by embarking on a succession of projects: *Cadillac Man* (1990), *Awakenings* (1990) opposite Robert De Niro, *The Fisher King* (1991), and *Hook* (1991) in which he plays Peter Pan.

WILLIS, Bruce

Born 19 March, 1955, Germany.

The oldest son of a United States Army serviceman, Willis was raised in New Jersey and attended Penns Grove High School where he performed in the drama club but also gained a reputation as a prankster and troublemaker.

Bruce Willis with Cybill Shepherd in *Moonlighting*

After school he drifted along for a while, working as a night security guard and playing in a blues band called Loose Goose, before enrolling in drama at Montclair State College.

He eventually left to make his off-Broadway debut in *Heaven and Earth* (1977). In between stints as a bartender, he diligently pursued his craft, working with a comedy improvisation group and graduating to lead roles in plays like *Railroad Bill* (1981) and *Bayside Boys* (1981).

Renowned as a self-confident extrovert, he appeared in a Levi Strauss commercial, won critical acclaim off-Broadway in *Fool for Love* (1984), was seen in a 1985 episode of ▷ *Miami Vice* and lost out on a role in *Desperately Seeking Susan* (1985) before being chosen from over 3,000 actors to play David Addison in the television series *Moonlighting* (1985–89).

Balding, with an impish grin, his combination of aggressive, wisecracking masculinity and off-the-cuff charm in the role, allied to a bantering rapport with co-star Cybill Shepherd (1949–) made him a star and won him an ▷ Emmy.

He moved in to leading roles for the cinema with the dim-witted but popular comedy *Blind Date* (1987) and the tedious western mystery *Sunset* (1988), but enjoyed a huge personal success with the comic thriller *Die Hard* (1988) as a resourceful cop who single-handedly saves a tower block from a Christmas Eve terrorist hi-jack.

Attempting a few character parts, he was the Vietnam veteran in *In Country* (1989) and a British journalist in *Bonfire at the Vanities* (1990) and recreated his original role, at a reported fee of $7.5 million, in the lavishly budgeted but more transparently implausible sequel *Die Hard 2: Die Harder* (1990).

He became one of the busiest stars in the film world; 1991 saw the release of *Billy Bathgate*, as well as *Hudson Hawk* in which he plays a suave cat burglar and *Mortal Thoughts* in which he co-stars with wife Demi Moore.

In 1987, he revealed his vocal abilities on the Motown album *Bruce Willis: The Return of Bruno*.

WIPE A transition effect where one picture image is replaced by another at a boundary line moving across the picture area and reminiscent of a wind-screen wiper. It may be sharply defined ('hard-edge') or diffuse ('soft-edge'). An expanding or contracting circular outline is termed an iris wipe. Popular during the heyday of the Hollywood studios in the 1930s and war years it still serves as a nostalgic device in films like the Indiana Jones series.

WISEMAN, Frederick

Born 1 January 1930, Boston, Massachusetts, USA.

A student at Yale Law School and Harvard, Wiseman completed his military service in 1956 and then practiced law in Paris where he developed an interest in film and began making his own 8mm studies of life. A lecturer in legal science at Boston University (1958–61), he produced an unsuccessful adaptation of the novel *The Cool World* (1963) before turning to the documentary format.

He established his reputation as the director of *Titicut Follies* (1967), a portrait of the State Prison for the Criminally Insane at Bridgewater, Massachusetts and followed this with *High School* (1968), *Law and Order* (1969) and *Hospital* (1970) for which he received an Emmy as Best Documentary Director.

His films offer an alternative social history of America, holding up a mirror to major institutions and using a ▷ cinéma-verité technique to amass many hours of footage that are subsequently edited into pungent feature-length reports. Unhindered by added commentaries or musical soundtracks, the viewer is dispassionately presented with the natural sights and sounds captured by Wiseman and his crew and allowed to bring his own value judgements to bear on the evidence of fly-on-the-wall observations and astonishingly unguarded self-incriminating testimonies.

As America's foremost documentarist, he has attacked every sacred cow in his native land and although often criticised for his decision not to inject more anger or polemic into his work, his many significant achievements include *Juvenile Court* (1973), *Welfare* (1975), *Meat* (1976) and a trio on aspects of military indoctrination and service practices – *Basic Training* (1971), *Manoeuvre* (1979) and *Missile* (1987).

His one attempt at fiction remains *Seraphita's Diary* (1982). *Near Death* (1989), his most recent work, is an exhaustive six-hour document set in the intensive care unit of a Boston hospital, probing the ties between staff and patients and the crucial interplay between man and machine.

THE WIZARD OF OZ

Victor Fleming (US 1939)
Judy Garland, Ray Bolger, Bert Lahr, Jack Haley, Margaret Hamilton, Frank Morgan.
Academy Awards: *Best Song* ('Over The Rainbow') Harold Arlen, E. Y. Harburg; *Best Original Score* Herbert Stothart; *Special Award* Judy Garland.
101 mins. col & b/w.

This elaborate fantasy is a tribute to the collective nature of film-making in

Judy Garland and Jack Haley in *The Wizard of Oz* (1939)

general, and to the quality of the team which M-G-M had gathered together in its studio by the end of the 1930s in particular. It would figure prominently in any definitive poll of the best-loved Hollywood movies, and made the astonishingly assured ▷ Judy Garland an instant star. It broke with convention in its integration of the musical numbers into the action, rather than working them in to carefully contrived performance situations within an otherwise naturalistic narrative, and gave full rein to the magical fantasy of L. Frank Baum's imaginary wonderland. Dorothy and her dog are transported from a monochrome Kansas to a vividly coloured and fantastically designed Oz, with its cowardly lion, scarecrow, tin man, wicked witch and the Wizard himself, the product of a very expensive and little used technique. The complex three-strip technicolor process which M-G-M adopted produced the astonishingly vivid and very carefully planned colour schemes which bring such a memorable visual quality to this ground-breaking classic. Even if the technique itself quickly became obsolete, *The Wizard of Oz* remains a lasting tribute to it, but more importantly, it is a lasting reminder of the capacity of Hollywood to capture the imagination with wonders (albeit of a more innocent kind than modern audiences would demand), and bring us back to earth with the final assertion that there is no place like home. It is a perfect encapsulation of the escapist function of the movies in its best sense, and remains a delight. It is a delight, though, very much of its time, as the attempt at an updated, black version, *The Wiz* (1978), convincingly demonstrated.

WOODWARD, Joanne

Born 27 February 1930, Thomasville, Georgia, USA.

Active in college drama, Woodward studied acting at the Neighbourhood Playhouse and Actors Studio in New York before securing an understudying job in the Broadway production of *Picnic* (1953). Appearances in scores of television dramas brought her to the attention of Buddy Adler (1906–60), then head of production at Twentieth-Century Fox, and she was signed to a long-term contract, making her film debut in the minor western *Count Three and Pray* (1955).

After one further role in *A Kiss Before Dying* (1956), she won the Best Actress Oscar for *The Three Faces of Eve* (1957) in which she was accomplished, against the odds, as a schizophrenic.

She married ▷ Paul Newman in 1958 and their careers have been inextricably intertwined ever since. An actress of compassion and range, she refused to play the star and a certain public self-effacement may have kept her from the choicest of roles. In the meantime, she worked productively bringing a sense of warmth, intelligence and reality to even the most cardboard of characters. Notable roles include the suburban housewife in *No Down Payment* (1957), the Southern belle in *The Long Hot Summer* (1958) and the blowsy title character in *The Stripper* (UK: *Woman of Summer*) (1963).

Her career appeared to flounder in the 1960s, but she returned to display a deft comic touch in *Big Hand for A Little Lady* (1966) and had her best chance in years with *Rachel, Rachel* (1968) which Newman directed. Her realistically observed portrait of a spinster schoolteacher transformed by love brought her a further Oscar nomination. In between more routine assignments, she has continued to delineate the fragility and vulnerability of the lonely, repressed and lovelorn, impressing as the crisis-beset middle-aged woman in *Summer Wishes, Winter Dreams* (1973) (Oscar nomination).

Disappointed by the commercial failure of the latter film and aware of the restricted cinema opportunities for mature actresses, she has more frequently appeared on stage and television in recent years earning a well-deserved reputation as the doyenne of TV-movie drama in projects like *Sybil* (1976), *See How She Runs* (1978) (Emmy), *Crisis at Central High* (1981) (Emmy nomination) and *Do You Remember Love?* (1985) (Emmy).

Recent cinema films include Newman's accomplished version of *The Glass Menagerie* (1987) in which she recreated her stage success as Amanda, and *Mr. and Mrs. Bridge* (1990) (Best Actress Oscar nomination).

THE WORLD AT WAR

UK 1973

Two years of effort and £1 million went into this 26-hour study of World War II, produced by Jeremy Isaccs (1932–). Narrated by ▷ Lord Olivier (▷ Burt Lancaster did the honours for the US market), it was a considered appraisal of the progress of the conflict. Three and a half million feet of film was viewed by the programme's researchers, and new interviews were conducted with the participants in the war, including Chief of Bomber Command, RAF Marshall Sir Arthur Harris, and Hitler's personal secretary Traudl Junge. Isaacs avoided reconstruction of scenes, and only used archive footage when he knew it was specifically appropriate. The music score was by Carl Davis (1936–). A book of the series sold 500,000 copies.

WYLER, William

Real Name Willy Wyler
Born 1 July 1902, Mulhouse, Alsace-Lorraine, Germany (now in France).
Died 27 July 1981, Beverly Hills, California, USA.

Invited to America by his cousin Carl Laemmle (1867–1939), the then head of Universal Pictures, Wyler began his lengthy career in the publicity department there but soon graduated to assistant director and made his directorial debut with *Crook Busters* (1925).

Over the next five years, he made numerous westerns before moving on to more prestigious productions, usually involving star actors and noted literary sources. Among his early successes were the punchy drama of a lawyer's self doubts *Counsellor-At-War* (1933), the romantic trifle *The Good Fairy* (1935) and an incisive account of the pernicious damage wrought by the spread of malicious tittle tattle in *These Three* (1936).

At the peak of his talents from 1936, he became renowned for his obsessively meticulous approach to composition, performance and narrative structure, insisting on countless ▷ takes of any given scene until his notion of dramatic perfection was attained. The end results generally justified the blood, sweat and tears testified to by his actors and technicians.

Ranking high among his greatest achievements are his films with ▷ Bette Davis *Jezebel* (1938), *The Letter* (1940) (Best Director Oscar nomination) and *The Little Foxes* (1941) (Oscar nomination) but his abilities and experimentation with the use of ▷ deep-focus photography also vividly enhanced such classics as *Dodsworth* (1936) (Oscar nomination), *Wuthering Heights* (1939) (Oscar nomination), the shameless wartime propaganda of *Mrs Miniver* (1942) (Oscar) and the sensitive account of returning veterans adjusting to changes at the Home Front in ▷ *The Best Years of Our Lives* (1946) (Oscar).

His virtually unrivalled success rate continued with a acutely acted version of *The Heiress* (1949) (Oscar nomination), *The Detective Story* (1950) (Oscar nomination) and the enchanting if now dated comedy romance *Roman Holiday* (1953) (Oscar nomination).

Later work showed him moving towards a more epic scale of filmmaking and, whilst popular at the box-office, increasingly lacked the passion in performance and direction that distinguished earlier work, settling instead for a high standard of technical accomplishment allied to a blander level of polished, escapist entertainment. His many popular credits include *Friendly Persuasion* (1956) (Oscar nomination), the sweeping western *The Big Country* (1958), the lengthy spectacle ▷ *Ben-Hur* (1959) (Oscar) and the jewel caper *How To Steal A Million* (1966).

He received his final Oscar nomination for *The Collector* (1965) and retired from direction after the overblown musical *Funny Girl* (1968) and *The Liberation of L. B. Jones* (1970), a study of racism in the American South.

His reputation endures as one of the American cinema's most versatile and accomplished directors and he was the 1976 recipient of the ▷ American Film Institute Life Achievement Award.

YES, MINISTER

UK 1980–87

Written by Antony Jay (1930–) and Jonathan Lynn (1943–), *Yes, Minister* was popular with public and politicians alike, surprisingly since it showed the latter as spineless incompetents whose success or failure was governed by the whims of an all-powerful civil service. It was highly praised for its detail, much of which was based on inside knowledge gleaned by Jay during a spell with the *Tonight* programme. The scripts were erudite, following the bumbling efforts of rising politician Jim Hacker (Paul Eddington (1927–)), whose well-meaning schemes were continually scuppered by his double-talking Permanent Under-Secretary Sir Humphrey Appleby (Nigel Hawthorne (1929–)) and private secretary Bernard Woolley (Derek Fowlds (1937–)), late of the

Basil Brush Show). Mrs Thatcher said the show was her favourite, and in 1984 enacted a scene with Eddington at an awards ceremony. Eddington and Hawthorne were awarded CBEs in the 1986 New Year's Honours List. In 1986, after three series Hacker was elevated to Prime Minister (the title changed accordingly, to *Yes, Prime Minister*), a switch which removed one of the central planks from the show – his fear of damaging his promotion prospects. Yet even towards the end it was head and shoulders above the competition. It attained a rare moment of topicality when an episode detailing a dispute between Hacker and his defence secretary was screened on the day Michael Heseltine walked out of Mrs Thatcher's cabinet.

YOUNG ONES, THE

UK 1982–84

The leading lights in the shortlived 'alternative comedy' movement of the early 1980s were well-presented in this bizarre, sometimes surreal sitcom, produced by Paul Jackson. The show had its roots in London's Comedy Store, where Rik Mayall (1958–), Adrian Edmondson (1957–) and Ben Elton (1959–) (collectively known as 20th-Century Coyote) performed alongside Nigel Planer (1953–), Alexei Sayle (1952–) and Arnold Brown. The group gelled into the Comic Strip with the addition of Dawn French (1958–) and Jennifer Saunders (1958–). All had some part in the series, which was written by Elton, Mayall and Lise Mayer.

Set in the squalid student household, the sometimes uneven writing was compensated for by exaggerated, cartoon-like characters. Mayall was Rik, a badge-wearing trendy, obsessed with insulting lentil-eating hippie Neil (Planer), and remaining right-on at all times. Threatening them both was the ultra-violent, abusive punk Vivien (Edmondson). And standing at the edge of the action was the unresolved character of Mike (Christopher Ryan). The show was loosely structured – a guest band played each week in an impromptu interval – and was enlivened by guest spots from Sayle. *The Young Ones* took its name from Rik's ironic reading of the Cliff Richard song, and much of its humour was derived from the inversion of sitcom rules.

The Comic Strip made a patchy series of films for Channel 4. Mayall, Edmondson, and Planer have moved in to mainstream television roles. Elton, while succeeding as a comic in his own right, also co-scripted the *Blackadder* series, starring Rowan Atkinson (1955–).

Z

Constantin Costa-Gavras (France, 1968)
Yves Montand, Jean-Louis Trintignant, Jacques Perrin, Francois Pertier, Irene Papas.
Academy Awards: *Best Foreign Film*; *Best Film Editing* Françoise Bonnot.
Cannes Film Festival: *Best Actor* Jean-Louis Trintignant.
New York Film Critics Awards: *Best Motion Picture*; *Best Direction.*
123 mins. col.

Z is widely held to be an exemplary model of the political thriller, and established the exiled Greek director ▷ Constantin Costa-Gavras as the leading film-maker in that genre. Its greatest strength lies in its simplicity, and in the universality of the message which it contains. ▷ Yves Montand plays Z, the leader of a Pacifist Opposition movement in an unidentified Mediterranean state, but one which clearly invokes the military government of Greece, and is based

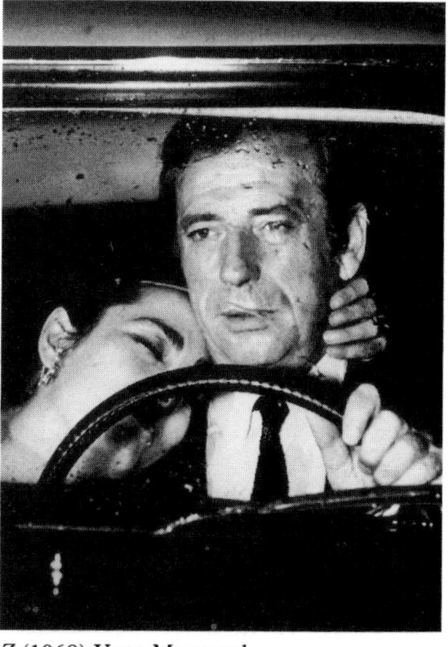

Z (1968) Yves Montand

upon an actual incident which occurred in that country. The charismatic Z is winning widespread popularity and threatening radical political change within the state, until he suddenly becomes the victim of a suspicious road accident, hit by a van. When he dies after undergoing brain surgery, the subsequent investigation uncovers the existence of a secret right-wing organisation dedicated to the overthrow of the democratic state. ▷ Jean Louis-Trintignant is excellent as the investigating magistrate attempting to expose the plot in the face of vicious intimidation, and the film's enduring appeal lies in this eternal conflict between individual and state machine. The topicality of many political thrillers tends to date them quickly, but Costa-Gavras's concern with human rights goes beyond specific applicability to the Colonels' regime, and is readily translated to any totalitarian regimes or rulers. Beyond that political dimension, however, it succeeds because it is also a gripping and suspenseful thriller, with actors of international box-office standing who grab and hold the audience's attention and sympathy. That unusual combination of serious political comment with highly entertaining commercial appeal has already conferred classic status on the film, and became a characteristic of the director's style in such subsequent films as *State of Siege* (1973) and *Missing* (1981).

Z CARS
UK 1962–78

The idea for this gritty drama series came after writer Troy Kennedy Martin (1932–) found himself with a dose of the mumps, bedridden and bored. To pass the time he listened in to police radio messages, and quickly realised that their work was a lot more interesting than the cosy world portrayed in ▷ *Dixon of Dock Green*.

The series was set in Liverpool – Kirby was re-named Newtown, Seaforth was called Seaport – and swapped Dixon's beat for the tougher urban realities encountered by police panda cars. Many of the innovations seem tame and obvious by today's standards, but even showing the police as being prone to human frailties was controversial. The first episode had Constable Lynch (James Ellis) asking for the result of a horse race, and revealed Constable Bob Steele (Jeremy Kemp (1934–)) to be a wife-beater. Viewers, among them the chairman of the Police Federation, complained, but before long the characters were established and popular, and the idea that the police were anything other than human seemed faintly ridiculous. Early stars Inspector Barlow (Stratford Johns (1925–)) and Sergeant Watt (Frank Windsor (1927–)) were given their own series, *Softly, Softly*, in 1966. *Z Cars* ran till 1978, by which time it was looking tame and dated against shows like ITV's *The Sweeney*. In 1985 Troy Kennedy Martin wrote *Edge of Darkness*, a chilling political thriller which won three BAFTA awards.

ZANUCK, Darryl Francis
Born 5 September 1902, Wahoo, Nebraska, USA.
Died 22 December 1979, Palm Springs, California, USA.

A member of the Nebraska National Guard who had fought in France towards the end of World War I, Zanuck drifted through a succession of uneventful jobs before having his writing published and making his way to Hollywood as a scriptwriter at Warner Brothers where his earliest efforts were canine adventures for wonder dog Rin Tin Tin.

Often writing under the pseudonyms of Mark Canfield, Melville Crossman or Gregory Rogers, his many scripts include *The Lighthouse By The Sea* (1924), *Across the Pacific* (1926) and *Tracked By The Police* (1927).

Part of the Warner Brothers executive from 1929, he was placed in charge of production between 1931 and 1933, a period that saw the studio enter its golden era of gangster yarns like ▷ *Public Enemy* (1931) and musicals like *Golddiggers of 1933*.

He left to join a partnership with Joseph Schenck (1878–1961) in Twentieth Century Productions and in 1935 this became Twentieth Century-Fox. With hard-working, dictatorial Zanuck as Vice-President in Charge of Production from 1935 to 1952, the studio became home to such stars as ▷ Shirley Temple, Betty Grable (1916–73), Alice Faye (1912–), ▷ Marilyn Monroe, Tyrone Power (1913–58) and Don

Ameche (1908–), with Zanuck's name seen on such films as *Alexander's Ragtime Band* (1938), *Young Mr. Lincoln* (1939), ▷ *The Grapes of Wrath* (1940) and *How Green Was My Valley* (1941).

After the war, he encouraged the studio in their fondness for realistic, documentary-like thrillers and developed a reputation as someone who courageously tackled a number of thorny social issues although the tales of racial bigotry in *Pinky* (1949) and *No Way Out* (1950) and the anti-Semitism in *Gentleman's Agreement* (1947) have not worn especially well.

When shifting allegiances made his position vulnerable, he moved to Europe as an independent producer on a series of uninspiring box-office flops like *The Roots of Heaven* (1958) and *Crack in the Mirror* (1960) but he did enjoy success with the all-star World War II account of D-Day, *The Longest Day* (1962).

Restored to favour, he returned to Fox as Executive President and was ensconced during the studio's box-office successes like ▷ *The Sound of Music* (1965) before being made Chairman of the Board in 1969 by his son Richard (1934–) who was then head of production and whose subsequent independent career has included *The Sting* (1973), *Jaws* (1975), *Cocoon* (1985) and *Driving Miss Daisy* (1989).

ZEFFIRELLI, Franco

Real Name Gianfranco Corsi
Born 12 February 1923, Florence, Italy.

A student of architecture at Florence, Zeffirelli began his long association with the arts as a stage actor who caught the eye of noted director ▷ Luchino Visconti and was hired to appear in a number of his productions.

On the basis of his sole screen performance in the film *L'Onorevole Angelina* (Angelina) (1948) he was offered a screen contract by RKO in Hollywood but rejected the offer to work as an assistant to Visconti on stage and in the cinema on such films as *La Terra Trema* (1948), *Bellissima* (1951) and *Senso* (1954).

He directed his own film *Camping* (1957) but concentrated on his illustrious and glittering career as an opulent set designer and director of opera, noted for

Leonard Whiting and Olivia Hussey in *Romeo and Juliet* (1968) directed by Franco Zeffirelli

his stormy collaborations with diva Maria Callas.

He eventually returned to the cinema with a boisterous version of *The Taming of the Shrew* (1967) that was one of the better pairings of then husband and wife ▷ Elizabeth Taylor and ▷ Richard Burton and declared his avowed intention to bring the classics to a young audience by making the texts relevant and accessible to a contemporary sensibility. Thus a beautifully photographed *Romeo and Juliet* (1968) (Best Director Oscar nomination) that cast actors of the correct age for the two central roles.

All of his subsequent ventures were less admired, betraying the hand of a far from fluid cinematic talent and a belief in prettiness and vulgarity to capture the attention. His several critical and artistic failures include a crass biography of Francis of Assissi, *Fratello Sole Sorella Luna* (Brother Sun, Sister Moon) (1972), an unnecessary and unashamedly tear-jerking remake of *The Champ* (1979) and a dreary wallow in the torments of young ardour *Endless Love* (1981) which he subsequently disowned and which introduced ▷ Tom Cruise to the screen.

His sincere, all-star television production *Jesus of Nazareth* (1977) was one of the highlights of small screen drama and his often flamboyant and static style finally seemed appropriate to the chocolate box prettiness of his cinema version of *La Traviata* (1982).

He stayed with opera for a film of *Otello* (1986) starring Placido Domingo and then took up the cudgels of his campaign to popularise the Bard with a version of *Hamlet* (1990) starring ▷ Mel Gibson.

ZEMECKIS, Robert

Born 14 May 1951, Chicago, Illinois, USA.

A graduate of the School of Cinema at the University of Southern California, Zemeckis had already shown promise with his short film *Field of Honor* (c. 1973) which had earned him a Best Student Film Oscar.

Subsequently an editor of television commercials, he made his directorial debut (with ▷ Steven Spielberg as ▷ executive producer) on the offbeat comedy-drama *I Wanna Hold Your Hand* (1978) (which he also co-wrote), focusing on the eruption of Beatlemania when the Fab Four visited New York in 1964.

There are similar elements in his script for *1941* (1979), which Spielberg directed, in which hysteria again erupts as California over-reacts to the threat unleashed by the attack on Pearl Harbor.

He then made the underrated *Used Cars* (1980), a brash satirical swipe at the business practices of twin brothers who are deadly rivals in the car resale business.

He showed his flair for sweeping high adventure on *Romancing the Stone* (1984) in which a dowdy writer is transformed into a latterday superhero as she encounters real-life intrigue, and emphasised his skill as a vivid and humorous storyteller with the exhilarating Oedipal time-travelling fantasy *Back to the Future* (1985).

His sense of timing, pace and characterisation were also much in evidence as the director who masterminded all the elements of humour, pastiche, performance and cartoon in the vastly entertaining *Who Framed Roger Rabbit* (1988) which expertly mixed live-action and animation in a breathtaking homage to the ▷ film noir and the heyday of the studio cartoon.

He followed this with the convoluted and meanspirited *Back to the Future, Part II* (1989) and the Western spoof of *Back to the Future, Part III* (1990), both significant box-office hits, and now rivals his mentor Spielberg as a master of popular entertainment.

ZETTERLING, Mai

Born 24 May 1925, Vasteras, Sweden.

Zetterling made her stage debut at the age of sixteen in *A Midsummer Night's Dream* at the Workhouse (1941) and studied at the Royal Dramatic Theatre School in Stockholm (1942–45) before appearing in productions of *Twelfth Night*, *The Cherry Orchard* and *The House of Bernarda Alba* over the next two years.

She made her film debut in *Lasse-Maja* (1941) and gave a striking performance as an alcoholic prostitute in *Hets* (Frenzy) (1944) the first script by ▷ Ingmar Bergman. She journeyed to Britain for *Frieda* (1947) and began an international career where her intelligence and wit enlivened some lacklustre roles. Notable films include *Quartet* (1948), *Knock On Wood* (1954) and *Only Two Can Play* (1961).

In the 1960s, she turned to direction with a series of hard-hitting documentaries that were often critical of her homeland; titles include *The Polite Invasion* (1960), *The Prosperity Race* (1962) and *The Do-It-Yourself Democracy* (1963). Her first feature *Alskandra Par* (Loving Couples) (1964) conveyed the different life experiences of three pregnant women at the turn of the century.

She followed this with *Nattlek* (Night Games) (1966), a sexually frank adaptation of her novel about a repressed man and the mother-dominated childhood at the root of his adult inadequacies.

Labelled a feminist director, she has created portraits of loneliness and obsession that were often boldly shocking in their depiction of sexuality and women dissatisfied with their subservient role in society. Her best known work includes *Flickorna* (The Girls) (1968) in which three actresses rebel against the deficiencies of their menfolk, *Scrubbers* (1982), a harrowing view of life in a girl's borstal, and *Amarosa* (1986) a biography of writer Agnes Von Krusentjerna, one of whose novels had inspired Zetterling's first feature.

Among her many documentaries are *Vincent the Dutchman* (1971), a segment on the Munich Olympics in *Vision of Eight* (1973) and *Of Seals and Men* (1980). She has also written a number of books including *The Cat's Tale* (1965), *Bird of Passage* (1976) and *Osminkat* (All Those Tomorrows) (1984).

After a long absence from the screen, she returned with small roles in *The Witches* (1990) and *Hidden Agenda* (1990).

ZINNEMANN, Fred
Born 29 April 1907, Vienna, Austria.

A law student at the University of Vienna, Zinnemann then studied at the Ecole Technique de Photographie et Cinématographie in Paris and worked as an assistant cameraman on films like *La Marche des Machines* (1927) and *Menschen am Sonntag* (People on Sunday) (1929).

He relocated to Hollywood in 1929, gaining employment as an extra in ▷ *All Quiet on the Western Front* (1930) and as an assistant on *The Spy* (1931) and *The Kid from Spain* (1932) before his association with documentarist ▷ Robert Flaherty led to his co-direction of the Mexican tale *Los Redes* (The Wave) (1934).

At M-G-M's short film department from 1937, he directed a number of the *Crime Does Not Pay* series and an Oscar win for the medical subject *That Mothers Might Live* (1938) accelerated his graduation to features with *Kid Glove Killer* (1942). The best of his early films, including *The Seventh Cross* (1944), *The Search* (1948) (Best Director Oscar nomination) and *The Men* (1950), used a documentary-like realism to explore the physical and emotional traumas of war and its after effects and also displayed his keen eye for spotting talent in the early performances of ▷ Marlon Brando and ▷ Montgomery Clift.

Often concerned with conflicts of conscience and the moral dilemmas of reluctant heroes, he illustrated these themes most eloquently in ▷ *High Noon* (1952) (Oscar nomination), *From Here To Eternity* (1953) (Oscar) and *A Man For All Seasons* (1966) (Oscar).

A meticulous craftsman of increasingly large-scale productions, he has enjoyed consistent box-office success with a wide range of topics and genres from the musical *Oklahoma!* (1955) to *The Nun's Story* (1959) (Oscar nomination), the warmhearted saga of Australian family life *The Sundowners* (1960) (Oscar nomination), the fact-based thriller *The Day of the Jackal* (1973) and the handsomely mounted 1930s memoir *Julia* (1977) which introduced ▷ Meryl Streep to the screen. His last film to date is *Five Days One Summer* (1982).

He received a further Oscar for the documentary *Benjy* (1951).

ZOOM A visual effect involving the enlarging or diminishing of the image as though rapidly approaching or receding from it. Originally made by actual camera movement, it is now provided by a zoom lens whose focal length can be continuously variable and thus create the illusion of movement, from say wide-angle to long focus, whilst the camera remains stationary.

ZUKOR, Adolph
Born 7 January 1873, Hungary.
Died 10 June 1976, Los Angeles, California, USA.

Emigrating to America from Hungary in 1889, Zukor first savoured financial success in the fur trade and later invested heavily in penny arcades and ▷ nickelodeons, building a small distribution network.

After producing and distributing *Queen Elizabeth* (1912) with Sarah Bernhardt (1844–1923), he founded the Famous Players Company and was among the first of the pioneering Hollywood moguls to recognise the potential of celebrity in attracting audiences to the cinema.

Promising 'famous players in famous plays', he was subsequently instrumental in establishing the star system, grooming the careers of individual personalities, notably ▷ Mary Pickford who was lured from Biograph in 1914 for the princely sum of $2,000 a week and name billing. Other stars created or substantially promoted by Zukor include ▷ Rudolph Valentino, ▷ Douglas Fairbanks Senior, and ▷ Gloria Swanson.

In 1916, he became President of the Famous Players-Lasky Corporation which was influential in furthering the careers of ▷ Cecil B. De Mille, ▷ Ernst Lubitsch and many others. In 1927, the corporation became Paramount Pictures. A clever and canny businessman, he concentrated more on corporate and fiscal matters than creative endeavours, helping to expand Paramount's national cinema ownership and developing their overseas business interests. However, his many significant hits with Famous Players and Paramount include *The Prisoner of Zenda* (1914), *The Sheik* (1921), *The Ten Commandments* (1923), *The Covered Wagon* (1923), *Trouble in Paradise* (1932), and *The Sign of the Cross* (1932).

In 1935, he became Chairman of the Board at Paramount, a position of influence that he retained for some decades and he was still chairman emeritus at the time of his death. His autobiography, *The Public Is Never Wrong* was published in 1945 and he received a special Oscar in March 1949 as 'the man who has been called the father of the feature film in America'.

PRIZEWINNERS
Oscars and Film Festivals

Oscar Winners

Awards of the American Academy of Motion Picture Arts Society

1927–28
Best Picture: *The Last Command; Wings*
Best Director: Frank Borzage, *Seventh Heaven*
Comedy Direction: Lewis Milestone, (*Two Arabian Nights*)
Best Actress: Janet Gaynor (*Seventh Heaven; Street Angel; Sunrise*)
Best Actor: Emil Jannings (*The Last Command; The Way of All Flesh*)
Best Cinematography: Charles Rosher and Karl Struss (*Sunrise*)
Best Interior Decorator: William Cameron Menzies (*The Dove; The Tempest*)

1928–29
Best Picture: *Broadway Melody*
Best Director: Frank Lloyd (*The Divine Lady; Weary River; Drag*)
Best Actess: Mary Pickford (*Coquette*)
Best Actor: Warner Baxter (*In Old Arizona*)
Best Cinematography: Clyde De Vinna (*White Shadows in The South Seas*)
Best Interior Decoration: Cedric Gibbons (*The Bridge of San Luis Rey*)

1929–30
Best Picture: *All Quiet on the Western Front*
Best Director: Lewis Milestone (*All Quiet on the Western Front*)
Best Actress: Norma Shearer (*The Divorcee*)
Best Actor: George Arliss (*Disraeli*)
Best Cinematography: Joseph T. Rucker and Willard Van Der Veer (*With Byrd at the South Pole*)
Best Interior Decoration: Herman Rosse (*King of Jazz*)

1930–31
Best Picture: *Cimarron*
Best Director: Norman Taurog (*Skippy*)
Best Actress: Marie Dressler (*Min and Bill*)
Best Actor: Lionel Barrymore (*A Free Soul*)
Best Cinematography: Floyd Crosby (*Tabu*)
Best Interior Decoration: Max Ree (*Cimarron*)

1931–32
Best Picture: *Grand Hotel*
Best Director: Frank Borzage (*Bad Girl*)
Best Actress: Helen Hayes (*The Sin of Madelon Claudet*)
Best Actor: Wallace Beery (*The Champ*); Fredric March (*Dr. Jekyll and Mr. Hyde*)
Best Cinematography: Lee Garmes (*Shanghai Express*)
Best Interior Decoration: Gordon Wiles (*Transatlantic*)

1932–33
Best Picture: *Cavalcade*
Best Director: Frank Lloyd (*Cavalcade*)
Best Actress: Katharine Hepburn (*Morning Glory*)
Best Actor: Charles Laughton (*The Private Life of Henry VIII*)
Best Cinematography: Charles Bryant Lang, Jr. (*A Farewell to Arms*)
Best Interior Decoration: William S. Darling (*Cavalcade*)

1934
Best Picture: *It Happened One Night*
Best Director: Frank Capra (*It Happened One Night*)
Best Actress: Claudette Colbert (*It Happened One Night*)
Best Actor: Clark Gable (*It Happened One Night*)
Best Cinematography: Victor Milner (*Cleopatra*)
Best Interior Decoration: Cedric Gibbons and Frederic Hope (*The Merry Widow*)

1935
Best Picture: *Mutiny on the Bounty*
Best Director: John Ford (*The Informer*)
Best Actress: Bette Davis (*Dangerous*)
Best Actor: Victor McLaglen (*The Informer*)

Best Cinematography: Hal Mohr (*A Midsummer Night's Dream*)
Best Interior Decoration: Richard Day (*The Dark Angel*)

1936
Best Picture: *The Great Ziegfeld*
Best Director: Frank Capra (*Mr. Deeds Goes to Town*)
Best Actress: Luise Rainer (*The Great Ziegfeld*)
Best Actor: Paul Muni (*The Story of Luis Pasteur*)
Best Supporting Actress: Gale Sondergaard (*Anthony Adverse*)
Best Supporting Actor: Walter Brennan (*Come and Get It*)
Best Cinematography: Gaetano Gaudio (*Anthony Adverse*)
Best Interior Decoration: Richard Day (*Dodsworth*)

1937
Best Picture: *The Life of Emile Zola*
Best Director: Leo McCarey (*The Awful Truth*)
Best Actress: Luise Rainer (*The Good Earth*)
Best Actor: Spencer Tracy (*Captains Courageous*)
Best Supporting Actress: Alice Brady (*In Old Chicago*)
Best Supporting Actor: Joseph Schildkraut (*The Life of Emile Zola*)
Best Cinematography: Karl Freund (*The Good Earth*)
Best Interior Decoration: Stephen Goosson (*Lost Horizon*)

1938
Best Picture: *You Can't Take It With You*
Best Director: Frank Capra (*You Can't Take It With You*)
Best Actress: Bette Davis (*Jezebel*)
Best Actor: Spencer Tracy (*Boy's Town*)
Best Supporting Actress: Fay Bainter (*Jezebel*)
Best Supporting Actor: Walter Brennan (*Kentucky*)
Best Cinematography: Joseph Ruttenberg (*Jezebel*)
Best Interior Decoration: Carl J. Weyl (*The Adventures of Robin Hood*)

1939
Best Picture: *Gone With the Wind*
Best Director: Victor Fleming (*Gone With the Wind*)
Best Actress: Vivien Leigh (*Gone With the Wind*)
Best Actor: Robert Donat (*Goodbye Mr Chips*)
Best Supporting Actress: Hattie McDaniel (*Gone With the Wind*)
Best Supporting Actor: Thomas Mitchell (*Stagecoach*)
Best Cinematography (Black & White): Gregg Toland (*Wuthering Heights*)
Best Cinematography (Colour): Ernest Haller and Ray Rennahan (*Gone With the Wind*)
Best Interior Decoration: Lyle Wheeler (*Gone With the Wind*)

1940
Best Picture: *Rebecca*
Best Director: John Ford (*The Grapes of Wrath*)
Best Actress: Ginger Rogers (*Kitty Foyle*)
Best Actor: James Stewart (*The Philadelphia Story*)
Best Supporting Actress: Jane Darwell (*The Grapes of Wrath*)
Best Supporting Actor: Walter Brennan (*The Westerner*)
Best Cinematography (Black & White): George Barnes (*Rebecca*)
Best Cinematography (Colour): George Perinal (*The Thief of Bagdad*)
Best Interior Decoration (Black & White): Cedric Gibbons and Paul Groese (*Pride and Prejudice*)
Best Interior Decoration (Colour): Vincent Korda (*The Thief of Bagdad*)

1941
Best Picture: *How Green Was My Valley*
Best Director: John Ford (*How Green Was My Valley*)
Best Actress: Joan Fontaine (*Suspicion*)
Best Actor: Gary Cooper (*Sergeant York*)
Best Supporting Actress: Mary Astor (*The Great Lie*)
Best Supporting Actor: Donald Crisp (*How Green Was My Valley*)
Best Cinematography (Black and White): Arthur Miller (*How Green Was My Valley*)
Best Cinematography (Colour): Ernest Palmer and Ray Rennahan (*Blood and Sand*)
Best Interior Decoration (Black & White): Richard Day, Nathan Juran and Thomas Little (*How Green Was My Valley*)

Best Interior Decoration (Colour): Cedric Gibbons, Urie McCleary and Edwin B. Willis (*Blossoms in the Dust*)

1942
Best Picture: *Mrs. Miniver*
Best Director: William Wyler (*Mrs. Miniver*)
Best Actress: Greer Garson (*Mrs. Miniver*)
Best Actor: James Cagney (*Yankee Doodle Dandy*)
Best Supporting Actress: Teresa Wright (*Mrs. Miniver*)
Best Supporting Actor: Van Heflin (*Johnny Eager*)
Best Cinematography (Black & White): Joseph Ruttenberg (*Mrs. Miniver*)
Best Cinematography (Colour): Leon Shamroy (*The Black Swan*)
Best Interior Decoration (Black & White): Richard Day, Joseph Wright and Fay Babcock (*This Above All*)
Best Interior Decoration (Colour): Richard Day, Joseph Wright and Thomas Little (*My Gal Sal*)

1943
Best Picture: *Casablanca*
Best Director: Michael Curtiz (*Casablanca*)
Best Actress: Jennifer Jones (*The Song of Bernadette*)
Best Actor: Paul Lukas (*Watch on the Rhine*)
Best Supporting Actress: Katina Paxinou (*For Whom the Bell Tolls*)
Best Supporting Actor: Charles Coburn (*The More the Merrier*)
Best Cinematography (Black & White): Arthur Miller (*The Song of Bernadette*)
Best Cinematography (Colour): Hal Mohr and W. Howard Greene (*The Phantom of the Opera*)
Best Interior Decoration (Black & White): James Basevi, William Darling and Thomas Little (*The Song of Bernadette*)
Best Interior Decoration (Colour): Alexander Golitzen, John B. Goodman, Russell A. Gausman and Ira S. Webb (*The Phantom of the Opera*)

1944
Best Picture: *Going My Way*
Best Director: Leo McCarey (*Going My Way*)
Best Actress: Ingrid Bergman (*Gaslight*)
Best Actor: Bing Crosby (*Going My Way*)
Best Supporting Actress: Ethel Barrymore (*None But the Lonely Heart*)
Best Supporting Actor: Barry Fitzgerald (*Going My Way*)
Best Cinematography (Black & White): Joseph La Shelle (*Laura*)
Best Cinematography (Colour): Leon Shamroy (*Wilson*)
Best Interior Decoration (Black & White): Cedric Gibbons, William Ferrari, Edwin B. Willis and Paul Huldschinsky (*Gaslight*)
Best Interior Decoration (Colour): Wiard Ihnen and Thomas Little (*Wilson*)

1945
Best Picture: *The Lost Weekend*
Best Director: Billy Wilder (*The Lost Weekend*)
Best Actress: Joan Crawford (*Mildred Pierce*)
Best Actor: Ray Milland (*The Lost Weekend*)
Best Supporting Actress: Anne Revere (*National Velvet*)
Best Supporting Actor: James Dunn (*A Tree Grows In Brooklyn*)
Best Cinematography (Black & White): Harry Stradling (*The Keys of the Kingdom*)
Best Cinematography (Colour): Leon Shamroy (*Leave Her To Heaven*)
Best Interior Decoration (Black & White): Wiard Ihnen and A. Roland Fields (*Blood on the Sun*)
Best Interior Decoration (Colour): Hans Dreier, Ernest Fegte and Sam Comer (*Frenchman's Creek*)

1946
Best Picture: *The Best Years of Our Lives*
Best Director: William Wyler (*The Best Years Of Our Lives*)
Best Actress: Olivia De Havilland (*To Each His Own*)
Best Actor: Fredric March (*The Best Years Of Our Lives*)
Best Supporting Actress: Anne Baxter (*Razor's Edge*)
Best Supporting Actor: Harold Russell (*The Best Years Of Our Lives*)
Best Cinematography (Black & White): Arthur Miller (*Anna and the King of Siam*)
Best Cinematography (Colour): Charles Rosher, Leonard Smith and Arthur Arling (*The Yearling*)

Best Interior Decoration (Black & White): Lyle Wheeler, William Darling, Thomas Little and Frank E. Hughes (*Anna and the King of Siam*)
Best Interior Decoration (Colour): Cedric Gibbons, Paul Groesse and Edwin B. Willis (*The Yearling*)

1947
Best Picture: *Gentleman's Agreement*
Best Director: Elia Kazan (*Gentleman's Agreement*)
Best Actress: Loretta Young (*The Farmer's Daughter*)
Best Actor: Ronald Colman (*A Double Life*)
Best Supporting Actress: Celeste Holm (*Gentleman's Agreement*)
Best Supporting Actor: Edmund Gwenn (*Miracle on 34th Street*)
Best Foreign Film: *Sciuscia* (Shoeshine)
Best Cinematography (Black & White): Guy Green (*Great Expectations*)
Best Cinematography (Colour): Jack Cardiff (*Black Narcissus*)
Best Art Direction-Set Decoration (Black & White): John Bryan and Wilfred Shingleton (*Great Expectations*)
Best Art Direction-Set Decoration (Colour): Alfred Junge (*Black Narcissus*)

1948
Best Picture: *Hamlet*
Best Director: John Huston (*Treasure of the Sierra Madre*)
Best Actress: Jane Wyman (*Johnny Belinda*)
Best Actor: Laurence Olivier (*Hamlet*)
Best Supporting Actress: Claire Trevor (*Key Largo*)
Best Supporting Actor: Walter Huston (*Treasure of the Sierra Madre*)
Best Foreign Film: *Monsieur Vincent*
Best Cinematography (Black & White): William Daniels (*The Naked City*)
Best Cinematography (Colour): Joseph Valentine, William V. Skall and Winton Hoch (*Joan of Arc*)
Best Art Direction-Set Decoration (Black & White): Roger K. Furse and Carmen Dillon (*Hamlet*)
Best Art Direction-Set Decoration (Colour): Hein Heckroth and Arthur Lawson (*Red Shoes*)

1949
Best Picture: *All The King's Men*
Best Director: Joseph L. Mankiewicz (*A Letter to Three Wives*)
Best Actress: Olivia De Havilland (*The Heiress*)
Best Actor: Broderick Crawford (*All the King's Men*)
Best Supporting Actress: Mercedes McCambridge (*All the King's Men*)
Best Supporting Actor: Dean Jagger (*12 O'Clock High*)
Best Foreign Film: *Ladri di Biciclette* (Bicycle Thieves)
Best Cinematography (Black & White): Paul C. Vogel (*Battleground*)
Best Cinematography (Colour): Winton Hoch (*She Wore A Yellow Ribbon*)
Best Art Direction–Set Decoration (Black & White): John Meehan, Harry Horner and Emile Kuri (*The Heiress*)
Best Art Direction-Set Decoration (Colour): Cedric Gibbons, Paul Groese, Edwin B. Willis and Jack D. Moore (*Little Women*)

1950
Best Picture: *All About Eve*
Best Director: Joseph L. Mankiewicz (*All About Eve*)
Best Actress: Judy Holliday (*Born Yesterday*)
Best Actor: Jose Ferrer (*Cyrano de Bergerac*)
Best Supporting Actress: Josephine Hull (*Harvey*)
Best Supporting Actor: George Sanders (*All About Eve*)
Best Foreign Film: *Au Delà des Grilles* (The Walls of Malapaga)
Best Cinematography (Black & White): Robert Krasker (*The Third Man*)
Best Cinematography (Colour): Robert Surtees (*King Solomon's Mines*)
Best Art Direction-Set Decoration (Black & White): Hans Dreier, John Meehan, Sam Comer and Ray Moyer (*Sunset Boulevard*)
Best Art Direction-Set Decoration (Colour): Hans Dreier, Walter Tyler, Sam Comer and Ray Moyer (*Samson and Delilah*)

1951
Best Picture: *An American in Paris*
Best Director: George Stevens (*A Place in the Sun*)
Best Actress: Vivien Leigh (*A Streetcar Named Desire*)

Best Actor: Humphrey Bogart (*The African Queen*)
Best Supporting Actress: Kim Hunter (*A Streetcar Named Desire*)
Best Supporting Actor: Karl Malden (*A Streetcar Named Desire*)
Best Foreign Film: *Rashomon*
Best Cinematography (Black & White): William C. Mellor (*A Place in the Sun*)
Best Cinematography (Colour): Alfred Gilks and John Alton (*An American in Paris*)
Best Art Direction-Set Decoration (Black & White): Richard Day; George James Hopkins (*A Streetcar Named Desire*)
Best Art Direction-Set Decoration (Colour): Cedric Gibbons and Preston Ames; Edwin B. Willis and Keogh Gleason (*An American in Paris*)

1952
Best Picture: *The Greatest Show On Earth*
Best Director: John Ford (*The Quiet Man*)
Best Actress: Shirley Booth (*Come Back, Little Sheba*)
Best Actor: Gary Cooper (*High Noon*)
Best Supporting Actress: Gloria Grahame (*The Bad and the Beautiful*)
Best Supporting Actor: Anthony Quinn (*Viva Zapata!*)
Best Foreign Film: *Jeux Interdits* (Forbidden Games)
Best Cinematography (Black & White): Robert Surtees (*The Bad and The Beautiful*)
Best Cinematography (Colour): Winton C. Hoch and Archie Stout (*The Quiet Man*)
Best Art Direction-Set Decoration (Black & White): Cedric Gibbons and Edward Carfagno; Edwin B. Willis and Keogh Gleason (*The Bad and The Beautiful*)
Best Art Direction-Set Decoration (Colour): Paul Sheriff, Marcel Vertes (*Moulin Rouge*)

1953
Best Picture: *From Here To Eternity*
Best Director: Fred Zinnemann (*From Here to Eternity*)
Best Actress: Audrey Hepburn (*Roman Holiday*)
Best Actor: William Holden (*Stalag 17*)
Best Supporting Actress: Donna Reed (*From Here to Eternity*)
Best Supporting Actor: Frank Sinatra (*From Here to Eternity*)
Best Foreign Film: Not awarded
Best Cinematography (Black & White): Burnett Guffey (*From Here to Eternity*)
Best Cinematography (Colour): Loyal Griggs (*Shane*)
Best Art Direction-Set Decoration (Black & White): Cedric Gibbons and Edward Carfagno; Edwin B. Willis and Hugh Hunt (*Julius Caesar*)
Best Art Direction-Set Decoration (Colour): Lyle Wheeler and George W. Davis; Walter M. Scott and Paul S. Fox (*The Robe*)

1954
Best Picture: *On The Waterfront*
Best Director: Elia Kazan (*On the Waterfront*)
Best Actress: Grace Kelly (*The Country Girl*)
Best Actor: Marlon Brando (*On the Waterfront*)
Best Supporting Actress: Eva Marie Saint (*On the Waterfront*)
Best Supporting Actor: Edmond O'Brien (*The Barefoot Contessa*)
Best Foreign Film: *Jigokumon* (Gate of Hell)
Best Cinematography (Black & White): Boris Kaufman (*On the Waterfront*)
Best Cinematography (Colour): Milton Krasner (*Three Coins in a Fountain*)
Best Art Direction-Set Decoration (Black & White): Richard Day (*On the Waterfront*)
Best Art Direction-Set Decoration (Colour): John Meehan; Emile Kuri (*20,000 Leagues Under the Sea*)

1955
Best Picture: *Marty*
Best Director: Delbert Mann (*Marty*)
Best Actress: Anna Magnani (*The Rose Tattoo*)
Best Actor: Ernest Borgnine (*Marty*)
Best Supporting Actress: Jo Van Fleet (*East of Eden*)
Best Supporting Actor: Jack Lemmon (*Mister Roberts*)
Best Foreign Film: *Myamoto Musashi* (Samurai, The Legend of Musashi)
Best Cinematography (Black & White): James Wong Howe (*The Rose Tattoo*)

Best Cinematography (Colour): Robert Burks (*To Catch A Thief*)
Best Art Direction-Set Decoration (Black & White): Hal Pereira and Tambi Larsen; Sam Comer and Arthur Krams (*The Rose Tattoo*)
Best Art Direction-Set Decoration (Colour): William Flannery and Jo Mielziner; Robert Priestley (*Picnic*)

1956
Best Picture: *Around the World in 80 Days*
Best Director: George Stevens (*Giant*)
Best Actress: Ingrid Bergman (*Anastasia*)
Best Actor: Yul Brynner (*The King and I*)
Best Supporting Actress: Dorothy Malone (*Written on the Wind*)
Best Supporting Actor: Anthony Quinn (*Lust for Life*)
Best Foreign Film: *La Strada*
Best Cinematography (Black & White): Joseph Ruttenberg (*Somebody Up There Likes Me*)
Best Cinematography (Colour): Lionel Lindon (*Around the World in 80 Days*)
Best Art Direction-Set Decoration (Black & White): Cedric Gibbons and Malcolm F. Brown; Edwin B. Willis and F. Keogh Gleason (*Somebody Up There Likes Me*)
Best Art Direction-Set Decoration (Colour): Lyle R. Wheeler and John De Cuir; Walter M. Scott and Paul S. Fox (*The King and I*)

1957
Best Picture: *Bridge on the River Kwai*
Best Director: David Lean (*Bridge on the River Kwai*)
Best Actress: Joanne Woodward (*The Three Faces of Eve*)
Best Actor: Alec Guinness (*Bridge on the River Kwai*)
Best Supporting Actress: Miyoshi Umeki (*Sayonara*)
Best Supporting Actor: Red Buttons (*Sayonara*)
Best Foreign Film: *Le Notti di Cabiria* (The Nights of Cabiria)
Best Cinematography: Jack Hildyard (*Bridge on the River Kwai*)
Best Art Direction-Set Decoration: Ted Hawarth; Robert Priestley (*Sayonara*)

1958
Best Picture: *Gigi*
Best Director: Vincente Minnelli (*Gigi*)
Best Actress: Susan Hayward (*I Want to Live!*)
Best Actor: David Niven (*Separate Tables*)
Best Supporting Actress: Wendy Hiller (*Separate Tables*)
Best Supporting Actor: Burl Ives (*The Big Country*)
Best Foreign Film: *Mon Oncle* (My Uncle)
Best Cinematography (Black & White): Sam Leavitt (*The Defiant Ones*)
Best Cinematography (Colour): Joseph Ruttenberg (*Gigi*)
Best Art Direction-Set Decoration: William A. Horning and Preston Ames; Henry Grace and Keogh Gleason (*Gigi*)

1959
Best Picture: *Ben-Hur*
Best Director: William Wyler (*Ben-Hur*)
Best Actress: Simone Signoret (*Room at the Top*)
Best Actor: Charlton Heston (*Ben-Hur*)
Best Supporting Actress: Shelley Winters (*The Diary of Anne Frank*)
Best Supporting Actor: Hugh Griffith (*Ben-Hur*)
Best Foreign Film: *Orfeu Negro* (Black Orpheus)
Best Cinematography (Black & White): William C. Mellor (*The Diary of Anne Frank*)
Best Cinematography (Colour): Robert L. Surtees (*Ben-Hur*)
Best Art Direction-Set Decoration (Black & White): Lyle R. Wheeler and George W. Davis; Walter M. Scott and Stuart A. Reiss (*The Diary of Anne Frank*)
Best Art Direction-Set Decoration (Colour): William A. Horning and Edward Carfagno; Hugh Hunt (*Ben-Hur*)

1960
Best Picture: *The Apartment*
Best Director: Billy Wilder (*The Apartment*)
Best Actress: Elizabeth Taylor (*Butterfield 8*)
Best Actor: Burt Lancaster (*Elmer Gantry*)
Best Supporting Actress: Shirley Jones (*Elmer Gantry*)

Best Supporting Actor: Peter Ustinov (*Spartacus*)
Best Foreign Film: *Jungfru Källan* (The Virgin Spring)
Best Cinematography (Black & White): Freddie Francis (*Sons and Lovers*)
Best Cinematography (Colour): Russell Metty (*Spartacus*)
Best Art Direction-Set Decoration (Black & White): Alexander Trauner; Edward G. Boyle (*The Apartment*)
Best Art Direction-Set Decoration (Colour): Alexander Golitzen and Eric Orbom; Russell A. Gausman and Julia Heron (*Spartacus*)

1961
Best Picture: *West Side Story*
Best Director: Robert Wise and Jerome Robbins (*West Side Story*)
Best Actress: Sophia Loren (*Two Women*)
Best Actor: Maximilian Schell (*Judgement at Nuremberg*)
Best Supporting Actress: Rita Moreno (*West Side Story*)
Best Supporting Actor: George Chakiris (*West Side Story*)
Best Foreign Film: *Såsom I En Spegel* (Through A Glass Darkly)
Best Cinematography (Black & White): Eugen Shuftan (*The Hustler*)
Best Cinematography (Colour): Daniel L. Fapp (*West Side Story*)
Best Art Direction-Set Decoration (Black & White): Harry Horner; Gene Callahan (*The Hustler*)
Best Art Direction-Set Decoration (Colour): Boris Leven; Victor A. Gangelin (*West Side Story*)

1962
Best Picture: (*Lawrence of Arabia*)
Best Director: David Lean (*Lawrence of Arabia*)
Best Actress: Anne Bancroft (*The Miracle Worker*)
Best Actor: Gregory Peck (*To Kill A Mockingbird*)
Best Supporting Actress: Patty Duke (*The Miracle Worker*)
Best Supporting Actor: Ed Begley (*Sweet Bird of Youth*)
Best Foreign Film: *Cybèle Ou Les Dimanches De Ville D'Avray* (Sundays and Cybele)
Best Cinematography (Black & White): Jean Bourgoin and Walter Wottitz (*The Longest Day*)
Best Cinematography (Colour): Fred A. Young (*Lawrence of Arabia*)
Best Art Direction-Set Decoration (Black & White): Alexander Golitzen and Henry Bumstead; Oliver Emert (*To Kill A Mockingbird*)
Best Art Direction-Set Decoration (Colour): John Box and John Stoll; Dario Simoni (*Lawrence of Arabia*)

1963
Best Picture: *Tom Jones*
Best Director: Tony Richardson (*Tom Jones*)
Best Actress: Patricia Neal (*Hud*)
Best Actor: Sidney Poitier (*Lilies of the Field*)
Best Supporting Actress: Margaret Rutherford (*The V.I.P.'s*)
Best Supporting Actor: Melvyn Douglas (*Hud*)
Best Foreign Film: *Otto e Mezzo* ($8\frac{1}{2}$)
Best Cinematography (Black & White): James Wong Howe (*Hud*)
Best Cinematography (Colour): Leon Shamroy (*Cleopatra*)
Best Art Direction-Set Decoration (Black & White): Gene Callahan (*America, America*)
Best Art Direction-Set Decoration (Colour): John De Cuir, Jack Martin Smith, Hilyard Brown, Herman Blumenthal, Elven Webb, Maurice Pelling and Boris Juraga; Walter M. Scott, Paul S. Fox and Ray Moyer (*Cleopatra*)

1964
Best Picture: *My Fair Lady*
Best Director: George Cukor (*My Fair Lady*)
Best Actress: Julie Andrews (*Mary Poppins*)
Best Actor: Rex Harrison (*My Fair Lady*)
Best Supporting Actress: Lila Kedrova (*Zorba, the Greek*)
Best Supporting Actor: Peter Ustinov (*Topkapi*)
Best Foreign Film: *Ieri, Oggi, Domani* (Yesterday, Today and Tomorrow)
Best Cinematography (Black & White): Walter Lassally (*Zorba, the Greek*)
Best Cinematography (Colour): Harry Stradling (*My Fair Lady*)
Best Art Direction-Set Decoration (Black & White): Vassilis Fotopoulos (*Zorba, the Greek*)
Best Art Direction-Set Decoration

(Colour): Gene Allen and Cecil Beaton; George James Hopkins (*My Fair Lady*)

1965

Best Picture: *The Sound of Music*
Best Director: Robert Wise (*The Sound of Music*)
Best Actress: Julie Christie (*Darling*)
Best Actor: Lee Marvin (*Cat Ballou*)
Best Supporting Actress: Shelley Winters (*A Patch of Blue*)
Best Supporting Actor: Martin Balsam (*A Thousand Clowns*)
Best Foreign Film: *Obchod Od na Korze* (The Shop on the High Street)
Best Cinematography (Black & White): Ernest Laszlo (*Ship of Fools*)
Best Cinematography (Colour): Freddie Young (*Dr. Zhivago*)
Best Art Direction-Set Decoration (Black & White): Robert Clatworthy; Joseph Kish (*Ship of Fools*)
Best Art Direction-Set Decoration (Colour): John Box and Terry Marsh; Dario Simoni (*Dr. Zhivago*)

1966

Best Picture: *A Man for All Seasons*
Best Director: Fred Zinnemann (*A Man for All Seasons*)
Best Actress: Elizabeth Taylor (*Who's Afraid of Virginia Woolf?*)
Best Actor: Paul Scofield (*A Man for All Seasons*)
Best Supporting Actress: Sandy Dennis (*Who's Afraid of Virginia Woolf?*)
Best Supporting Actor: Walter Matthau (*The Fortune Cookie;* UK: *Meet Whiplash Willie*)
Best Foreign Film: *Un Homme et une Femme* (A Man and a Woman)
Best Cinematography (Black & White): Haskell Wexler (*Who's Afraid of Virginia Woolf?*)
Best Cinematography (Colour): Ted Moore (*A Man for All Seasons*)
Best Art Direction-Set Decoration (Black & White): Richard Sylbert; George James Hopkins (*Who's Afraid of Virginia Woolf?*)
Best Art Direction-Set Decoration (Colour): Jack Martin Smith and Dale Hennesy; Walter M. Scott and Stuart A. Reiss (*Fantastic Voyage*)

1967

Best Picture: *In The Heat of the Night*
Best Director: Mike Nichols (*The Graduate*)
Best Actress: Katharine Hepburn (*Guess Who's Coming to Dinner*)
Best Actor: Rod Steiger (*In the Heat of the Night*)
Best Supporting Actress: Estelle Parsons (*Bonnie and Clyde*)
Best Supporting Actor: George Kennedy (*Cool Hand Luke*)
Best Foreign Film: *Ostře Sledované Vlaky* (Closely Observed Trains)
Best Cinematography: Burnett Guffey (*Bonnie and Clyde*)
Best Art Direction-Set Decoration: John Truscott and Edward Careere; John W. Brown (*Camelot*)

1968

Best Picture: *Oliver!*
Best Director: Carol Reed (*Oliver!*)
Best Actress: Katharine Hepburn (*The Lion in Winter*); Barbra Streisand (*Funny Girl*)
Best Actor: Cliff Robertson (*Charly*)
Best Supporting Actress: Ruth Gordon (*Rosemary's Baby*)
Best Supporting Actor: Jack Albertson (*The Subject Was Roses*)
Best Foreign Film: *Voina I Mir* (War and Peace)
Best Cinematography: Pasqualino De Santis (*Romeo and Juliet*)
Best Art Direction-Set Decoration: John Box and Terence Marsh; Vernon Dixon and Ken Muggleston (*Oliver!*)

1969

Best Picture: *Midnight Cowboy*
Best Director: John Schlesinger (*Midnight Cowboy*)
Best Actress: Maggie Smith (*The Prime of Miss Jean Brodie*)
Best Actor: John Wayne (*True Grit*)
Best Supporting Actress: Goldie Hawn (*Cactus Flower*)
Best Supporting Actor: Gig Young (*They Shoot Horses Don't They?*)
Best Foreign Film: *Z*
Best Cinematography: Conrad Hall (*Butch Cassidy and The Sundance Kid*)
Best Art Direction-Set Decoration: John De Cuir, Jack Martin Smith and Herman Blumenthal; Walter M. Scott, George Hopkins and Raphael Bretton (*Hello, Dolly!*)

1970

Best Picture: *Patton*
Best Director: Franklin J. Schaffner (*Patton*)

Best Actress: Glenda Jackson (*Women in Love*)
Best Actor: George C. Scott (*Patton*)
Best Supporting Actress: Helen Hayes (*Airport*)
Best Supporting Actor: John Mills (*Ryan's Daughter*)
Best Foreign Film: *Indagine su un Cittadino al di Sopra di ogni Sospetto* (Investigation of a Citizen Above Suspicion)
Best Cinematography: Freddie Young (*Ryan's Daughter*)
Best Art Direction-Set Decoration: Urie McCleary and Gil Parrondo; Antonio Mateos and Pierre-Louis Thevent (*Patton*)

1971
Best Picture: *The French Connection*
Best Director: William Friedkin (*The French Connection*)
Best Actress: Jane Fonda (*Klute*)
Best Actor: Gene Hackman (*The French Connection*)
Best Supporting Actress: Cloris Leachman (*The Last Picture Show*)
Best Supporting Actor: Ben Johnson (*The Last Picture Show*)
Best Foreign Film: *Il Giardino dei Finzi-Contini* (The Garden of the Finzi-Continis)
Best Cinematography: Oswald Morris (*Fiddler On The Roof*)
Best Art Direction-Set Decoration: John Box, Ernest Archer, Jack Maxsted and Gil Parrondo; Vernon Dixon (*Nicholas and Alexandria*)

1972
Best Picture: *The Godfather*
Best Director: Bob Fosse (*Cabaret*)
Best Actress: Liza Minnelli (*Cabaret*)
Best Actor: Marlon Brando (*The Godfather*)
Best Supporting Actress: Eileen Heckart (*Butterflies Are Free*)
Best Supporting Actor: Joel Grey (*Cabaret*)
Best Foreign Film: *Le Charme Discret de la Bourgeoisie* (The Discreet Charm of the Bourgeoisie)
Best Cinematography: Geoffrey Unsworth (*Cabaret*)
Best Art Direction-Set Decoration: Rolf Zehetbauer and Jurgen Kiebach; Herbert Strabel (*Cabaret*)

1973
Best Picture: *The Sting*
Best Director: George Roy Hill (*The Sting*)
Best Actress: Glenda Jackson (*A Touch of Class*)
Best Actor: Jack Lemmon (*Save the Tiger*)
Best Supporting Actress: Tatum O'Neal (*Paper Tiger*)
Best Supporting Actor: John Houseman (*The Paper Chase*)
Best Foreign Film: *La Nuit Americaine* (Day for Night)
Best Cinematography: Sven Nykvist (*Viskingar och Rop* (Cries and Whispers))
Best Art Direction-Set Decoration: Henry Bumstead; James Payne (*The Sting*)

1974
Best Picture: *The Godfather Part II*
Best Director: Francis Ford Coppola (*The Godfather Part II*)
Best Actress: Ellen Burstyn (*Alice Doesn't Live Here Anymore*)
Best Actor: Art Carney (*Harry and Tonto*)
Best Supporting Actress: Ingrid Bergman (*Murder on the Orient Express*)
Best Supporting Actor: Robert De Niro (*The Godfather Part II*)
Best Foreign Film: *Amarcord*
Best Cinematography: Fred Koenekamp and Joseph Biroc (*The Towering Inferno*)
Best Art Direction-Set Decoration: Dean Tavoularis and Angelo Graham; George R. Nelson (*The Godfather Part II*)

1975
Best Picture: *One Flew Over the Cuckoo's Nest*
Best Director: Milos Forman (*One Flew Over the Cuckoo's Nest*)
Best Actress: Louise Fletcher (*One Flew Over the Cuckoo's Nest*)
Best Actor: Jack Nicholson (*One Flew Over the Cuckoo's Nest*)
Best Supporting Actress: Lee Grant (*Shampoo*)
Best Supporting Actor: George Burns (*The Sunshine Boys*)
Best Foreign Film: *Derzu Uzala*
Best Cinematography: John Alcott (*Barry Lyndon*)

Best Art Direction-Set Decoration: Ken Adam and Roy Walker; Vernon Dixon (*Barry Lyndon*)

1976

Best Picture: *Rocky*
Best Director: John G. Avildsen (*Rocky*)
Best Actress: Faye Dunaway (*Network*)
Best Actor: Peter Finch (*Network*)
Best Supporting Actress: Beatrice Straight (*Network*)
Best Supporting Actor: Jason Robards (*All the President's Men*)
Best Foreign Film: *La Victoire en Chantant* (Black and White in Colour)
Best Cinematography: Haskell Wexler (*Bound for Glory*)
Best Art Direction-Set Decoration: George Jenkins; George Gaines (*All the President's Men*)

1977

Best Picture: *Annie Hall*
Best Director: Woody Allen (*Annie Hall*)
Best Actress: Diane Keaton (*Annie Hall*)
Best Actor: Richard Dreyfuss (*The Goodbye Girl*)
Best Supporting Actress: Vanessa Redgrave (*Julia*)
Best Supporting Actor: Jason Robards (*Julia*)
Best Foreign Film: *La Vie Devant Soi* (Madame Rosa)
Best Cinematography: Vilmos Zsigmond (*Close Encounters Of The Third Kind*)
Best Art Direction-Set Decoration: John Barry, Norman Reynolds and Leslie Dilley; Roger Christian (*Star Wars*)

1978

Best Picture: *The Deer Hunter*
Best Director: Michael Cimino (*The Deer Hunter*)
Best Actress: Jane Fonda (*Coming Home*)
Best Actor: Jon Voight (*Coming Home*)
Best Supporting Actress: Maggie Smith (*California Suite*)
Best Supporting Actor: Christopher Walken (*The Deer Hunter*)
Best Foreign Film: *Préparez vos Mouchoirs* (Get Out Your Handkerchiefs)
Best Cinematography: Nestor Almendros (*Days of Heaven*)
Best Art Direction-Set Decoration: Paul Sylbert and Edwin O'Donovan; George Gaines (*Heaven Can Wait*)

1979

Best Picture: *Kramer Vs. Kramer*
Best Director: Robert Benton (*Kramer Vs. Kramer*)
Best Actress: Sally Field (*Norma Rae*)
Best Actor: Dustin Hoffman (*Kramer Vs. Kramer*)
Best Supporting Actress: Meryl Streep (*Kramer Vs. Kramer*)
Best Supporting Actor: Melvyn Douglas (*Being There*)
Best Foreign Film: *Die Blechtrommel* (The Tin Drum)
Best Cinematography: Vittorio Storaro (*Apocalypse Now*)
Best Art Direction-Set Decoration: Philip Rosenberg and Tony Walton; Edward Stewart and Gary Brink (*All That Jazz*)

1980

Best Picture: *Ordinary People*
Best Director: Robert Redford (*Ordinary People*)
Best Actress: Sissy Spacek (*Coalminer's Daughter*)
Best Actor: Rober De Niro (*Raging Bull*)
Best Supporting Actress: Mary Steenburgen (*Melvin and Howard*)
Best Supporting Actor: Timothy Hutton (*Ordinary People*)
Best Foreign Film: *Moskava Slezam Ne Verit* (Moscow Distrusts Tears)
Best Cinematography: Geoffrey Unsworth and Ghislain Cloquet (*Tess*)
Best Art Direction-Set Decoration: Pierre Guffroy and Jack Stevens (*Tess*)

1981

Best Picture: *Chariots of Fire*
Best Director: Warren Beatty (*Reds*)
Best Actress: Katharine Hepburn (*On Golden Pond*)
Best Actor: Henry Fonda (*On Golden Pond*)
Best Supporting Actress: Maureen Stapleton (*Reds*)
Best Supporting Actor: Sir John Gielgud (*Arthur*)
Best Foreign Film: *Mephisto*
Best Cinematography: Vittorio Storaro (*Reds*)
Best Art Direction-Set Decoration: Norman Reynolds and Leslie Dilley; Michael Ford (*Raiders of the Lost Ark*)

1982

Best Picture: *Gandhi*
Best Director: Richard Attenborough (*Gandhi*)

Best Actress: Meryl Streep (*Sophie's Choice*)
Best Actor: Ben Kingsley (*Gandhi*)
Best Supporting Actress: Jessica Lange (*Tootsie*)
Best Supporting Actor: Louis Gossett, Jr. (*An Officer and a Gentleman*)
Best Foreign Film: *Volver A Empezar* (*To Begin Again*)
Best Cinematography: Billy Williams and Ronnie Taylor (*Gandhi*)
Best Art Direction-Set Decoration: Stuart Craig and Bob Laing; Michael Seirton (*Gandhi*)

1983
Best Picture: *Terms of Endearment*
Best Director: James L. Brooks (*Terms of Endearment*)
Best Actress: Shirley MacLaine (*Terms of Endearment*)
Best Actor: Robert Duvall (*Tender Mercies*)
Best Supporting Actress: Linda Hunt (*The Year of Living Dangerously*)
Best Supporting Actor: Jack Nicholson (*Terms of Endearment*)
Best Foreign Film: *Fanny och Alexander* (Fanny and Alexander)
Best Cinematography: Sven Nyvist (*Fanny och Alexander*) (Fanny and Alexander)
Best Art Direction-Set Decoration: Susanne Lingheim (*Fanny och Alexander*, (Fanny and Alexander))

1984
Best Picture: *Amadeus*
Best Director: Milos Forman (*Amadeus*)
Best Actress: Sally Field (*Places in the Heart*)
Best Actor: F. Murray Abraham (*Amadeus*)
Best Supporting Actress: Peggy Ashcroft (*A Passage to India*)
Best Supporting Actor: Haing S. Ngor (*The Killing Fields*)
Best Foreign Film: *La Diagonale du Fou* (Dangerous Moves)
Best Cinematography: Chris Menges (*The Killing Fields*)
Best Art Direction-Set Decoration: Patrizia Von Brandenstein; Karel Cerny (*Amadeus*)

1985
Best Picture: *Out of Africa*
Best Director: Sydney Pollack (*Out of Africa*)
Best Actress: Geraldine Page (*The Trip To Bountiful*)
Best Actor: William Hurt (*The Kiss of the Spiderwoman*)
Best Supporting Actress: Anjelica Huston (*Prizzi's Honor*)
Best Supporting Actor: Don Ameche (*Cocoon*)
Best Foreign Film: *La Historia Official* (The Official Version)
Best Cinematography: David Watkin (*Out of Africa*)
Best Art Direction-Set Decoration: Herbert Westbrook, Colin Grimes and Cliff Robinson; Josie MacAvin (*Out of Africa*)

1986
Best Picture: *Platoon*
Best Director: Oliver Stone (*Platoon*)
Best Actress: Marlee Matlin (*Children of a Lesser God*)
Best Actor: Paul Newman (*The Color of Money*)
Best Supporting Actress: Dianne Wiest (*Hannah and Her Sisters*)
Best Supporting Actor: Michael Caine (*Hannah and Her Sisters*)
Best Foreign Film: *De Aanslag* (The Assault)
Best Cinematography: Chris Menges (*The Mission*)
Best Art Direction-Set Decoration: Gianni Quaranta, Brian Ackland-Snow, Brian Savegar, Elio Altramusa (*A Room with a View*)

1987
Best Picture: *The Last Emperor*
Best Director: Bernardo Bertolucci (*The Last Emperor*)
Best Actress: Cher (*Moonstruck*)
Best Actor: Michael Douglas (*Wall Street*)
Best Supporting Actress: Olympia Dukakis (*Moonstruck*)
Best Supporting Actor: Sean Connery (*The Untouchables*)
Best Foreign Film: *Babettes Gaestebud* (Babette's Feast)
Best Cinematography: Vittorio Storaro (*The Last Emperor*)
Best Art Direction-Set Decoration: Ferdinando Scarfiotti (*The Last Emperor*)

1988
Best Picture: *Rain Man*
Best Director: Barry Levinson (*Rain Man*)

Best Actress: Jodie Foster (*The Accused*)
Best Actor: Dustin Hoffman (*Rain Man*)
Best Supporting Actress: Geena Davis (*The Accidental Tourist*)
Best Supporting Actor: Kevin Kline (*A Fish Called Wanda*)
Best Foreign Film: *Pelle Erobreren* (Pelle The Conqueror)
Best Cinematography: Peter Biziou (*Mississippi Burning*)
Best Art Direction-Set Decoration: Stuart Craig, Gerard James (*Dangerous Liaisons*)

1989
Best Picture: *Driving Miss Daisy*
Best Director: Oliver Stone (*Born on the 4th of July*)
Best Actress: Jessica Tandy (*Driving Miss Daisy*)
Best Actor: Daniel Day-Lewis (*My Left Foot*)
Best Supporting Actress: Brenda Fricker (*My Left Foot*)
Best Supporting Actor: Denzel Washington (*Glory*)
Best Foreign Film: *Nuovo Cinema Paradiso* (Cinema Paradiso)
Best Cinematography: Freddie Francis (*Glory*)
Best Art Direction-Set Decoration: Anton Furst, Peter Young (*Batman*)

1990
Best Picture: *Dances with Wolves*
Best Director: Kevin Costner (*Dances with Wolves*)
Best Actress: Kathy Bates (*Misery*)
Best Actor: Jeremy Irons (*Reversal of Fortune*)
Best Supporting Actress: Whoopi Goldberg (*Ghost*)
Best Supporting Actor: Joe Pesci (*Goodfellas*)
Best Foreign Film: *Reise der Hoffnung* (Journey of Hope)
Best Cinematography: Dean Semler (*Dances with Wolves*)
Best Art Direction-Set Decoration: Richard Sybert and Rick Simpson (*Dick Tracy*)

Berlin Film Festival

1956
Best Film: First prize *Invitation to the Dance*
Second Prize *Richard III*
Best Director: Robert Aldrich (*Autumn Leaves*)
Best Actress: Elsa Martinelli (*Donatella*)
Best Actor: Burt Lancaster (*Trapeze*)

1957
Best Film: *Twelve Angry Men*
Best Director: Mario Monicelli (*Padri e Figli* (Fathers and Sons))
Best Actress: Yvonne Mitchell (*Woman in a Dressing Gown*)
Best Actor: Pedro Infante (*Tizoc*)

1958
Best Film: *The End of the Day*
Best Director: Tadashi Imai, *Junai Monogatari* (Story of True Love)
Best Actress: Anna Magnani (*Wild is the Wind*)
Best Actor: Sidney Poitier (*The Defiant Ones*)

1959
Best Film: *Les Cousins* (The Cousins)
Best Director: Akira Kurosawa (*Kakushi Toride No San-Akunin* (The Hidden Fortress))
Best Actress: Shirley Maclaine (*Ask Any Girl*)
Best Actor: Jean Gabin (*Archimede the Tramp* (Archimède le Clochard))

1960
Best Film: First Prize *Lazarillo de Tormes*
Second Prize *The Love Game*
Best Director: Jean-Luc Godard (*A Bout de Souffle* (Breathless))
Best Actress: Juliette Mayniel (*Country Fair*)
Best Actor: Fredric March (*Inherit the Wind*)

1961
Best Film: *La Notte* (The Night)
Best Director: Bernhard Wicki (*Das Wunderdes Malachias* (The Miracle of Father Malachias))
Best Actress: Anna Karina (*Une Femme est une Femme* (A Woman is a Woman))
Best Actor: Peter Finch (*No Love for Johnnie*)

1962
Best Film: *A Kind of Loving*
Best Director: Francesco Rosi (*Salvatore Guiliano*)
Best Actress: Rita Gam and Viveca Lindfors (*No Exit*)
Best Actor: James Stewart (*Mr. Hobbs Takes a Vacation*)

1963
Best Film: *Oath of Obedience; Il Diavolo* (The Devil)
Best Director: Nikos Koundouros (*Little Aphrodite*)
Best Actress: Bibi Andersson (*The Lovers*)
Best Actor: Sidney Poitier (*Lilies of the Field*)

1964
Best Film: *Susuz Yaz* (Waterless Summer)
Best Director: Satyajit Ray (*Mahanager* (The Big City))
Best Actress: Sachiko Hidari (*Kanojo To Kare* (She and He))
Best Actor: Rod Steiger (*The Pawnbroker*)

1965
Best Film: *Alphaville*
Best Director: Satyajit Ray (*Charluta* (The Lonely Wife))
Best Actress: Mahdur Jaffrey (*Shakespeare Wallah*)
Best Actor: Lee Marvin (*Cat Ballou*)

1966
Best Film: *Cul de Sac*
Best Director: Carlos Saura (*La Caza* (The Hunt))
Best Actress: Lola Albright (*Lord Love a Duck*)
Best Actor: Jean-Pierre Léaud (*Masculin-Féminin* (Masculine-Feminine))

1967
Best Film: *Le Départ*
Best Director: Zivojin Pavlovic (*The Rats Awaken*)
Best Actress: Edith Evans (*The Whisperers*)

Best Actor: Michel Simon (*La Vieil Homme et L'Enfant* (The Old Man and the Boy))

1968

Best Film: First Prize *Ole Dole Doff*
Second Prize *Nevinost Bez Zastite* (Innocence Unprotected); *Come L'Amore*
Best Director: Carlos Saura (*Peppermint Frappé*)
Best Actress: Stephane Audran (*Les Biches* (The Does))
Best Actor: Jean-Louis Trintignant (*L'Homme qui Ment* (The Man Who Lies))

1969

Best Film: First Prize *Early Years*
Second Prize *Brazil Year 2000; Made in Sweden; I am a Elephant, Madame; Greetings; Un Tranquillo Posto di Campagna* (A Quiet Place in the Country)

1970

Prizes suspended

1971

Best Film: First Prize *Il Giardino dei Finzi-Contini* (The Garden of the Finzi-Continis)
Second Prize *Il Decamerone* (The Decameron)
Best Director: (Not awarded)
Best Actress: Shirley MacLaine (*Desperate Characters*)
Simone Signoret (*Le Chat* (The Cat))
Best Actor: Jean Gabin (*Le Chat* (The Cat))

1972

Best Film: First Prize *Il Racconti di Canterbury* (The Canterbury Tales)
Second Prize *The Hospital*
Best Director: Jean-Pierre Blanc (*The Spinster*)
Best Actress: Elizabeth Taylor (*Hammersmith is Out*)
Best Actor: Albert Sordi (*Detenuto in Attesa di Giudizio* (Detained Whilst Waiting for Justice))

1973

Best Film: First Prize *Distant Thunder*
Best Film: Special Jury Prize *Where There's Smoke There's Fire*
Best Film: Second Prize *The Revolution of the Seven Madmen; Le Grand Blond avec une chaussure Noire* (The Tall Blond Man with One Black Shoe); *The Experts; All Nudity Will be Punished; The 14*
Best Director: (Not awarded)
Best Actress: (Not awarded)
Best Actor: (Not awarded)

1974

Best Film: First Prize *The Apprenticeship of Duddy Kravitz*
Best Film: Special Jury Prize *L'Horloger de St. Paul* (The Watchmaker of St. Paul)
Best Film: Second Prize *In the Name of the People; Little Malcolm; Still Life; Pane e Cioccolata* (Bread and Chocolate); *Rebellion in Patagonia*
Best Director: (Not awarded)
Best Actress: Marta Vancourova (*The Lovers of the Year 1*)
Best Actor: Antonio Ferrandiz (*Next of Kin*)

1975

Best Film: First Prize *Orkobefogadas*
Best Film: Special Jury Prizes *Overlord; Dupont Lajoie*
Best Director: Sergey Solovyov (*A Hundred Days After Childhood*)
Best Actress: Kinuyo Tanaka (*Sandakan, House 8*)
Best Actor: Vlastimil Brodsky (*Jakob der Lügner*)

1976

Best Film: First Prize *Buffalo Bill and the Indians* (prize declined)
Best Film: Special Jury Prize *Canoa*
Best Director: Mario Monicelli (*Caro Michele*)
Best Actress: Jadwiga Baranska (*Night and Days*)
Best Actor: Gerhard Olschewski (*Lost Life*)

1977

Best Film: First Prize *Voskhozzhdenie* (The Ascent)
Second Prize *Le Diable, Probablement* (The Devil Probably); *The Bricklayers; A Strange Role*
Best Director: Manuel Gutierrez (*Black Litter*)
Best Actress: Lily Tomlin (*The Late Show*)
Best Actor: Fernando Fernan Gomez (*The Anchorite*)

1978

Best Film: 'the entire Spanish program', which included *Max's Words* and *The Trouts*
Best Film: Special Jury Prize *A Queda*
Best Director: Georgi Dyulgerov (*Advantage*)
Best Actress: Gena Rowlands (*Opening Night*)
Best Actor: Craig Russell (*Outrageous*)

1979

Best Film: First Prize *David*
Best Film: Special Jury Prize *Alexandria–Why*
Best Director: Astrid Henning Jensen (*Vinterbørn* (Winter Children))
Best Actress: Hanna Schygulla (*Die Ehe der Maria Braun* (The Marriage of Maria Braun))
Best Actor: Michele Pacido (*Ernesto*)

1980

Best Film: First Prizes *Heartland; Palermo Oder Wolfsburg*
Best Film: Special Jury Prize *Chiedo Asilo*
Best Director: Istvan Szabo (*Bizalom* (Confidence))
Best Actress: Renate Krossner (*Solo Sunny*)
Best Actor: Andrzej Seweryn (*Dyrygent* (The Conductor))

1981

Best Film: *Di Presa, Di Presa* (Hurry, Hurry)
Best Director: Markus Imhoof (*Das Boot ist Voll* (The Boat Is Full))
Best Actress: Barbara Grabowska (*Goracza* (Fever))
Best Actor: Jack Lemmon (*Tribute*); Anatoli Solonitsyn (*Dwadzat Schest Dne is Shisni Dostojewskogo* (26 Days In The Life Of Dostoevsky))

1982

Best Film: *Die Sehnsucht der Veronika Voss* (Veronika Voss)
Best Director: Mario Monicelli (*The Marquis Of Grillo*)
Best Actress: Katrin Saaa (*On Probation*)
Best Actor: Michel Piccoli (*Une Etrange Affaire* (A Strange Affair))

1983

Best Film: *Ascendency; The Beehive*
Best Director: Eric Rohmer, *Pauline à la Plage* (*Pauline at the Beach*)
Best Actress: Jewgenija Gluschenko (*Vlublen Po Sobstvennomu*) (Love By Request))
Best Actor: Bruce Dern (*The Championship Season*)

1984

Best Film: *Love Streams*
Best Director: Ettore Scola (*Le Bal*)
Best Actress: Monica Vitti (*The Flirt*)
Best Actor: Albert Finney (*The Dresser*)

1985

Best Film: *Wetherby; The Woman and the Stranger*
Best Director: Robert Benton (*Places in the Heart*)
Best Actress: Jo Kennedy (*Wrong World*)
Best Actor: Fernando Fernan Gomez (*Stico*)

1986

Best Film: *Stammhein*
Best Director: Georgi Shengelaya (*A Young Composer's Odyssey*)
Best Actress: Charlotte Valandrey (*Red Kiss*); Marcella Cartaxo (*The Hour of the Star*)
Best Actor: Tuncel Kurtiz (*The Smile of The Lamb*)

1987

Best Film: *Thema* (The Theme)
Best Director: Oliver Stone, *Platoon*
Best Actress: Ana Beatriz Nogueira (*Vera*)
Best Actor: Gian Maria Volonte (*The Moro Case*)

1988

Best Film: *Red Sorghum*
Best Director: Norman Jewison (*Moonstruck*)
Best Actress: Anne Bancroft (*84 Charing Cross Road*)
Best Actor: Jorg Pose and Manfred Mock (*Bear Ye One Another's Burdens*)

1989

Best Film: *Rain Man*
Best Director: Dusan Hanak (*Ja Milujem, Ty Milujes* (*I Love, You Love*))
Best Actress: Isabelle Adjani (*Camille Claudel*)
Best Actor: Gene Hackman (*Mississippi Burning*)

1990

Best Film: *Skřivànci na Niti* (Larks on a String); *The Music Box*

Best Director: Michael Verhoeven (*Das Schreckliche Mädchen* (The Nasty Girl))
Best Actor: Iain Glen (*The Silent Scream*)
Best Joint Performance: Jessica Tandy, Morgan Freeman (*Driving Miss Daisy*)

1991

Best Film: *La Casa de Sorriso* (The House of Smiles)
Best Director: Ricky Tognazzi (*Ultra*) Jonathan Demme (*Silence of the Lambs*)
Best Actress: Victoria Abril (*Amantes* (Lovers))
Best Actor: Maynard Eziashi (*Mr Johnson*)
Outstanding Single Achievement as Actor, Producer and Director: Kevin Costner (*Dances with Wolves*)

Cannes Film Festival

1946
Best Films: *La Bataille du Rail* (Battle of the Rails)*; Symphonie Pastorale; The Lost Weekend; Brief Encounter; Roma, Città A perta* (Rome, Open City)*; Maria Candelaria; The Last Chance*
Best Director: René Clément (*La Bataille du Rail* (Battle of the Rails))
Best Actress: Michèle Morgan (*Symphonie Pastorale*)
Best Actor: Ray Milland (*The Lost Weekend*)

1947
Best Films: *Antoine et Antoinette; Les Maudits* (The Damned)*; Crossfire; Dumbo; Ziegfeld Follies*
Best Director: (Not awarded)
Best Actress: (Not awarded)
Best Actor: (Not awarded)

1948
No festival

1949
Best Film: *The Third Man*
Best Director: René Clément (*Au Delà des Grilles* (The Walls of Malapaga))
Best Actress: Isa Miranda (*Au Delà des Grilles* (The Walls of Malaplaga))
Best Actor: Edward G. Robinson (*House of Strangers*)

1950
No festival

1951
Best Film: *Miracolo a Milano* (Miracle in Milan)*; Miss Julie*
Best Director: Luis Buñuel (*Los Olvidados* (The Young and the Damned))
Best Actress: Bette Davis (*All About Eve*)
Best Actor: Michael Redgrave (*The Browning Version*)

1952
Best Film: *Othello; Two Cents Worth of Hope*
Best Director: Christian-Jaque (*Fanfan la Tulipe* (Fanfan the Tulip))
Best Actress: Lee Grant (*Detective Story*)
Best Actor: Marlon Brando (*Viva Zapata!*)

1953
Best Film: *La Salaire de la Peur* (Wages of Fear)
Best Director: Walt Disney, for all his work
Best Actress: Shirley Booth (*Come Back, Little Sheba*)
Best Actor: Charles Vanel (*La Salaire de la Peur* (Wages of Fear))

1954
Best Film: *Jigokumon* (Gate of Hell)
Best Director: René Clément (*Monsieur Ripois* (Knave of Hearts))
Best Actress: (Not awarded)
Best Actor: (Not awarded)

1955
Best Film: *Marty*
Best Director: Jules Dassin (*Rififi*)*;* Serge Vasiliev (*Geroite na Shipka* (Heroes of Shipka)
Best Actress: Betsy Blair (*Marty*)
Best Actor: Spencer Tracy (*Bad Day at Black Rock*); Ernest Borgnine (*Marty*)

1956
Best Film: *World Of Silence*
Best Director: Serge Youtkevitch (*Otello* (*Othello*))
Best Actress: Susan Hayward (*I'll Cry Tomorrow*)
Best Actor: (Not awarded)

1957
Best Film: *Friendly Persuasion*
Best Director: Robert Bresson (*Un Condamné à Mort s'est Échappé* (A Man Escaped))
Best Actress: Guilietta Massina (*Nights of Cabiria*)
Best Actor: John Kitzmiller (*Valley of Peace*)

1958
Best Film: (*Letyat Zhuravli* (The Cranes Are Flying))
Best Director: Ingmar Bergman (*Nara Livet* (So Close to Life))
Best Actress: (collective prize) Eva Dahlbeck, Ingrid Thulin, Bibi Andersson, Babro Ornas (*Nara Livet* (So Close to Life))
Best Actor: Paul Newman (*The Long Hot Summer*)

1959
Best Film: *Orfeu Negro* (Black Orpheus)
Best Director: François Truffaut (*Les Quatres Cents Coups* (The 400 Blows))
Best Actress: Simone Signoret (*Room at the Top*)
Best Actor: (collective prize) Dean Stockwell, Bradford Dillman, Orson Welles (*Compulsion*)

1960
Best Film: *La Dolce Vita* (*Jungfrukällan* (The Virgin Spring) and *The Young One* were declared too good to be judged)
Best Director: (Not awarded)
Best Actress: Melina Mercouri (*Never on Sunday*); Jeanne Moreau (*Moderato Cantabile*)
Best Actor: (Not awarded)

1961
Best Film: *Viridiana; Une Aussi Longue Absence*
Best Director: Yulia Solntseva, *History of the Flaming Years*
Best Actress: Sophia Loren (*La Ciociara* (Two Women))
Best Actor: Anthony Perkins (*Goodbye Again*)

1962
Best Film: *O Pagador de Promessas* (The Given Word)
Best Director: (Not awarded)
Best Acting: (collective prize) Katharine Hepburn, Ralph Richardson, Jason Robards, Jr., Dean Stockwell (*Long Day's Journey Into Night*); Rita Tushingham, Murray Melvin (*A Taste of Honey*)

1963
Best Picture: *Il Gattopardo* (The Leopard)
Best Director: (Not awarded)
Best Actress: Marina Vlady (*Una Storia Moderna: L'Ape Regina* (The Queen Bee))
Best Actor: Richard Harris (*This Sporting Life*)

1964
Best Film: *Les Parapluies de Cherbourg* (The Umbrellas of Cherbourg)
Best Director: (Not awarded)
Best Actress: Anne Bancroft (*The Pumpkin Eater*); Barbara Barrie (*One Potato, Two Potato*)
Best Actor: Antal Pager (*Pacsirta*); Saro Urzi (*Sedotta e Abandonata* (Seduced and Abandoned)

1965
Best Film: *The Knack*
Best Director: Liviu Ciulei (*Padurea Spinzuratibr* (The Lost Forest))
Best Acting: (awarded collectively) Samantha Eggar and Terence Stamp (*The Collector*)

1966
Best Film: *Un Homme et une Femme* (A Man and a Woman)*; Signore e Signori* (The Birds, the Bees and the Italians)
Best Director: Serge Youtkevitch (*Lenin V Polshe* (Lenin in Poland))
Best Actress: Vanessa Redgrave (*Morgan*)
Best Actor: Per Oscarsson (*Sult* (Hunger)

1967
Best Film: *Blow-Up*
Best Director: Ferenc Kosa (*Ten Thousand Suns*)
Best Actress: Pia Degermark (*Elvira Madigan*)
Best Actor: Odded Kotler (*Shlosha Yamin ve Yeled* (Three Days and a Child))

1968
Festival closed

1969
Best Film: *If*
Best Director: Glauber Rocha (*Antonio Das Mortes*); Vojtech Jasny (*My Dear*)
Best Actress: Vanessa Redgrave (*Isadora*)
Best Actor: Jean-Louis Trintignant (*Z*)

1970
Best Film: *M*A*S*H*
Best Director: John Boorman (*Leo the Last*)
Best Actress: Ottavio Piccolo (*Metelo*)
Best Actor: Marcello Mastroianni (*Drama of Jealousy*)

1971
Best Film: *The Go-Between*
Best Director: (Not awarded)
Best Actress: Kitty Winn (*Panic in Needle Park*)
Best Actor: Ricardo Cucciola (*Sacco and Vanzetti*)

1972
Best Film: *La Classe Operaia va in Paradiso* (The Working Class Goes to Paradise); *Il Caso Mattei* (The Mattei Affair)
Best Director: Miklós Jancsó (*Meg Kér a Nép* (The Red Psalm))
Best Actress: Susannah York (*Images*)
Best Actor: Jean Yanne (*Nous ne Vieillirons pas Ensemble* (We Will Not Grow Old Together))

1973
Best Films: *Scarecrow; The Hireling*
Best Director: (Not awarded)
Best Actress: Joanne Woodward (*The Effect of Gamma Rays on Man In The Moon Marigolds*)
Best Actor: Giancarlo Giannini (*Film d'Amore e d'Anarchia* (Love and Anarchy))

1974
Best Film: *The Conversation*
Best Director: (Not awarded)
Best Actress: Marie-Jose Nat (*Les Violons du Bal*)
Best Actor: Jack Nicholson (*The Last Detail*)

1975
Best Film: *Ahdat Sanawouach El-Djamr* (Chronicle of the Burning Years)
Best Director: Constantine Costa-Gavras (*Section Spéciale* (Special Section)); Michel Brault (*Les Ordes*)
Best Actress: Valerie Perrine (*Lenny*)
Best Actor: Vittorio Gassman (*Profumo di Donna* (Scent of Woman))

1976
Best Film: *Taxi Driver*
Best Director: Ettore Scola (*Brutti, Sporchi, Cattivi* (Down and Dirty))
Best Actress: Mari Torocsik (*Deryne, Hol Van*); Dominique Sanda (*L'Eredità Ferramonti* (The Inheritance))
Best Actor: José-Luis Gómez (*La Familia de Pascual Duarte*)

1977
Best Film: *Padre Padrone*
Best Director: (Not awarded)
Best Actress: Shelley Duvall (*Three Women*); Monique Mercure (*J. A. Martin, Photographer*)
Best Actor: Fernando Rey (*Elisa, Vida Mia* (Elisa My Love))

1978
Best Film: *Albero Degli Zoccoli* (The Tree of Wooden Clogs)
Best Director: Nagisa Oshima (*Ai No Borei* (Empire of Passion))
Best Actress: Jill Clayburgh (*An Unmarried Woman*); Isabelle Huppert (*Violette Nozière*)
Best Actor: Jon Voight (*Coming Home*)

1979
Best Film: *Apocalypse Now; Die Blechtrommel* (The Tin Drum)
Best Director: Terence Malick (*Days of Heaven*)
Best Actress: Sally Field (*Norma Rae*)
Best Actor: Jack Lemmon (*The China Syndrome*)

1980
Best Film: *Kagemusha; All That Jazz*
Best Director: (Not awarded)
Best Actress: Anouk Aimée (*Salto Nel Vuoto* (Leap into The Void))
Best Actor: Michel Piccoli (*Salto Nel Vuoto* (Leap into The Void))

1981
Best Film: *Czlowiek Z Zelaza* (Man of Iron)
Best Director: (Not awarded)
Best Actress: Isabelle Adjani (*Possession*)
Best Actor: Ugo Tognazzi (*La Tragedia di un Uomo Ridicolo* (The Tragedy of a Ridiculous Man))

1982
Best Film: *Missing*
Best Director: Werner Herzog (*Fitzcarraldo*)
Best Actress: Jadwiga Jankowska-Cieslak (*Un Autre Regard* (Another Kind of Look))
Best Actor: Jack Lemmon (*Missing*)

1983
Best Film: *Narayama Bushi-Ko* (The Ballad of Narayama)
Best Director: Commendations to Andrei Tarkovsky (*Nostalghia* (Nostalgia)); Robert Bresson (*L'Argent*)
Best Actress: Hanna Schygulla (*La Storia di Piera* (The Story of Piera))
Best Actor: Gian Maria Volonte (*La Mort de Mario Ricci* (The Death of Mario Ricci))

1984
Best Film: *Paris Texas*

Best Director: Bertrand Tavernier (*Un Dimanche à la Campagne* (A Sunday in The Country))
Best Actress: Helen Mirren (*Cal*)
Best Actor: Francisco Rabal (*Les Saints Innocents* (The Holy Innocents))

1985
Best Film: *Octa Na Slŭzbenom Putu* (When Father Was Away On Business)
Best Director: André Techne (*Rendez-vous*)
Best Actress: Cher (*Mask*); Norma Alessandro (*La Historia Official* (The Official Version))
Best Actor: William Hurt (*The Kiss of the Spiderwoman*)

1986
Best Film: *The Mission*
Best Director: Martin Scorsese (*After Hours*)
Best Actress: Barbara Sukowa (*Rosa Luxemburg*); Fernanda Torres (*Parle-moi d'Amour*)
Best Actor: Michel Blanc (*Tenue de soirée* (Evening Dress)); Bob Hoskins (*Mona Lisa*)

1987
Best Film: *Sous Le Soleil de Satan* (Under Satan's Sun)
Best Director: Wim Wenders (*Der Himmel über Berlin* (Wings of Desire))
Best Actress: Barbara Hershey (*Shy People*)
Best Actor: Marcello Mastroianni (*Oci Ciornie* (Dark Eyes))

1988
Best Film: *Pelle Erobreren* (Pelle the Conqueror)
Best Director: Fernando Solanas (*Sur* (The South))
Best Actress: Barbara Hershey (*A World Apart*)
Best Actor: Forest Whitaker (*Bird*)

1989
Best Film: *sex, lies and videotape*
Best Director: Emir Kusturica (*Dom Za Vesanje* (Time of the Gypsies))
Best Actress: Meryl Streep (*Cry in the Dark*)
Best Actor: James Spader (*sex, lies and videotape*)

1990
Best Film: *Wild at Heart*
Best Director: Pavel Lounguine (*Taxi Blues*)
Best Actress: Krzystyna Janda (*Przesluchanie* (The Interrogation))
Best Actor: Gerard Depardieu (*Cyrano De Bergerac*)

1991
Best Film: *Barton Fink*
Best Director: Joel and Ethan Coen (*Barton Fink*)
Best Actress: Irene Jacob (*La Double Vie de Véronique*)
Best Actor: John Turturro (*Barton Fink*)

Venice Film Festival

1934
Best Foreign Film: *Man of Aran*
Best Italian Film: *Teresa Gonfalonieri*
Best Director: Gustave Machaty (*Extase*); J Rovensky (*Young Love*); Tomas Trnka (*Hurricane in the Tatras*); Karel Plicka (*Zem Spieva*)
Best Actress: Katharine Hepburn (*Little Women*)
Best Actor: Wallace Beery (*Viva Villa*)

1935
Best Foreign Film: *Anna Karenina*
Best Italian Film: *Casta Diva*
Best Director: King Vidor (*Wedding Night*)
Best Actress: Paula Wessely (*Episode*)
Best Actor: Pierre Blanchar (*Crime and Punishment*)

1936
Best Foreign Film: *Der Kaiser von Kalifornien*
Best Italian Film: *Squadrone Bianco*
Best Director: Jacques Feyder (*La Kermesse Héroïque* (Carnival in Flanders)
Best Actress: Annabella (*Veille d'Armes*)
Best Actor: Paul Muni (*The Story of Louis Pasteur*)

1937
Best Foreign Film: *Un Carnet de Bal*
Best Italian Film: *Scipione l'Africano*
Best Director: Robert Flaherty, Zoltan Korda (*Elephant Boy*)
Best Actress: Bette Davis (*Marked Woman; Kid Galahad*)
Best Actor: Emil Jannings (*Der Herrscher*)

1938
Best Foreign Film: *Olympia*
Best Italian Film: *Lucianno Serra Pilota*
Best Director: (Not awarded)
Best Actress: Norma Shearer (*Marie Antoinette*)
Best Actor: Leslie Howard (*Pygmalion*)

1939
Best Foreign Film: (Not awarded)
Best Italian Film: *Abuna Messias*
Best Actress: (Not awarded)
Best Actor: (Not awarded)

1940
Best Foreign Film: *Der Postmeister*
Best Italian Film: *L'Assedio dell'Alcazar*
Best Director: (Not awarded)
Best Actress: (Not awarded)
Best Actor: (Not awarded)

1941
Best Foreign Film: *Ohm Kruger*
Best Italian Film: *La Corona di Ferro*
Best Director: (Not awarded)
Best Actress: Luise Ullrich (*Annelie*)
Best Actor: Ermete Zacconi (*Don Buonaparte*)

1942
Best Foreign Film: *Der Grosse König*
Best Italian Film: *Bengasi*
Best Director: (Not awarded)
Best Actress: Kristina Soderbaum (*Der Grosse König; Die Goldene Stadt*)
Best Actor: Fosco Giachetti (*Un Colpo do Pistola; Bengasi; Noi Vivi*)

1943
No festival

1944
No festival

1945
No festival

1946
Best Film: *The Southerner*
Best Director: (Not awarded)
Best Actress: (Not awarded)
Best Actor: (Not awarded)

1947
Best Film: International Grand Prize *Sirena*
Best Director: Henri-Georges Clouzot (*Quai des Orfèvres*)
Best Actress: Anna Magnana (*Honorable Angelina*)
Best Actor: Pierre Fresnay (*Monsieur Vincent*)

1948
Best Film: International Grand Prize *Hamlet*
Best Director: G. W. Pabst (*Der Prozess* (The Trial))
Best Actress: Jean Simmons (*Hamlet*)
Best Actor: Ernest Deutsch (*Der Prozess* (The Trial))

1949
Best Film: *Manon*
Best Director: Augusto Genina (*Cielo sulla Palude*)
Best Actress: Olivia de Havilland (*The Snake Pit*)
Best Actor: Joseph Cotten (*Portrait of Jennie*)

1950
Best Film: *Justice is Done*
Best Director: (Not awarded)
Best Actress: Eleanor Parker (*Caged*)
Best Actor: Sam Jaffe (*The Asphalt Jungle*)

1951
Best Film: *Rashomon*
Best Director: (Award dropped)
Best Actress: Vivien Leigh (*A Streetcar Named Desire*)
Best Actor: Jean Gabin (*La Nuit est Mon Royaume*)

1952
Best Film: *Jeux Interdits* (Forbidden Games)
Best Actress: (Not awarded)
Best Actor: Fredric March (*Death of a Salesman*)

1953
Best Film: (Not awarded)
Best Actress: Lilli Palmer (*The Fourposter*)
Best Actor: Henri Vilbert (*Absolution*)

1954
Best Film: *Romeo and Juliet*
Best Actress: (Not awarded)
Best Actor: Jean Gabin (*Touchez pas au Grisbi* (Grisbi); *The Air of Paris*)

1955
Best Film: *Ordet* (The Word)
Best Actress: (Not awarded)
Best Actor: Kenneth More (*The Deep Blue Sea*); Curt Jurgens (*Des Teufels General* (The Devil's General); *Les Héros sont Fatigués* (Heroes and Sinners))

1956
Best Film: (Not awarded)
Best Actress: Maria Schell (*Gervaise*)
Best Actor: Bourvil (*La Traversée de Paris* (A Pig Across Paris))

1957
Best Film: *Aparajito* (The Unvanquished)
Best Actress: Dzidra Ritenbery (*Malva*)
Best Actor: Anthony Franciosa (*A Hatful of Rain*)

1958
Best Film: *Muhomatsu no Issho*
Best Actress: Sophia Loren (*The Black Orchid*)
Best Actor: Alec Guinness (*The Horse's Mouth*)

1959
Best Film: *Il Generale della Rovere; La Grande Guerre*
Best Actress: Madeline Robinson (*Double Tour*)
Best Actor: James Stewart (*Anatomy of a Murder*)

1960
Best Film: *Le Passage du Rhin*
Best Actress: Shirley MacLaine (*The Apartment*)
Best Actor: John Mills (*Tunes of Glory*)

1961
Best Film: *L'Année Dernière à Marienbad* (Last Year at Marienbad)
Best Actress: Suzanne Flon (*Non Uccidere* (Thou Shalt Not Kill))
Best Actor: Toshiro Mifune (*Yojimbo*)

1962
Best Film: *Ivanovo Detstro* (Ivan's Childhood)
Best Actress: Emmanuele Riva (*Thérèse Desqueyroux*)
Best Actor: Burt Lancaster (*Bird Man of Alcatraz*)

1963
Best Film: *Le Mani sulla Città*
Best Actress: Delphine Seyrig (*Muriel*)
Best Actor: Albert Finney (*Tom Jones*)

1964
Best Film: *Il Deserto Rosi* (Red Desert)
Best Actress: Harriet Andersson (*Att Älska* (To Love))
Best Actor: Tom Courtenay (*King and Country*)

1965
Best Film: *Vaghe Stelle dell' Orsa* (Of a Thousand Delights)

Best Actress: Annie Girardot (*Trois Chambres à Manhattan*)
Best Actor: Toshiro Mifune (*Akahige* (Red Beard))

1966
Best Film: *La Battaglia di Algeri* (Battle of Algiers)
Best Actress: Natalia Arinbasavora (*The First Schoolteacher*)
Best Actor: Jacques Perrin (*Quest; Half a Man*)

1967
Best Film: *Belle de Jour*
Best Actress: Shirley Knight (*Dutchman*)
Best Actor: Ljubisa Samardzic (*Dawn*)

1968
Best Film: *Die Artisten in der Zirkuskuppel*
Best Actress: Laura Betti (*Teorema* (Theorem))
Best Actor: John Morely (*Faces*)

1969–79
Jury and Award system discontinued

1980
Best Film: *Gloria, Atlantic City*

1981
Best Film: *Die Bleierne Zeit* (The German Sisters)

1982
Best Film: *Der Stand der Dinge* (The State of Things)

1983
Best Film: *Prenom Carmen* (First Name Carmen)
Best Actress: Darling Legitimus (*Rue Cases Negres* (Black Shack Alley))
Best Actor: (collective award) Matthew Modine, Michael Wright, Mitchell Lichtenstein, David Alan Grier, Guy Boyd and George Dzundza (*Streamers*)

1984
Best Film: *Rok Spokojnego Sloncá* (A Year Of The Quiet Sun)
Best Actress: Pascale Ogier (*Les Nuits de la Pleine Lune* (Full Moon in Paris))
Best Actor: Naseeruddin Shaf (*The Crossing*)

1985
Best Film: *Sans Toit, Ni Loi* (*Vagabonde*)
Best Actress: Sandrine Bonnaire (*Sans Toit, Ni Loi* (Vagabonde)); Jane Birkin (*Dust*); Themis Bazaka (*Petrina Chronia*); Galija Novents (*Tango Nasego*); Sonja Savic (*Zivot Je Lep*)
Best Actor: Gerard Depardieu (*Police*)

1986
Best Film: *Le Rayon Vert* (The Green Ray)
Best Actress: Valeria Golino (*Storia D'Amore*)
Best Actor: Carlo Delle Piane (*Regalo di Natale*)

1987
Best Film: *Au Revoir Les Enfants*
Best Actress: Kang Soo-Yeon (*Contract Mother*)
Best Actor: James Wilby and Hugh Grant (*Maurice*)

1988
Best Film: *La Leggenda del Santo Bevitore* (The Legend of The Holy Drinker)
Best Actress: Shirley MacLaine (*Madame Sousatzka*)
Best Actor: Don Ameche (*Things Change*)

1989
Best Film: *Beiging Chengshi* (A City Of Sadness)
Best Actress: Peggy Ashcroft and Geraldine James (*She's Been Away*)
Best Actor: Marcello Mastroianni and Massimo Troisi (*Che Ora È?;* (What Time Is It?))

1990
Best Film: *Rosencrantz and Guildenstern Are Dead*
Best Actress: Gloria Munchmeyer (*La Lune en el Espejo* (Moon in the Mirror))
Best Actor: Oleg Borisov (*Edinstvenijat Svidetel* (The Only Witness))
Best Director: Martin Scorsese (*Goodfellas*)